Origami⁵

Origami^5

Fifth International Meeting of Origami Science, Mathematics, and Education

Edited by
Patsy Wang-Iverson
Robert J. Lang
Mark Yim

 **CRC Press**
Taylor & Francis Group
Boca Raton London New York

CRC Press is an imprint of the
Taylor & Francis Group, an **informa** business

AN A K PETERS BOOK

Cover Illustrations: See "A Systematic Approach to Twirl Design," Figure 15 (p. 120), and "Reconstructing David Huffman's Legacy in Curved-Crease Folding," Figure 14 (p. 49).

CRC Press
Taylor & Francis Group
6000 Broken Sound Parkway NW, Suite 300
Boca Raton, FL 33487-2742

© 2011 by Taylor & Francis Group, LLC
CRC Press is an imprint of Taylor & Francis Group, an Informa business

No claim to original U.S. Government works

Printed in the United States of America on acid-free paper
Version Date: 20110526

International Standard Book Number: 978-1-56881-714-9 (Paperback)

Library of Congress Cataloging-in-Publication Data

International Meeting of Origami Science, Mathematics, and Education (5th : 2010 : Singapore)
Origami 5 : Fifth International Meeting of Origami Science, Mathematics, and Education / edited by Patsy Wang-Iverson, Robert J. Lang, Mark Yim.
p. cm.
Includes bibliographical references and index.
ISBN 978-1-56881-714-9 (pbk.)
1. Origami--Mathematics--Congresses. 2. Origami in education--Congresses. I. Wang-Iverson, Patsy. II. Lang, Robert J. III. Yim, Mark. IV. Title. V. Title: Origami five.

QA491.I55 2011
736'.982--dc22   2011008391

Visit the Taylor & Francis Web site at
http://www.taylorandfrancis.com

and the CRC Press Web site at
http://www.crcpress.com

Contents

Preface

Origami[5] follows in the large footprints of four volumes,[1] each documenting work presented at an extraordinary series of meetings that have explored the connections between origami, mathematics, science, technology, education, and other academic fields. The idea for these meetings originated with Professor Humiaki Huzita, who organized the First International Meeting of Origami Science and Technology, held December 6–7, 1989, at the Casa di Ludovico Ariosto in Ferrara, Italy. Five years later, under the leadership of Professor Koryo Miura, the Second International Meeting of Origami Science and Scientific Origami took place November 29–December 2, 1994, at Seian University of Art and Design, Otsu, Shiga, Japan. This meeting officially expanded beyond origami mathematics and science to include origami design, origami in education, and the history of origami. The third meeting, held at Asilomar, Pacific Grove, California, March 9–11, 2001, was organized by Professor Thomas Hull; titled The Third International Meeting of Origami Science, Mathematics and Education, it became known by

---

[1][Huzita 89] *Proceedings of the First International Meeting of Origami Science and Technology*, edited by Humiaki Huzita. Padova, Italy: Dipartimento di Fisica Galileo Galilei dell'Universita degli studi di Padova, 1989.

[Miura et al. 97] *Origami Science and Art: Proceedings of the Second International Meeting of Origami Science and Scientific Origami*, edited by Koryo Miura, Tomoko Fuse, Toshizaku Kawaski, and Jun Maekawa. Otsu, Shiga, Japan: Seian University of Art and Design, 1997.

[Hull 02] *Origami*[3]: *Third International Meeting of Origami Science, Mathematics and Education*, edited by Thomas Hull. Natick, MA: A K Peters, 2002.

[Lang 09] *Origami*[4]: *Fourth International Meeting of Origami Science, Mathematics and Education*, edited by Robert J. Lang. Natick, MA: A K Peters, 2009.

its acronym, 3OSME, which inspired the format (and name) for subsequent meetings. The fourth meeting, 4OSME, organized by Robert J. Lang, took place September 6–10, 2006, at the California Institute of Technology in Pasadena, California. Most recently, the fifth such meeting, 5OSME, co-organized by Eileen Tan, Benjamin Tan, and Patsy Wang-Iverson, was held on July 13–17, 2010, at the Singapore Management University in Singapore. The majority of papers in this book are based on presentations from this meeting (with a few post-meeting contributions as well).

Origami[5] follows the precedent set by the second meeting and continued at 3OSME and 4OSME to expand the interdisciplinary connections to the world of origami. This book begins with a section on origami history, art, and design. It is followed by sections on origami in education, origami science, engineering, and technology, and it culminates with a section on origami mathematics—the pairing that inspired the first such meeting. The scope of the collected papers is broad; within this one volume, one can find historical information, artists' descriptions of their processes, various perspectives and approaches to the use of origami in education, mathematical tools for origami design, applications of folding in engineering and technology, and original and cutting-edge research on the mathematical underpinnings of the field.

We begin with a section on *origami history, art, and design*, in which the first two papers contrast Eastern and Western aspects of origami. Koshiro Hatori revisits the history of origami and identifies its origin in both the West and the East. Koichi Tateishi examines differences between current Japanese and western origami practices and attributes them to cross-linguistic differences between the two cultures. Then Arnold Tubis and Crystal Mills discuss the surprising role of origami in the creation of the original American flag by Betsy Ross. Erik Demaine et al. integrate history and design in their examination of the work of curved-crease folding pioneer David Huffman and, for the first time, unlock his secrets and reconstruct several of his most famous works.

Origami artists follow many different paths to create their art, often with a mathematical bent, always with a strong emphasis on aesthetic considerations. In a series of papers on art and design, several authors present the reader with eclectic but engaging papers that describe their approaches to the creation of origami art, all with a strong mathematical or technological flavor. Cheng Chit Leong describes curved folding and ties together aesthetical and mathematical considerations. Christine Edison explores the properties of twisted curved corrugations. Andrew Hudson presents a novel algorithm that can be generalized for construction of origami figures from regular polygons based on their rotational symmetry. Miyuki Kawamura and Hiroyuki Moriwaki straddle design and technology with their new technique for illuminating origami models. Faye Goldman demonstrates how an idea can be adapted for use with new material, leading to esthetically pleasing polyhedral models. Krystyna Burczyk and Wojciech Burczyk discuss their systematic approach to using twirls and spirals with differently shaped papers to produce an almost limitless array of new creations. Building upon his work presented at 4OSME, Matthew Gardiner describes advances he has made with oribotics with the aid of technology and inspiration from surprising sources.

The second section focuses on *origami in education*. Education is a shared global enterprise, with increasing numbers of countries using international assessments, such as *TIMSS: Trends in International Mathematics and Science Study* and *PISA: Programme for International Student Assessment*, in reading, mathematics, and science as benchmarks for measuring progress in educating their students. The chapters in this education section offer insight into educational systems in different countries and cultures through the lens of origami. Miri Golan shows how the teaching of origami is aligned with the van Hiele theory of teaching geometry. Maria Lluïsa Fiol et al. rediscover Froebel's recognition of the value of origami in the education of young children and describe the care and time necessary to educate and prepare teachers in the use of origami to stimulate student learning and creativity. Christine Edison presents cases of the effectiveness of origami in re-engaging students with school in an economically deprived environment.

We then turn to an examination of origami's effectiveness and descriptions of specific ways that origami may be used in education. Norma Boakes studies whether constructing origami models can enhance college students' spatial skills. James Morrow and Charlene Morrow offer a detailed approach to learning origami through reverse engineering of models. Sue Pope and Tung Ken Lam present ways in which novices can be introduced to origami and describe how origami can be used to teach challenging mathematics concepts and to avoid common misconceptions. Michael Winckler et al. describe the use of origami to teach geometry to grade eight students in a gymnasium. Shi-Pui Kwan closes out the education section with a discussion of his use of origami in high school geometry.

Origami science, engineering, and technology are the focus of the next group of papers. It is exciting to see the development of what is considered by many as an art form have more direct impact on society through applications to science and technology. This section includes methods for patterning origami mazes and cylinders, origami used as engineered structures, computer aided origami, and fantastical origami-inspired devices.

The section starts with two papers on rigid origami: first, a design technique for rigid folding using thick panels by Tomohiro Tachi, followed by a kinematic analysis of patterned cylinders from Kunfeng Wang and Yan Chen. We then see how pattern folding techniques can be used for

their mechanical structural characteristics in papers from Jiayao Ma and
Zhong You, Mark Schenk and Simon Guest, and Yves Klett and Klaus
Drechsler. These papers demonstrate how origami can be useful for designing structures such as bumpers in cars and airplane fuselages. Steven
Gray et al. describe software for aiding origami construction of a self-folding
origami robot, while Naoya Tsurutaet al. and Hugo Akitaya et al. describe
systems and algorithms for automated diagramming of origami instruction. Finally, Noy Bassiket al. and Kaori Kuribayashi-Shigetomi and Shoji
Takeuchi show self-folding micro origami structures and bio-compatible
origami structures that could lead to bio-implantable devices.

The *mathematics of origami* has always involved a creative pairing between "origami-math"—the fundamental laws underlying mathematics—
and "computational origami"—the mathematics specific to the origami design problem. The final section, devoted to origami mathematics, starts
off with a series of papers on mathematical design algorithms beginning
with Jun Maekawa's analysis of the design of knots folded from strips of
paper. Nadia Benbernou et al. provide patterns for folding arbitrary 3D
shapes composed of cubes; Herng Yi Cheng describes how to construct arbitrary biplanar three-dimensional solids; Jun Mitani gives an algorithm
for polyhedral and/or curved rotationally symmetric solids.

Then we have a group of papers giving design algorithms for generally
planar structures. Erik Demaine et al. present a recipe for construction of
arbitrary mazes. Robert J. Lang and Alex Bateman show how a large class
of twist tessellations can be designed, solving a longstanding problem in
the process, and Lang gives a prescription for designing woven tessellations,
introducing a new law of "flat-unfoldability" along the way.

We then shift focus to the mathematical underpinnings of origami design. Tomohiro Tachi and Erik Demaine give a surprising explanation for
the utility of highly symmetric crease patterns in the design world based on
the number fields of the points constructed thereby, and this leads into a
series of works that describe geometric constructions using origami. Robert
Orndorff introduces a new construction of square roots, while Emma Frigerio and Kazuo Haga describe two families of geometric construction based
on earlier work by Haga, one of the pioneers of origami constructions.

Continuing the theme of geometric constructions, we have several studies of axiomatic origami, i.e., the foundations of the geometric constructions
presented earlier in this section. Eulàlia Tramuns creates formal measures
of the number of steps required in origami constructions and compares
them to other geometric construction tools. Toshikazu Kawasaki and Roger
Alperin examine origami constructions in other spaces: spherical and hyperbolic, respectively. We then move to the area of flat-foldability: Hidefumi Kawasaki presents a new proof of a flat-foldability result. Ryuhei
Uehara gives new results on configuration counting in the one-dimensional

stamp folding problem, followed by Thomas Hull and Eric Chang's work establishing bounds on the number of crease assignments at a single vertex. The section closes with a proof by Erik Demaine et al. that one of the fundamental steps in circle/river packing based origami design is NP-hard.

There is an accompanying website to $Origami^5$, http://www.origami-usa .org/origami5, that provides additional resources and links related to the papers.

Acknowledgments

This book, $Origami^5$, is the result of a great deal of work by many people that began with the original 5OSME meeting. We editors wish to thank the program committee members who reviewed the abstracts accepted for presentation at the meeting: Philip Chapman-Bell, Martin Demaine, Christine Edison, Emma Frigerio, Thomas Hull, Cheng Chit Leong, Neil Sloane, Tomohiro Tachi, Benjamin Tan, Eileen Tan, Koichi Tateishi, and Seng Kai Wong.

We also thank the individuals working behind the scenes to evaluate the presentations and to review and improve the submitted papers: Roger Alperin, Marshall Bern, Jin-Yi Cai, Martin Demaine, Ken Fan, Emma Frigerio, Simon Guest, Barry Hayes, Thomas Hull, Jason Ku, Daniel Kwan, David Lister, Perla Myers, Jun Mitani, Charlene Morrow, Joseph O'Rourke, Aviv Ovadya, Ryda Rose, Tomohiro Tachi, Arnold Tubis, Ryuhei Uehara, Tamara Veenstra, and those who chose to remain anonymous.

The 5OSME meeting was made possible through the collaborative efforts of many organizations/institutions and people. Partners included the Gabriella and Paul Rosenbaum Foundation, Origami-USA, Principals Academy, and Singapore Management University. Additionally, several sponsors contributed to making the meeting a reality through the sponsorship of student and teacher registrants: BioSym (Biosystems and Micromechanics), Censam (Center for Environmental Sensing and Modeling), Marshall Cavendish, the SMART Centre (Singapore-MIT Alliance for Research and Technology).

In addition to institutional recognition, we must thank the individuals within the institutions who were responsible for the generous contributions: Rohan Abeyaratne, SMART Centre Director; Duriya Aziz, Marshall Cavendish; George Barbastathis, SMART Centre; Madge Goldman, Gabriella and Paul Rosenbaum Foundation; Roger Kamm, BioSyM IRG PI; Les Norford, CENSAM IRG PI; and Kirpal Singh and Sumathi Nair, Wee Kim Wee Centre, SMU.

We thank Cheng Chit Leong for procuring the SMU site for the meeting, Kirpal Singh for offering SMU as the venue, and Sumathi Nair for all her work and arrangements. We also thank two most important people, without whom the meeting in Singapore would not have occurred: Eileen Tan and Benjamin Tan.

Most of the papers in this book grew out of the meeting presentations, but these papers are original to this book; the authors wrote, re-rewrote, responded to reviews and editing, and in many cases worked through numerous revision cycles. Thanks to their hard work, you have before you, a collection that represents the leading edge of academic origami developments. Without their work, this book would not exist, and so we thank the authors of these papers most of all.

We hope you will enjoy *Origami*^{5}.

Patsy Wang-Iverson
Robert J. Lang
Mark Yim

March, 2011

Part I

Origami History, Art, and Design

History of Origami in the East and the West before Interfusion

Koshiro Hatori

1 Introduction

Origami used to be, and still sometimes is, said to originate in the second century in China. This conjecture was first stated, as far as I know, by Lillian Oppenheimer when she wrote a foreword for Isao Honda's book *How to Make Origami* [Honda 59]. It assumes two things: paper was invented in the second century in China, and origami started just after the invention of paper. I am going to argue that both assumptions are unacceptable.

2 Origin of Origami: Many Misunderstandings and Some Suppositions

When Oppenheimer wrote the foreword to *How to Make Origami*, it was widely believed that paper was invented by the Chinese eunuch Cai Lun (also alphabetized as T'sai Lun) in 105 CE. However, much older paper has been unearthed from some tombs of the Western Han Dynasty (206 BCE–8 CE). The oldest piece, which was discovered at Fan Ma Tan in 1986, is estimated to have been made in the middle of the second century BCE [Komiya 01].

Moreover, recent studies show that high-quality bark paper, called *amate* in Meso-America, *kapa* in Hawaii, and *tapa* in Southeast Asia, dates back to 5000 BCE [Sakamoto 09]. Such ancient paper was so sophisticated that some even had watermarks. The beaten bark paper has a texture that is close to *washi* or Japanese handmade paper. In fact, it has been made from the mulberry tree in Southeast Asia. Even though this bark paper is sometimes regarded as cloth rather than paper, it may well be the origin of *washi* [Sakamoto 08].

The ancient bark paper had such high quality that I believe it could be folded. Then, one may ask whether origami dates back to 7,000 years ago, and my answer would be "no." When one examines the origin of origami, the question should be, in my opinion, about how origami has emerged and developed instead of who folded paper first. It is likely that ancient people folded paper, but such paper folding would have no relationship with today's origami. We cannot trace the history of origami more than a few hundred years.

Some say origami started in the Heian period (794–1185) in Japan. One of the stories they refer to is an anecdote of Abe no Seimei, the most famous *onmyoji* (specialist of a Japanese traditional spiritual cosmology) of the tenth century. The story says that he took a piece of paper and turned it into a real heron to search for his most formidable rival Ashiya Doman.

According to Masao Okamura, however, this tale of Seimei has nothing to do with origami. Some books say he tied a piece of paper to make a knot, some say he cut paper in a shape of a bird, some say he drew a heron on paper, but no book says he folded paper. Okamura's extensive studies revealed that there is no evidence of origami in the Heian period [Okamura 99].

Others mention the *shide*, cut and folded zigzag paper strips used in Shinto rituals, as an example of ancient paper folding. They were originally, however, pieces of cloth offered to the gods. Although it is possible that the *shide* came to be made of paper in the Heian period, I do not see any connection between the *shide* and today's origami.

The origin of origami in Japan is thought to be ceremonial wrappers as represented by *noshi*. *Noshi* was originally a form of folded wrappers for *noshi-awabi*, or stripped and dried abalone meat, although today it is just attached or printed on wrapping paper as a token of good fortune. Another example is a pair of paper butterflies known as *ocho* and *mecho* (Figure 1). They are, in fact, wrappers for sake bottles, although today they are just attached onto the neck of the bottle and used mainly in wedding ceremonies. Some say such wrappers date back to the Heian period, for which I have never seen any evidence.

The *samurai* warriors of the Edo period (1603–1868) were supposed to fold wrapping paper in a specific way according to what was inside when they sent a gift. It is part of the etiquette of the *samurai* class, which was carried down from generation to generation in some houses, most notably

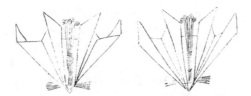

Figure 1. *Ocho* and *mecho* illustrated in the nineteenth century.

Ogasawara, Ise, and Kira. According to Ise Sadatake, who wrote a book of the ceremonial origami *Tsutsumi-no Ki* in 1764, such paper folding was established in the Muromachi period (1333–1573) [Araki 03].

In contrast, the origin of Western origami is thought to be baptismal certificates folded in a "double blintz," that is, folding all the four corners of a square to the center and repeating the same folds on the smaller square. According to Ann Herring, the baptismal certificate of Friedrich Froebel, whom we shall meet in the next section, was also folded [Herring 99].

This custom of folding baptismal certificates seems to have been popular in Central Europe in the seventeenth and eighteenth centuries (Figure 2). Herring suggested that such paper folding may have started before the Protestant Reformation [Herring 99]. So, origami in the West probably dates back to the sixteenth century.

David Lister has noticed that the crease pattern of baptismal certificates is closely similar to the design of old European astrological horoscopes. According to Vicente Palacios, such an "astrological square" was introduced into Spain in the twelfth century [Lister 97]. I must, however, point out that there is no evidence that horoscopes in either Spain or Germany were folded.

Comparing Japanese wrappers with European baptismal certificates, we can observe the difference in the folding styles of the East and the West. The crease lines of *samurai* wrappers run in arbitrary angles, whereas those of baptismal certificates are limited to square grids and diagonals.

Figure 2. German baptismal certificate from the eighteenth century.

As we shall see in the following section, the difference between Eastern and Western folding stayed with almost all of the origami models until the second half of the nineteenth century, when Japan opened its border and started to exchange cultures with Europe. This fact suggests that the two traditions of origami in Japan and Europe emerged and developed independently of each other.

3   The East and the West: Different Styles, Different Traditions

The most typical European origami model is perhaps the little bird called *pajarita* in Spain and *cocotte* in France. Although its origin is rather vague, I suspect that pajarita must have existed in the late eighteenth century, because horses and riders (Figure 3) in the collection of the German National Museum, which were made around the time of the War of the Sixth Coalition (1812–1814) [Kono 58], appear to have derived from it.

Another popular model in Europe is the boat. Vicente Palacios argued that the boat is recorded in an edition of *Tractatus de Sphaera Mundi* published in Venice in 1490 [Lister 97]. This book was written by Johannes de Sacrobosco (John of Holywood) in the thirteenth century and reprinted more than 60 times through the middle of the seventeenth century. Even though the illustrated boats look like folded models, I would refrain from making judgment until we have more evidence of origami before the nineteenth century.

Also well known is the hat, which John Tenniel has depicted in artwork for Lewis Carroll's *Through the Looking Glass* published in 1872 [Carroll 03]. Although both the boat and the hat are made from rectangular sheets of paper, most of the European traditional models are made from square sheets. Those models are well documented in the context of the Froebelian education system.

Figure 3. Horses and riders from the early nineteenth century.

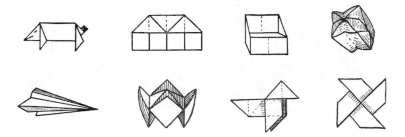

Figure 4. Examples of European traditional models.

Friedrich Wilhelm August Froebel was a German educator who founded the first modern kindergarten in 1837. His education system contained a set of toys called "Gifts" and a set of plays called "Occupations," and one of the most important Occupations was origami. Maria Kraus-Boelté recorded nearly one hundred origami models in her book *The Kindergarten Guide* [Kraus-Boelté and Kraus 82].

Many of the European origami models contained in Kraus-Boelté's book are not included in contemporary Japanese records. The pig, house, sofa (also known as piano or organ), balloon (waterbomb), arrow (paper plane), salt cellar (cootie catcher), bird (*pajarita* or *cocotte*), and windmill shown in Figure 4 were all born in Europe and imported into Japan along with the kindergarten system.

The balloon may be the same model as the "paper prison" mentioned in John Webster's play *The Duchess of Malfi*, which premiered in 1614. That may be the oldest reference to European origami, but again, I would suspend my conclusion for now. Moreover, I do not consider just a few unconvincing references scattered over two centuries enough to prove the existence of origami.

Looking through the European models, one can easily notice that most of them have only creases that are either square grids or diagonals. This is true even in the Chinese junk and the gondola, which are similar to the Japanese *takara-bune*, or treasure ship (Figure 5). The treasure ship has a pointier bow that is folded with sharp-angled creases. This difference is, in my opinion, so critical that I am sure the Chinese junk and *takara-bune* developed independently on the opposite sides of the world.

Figure 5. Chinese junk, gondola, and *takara-bune* (left to right).

Figure 6. Examples of Japanese traditional models from the early nineteenth century.

The origami history researcher Satoshi Takagi one day bought a box containing many origami pieces. They are considered to have been folded by several persons in the house of Moriwaki from the middle eighteenth century through the nineteenth century. The older pieces are ceremonial wrappers, including *ocho* and *mecho*, and the newer ones are the traditional models we know well, such as the *orizuru* (crane) and *yakko-san* (servant) [Takagi 99].

Most interesting among these pieces are those estimated to have been made in the early nineteenth century (Figure 6), around the same time as the horses and riders were made in Germany. The pieces were made of sheets in different shapes with many cuts, many of which were painted. These are indeed the characteristics of Japanese origami. In contrast, the models in Europe were mostly made of square sheets without cuts.

Many of the Moriwaki models closely resemble those illustrated in contemporary books such as *Kayara-gusa* and *Chushingura Orikata*, but they are slightly different. The variations suggest that many people at that time were making the models, with or without looking at books. In fact, several versions of the *Chushingura Orikata* booklet were published throughout the nineteenth century.

Kayara-gusa was compiled by Adachi Kazuyuki, about whom we know virtually nothing. He copied numerous books to make his own encyclopedia, and completed more than two hundred volumes in 1845 without giving a title to the whole. He reproduced at least three origami books in volumes 27 and 28. We call the origami-related part of this book *Kayara-gusa* because the volumes from 21 to 30 are so titled [Okamura 94].

Kayara-gusa contains 15 patterns of wrappers, diagrams of 25 models, and colored crease patterns for 6 figures [Brossman and Brossman 61]. It also includes two paragraphs, one saying he did not diagram *sembazuru* (connected cranes), boat, *kago* (Japanese palanquin), lotus flower, *sanpo* (tray with stand), box, *komoso* (also known as *komuso*, Zen Buddhist monk), thread holder, or *kabuto* (helmet) because everyone knew how to fold them [Okamura 94]. These models are still popular in today's origami world. It should be noted that most of them were unique to Japan.

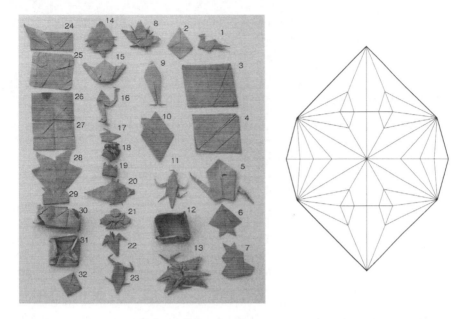

Figure 7. More examples of Japanese traditional models (left) and the crease pattern for the tortoise (right).

The other paragraph says that there were origami models of a peacock, praying mantis, *fukura-suzume* (sparrow ruffling feathers), blowfish (also known as catfish), and *kitsune-no yomeiri* (marriage of fox). They were, and still are, not well-known models, and Adachi himself did not know how to fold them [Okamura 94]. They have been, as it were, lost traditional models. Therefore, it would be surprising that all of the models are contained in the recently discovered origami collection of the Kanchazan Museum (Figure 7, left).

Kuzuhara Koto, a blind *koto* (Japanese stringed musical instrument) teacher born in 1812, made most of the origami pieces in the collection. They include some ceremonial wrappers; some connected cranes similar to those illustrated in *Sembazuru Orikata* published in 1797; many models recorded in *Kayara-gusa*; *tamate-bako* or treasure box, which is perhaps the oldest modular origami model, also depicted in *Ranma Zushiki* in 1734; and even models that have not been found in any other sources, such as the tortoise with the point-splitting crease pattern (Figure 7, right) and the box with the twist fold [Okamura 08].

All of the 66 pieces are unfortunately squashed flat to the extent that some cannot be recognized. Still, the collection is a comprehensive showcase of the diversity of Japanese traditional origami. Although some were made with the same square-grid pattern as the European traditional models, many Japanese models were highly complicated with advanced folding techniques as well as many cuts.

4   Conclusion

When comparing hundreds of traditional models recorded in the eighteenth and nineteenth century, one may be astonished to realize that only a few models were common to Europe and Japan at that time. Moreover, one can notice some differences even between models that appear to be shared between the East and the West.

Not only did the repertoires have little overlap, the folding styles also differed completely between the East and West. The Japanese origami models before the middle of the nineteenth century were made of sheets in various shapes: squares, rectangles, hexagons, octagons, and even many eccentric shapes. They were also folded with many cuts as well as with sophisticated folding techniques, and often were painted. Their European counterparts were made mainly from squares, sometimes from rectangles, and had few cuts. In addition, their crease lines were mostly limited to square grids and diagonals.

The difference has its root in the origin of origami—ceremonial wrappers of the fourteenth century in Japan and baptismal certificates of the sixteenth century in Europe. The crease lines for the wrappers run at different angles, whereas the folds in the baptismal certificates were the double blintz. This fact strongly suggests that Japanese and European origami arose and evolved independently.

In the first years of the Meiji Restoration, in the 1860s and 1870s, the European education system was introduced and adopted in Japan. As a result, European origami was imported to Japan as a part of the kindergarten curriculum. In addition, as people traveled internationally, Japanese origami spread over the Western world. The state of origami as we know it today has been developed as a consequence of such a cultural exchange. Thus, origami has never been a "Japanese" art.

Bibliography

[Araki 03] Makio Araki. *Fukkoku Ise Sadatake "Houketsuki"* (in Japanese). Kyoto: Tankosha, 2003.

[Brossman and Brossman 61] Julia Brossman and Martin Brossman. *A Japanese Paper-Folding Classic: Excerpt from the "Lost" Kan no Mado* (reprinted version). Washington, DC: The Pinecone Press, 1961.

[Carroll 03] Lewis Carroll. *Through the Looking Glass* (reprinted version). London: Penguin Books, 2003.

[Herring 99] Ann Herring. "Origami-no Bunkashiko: Origami Shinwa, Taiken, Rekishiteki Shogen-wo Chushin-ni" (in Japanese). In *Oru Kokoro*, pp. 84–89. Tatsuno: Tatsuno City Museum of History and Culture, 1999.

[Honda 59] Isao Honda. *How to Make Origami.* New York: McDowell, Obolensky, 1959.

[Komiya 01] Hidetoshi Komiya. "Kami-no Tanjo-to Sono Rekishi" (in Japanese). In *Kami-no Daihyakka*, pp. 38–49. Tokyo: Bijutsu Shuppan-sha, 2001.

[Kono 58] Yoichi Kono. *Gakumon-no Magarikado* (in Japanese). Tokyo: Iwanami Shoten, 1958.

[Kraus-Boelté and Kraus 82] Maria Kraus-Boelté and John Kraus. *The Kindergarten Guide Volume Two: The Occupations* (reprinted version). New York: E. Steiger & Co., 1882.

[Lister 97] David Lister. "Some Observations on the History of Paperfolding in Japan and the West - a Development in Parallel." In *Origami Science and Art: Proceedings of the Second International Meeting of Origami Science and Scientific Origami*, edited by K. Miura, pp. 511–524. Shiga, Japan: Seian University of Art and Design, 1997.

[Okamura 94] Masao Okamura. "Koten Kenkyu 'Karayagusa'" (in Japanese). *Oru* 5 (1994), 58–63, and 8 (1995), 59–65.

[Okamura 99] Masao Okamura. "Origami-no Nagare" (in Japanese). In *Oru Kokoro*, pp. 4–15. Tatsuno: Tatsuno City Museum of History and Culture, 1999.

[Okamura 08] Masao Okamura. "Koto-san-no Origami" (in Japanese). *Monthly Origami Magazine* 398 (2008), 30–31, and 409 (2009), 20–21.

[Sakamoto 08] Imamu Sakamoto. "Juhishi-no Umoreta Rekishi" (in Japanese). *Hyakumantoh* 130 (2008), 52–71.

[Sakamoto 09] Imamu Sakamoto. "Kami-to Hito-wo Tsunagu Juhishi" (in Japanese). *Hyakumantoh* 134 (2009), 63–86.

[Takagi 99] Satoshi Takagi. "Moriwaki-ke Kyuzo-no Origami Shiryo-nitsuite" (in Japanese). In *Oru Kokoro*, pp. 67–74. Tatsuno: Tatsuno City Museum of History and Culture, 1999.

Deictic Properties of Origami Technical Terms and Translatability: Cross-Linguistic Differences between English and Japanese

Koichi Tateishi

1 Introduction

The hardship that diagram translators and interpreters often have between Japanese and English, and also with other European languages, is due to deictic differences between the languages, resulting in some terms without proper translations. For example, the English direction, "Fold corner to corner and unfold" uses two expressions in Japanese: "Corner-OBJECT corner-*e* fold" and "Corner-OBJECT corner-*ni* fold," where both *e* and *ni* mean *to* in English. When unfolding is the next step, more people tend to use *ni* than *e*. The only difference between *ni* and *e* is that *e* implies gone forever and *ni* indicates staying only tentatively. English does not make this distinction.

There are two possible explanations for this difference:

1. Japanese viewpoints stop at where the paper has been, and then wait and see what happens to the paper, whereas English-speaking people's viewpoints move while they are *making actions* to paper, according to where they are doing so (of course, with *fixed reference points*);

2. English lacks such a semantic distinction and English-speaking people do not care for it. I take the former view, that Japanese diagrams refer only to *similarity*.

The reason that I do not take the second view, at least as the primary reason, is that the second view is a type of strict linguistic determinism, as strongly argued by Sapir and Whorf [Sapir 83]. They often cite an example of a language called Pirahã, an Amazonian language. Pirahã people do not have numeral terms except for *few, a few, many*. Some scholars, including Everett [Everett 05], often cite these as *one, two, many* for propagandistic purposes, which is wrong; the languages' numerical distinction is basically (almost) none, a few, versus many (e.g., [Pinker 07]), and, when asked in Pirahã, they cannot distinguish numerically between three fish and four fish. Everett and his followers argue that words determine concepts and thoughts [Everett 05]. However, Pirahã men (but not women) speak nearly perfect Portuguese for trade with other tribes, and, in such a situation, they understand numbers because they have to understand money and goods to be exchanged. This, as Pinker and Chomsky (although quite indirectly) point out [Pinker 07, Chomsky 95], shows that the meaning does not directly affect thoughts. In the "language" of thoughts, we do have the ability to distinguish between minute differences. However, in real language (a "meta-"language of the thought language), the distinction is categorized in different ways according to the conventions of linguistic communities, as Saussure and Chomsky suggest (the difference between Saussure and Chomsky is whether they take such conventions as social or biological) [Saussure 83, Chomsky 95]. Thus, I would say that the *-e/-ni* fact accords with the linguistic determinism, but only indirectly, because it does not affect our thoughts, which is why English readers can understand the distinctions when explained.

This paper discusses such viewpoint differences in verbal instructions in origami diagrams in Japanese and English. After semantic considerations, the linguistic and/or psychological causes and effects of the differences are considered. Origami is taken for granted in Japanese society outside the world of folders, as easily shown by the following facts: first, the oldest published origami book, *Hiden Senbazuru Orikata*, was published in Japan; second, laymen generally consider origami something Japanese; finally, Japanese frequently answer "origami" when they are asked, "Name whatever you think is a feature of Japanese culture"—along with kimono, tea ceremony, flower arrangement, and so on.[1] This perception has made it hard to construct rigid methods and devices for teaching or telling how to fold, and maybe results in the folders' attitudes toward origami as artwork. In particular, the differences alter creators' psychological attitudes. Because they have not studied anything taken for granted, discrepancies

---

[1] In this sense, I do not consider my claiming origami being taken for granted an "attitude," but as fact. We do see tons of origami books, from children's to adults', in regional bookstores, and, as a college professor, I encounter students planning to study abroad who are ashamed of not being able even to fold a crane. These factors all show that origami is something presupposed to be a part of Japanese culture.

are created between those who *know* origami and those who do not.[2] Thus, Japanese creators tend to be satisfied with creating only crease patterns, not caring for diagrams. Paper becoming something else *is* a model for Japanese, especially for younger creators of complex models.

2 Previous Studies on Origami Terms

Attempts have been made with regard to fixing the technical terms of origami worldwide, mostly from the Western world. On the Origami-L email list for folders, such a proposal or two (or more) have been made. Lang presents a long list of origami technical terms in the appendix of his seminal book on the mathematical foundations of origami design [Lang 03]. However, such attempts have hardly been made in Japan. The only thing that can be seen in Japan is a list of technical terms with diagrams on the front page of origami books. Given that there are no records of it so far, it seems that Japanese have not paid attention to fixing the terms, or, even if they have, they could not find an effective way of doing so,

Interestingly, though, the Japanese have always shown interest in base forms, i.e., patterns on which various models can be based. Uchiyama Kosho, for example, lists patterns, but not technical terms, on the front pages of his famous epoch-making book [Uchiyama 62], and his father, Uchiyama Mitsuhiro, is quite famous for the patterns themselves, as shown in his *Tatou* book [Yanagi 88]. Moreover, one of the very first books on origami in Japan and in the world, *Hiden Senbazuru Orikata* [Akisato 97], basically shows only patterns. With regard to origami in particular, the Japanese seem to be disinterested in already established methods of folding per se, even in a book format where instruction occupies a very important part. Where have all the steps gone?

I have noted this issue previously [Tateishi 09], pointing out that, as diagrams and verbal instructions help each other, there are cases where too much verbal instruction can cause confusion for readers. I also pointed out that verbal instructions below the diagrams are only for security, so readers are not left alone in a sea of geometrical diagrams. Komatsu presents a significant article on how diagrams can be developed to avoid confusion

---

[2]All the facts that I will show appear to demonstrate that the Japanese are not good at instructions of origami, but that is not my point. Actually, what I am claiming in this paper is that Japanese folders today are not good at making truly effective origami instructions and diagrams, and I am searching for the reasons. I am not suggesting that people such as Akira Yoshizawa [Yoshizawa 96, Yoshizawa 99] and Kosho Uchiyama [Uchiyama 62] were bad in this respect, because their choices of diagrams, words, and diagrams were neatly done, given the poverty of methods for showing origami models in publication at that time. I am only speaking of recent folders and nonfolders in this paper, although I do not deny that these old masters influenced recent folders.

[Komatsu 03]. However, we both are interested in how the diagrams, not the terms, can be developed.

In this paper, I will add another aspect of the issue: Japanese origami terms are made vague in the first place. Whereas instructions written in English are very referential, those in Japanese are mostly either figurative or metaphorical. For English-speaking people, instructions below diagrams are clearly written instructive advice. Readers will read diagrams and see enough information on how to reproduce the models.[3] However, for the Japanese, instructions are only for help. This paper points the readers to some possible linguistic and pedagogical backgrounds of such differences. Finally, the paper contends that the vagueness in the Japanese origami world may lead to the recent shortage of diagrams of those complex models created by young Japanese origamists.

One may speculate that the competition on notational tools by Kawai, Uchiyama, Yoshizawa, and others, in the 1950s and 1960s might have contributed to the confusion in Japanese diagrams [Kawai 70, Uchiyama 62, Yoshizawa 96, Yoshizawa 99]. This possibility is, of course, one of the factors; however, the fact that this has continued even after the birth of complex origami models, initially perhaps triggered by Maekawa [Maekawa 83] is what I am trying to point out. Maekawa regained the status of crease patterns through the aid of Kunihiko Kasahara [Kasahara 96], and the origami instructions then made a shift in Japan. In the Western world, Peter Engel played a comparable role in reestablishing crease patterns in origami [Engel 89]. Even so, the diagrams had not changed, despite the efforts of people, including Makoto Yamaguchi and Kasahara, whose instructions are truly referential and clear [Yamaguchi 04, Kasahara 96]. This discrepancy *is* what I would like to point out and for which I seek reasons in this paper.

3   Theoretical Backgrounds

Even though I do not blindly accept the Sapir-Whorf hypothesis [Sapir 83], which says that language determines thoughts (and not vice versa), it cannot be denied that the way people speak somehow controls the way they think, exemplified in the frequently cited allegory of Japanese not speaking logically because their language is not logical. Suzuki, for example, points out that various words in Japanese have different ranges of meanings as

---

[3] Of course, I do not say that everyone can fold the models by this method. If there is an attitude on this point, it is perhaps my scientific attitude as a Chomskyan-theoretic linguist, which always sets an "ideal listener/speaker." I do not deny the existence of performance-dependent differences of individuals, but I will continue to assume ideal creators/authors throughout this paper, as this is the attitude linguists in my field must take.

compared with other languages, including English [Suzuki 73]. The word *lip*, unlike its correspondent *kuchibiru*, can have a mustache on it, for example. As language is a reflection of its cultural background, there is always a problem of translatability. It is the *mouth* (*kuchi*) that has a mustache on it in Japanese.

It is not clear whether the Sapir-Whorf hypothesis or its developments thereof have been accepted in the linguistic analysis of semantics in recent periods. Generative linguistics (e.g., [Chomsky 95]) has never been clear about it. Generativists have put semantics outside their scope, which is genetically endowed inside human beings. Cognitive semantics (e.g., [Lakoff 87]) is not clear on the issue either, because, even though proponents of cognitive semantics quote Whorf [Sapir 83], language for them is a matter of categorization and learning, behind which thoughts and inferences lie. However, we cannot deny that vagueness in verbal expressions is often caused by cross-linguistic differences of categorization of entities, materials, states, and events around us, although, as a linguist, I do not take it as the aforementioned strict linguistic determinism. Language may affect our thought, but not directly. Categorization of senses may affect our cognitive behavior, but this is outside of language; i.e., if categorization of senses by a word is clear-cut, thought will not seek for disambiguation, but if not, we often have to do so—this is not a necessity, on the contrary.

One example that shows the vagueness of Japanese with regard to origami is the verb *oru* (fold). *Oru* can be translated in various ways into English: *fold, bend, twist, break,* and so on. The only common feature of those various *oru*s is that it must be applied to a thin and/or flat base. In contrast, the English verb *fold* is not that ambiguous, and most variations of its meaning pertain to putting a layer onto something else. I will show later that the Japanese verbal noun *ori*, a nominal form of *oru*, can be used even for a case that by no means is folding for English-speaking people, and it is *this* vagueness that causes Japanese origami terms to become vague.

However, given the context of origami folding, *oru/ori* is, in most cases, interpreted as folding paper, with no necessity of disambiguation, which shows that the relationship between language and thought is at best indirect. What is at issue is that this disambiguation often "leaks."

4   Maze of (Un)Translatability

This section introduces various cases that illustrate the difficulties in translation between English and Japanese. As this covers most of the terms in the world of origami, the readers can see how difficult it is to translate one series of diagrams into another language. The terms differ in their degree of (un)translatability; the following sections show them classified into four distinct categories.

4.1   Translatable but Too Generic

Yama-ori and *tani-ori* are the only cases I could find for which direct translation is possible: *valley fold* and *mountain fold*, respectively. These terms have two notable features. First, the term could not possibly be a synthetic compound word in either of the two languages.[4] A synthetic compound word is a word composed of two words, the first of which is semantically the grammatical object of the second. For example, the word *origami creator* is a kind of synthetic compound word, as it refers to one who creates origami (models). In that sense, nobody can fold a mountain or a valley in English, nor in Japanese either. *Tani-o oru* (fold a valley) is simply meaningless. Thus, the meaning of these compound words can be interpreted only figuratively or metaphorically. "Fold like a valley/mountain" is perhaps a proper interpretation, which by itself is rather vague and allows for various interpretations.

This leads us to the second feature of these terms. Even though we find explanations of valley and mountain folds in the front or back pages of origami books, we hardly ever find them in actual diagrams. Rather, there are more cases of "fold and unfold," "fold corner-to-corner," "fold edge-to-edge," "fold in half," and so on. This means that, even though valley and mountain folds are the most popular folds in the world of origami, we seldom use the terms for them in actual instructions. The fact that they indicate the direction of folding does not refute my claim, because, in most such cases, the direction is the only issue that must be disambiguated, which perhaps is why the two terms are so widespread. This is also why the terms are often useful in oral instructions in origami classes, too, but this is off the point because I am speaking of the issue of translatability. The terms should rather be taken as "icons" of directions of folding, whose functions are mostly absorbed into crease lines on diagrams.[5]

4.2   Translatable but with Significant Differences

Next, two types of very popular folding techniques are taken up: inside and outside reverse folds and sinks. Reverse folds, inside and outside, have one

---

[4]A *compound word* in linguistics, which is my field, means two (or more) words joined to form a unit that functions as a word [Fromkin et al. 07]. They need not be joined without a space between them nor joined by a hyphen. In this sense, the word *mountain* in *mountain fold* is not an adjectival use of a noun, because, if this is the case we should have such forms as *completely mountain fold, mountain enough fold*, and so on, in which some other word(s) modify the "adjective."

[5]In this sense, they are good examples of vague origami terms well disambiguated, both in Japanese and in English.

feature in common: when it is done, it makes a reverse fold—one partially turns a layer over a folded edge. This is why the two folding techniques, inside and outside, are classified into one category, the reverse fold. Actually, in diagrams written in English, one does see many instructions with "outside reverse fold" or "inside reverse fold" used as a compound verb, perhaps because what the instruction means is strict and unambiguous.

In Japanese, however, the two folding techniques, which do the same action in different directions, are not categorized together; they are linguistically completely distinct. *Naka-wari ori* (inside break-in *ori*) is the term for inside reverse fold, and *kabuse ori* (cover *ori*) is the term for outside reverse fold. *Wari* in Japanese usually means completely dividing a piece into two, which *naka-wari ori* does not. In addition, *kabuse* in Japanese usually means covering and hiding something completely, which *kabuse ori* does not, either. Not only do the Japanese terms miss a very important common feature of the two folding techniques, but they also do not give us unambiguous instructions. *Naka-wari ori* means "fold like breaking pieces" and *kabuse-ori* means "fold like covering," other cases of figurative terms, unlike English.

In addition, the reverse folds and *nakawari-* or *kabuse-ori* have other significant differences. In English diagrams, reverse folds do not necessarily mean only folding by reversing the sides of the sheet halfway. Sometimes, reversing completely can be called reverse folds. However, at least for many Japanese authors, *nakawari-* or *kabuse-ori* refers only to those folds whose outputs look like a beak, like the origami crane's beak. Then, what do the Japanese call those reverse folds that reverse the flap 180 degrees? They say, "Fold like *kabuse-ori*," or "Put the flap inside." These complete reversing processes are what *nakawari* and *kabuse* mean literally, but the Japanese use the terms only for something else. The two terms are already only figurative in Japanese.

The case of *sink* is more serious in Japanese; the English word *sink* means what it does: sink the tip inside layers along the crease lines. Because the original meaning of the instruction is clear enough, we can understand what *unsink* and *spread-sink* mean (*spread-sink* also *does* sink the tip onto the flat surface). What does Japanese use for *sink*? The most frequent case I see is to use the English word *sink*, no translation. That is, Japanese does not have a word for *sink*. The Japanese occasionally use the term *shizume-ori* (sink fold), but this term is problematic for two reasons. First, sinking does not usually create any new crease line, so that it is *tatami* (fold along the crease lines), not *ori* (dividing a layer into two by a crease line), with the meaning of *ori* being figurative again. Second, because the meaning of *ori* is figurative, the meaning of *shizume* also must be (if *sink* is *ori*, then the focus must be on the crease lines created, which do not exist; with a nonexistent focus, one cannot make any action of sinking, strictly speaking). *Shizume-ori* means "do something like sinking by doing something like creating new crease lines."

4.3   Directly Untranslatable Cases

Unlike the clear distinctions in English between *pleat* and *crimp* [Lang 03], Japanese does not usually distinguish between the two, and both are called *dan-ori* (step *ori*). As a result, *dan-ori* is not usually a series of steps; it is again a very figurative and vague expression.

Japanese, however, do use the term *jabara-ori* (bellows *ori*) for box-pleating or pleating through the whole edge of the sheet (again, the result of *jabara-ori* is not bellows). This term corresponds to either of Lang's pleat or crimp, depending on how the surface layer is used, making the direct translation between English and Japanese impossible.

The English *swivel* and Japanese *hikiyose-ori* (pull-onto *ori*) differ in what part in the action is the focus. *Swivel* focuses on the first motion on the layer and on the pivot on which another layer rotates, but *hikiyose-ori* focuses on the second layer pulled up onto the initially folded layer. In terms of an action of *hikiyose* (pulling onto), Japanese *hikiyose-ori* establishes the desired result only indirectly. All of these points, again, make the translation between the two languages extremely difficult. When the English-speaking say, "swivel fold," we often have to make an explanation such as, "fold a layer and then flatten the raised-up layer onto it."

4.4   Lost in Translation

Although it is a special case, the term *Elias-stretch*, named after Neal Elias [Lang 03, p. 464], does not have a Japanese translation, even though the term used for pinching and stretching the pleat-folded layers into a long flap is a frequently used technique in current complex models. When the English-speaking say, "Elias-stretch," we have to divide the process into ten or more steps of pinching the layers and folding the pleat inside. Of course, this is a truly special term named after the individual who introduced the technique. However, English managed to name it, and Japanese did not.

5   Referentiality/Deictic versus Similarity: The Role of a Japanese Verbal Noun *Ori*

This paper thus far has described cases for which translation between English and Japanese is either difficult or impossible. It is surprising that these cover almost all terms of origami. I hypothesize that the difference lies in Japanese terms being ambiguous or vague and in English terms having a reference- or action-based nature. Take, for example, the Elias-stretch described above: even though it is a term for experts, when advanced folders look at the crease patterns and see the instruction "Elias-stretch," they know exactly what they should do.

The point can be extended to the word *sink*, discussed in Section 4.2. "Sinking" presupposes the existence of the pre-creased lines, and the action to be completed is to sink the tip of the corner into clusters of layers along those lines. This is exactly what Japanese folders do not have words to express. *Shizume-ori* is just a simile of what *sink* does: "(Make actions like) fold(ing) like sinking the tip." English origami terms are referentially and deictically unambiguous, given appropriate contexts in diagrams (or even without it); however, when we see diagrams in Japanese, there are still details to fix and struggling with diagrams.

Why are Japanese terms so indirect and figurative? The author hypothesizes that the secret lies in the term *ori* itself. Unlike English nouns such as *sink, pleat, crimp, fold,* and *reverse* that can also be verbs, Japanese *ori* is a noun, derived from the verb *oru* (to fold). The fact that Japanese origami terms can never be verbs can be shown by the lack of expressions like *Shizume-ori-nasai!* (Do sink!), the imperative form, or *shizume-otta* (Sank), the past tense form. This means that terms such as *tani-ori, yama-ori, kabuse-ori, nakawari-ori, dan-ori, jabara-ori, shizume-ori,* and so on are all compound nouns. It is well known, even in English, that already-established compound nouns are not interpreted synthetically but only figuratively. For example, a *blackboard* is never black these days, a *flatfoot* may put on high shoes, and a *jumping bean* may not be jumping now. The same situation applies to Japanese origami terms: they are only figuratively interpreted. Being figurative, Japanese origami terms cannot point the readers to the fixed reference points, so they rely on diagrams and/or crease patterns. It is the combination of diagrams and verbal or written words that have the readers follow the instructions. No over-the-phone teaching is possible with Japanese origami terms.

Such vagueness and figurative nature are seen in other origami terms, too. For example, the meaning of the Japanese term *kado* (literally, corner) is ambiguous between *corner* and *flap*, and *fuchi* (literally, edge) between *edge* and *layer*. Here again, Japanese terms do not have fixed points as reference.

Ori is a very convenient term because it can be attached to anything and create a new folding method. However, because of its convenience and its figurative nature, terms with *ori* are not that stable and easily become obsolete. For example, the term *fukuro-ori* (bag *ori*) means "open and squash" and actually is very convenient to express this series of folding actions. However, it is hardly used these days, and instead, a rather lengthy expression such as *uchigawa-o hirai-te tubusu-youni oru* (fold so that the output will look like it is opened inside and squashed) is used.

Similarly, *kannon-ori* (fold edges to the center line), *kaben-ori* (petal fold), and *zabuton-ori* (blintz fold) are rarely used these days.

In sum, English origami terms refer to actions so that readers know what to do immediately, whereas Japanese origami terms are often vague and/or figurative, or ad hoc, so that, even though they are convenient, they cannot be used as fixed technical terms.

6   Why Don't the Japanese Use Verbs?

Then, the natural question is, why do the Japanese not use compound verbs instead of nouns for origami terms? The reason lies in the linguistic system of Japanese.

First, unlike English, Japanese is a head-final language, which means that the core term of the phrase always comes at the end. For example, in a phrase *origami-o oru* "fold origami," which is about action, the term of the action, the verb, comes at the end. That is why, when English-speaking people say "Elias-stretch," the Japanese have to say:

> [[[[[*kado-o*] *tsubusu*]-*you-ni*] [[[*dan-o*] *tsuman*]-*de*]] [[[*naga-i*] *kado-o*] *tsukuru*]]
>
> (literally,   corner-OBJECT-squash-DATIVE   pleat-OBJECT pinch-and long-PRESENT corner-OBJECT make),

where the brackets show grammatical meaning units). In this expression, all the disambiguating elements of the phrase and the phrases therein are all at the end in the form of grammatical particles. This means that when one expresses the same concepts, it necessarily takes more words in Japanese than in English to do so.

Then, why not "Elias-stretch"? Here, the restriction on Japanese compound verbs is relevant. Japanese does not make much use of compound verbs in the first place, except for those ending with aspectual verbs like *-hajimeru* (start –ing), *-owaru* (finish –ing), *-tsuzukeru* (continue –ing), *-teiru* (be –ing), and so on. With other common verbs, Japanese has only fixed forms like *naname-yomu* (literally, slant-read meaning scan through). Why? Because Japanese has a denominal verbal suffix *-suru* (do). In Japanese, *-suru* can attach to virtually all nouns and make meaningful verbs. For example, *pasokon* (personal computer) can create with this suffix a compound verb *pasokon-suru* (use a personal computer) and similarly *doroboo* (thief) can create *doroboo-suru* (do robbery). As these two examples show, the semantic relations between a noun and a compound verb with *suru* are completely arbitrary. Of course, one can create words, such as *taniori-suru*, *nakawariori-suru*, and so on. One can even make *sink-suru*. Because of this very convenient term, the Japanese do not have to rely on compound verbs or any verb at all, but can just make compound nouns for origami. This fact, in turn, leaves Japanese origami terms vague, as discussed above, because compound nouns universally tend to be vague.

I have pointed out that, even though written instructions appear below almost all American diagrams, Japanese diagrams have written instructions only where necessary (for about 60% of all diagrams) [Tateishi 09].[6] Instructions below diagrams are not actually instructions; the real instructions are on the diagrams, with separate written instructions only providing security for readers in the sea of diagrams. I believe this explanation is still correct. Japanese reluctance to use words below diagrams most probably originates from the vagueness of Japanese origami terms, which in turn makes them rely on diagrams or crease patterns.

7 Further Considerations

Even though the Japanese linguistic structure—its morphological structure in particular—prevents Japanese diagrams from being truly explicit and explanatory, the Japanese do have a way around this problem; for example, they could use explanations such as, "Find the crease line just below the top corner, and find also the clusters of layers around the top corner. Now, you slightly open them and push the corner inside. The result must lock the clusters of layers," in place of "Closed sink-*suru*." Why do they not care for it?[7]

The main explanation for this lack of enthusiasm is that the Japanese these days cannot fold.[8] In recent years, Japanese kindergartens and elementary schools hardly use origami in their art and/or math classes. There are several possible reasons: (1) teachers and parents cannot fold origami well these days, perhaps not even a crane; (2) children raised by such origami-ignorant people cannot fold, either. In my semester-based origami class in college, students are certainly interested in origami, but when I ask one or two students to explain and/or teach how to fold a crane, they do not have words to explain it, even in Japanese. Only about half of the students can fold a crane properly. If a crane, the most famous Japanese origami model, produces such disastrous results, outcomes with other models cannot be better; most students cannot fold a helmet, a waterbomb, twin boats, or other popular Japanese models.

---

[6]I counted origami convention books, because there are individual differences in every country. Satoshi Kamiya, for example, uses words below most diagrams he draws (e.g., [Kamiya 05]).

[7]Actually, Akira Yoshizawa cared for such matters in many of his books. His instructions have been very clear both graphically and verbally [Yoshizawa 96, Yoshizawa 99].

[8]Here, I am speaking of the general public, not of those who attend origami conventions.

I have been teaching origami and its history for five years to students at a women's college, with about 200 students in total. At the beginning of the semester, I give a questionnaire to students to assess their "origami history." Out of 200 students, only 5 said that they had learned something using origami in their school days, no one recalled learning origami in kindergarten, 15 learned origami from their relatives, 158 did not own an origami book, and 126 said the course offered the first opportunity to fold an origami model. Competing activities, such as watching TV, playing video games, and so on, might have contributed to their ignorance of folding origami models. It should be noted that male complex origami folders in Japan are mostly game and/or anime obsessed, which may contribute to male/female differences. Unless they are interested in something tricky and puzzle-like, Japanese children do not even think of folding, and educational efforts in this direction, which is very important to develop their spatial-perceptual faculty, are not emphasized in Japan.

For the Japanese, origami had long been, and still is, taken for granted too much, even though it has practically been ignored in educational institutions today. I often hear the following claims: "Origami books are too difficult to read," "It is impossible to decipher series of diagrams with virtually no instruction at all," and so on. From these statements, I deduce that origami books in Japan, except for some truly good ones, are not for teaching how to fold. Because origami is taken for granted in Japanese society, schools do not even think of teaching folding; this oversight must be corrected. In Japan, origami books mostly seem to be for telling how creators have invented new art forms.[9]

Because there is virtually no origami-based education in Japanese society today, even origami enthusiasts are not taught anything about origami. They somehow manage to decipher ways to fold from hard-to-read diagram books. Because they are not taught, brand-new creators are not generally interested in teaching. They may teach a class, but most of them just hand out a sheet or two with diagrams and/or crease patterns, and say, "Raise your hand if you have questions," which in no sense is teaching. Perhaps, their interest is only in showing their brand-new models, but not sharing them with anybody in the sense of "reproduction," which sharing origami models often means. They therefore do not and cannot work as a resource for the development of the world of folding.

In 1999, when Origami Tanteidan changed its name to JOAS (Japan Origami Academic Society), *The Origami Tanteidan Magazine* started a series called "Crease Pattern Challenge," which shows only crease patterns

---

[9]Of course, from the creators' viewpoints, origami books publicize their own origami artwork. Creators decide what to put in their own books and how to tell readers what they are (and quite often how to explain how to fold them). From the readers' viewpoints, the authors only state how all these work.

and completed models, and readers are supposed to fold the models by deciphering the complex crease patterns. This, in the end, became reinvention, with the different name of "base forms." From the varieties of crease patterns shown in this series, new folders discover their own "base forms," even though it may be just 40×40 box pleats, and create their new models from them. Because crease patterns cannot be "taught," the new creators are not interested in teaching. The result will be a shortage of good origami model resources in the form of well-written diagrams, in which new generations of creator/folders are not necessarily interested.[10] Even though it may be a recent occurrence, it actually may be a reincarnation of old Japanese creators' reliance on "base forms." This approach has been prominent with relatively older generation complex model folders, but it is increasingly becoming true among younger generations.

Take, for example, a comparison of authors of Crease Pattern Challenges and the centerfold diagrams in *The Origami-Tanteidan Magazine*. After the academic year 2005 (starting from April 2005), there have been 33 issues of *The Origami Tanteidan Magazine* with 32 Crease Pattern Challenges and 36 centerfold diagrams. Among Crease Pattern Challenges, 13 crease patterns (CPs) are created by folders younger than Satoshi Kamiya (the reason why the author chose Satoshi Kamiya is only arbitrary, but the fact that he was featured in the now historic origami magazine *Oru* from Sojusha may justify that—he is the youngest of the "older generation") as opposed to 13 CPs by the older generation and 6 CPs by non-Japanese folders. However, in the centerfold diagrams, only 4 sets of diagrams are contributed by the younger generation, 28 by the older generations, and 4 by non-Japanese folders. I will not even try to draw statistical conclusions from these numbers, but these facts may show that younger creaters do have models, but they do not try to put them in diagram-form instructions. The complexity of their models is, of course, relevant. The more complex the model, the harder it is to draw sequential diagrams. I would like to emphasize the fact that it is harder to see models created by younger generations, except in exhibitions of origami artwork. Origami, on the one hand, is an established form of art, but the world of origami, on the other hand, needs to publicize its works to develop new prospective artists, because most of those who enter the world of origami start by mimicking their predecessors' works. Note that their models are not like those by Eric Joisel [Joisel 10], whose creations are unique.

As origami is taken too much for granted in Japan, creators have no idea of how to start folding from scratch, and they attribute their models to

---

[10]I do not mean that the Crease Pattern Challenge is harmful. It just shows a dichotomy of origami enthusiasts: those who can be nurtured by self-oriented learning and those who must be taught. I personally am interested in the effects of the spread of crease pattern disposition of models exported from Japan into the Western world these days, which, of course, takes time to investigate.

base forms and/or crease patterns. There has been no successful attempt to reconsider this cross-generational imprinting, so the Japanese cannot think of words for diagrams.

English, however, happens to be well suited as an instructional language. In addition to its relatively fixed word order, it has fewer inflectional systems than other Western languages—due to its historical contact with French, Latin, and Greek and to the United Kingdom's past colonialism—making it an across-the-board *lingua franca* of the world. (In other words, it is easier to learn than other western languages, grammar-wise.) In contrast, Japanese relies on coining new words when one has to give instruction, as in the *-ori* case discussed in Section 5. Furthermore, all the grammatical morphemes as modals, tense, aspects, voice, and so on, are stuck at the end of the sentence as a part of a single verb/auxiliary, which makes it harder to decipher instructions, even for the Japanese. This, in addition to the attitudinal points identified in this section, may have made the Japanese not rely on written instructions.

8   Conclusion

The Japanese cannot think of unambiguous terms for origami partly due to their traditional attitudes toward origami and partly due to the influences of the Japanese language's linguistic structure. With base forms and/or crease patterns as an escape hatch, the Japanese did not, and do not, seriously think about terms for instruction. The situation was different in the Western world because enthusiasts had to teach each other in the most effective ways. For creator-folders, such a situation is harmless. For novices in Japan, however, origami is now a form of art with no good instructive media, with a few exceptions, and they are hesitant to jump into the world of the rich varieties of art forms. I strongly feel that the teaching methods for origami as well as its terms must be seriously studied.

Bibliography

[Akisato 97] Rito Akisato. *Hiden Senbazuru Orikata.* Kyoto: Yoshino-ya Tame-hachi, 1797.

[Chomsky 95] Noam Chomsky. *The Minimalist Program.* Cambridge, MA: The MIT Press, 1995.

[Engel 89] Peter Engel. *Folding the Universe.* New York: Random House, 1989.

[Everett 05] Daniel L. Everett. "Cultural Constraints on Grammar and Cognition in Piraha: Another Look at the Design Features of Human Language." *Current Anthropology* 46:4 (August–October 2005), 621–646.

[Fromkin et al. 07] Victoria Fromkin, Robert Rodman, and Nina Hyams. *An Introduction to Language*, eighth edition. Boston: Thomson/Wadsworth, 2007.

[Joisel 10] Eric Joisel. *Eric Joisel: The Magician of Origami*. Tokyo: Origami House, 2010.

[Kamiya 05] Satoshi Kamiya. *Works of Satoshi Kamiya 1995–2003*. Tokyo: Origami House, 2005.

[Kasahara 96] Kunihiko Kasahara. *Joy of Origami*. Tokyo: Sojusha, 1996.

[Kawai 70] Toyoaki Kawai. *Origami*. Osaka: Hoikusha, 1970.

[Komatsu 03] Hideo Komatsu. "Orizu Hyogenni Tsuite" (Expressions in Diagrams). *Origami Tanteidan Magazine* 78 (2003), 11–13.

[Lakoff 87] George Lakoff. *Women, Fire, and Dangerous Things*. Chicago: The University of Chicago Press, 1987.

[Lang 03] Robert J. Lang. *Origami Design Secrets*. Natick, MA: A K Peters, 2003.

[Maekawa 83] Jun Maekawa. *Viva! Origami*. Tokyo: Sanrio, 1983.

[Pinker 07] Steven Pinker. *The Stuff of Thought: Language as a Window into Human Nature*. New York: Viking, 2007.

[Sapir 83] Edward Sapir. *Selected Writing of Edward Sapir in Language, Culture and Personality*. Berkeley: University of California Press, 1983.

[Saussure 83] Ferdinand de Saussure. *Course in General Linguistics*. Eds. Charles Bally and Albert Sechehaye. Trans. Roy Harris. La Salle, IL: Open Court, 1983.

[Suzuki 73] Takao Suzuki. *Kotobato Bunka* (Language and Culture). Tokyo: Iwanami-Shoten, 1973.

[Tateishi 09] Koichi Tateishi. "Redundancy of Verbal Instructions in Origami Diagrams." In *Origami⁴: Fourth International Meeting of Origami Science, Mathematics, and Education*, edited by Robert J. Lang, pp. 525–532. Wellesley, MA: A K Peters, 2009.

[Uchiyama 62] Kosho Uchiyama. *Origami*. Tokyo: Kokudosha, 1962.

[Yamaguchi 04] Makoto Yamaguchi. *The A to Z of Origami in English*. Tokyo: Natsumesha, 2004.

[Yanagi 88] Soetsu Yanagi. *Origami Flower Patterns: World of Uchiyama*. Tokyo: Geisodo, 1988.

[Yoshizawa 96] Akira Yoshizawa. *Inochi Yutakana Origami* (Origami with Life).
    Tokyo: Sojusha, 1996.

[Yoshizawa 99] Akira Yoshizawa. *Origami Tokuhon I* (Readers on Origami I).
    Tokyo: New Science-Sha, 1999.

Betsy Ross Revisited: General Fold and One-Cut Regular and Star Polygons

Arnold Tubis and Crystal Elaine Mills

1 Introduction

The five-pointed star of the American flag has been linked to a meeting in Philadelphia in May or June, 1776, between Betsy Ross and a committee headed by George Washington. Ross advocated the use of a five-pointed star pentagon (*pentagram*) instead of a six-pointed star in the flag, and demonstrated how easily a five-pointed star could be made by a fold and one-cut technique. Although historically controversial, this incident has been widely cited in testimonials, articles, books, and on the Internet, and mentioned in introductions to modern fold and one-cut algorithms (e.g., [Demaine and Demaine 04, Demaine and O'Rourke 07]).

2 Historical Sources for the Story

There are several recent compilations of, and commentary on, the historical bases for the Betsy Ross flag story [Timmins and Yarrington 83, Harker 05, Miller 10, Independence Hall 10]. The following summary is largely based on these sources.

Elizabeth Griscom Ross Ashburn Claypoole (1752–1836) (known as Betsy to family and friends, and thrice married and widowed) was for many years an upholsterer and flag maker in Philadelphia. Documents

do indeed exist showing that she made a flag for the Navy during the
Revolutionary War. An account of her meeting with then Colonel George
Washington, Robert Morris, and Colonel George Ross in 1776 (for which
no known official documentation exists) was first publicly delivered in a
March 1870 speech by one of her grandsons, William B. Canby (1825–
1890), before The Historical Society of Pennsylvania. This speech was
followed shortly afterward by affidavits concerning the meeting by one of
Betsy's daughters, Rachel Fletcher (1789–1823), in July 1871; one of her
granddaughters, Sophia B. Hildebrandt (1806–1891), in May 1870; and
one of her nieces, Margaret Donaldson Boggs (1776–1876), in June 1870.
Harker points out another written source for the flag story: a letter in
1903 from Rachel Albright (a granddaughter of Betsy) to a friend, Nellie
E. Chaffee, whose daughters were interested in Betsy's life story [Harker
05]. The letter is archived in the American Flag House and Betsy Ross
Memorial in Philadelphia. Collectively, the Canby speech, the three affidavits, and Rachel Albright's letter state that the meeting occurred in the
year 1776, shortly before the signing of the Declaration of Independence,
and that Betsy suggested changes in the flag design initially proposed by
Washington. Boggs, Canby, and Albright all explicitly refer to Betsy's
fold and one-cut five-pointed star. Fletcher's affidavit states that Betsy
proposed a 4 × 3 rectangular rather than a square flag shape and an arrangement for the stars in lines or in some adopted form as a circle, or
a star. The full texts of the Canby speech and the three affidavits are
available online [Independence Hall 10]. Demaine and Demaine [Demaine
and Demaine 04] and Demaine and O'Rourke [Demaine and O'Rourke 07]
reference an article [Wilcox 73] in the July 1873 issue of *Harper's New
Monthly Magazine* containing the Betsy Ross story described above, with
the year of the meeting given as 1777 instead of 1776. Wilcox was probably influenced by the date (June 14, 1777) of the so-called Flag Resolution of the Continental Congress. Harker offers a critical discussion
of the significance of this resolution [Harker 05]. It is not evident from
the article that the author was aware of the Canby speech and the three
affidavits.

Harker has compiled a considerable body of pre-1870 evidence in support of the 1776 events as described by Canby, Fletcher, Hildebrand, Boggs,
and Albright [Harker 05]. His basic assertion is that Betsy Ross sewed one
or more flags for Washington that featured thirteen five-pointed stars arranged in a circle, and that the Army carried these flags at the Battles of
Trenton (December 1776) and Princeton (January 1777). The key pieces
of evidence that Harker presents for these assertions are the following:

1. Roberts cites an article in the 1909 publication, *The Journal of American History* by Mrs. Katherine (Wright) Bennett (1854–1944) that
   recounts the life of Rebecca Prescott Sherman (1743–1813), the sec-

ond wife of Roger Sherman of Connecticut (1721–1793), signer of the Declaration of Independence [Roberts 04]. The article states that Rebecca heard from Roger about a flag, ordered by George Washington, being made and that she visited Betsy Ross and actually assisted her in sewing stars *on the very first flag of the young nation.*

2. In a February 29, 1896, interview in the *Harrisburg Telegraph,* retired General Edward C. Williams (1820–1900) described his experiences in the war with Mexico. His story is confirmed in all specific details by a report of Major William Brindle, Commanding Second Brigade, Volunteer Division, prepared September 15, 1847, the day after the taking of the Mexican Fortress of Chapultepec on the outskirts of Mexico City. Major Brindle reported that after the surrender of Mexican General Bravo, then Captain Williams *ascended to the top of* [the Fortress] *with the first flag made by Betsy Ross, of Philadelphia, which was presented to Washington before the Battle of Trenton, during the Revolution of 1776, which Captain Williams had obtained from the State Library in Harrisburg.* General Williams was born in Philadelphia in 1820 and lived there until 1838. He must have been familiar with the story of Betsy Ross and the American flag at about the time of the fiftieth anniversary of the signing of the Declaration of Independence and the American Revolution.

3. A 1784 painting by Charles Wilson Peale (1741–1827) titled *Washington at the Battle of Princeton, Jan. 3, 1777,* shows Washington having epaulets with five-pointed stars on his shoulders and a flag in the background with five-pointed stars, presumably arranged in a circle. Peale participated in the Battles of Trenton and Princeton, and was known for the painstaking attention to detail in his paintings.

4. A portrait (1832) of Betsy Ross by the famed artist Samuel L. Waldo (1783–1861) suggests that Betsy was considered a person of significance in Philadelphia.

5. A painting (1851) by the artist Ellie Wheeler (1816–1896) shows Betsy Ross and three men, with Ross having a flag on her lap with five-pointed stars in a circle. Wheeler grew up in the neighborhood where Betsy lived and was 20 years old when Betsy died in 1836. It is thus reasonable to assume that the story of the "committee" and Betsy's role in flag designing/making was common knowledge in Philadelphia.

6. A painting (Germany, 1851) by the artist Emanuel Leutze (1816–1868) entitled *Washington Crossing the Delaware* features a flag with a circle of five-pointed stars. Leutze also lived in Philadelphia in the 1820s and 1830s.

7. In a draft of three proposed frescoes for the Ladies Waiting Room (1856) by an architectural draftsman for the consideration of Constantino Brunidi (an Italian artist in the payroll of the Capital Building), one fresco shows a woman (Betsy Ross?) presenting a flag to three men (the committee?), with more uniformed men in the background.

8. The main focus of the remainder of this paper is the Pattern for Stars artifact [Harker 05, Timmins and Yarrington 83]. (See Figure 1.) At a 1963 luncheon meeting of the Women's Committee of the Philadelphia Flag Day Association, Reeves Wetherill of the Society of Free Quakers presented a sample folded five-pointed paper star pattern (a folded 5×8 piece of paper with a partial cut that shows how to obtain a pentagram). We will refer to it as the *Pattern for Stars artifact.* Wetherill explained that the artifact came from an old safe, which his father, Abel Wetherill, had caused to be opened in 1922. Other contents of the safe included a pistol and an old deed signed by John Penn (1729–1795), colonial governor of Pennsylvania. Reeves and Abel Wetherill were descendants of Samuel Wetherill (1736–1816), one of the founders of the Society of Free Quakers.[1] John and Elizabeth (Ross) Claypoole became members of the society in 1785. Harker reports that, according to the present-day clerk of the society, the safe did *not* belong to Samuel Wetherill, but to a Wetherill of a later generation, but that it was clear that the artifact had been in possession of someone associated with the society since the pattern was created [Harker 05]. There are four lines of writing on the corner of the paper in lead pencil as follows:

H.C. Wilson
Betsey Ross
Pattern for
Stars

Betsy is spelled as "Betsey" and the "H" appears to overwrite a "W." "Wilson" might possibly refer to Clarissa Claypoole Wilson (1785–1864), Betsy's daughter, who was widowed in 1812 and moved from Baltimore to live with Betsy in Philadelphia.

Although the identity of the person(s) who made the artifact and penciled in the four lines will probably never be determined, *the artifact was apparently considered important enough to be stored for safekeeping along with other items of historical significance.* Photographs of the artifact are found in two publications [Harker 05, Timmins and Yarrington 83]. For

---

[1] The Society of Free Quakers continues to exist as a philanthropic organization.

Figure 1. Pattern for Stars artifact with the four-line penciled inscription, Society of Free Quakers, Philadelphia (undated). The partial cut slants downward from the lower portion of the top edge. Photograph by John Balderston Harker, used with permission.

many years, the artifact was on display at the Society of Free Quakers Meeting House in Philadelphia.

3   Replicating the Pattern for Stars Artifact

Figure 2 presents fully land-marked steps for folding a $5'' \times 8''$ piece of paper (the same size paper used in the Pattern for Stars artifact in Figure 1) and obtaining a pentagram with a single cut, just as Betsy Ross supposedly demonstrated in 1776. It is not evident from photos of the Pattern for Stars artifact which, if any, folding landmarks were actually used.

An exact procedure for a fold and one-cut pentagram requires the division of a straight angle into five angles of measure $180°/5 = 36°$. This construction can, of course, be done by folding a golden-ratio triangle (e.g., [Row 05]). However, the required folding is hard to implement accurately with ordinary paper. We therefore use instead, in Steps 3 through 10, the first stage of the Fujimoto iterative method for division of a length or angle into an odd number of equal portions (e.g., [Huzita and Fujimoto 97; Hull 06, pp. 15–26]). This procedure is sufficiently accurate for most practical purposes. For the idealized case of infinitely thin and flexible paper, it gives the division of a straight angle of measure $180°$ into three angles of measure $35.78° = (180° - \arctan(0.75))/4$ and two of measure $36.32° = (180° - 3 \cdot 35.78°)/2$. All other folding steps are theoretically exact. Very thin paper is required if the folding of all layers in Step 11 is to be done accurately. If the paper is too thick, the crease lines of the regular pentagon shape in Step 12 should be made with the paper unfolded.

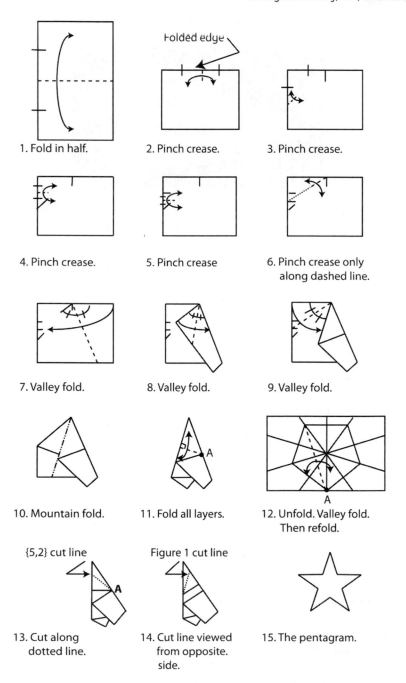

1. Fold in half.

2. Pinch crease.

3. Pinch crease.

4. Pinch crease.

5. Pinch crease

6. Pinch crease only along dashed line.

7. Valley fold.

8. Valley fold.

9. Valley fold.

10. Mountain fold.

11. Fold all layers.

12. Unfold. Valley fold. Then refold.

{5,2} cut line

Figure 1 cut line

13. Cut along dotted line.

14. Cut line viewed from opposite. side.

15. The pentagram.

Figure 2. Folding and one cutting the pentagram. The partial cut line in the Pattern for Stars artifact (Figure 1) overlays the cut line shown in Step 14.

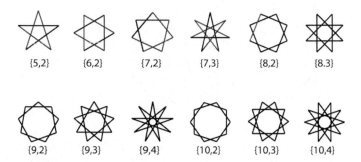

Figure 3. Star polygons, $\{p, x\}$, for $p = 5, 6, 7, \ldots, 10$. The corresponding regular polygons are obtained by joining the neighboring vertices with straight lines.

4 Generalizing the Betsy Ross Method to Fold and One-Cut Any Regular or Star Polygon

A straightforward generalization of the method used to obtain the Betsy Ross Pattern for Stars artifact gives *any* arbitrary regular polygon and its associated star polygon(s). Tubis and Mills previously reported a similar general procedure [Tubis and Mills 09]. It should be noted that the problem of determining the relevant folding for a fold and one-cut polygon (or group of polygons) starting with an *unmarked* piece of paper with no initial creases is different from that solved by Demaine and O'Rourke [Demaine and O'Rourke 07]. In the latter, the polygonal shapes are assumed to be initially inscribed on the paper, and then a general algorithm is derived for folding the paper so that *all* of the polygonal lines are superimposed on top of one another.

A regular polygon is one in which all of the side lengths and interior angles are equal. The sides and bisector lines of the interior angles of the polygon divide a p-sided polygon into p congruent isosceles triangle sections with vertex angle $(180/p)°$. The (regular) star polygon $\{p, x\}$ is a p-pointed star defined by line segments starting at each vertex of the regular p-sided polygon and ending at another vertex, with $x - 1$ vertices in between (e.g., [Coxeter 73, pp. 93–94; Caglayan 08]). It is easily seen that $2 \leq x < p/2$. Thus the $\{5, 2\}$ star polygon (pentagram) is the one with the fewest number of points. Star polygons for $p = 5$–10 are shown in Figure 3.

The first part of the general procedure is to fold the paper in half lengthwise, then to form a triangular section with an internal angle of measure $180°/p$ and with the vertex at the center of the folded edge and one ray along the folded edge, and finally to use this section as a template to fold the piece as in Steps 10 and 11 of Figure 2. Theoretically, this can be done exactly by folding for $p = 3$ through 10 (e.g., [Geretschläger 08]).

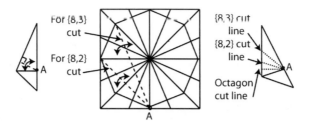

Figure 4. Correspondents of Steps 11–13 of Figure 2 for the case of a regular ($p = 8$) octagon and its star octagons, $\{8, 2\}$ and $\{8, 3\}$, from a starting square.

The method is particularly simple and well known for $p = 3$, 4, 6, and 8. A practical approximate method for $p = 5$ has already been given in Section 3, and may be applied to the case of $p = 10$ by angle bisection. For $p = 9$, one can use a folding procedure to trisect an angle of measure 60°. For $p = 7$, one can use a methodology similar to that for $p = 5$ by noting that $\arctan(0.5)$ ($= 26.56°$) is fairly close to $180°/7$ ($= 25.71°$), and again using an early stage of the Fujimoto procedure to achieve a better approximation to $180°/7$. Also, a square is perfectly adequate as the starting paper shape instead of the 8×5 shape used in Section 3.

To make the general method clear, the correspondents of Steps 11–13 of Figure 2 for the case of a regular ($p = 8$) octagon and its two star octagons, $\{8, 2\}$ and $\{8, 3\}$, from a starting square are given in Figure 4.

5   Discussion

Although a considerable body of circumstantial evidence, such as that outlined in this paper, exists today in support of the Betsy Ross flag story, historians will probably never be able to definitively resolve the various controversies and issues (concerning, e.g., the true provenance of the Pattern for Stars artifact) that still surround it. Nevertheless, the story encompasses an extremely interesting confluence of history, origami, and mathematics (geometry and trigonometry) and, as such, provides the basis for many great educational opportunities in mathematics and history classrooms.

Acknowledgment. We wish to thank John Balderston Harker, a fifth generation descendant of Betsy Ross, for sharing with us his extensive research on her life and the history of the American flag, for encouraging us to write this paper, and for providing us with photos of the Pattern for Stars artifact—a possible direct link to his famous ancestor.

Bibliography

[Caglayan 09] Gunhan Caglayan. "Mathematical Lens: Star Polygons." *Mathematics Teacher* 101:6 (February 2008), 432–438.

[Coxeter 73] Harold Scott MacDonald Coxeter. *Regular Polytopes.* New York: Dover Publications, 1973.

[Demaine and Demaine 04] Erik D. Demaine and Martin L. Demaine. "Fold-and-Cut Magic." In *Tribute to a Mathemagician*, edited by Barry Cipra, pp. 23–30. Natick, MA: A K Peters, 2004.

[Demaine and O'Rourke 07] Erik D. Demaine and Joseph O'Rourke. *Geometric Folding Algorithms: Linkages, Origami, Polyhedra.* Cambridge, UK: Cambridge University Press, 2007.

[Geretschläger 08] Robert Geretschläger. *Geometric Origami.* Shipley, UK: Arbelos, 2008.

[Harker 05] John Balderston Harker. *Betsy Ross's Five Pointed Star:Elizabeth Claypoole, Flag Maker—A Historical Perspective.* Melbourne Beach, FL: Canmore Press, 2005.

[Hull 06] Thomas Hull. *Project Origami: Activities for Exploring Mathematics.* Wellesley, MA: A K Peters, 2006.

[Huzita and Fujimoto 97] Humiaki Huzita and Shuzo Fujimoto. "Fujimoto Successive Method to Obtain Odd-Number Section of a Segment or an Angle by Folding Operations." In *Origami Science and Art. Proceedings of the Second International Meeting of Origami Science and Scientific Origami*, edited by K. Miura, pp. 1–13. Shiga, Japan: Seian University of Art and Design, 1997.

[Independence Hall 10] Independence Hall Association. "Betsy Ross Homepage Resources: Historic Analysis." Available at www.ushistory.org/betsy/flagpcp.html, 2010.

[Miller 10] Marla R. Miller. *Betsy Ross and the Making of America.* New York: Henry Holt & Publishers, 2010.

[Roberts 04] Cokie Roberts. *Founding Mothers.* New York: Harper Collins Company, Inc., 2004.

[Row 05] Tandalam Sundara Row. *Geometric Exercises in Paper Folding.* Chicago: Open Court Publishing, 1905; reprinted by New York: Dover Publications, Inc., 1966.

[Timmons and Yarrington 83] William D. Timmins and Robert W. Yarrington. *Betsy Ross, the Griscom Legacy.* Salem County, NJ: Salem County Cultural and Heritage Commission, 1983.

[Tubis and Mills 05] Arnold Tubis and Crystal Elaine Mills. "The Betsy Ross Star Pentagon (Pentagram)." In *PCOC (Pacific Coast Origami Convention) Play*, edited by Boaz Shuval, pp. 240–241. New York: OrigamiUSA, 2005.

[Tubis and Mills 09] Arnold Tubis and Crystal Elaine Mills. "Paper Folding One-Cut Regular Polygons and Star Polygons from 8 1/2″ × 11″ Paper." In *Activities Across the Strands, Grades K–12*, 2009–2010 special edition, edited by Janet Trentacosta, pp. 44, 46–51. Clayton, CA: California Mathematics Council, 2009.

[Wilcox 73] H. W. K. Wilcox. "National standards and emblems." *Harpers Magazine* (July 1873), 171–181. (Available at http://www.harpers.org/archive/1873/07/0057479.)

Reconstructing David Huffman's Legacy in Curved-Crease Folding

Erik D. Demaine, Martin L. Demaine, and Duks Koschitz

1 Introduction

David Huffman's curved-crease models (see Figure 1) are elegant, beautiful, and illustrative of Huffman's fascination with curved creases. Huffman's death in 1999 left us without his deep understanding, but his many models and notes provide a glimpse into his thinking. This chapter presents reconstructions of some of David Huffman's curved crease patterns and models, aiming to recover his insight and uncover the mathematical beauty underlying the artistic beauty. These initial reconstructions represent the beginning of an ongoing project with the Huffman family to study and document David Huffman's work in folding.

The first known reference of curved-crease folding is the work of a Bauhaus student in a course by Josef Albers in 1927–1928 [Wingler 69]. This model has creases in concentric circles and a hole in the center. Since the 1930s, Irene Schawinsky, Thoki Yenn, and Kunihiko Kasahara have built similar models with variations on the pleats and the size of the hole [Demaine and Demaine 08]. Other intricate curved-crease origami sculpture was designed by Ronald Resch in the 1970s. From the 1970s to the 1990s, Huffman created hundreds of models, which represent the majority of the work done in this field [Wertheim 04]. Huffman inspired further work on curved creases [Fuchs and Tabachnikov 99] and research on finding the nearest proper folding that approximates a three-dimensional scanned physical model [Kilian et al. 08].

Figure 1. David Huffman and his "Hexagonal Column with Cusps." (Photograph courtesy of University of California, Santa Cruz.)

David Huffman was simultaneously studying the mathematics and the art of curved-crease origami. He analyzed the local mathematical behavior of curved creases in his paper "Curvature and Creases: A Primer on Paper" [Huffman 76] and made sculptures to further study this special kind of folding. Our goal is to better understand the behavior of curved creases demonstrated in his models, given the lack of mathematical and algorithmic tools for designing curved-crease origami.

2 Approach

We are experimenting with both physical models and computer models to reconstruct Huffman's work. We analyze Huffman's designs by carefully studying photographs, measuring his models, and studying features that occur frequently in his designs. Based on personal communication with David Huffman in 1998, we assume that almost all of Huffman's creases are conic section (quadratic) curves. His work spans many areas of mathematical origami; here we select a few designs that use several quadratic curves within a single design or combine curved and straight creases.

Figure 2. Huffman used a spring-loaded ball burnisher similar to this one.

2.1 Folding Methods

The Huffman family provided us with an opportunity to see Huffman's estate, which they manage, and to see a variety of working models and crease patterns. We studied his techniques and work methods by looking at his drawing tools, templates, and model-making equipment. Huffman transferred his crease patterns onto sheets of white, matte PVC ("vinyl") that is 0.01″ thick using French curves. He then traced the creases with a spring-loaded ball burnisher similar to the one shown in Figure 2 to precrease the material. He slowly bent the material to the maximal angle without kinking the uncreased areas. The careful and time-consuming folding technique attests to his incredible patience and love for craft when producing his art.

2.2 Reconstruction Methods

As part of this reconstruction effort, we have decided to stay close to Huffman's way of making models. We deviate from his manual drawing methods in order to create digital files of the crease patterns. The reconstructions are drawn with computer-aided design (CAD) software, Rhinoceros 3D, which allows the use of quadratic curves. Most designs are drawn in two dimensions, folded, and then visually analyzed. Some designs have been recreated virtually in three dimensions and then made into physical models. We made paper versions before producing the final versions from the same material Huffman used in the 1970s.

We use an industrial vinyl cutter/printer to precrease the patterns, which requires format translation to Adobe Illustrator. The i-Cut i-XL-24M flatbed cutter and router has a 65″ by 120″ vacuum plate and 1″ cutting depth. The machine is furnished with two heads and can crease and cut at the same time. The drop-in tool slots allow the use of a special tool, called a creasing wheel, that gently pushes down onto a surface. The precreased vinyl then needs to be folded into its final shape by hand. Most of the reconstructions are close to the size of the originals. We decided to enlarge some examples to ensure better results, as tolerances become less of an issue.

3   Reconstructions

Huffman created a wide variety of models, most of which were never shown. The approximate reconstructions selected for this paper are grouped into categories that highlight some of the observations we have made during our study. Of course, aesthetic qualities and beauty are big factors, and it is impossible to know Huffman's exact motivations for each design.

We believe that David Huffman was interested in studying vertices of various degrees and that he made some of his models to show how they can be used. He studied vertices that are exploded and separated by polygons. Tessellations represent a large portion of his work, but we include only some that use curved creases. Huffman further studied foldings that have the characteristic of describing a volume. The crease patterns we show in Figures 3 through 15 have tags that identify circles (ci), ellipses (el), and parabolas (pa).

3.1   Degree-1 and -2 Vertices

The model in Figure 3 shows two mountain creases ending within the area of the paper. These vertices are located on the major axis of the ellipse of

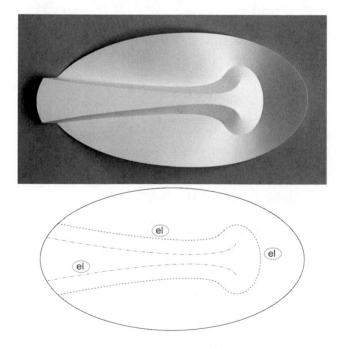

Figure 3. Huffman design using ellipses with two degree-1 vertices.

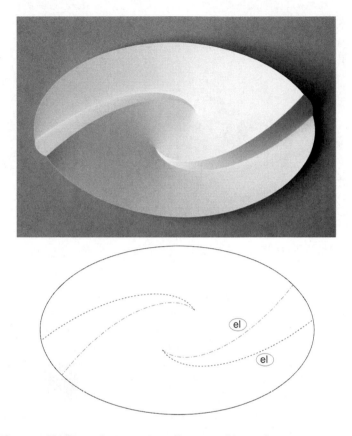

Figure 4. Huffman design using ellipses with two degree-2 vertices.

the valley crease, and the rulings around each degree-1 vertex describe a cone surface. The valley crease is drawn by splicing ellipses together. This design was published on the website of *Grafica Obscura* [Haeberli 96].

The model shown in Figure 4 [Haeberli 96] uses two degree-2 vertices and is drawn using ellipses. There is a ruling line between the two points, but no crease is necessary. The cut-out shape is an ellipse and the mountain and valley assignments alternate radially.

3.2 Inflated Vertices

We observe that Huffman was studying crease patterns with "exploded" vertices of varying degrees. The next series of models shows vertices that are exploded or inflated into flat polygons. Figure 5 shows a noninflated degree-4 vertex with creases that have pairwise common tangents. The creases are all circular arcs and alternate mountain and valley.

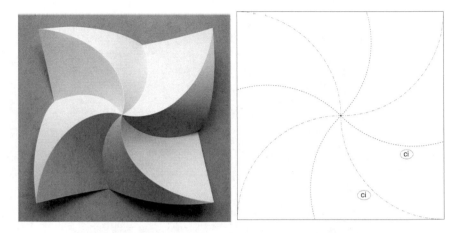

Figure 5. Huffman design using circles (unexploded vertex).

Figure 6 shows a degree-4 vertex inflated into two degree-3 vertices. The connecting element in this case is a straight line. The crease pattern uses ellipses that result in more dramatic curvature changes than in Figure 5.

In Figure 7, a degree-4 vertex has been inflated into four degree-3 vertices. The connecting flat square rotates very little when the design is folded into shape. This example is made of ellipses and is featured on *Grafica Obscura* [Haeberli 96].

Figure 8 shows a crease pattern that displays structural properties similar to the so-called "flashers." Simon Guest studied these shapes in terms

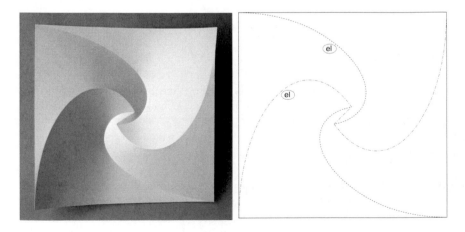

Figure 6. Huffman design using ellipses and a line (exploded vertex).

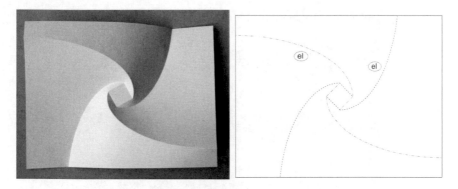

Figure 7. Huffman design using ellipses and a square (exploded vertex).

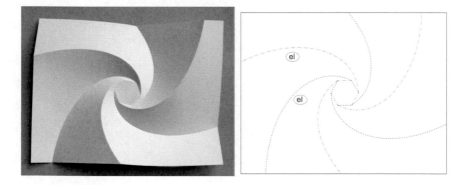

Figure 8. Huffman design using ellipses and a hexagon (exploded vertex).

of how they can curl up into themselves [Guest and Pellegrino 92], and Huffman made several of them. This design is made of ellipses that converge in a degree-6 vertex, which was inflated into six degree-3 vertices. The connecting element is a flat hexagon that rotates very little in comparison to Guest's shapes, and we believe that Huffman was not studying their kinetic behavior, but rather the inflated vertices.

3.3 Tessellations

Figure 9 shows a tessellation with reflectionally and 180° rotationally symmetric tiles. The crease pattern uses circles, and mountain and valley assignments alternate from row to row.

Huffman called the tessellation in Figure 10 "Arches" and used parabolas or ellipses that are connected at their focal points. The resulting arches have parallel rulings. The model shown here was constructed of paper and uses parabolas.

Figure 9. Huffman design using circles.

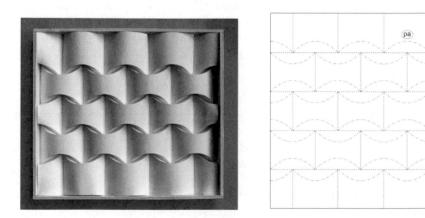

Figure 10. Huffman's "Arches" design using parabolas and lines. (See Color Plate I.)

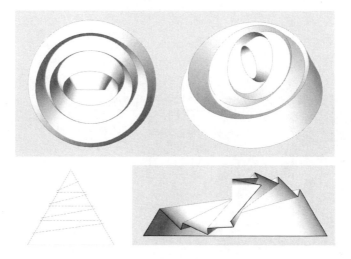

Figure 11. "Cone Reflected 7 Times," top and perspective views (top); section through entire cone and mirrored design (bottom).

3.4 Cones

Figure 11 shows Huffman's "Cone Reflected 7 Times" reconstructed as a digital three-dimensional model. A similar shape was designed by Ron Resch in 1969, which he called "Yellow Folded Cones: Kissing" [Resch 71], where he truncated a cone twice and mirrored it.

Here eight truncated cones are mirrored in an alternating way. The perpendicular cuts to the main axis of the cone result in a circular arc in the crease pattern similar to a design by Hiroshi Ogawa [Ogawa 71]. In the rotated cuts, increasing angles go up from the bottom. The truncated cones need to be unrolled to construct the curves for the crease pattern. These curves are a rare example of Huffman using non-quadratic curves; see Figure 12.

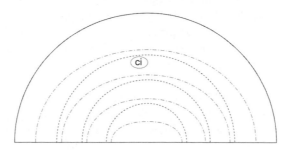

Figure 12. Crease pattern design using circles and non-quadratic curves.

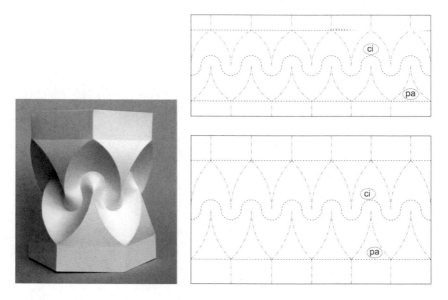

Figure 13. "Hexagonal Column with Cusps" design using circles and lines. The photograph corresponds to the upper crease pattern.

3.5   Complex Shapes

Huffman's "Hexagonal Column with Cusps," shown in Figure 13, is remarkable as two sides of the paper meet and create a continuous shape. This stunning and aesthetically very well received example of Huffman's designs has been reconstructed by Saadya Sternberg [Sternberg 09] and Robert Lang [Lang 10].

This crease pattern combines half-circles, parabolas, and straight lines. Huffman made many versions of this model with different proportions and sometimes even repeated the entire shape twice in a single crease pattern. Figure 13 shows the crease pattern for two differently proportioned versions. The model shown here was made out of paper.

Huffman's "4-Lobed Cloverleaf Design" is shown in Figure 14. It is symmetric along two axes and comprises lines and ellipses. The inner square is folded such that the triangular faces touch one another—a common way to hide material in straight-crease origami, but a rare characteristic among Huffman's designs.

The model in Figure 15, named "One Column," was recreated here from white Zanders elephant hide paper. Huffman joined parabolas together to create two wavy extrusions at different scales. This model is particularly striking. It starts with a very simple crease pattern yet creates dramatic features from folding those creases very tightly.

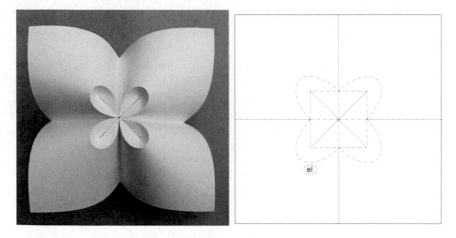

Figure 14. Huffman's "4-Lobed Cloverleaf Design" using ellipses and lines.

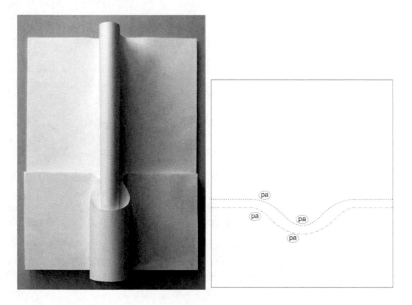

Figure 15. Huffman's "One Column" design using parabolas and lines.

4 Conclusion

Our goal is to expose to the world David Huffman's beautiful artwork and the underlying mathematics that he used to create it. We believe that much can be learned from reconstructing and analyzing his final models, which is the focus of this paper. This reconstruction project is ongoing, with many more models to be studied. We also believe that there is much to be learned from Huffman's notes, sketches and working models, a study which is just beginning.

Ultimately, we aim to develop a theory for how David Huffman designed his curved-crease foldings, to enable future origami artists and mathematicians to build upon his knowledge and expertise. We regret not being able to develop this theory in direct communication with David, but we are confident that the legacy he left behind will enable a fruitful collaboration.

Acknowledgments. We thank the Huffman family—in particular, Elise, Linda, and Marilyn Huffman, and Jeff Grubb—for their ongoing collaboration on this project and for their kind hospitality.

We are grateful to Peter Wilson, with Makepeace Digital Imaging in Boston, Massachusetts, who has provided us with generous support in making the models, and to Jenny Ramseyer for her help in making some of the models.

Bibliography

[Demaine and Demaine 08] Erik D. Demaine and Martin L. Demaine. "History of Curved Origami Sculpture." Available at http://erikdemaine.org/curved/history/, 2008.

[Fuchs and Tabachnikov 99] Dmitry Fuchs and Serge Tabachnikov. "More on paperfolding." *The American Mathematical Monthly* 106:1 (January 1999), 27–35.

[Guest and Pellegrino 92] Simon D. Guest and Sergio Pellegrino. "Inextensional Wrapping of Flat Membranes." In *Structural Morphology/Morphologie Structurale: Proceedings of the First International Seminar on Structural Morphology*, edited by R. Motro and T. Wester, pp. 203–215. Madrid: IASS, 1992.

[Haeberli 96] Paul Haeberli. "Geometric Paper Folding: Dr. David Huffman." *Grafica Obscura*. Available at http://www.graficaobscura.com/huffman/, November 1996.

[Huffman 76] David A. Huffman. "Curvature and Creases: A Primer on Paper." *IEEE Transactions on Computers* C-25:10 (October 1976), 1010–1019.

[Kilian et al. 08] Martin Kilian, Simon Flory, Zhonggui Chen, Niloy J. Mitra, Alla Sheffer, and Helmut Pottmann. "Curved Folding." *ACM Transactions on Graphics* 27:3 (2008), 1–9.

[Lang 10] Robert J. Lang. "Flapping Birds to Space Telescopes: The Modern Science of Origami." Guest Lecture, Massachusetts Instittue of Technology, Cambridge, MA, April 26, 2010.

[Ogawa 71] Hiroshi Ogawa. *Forms of Paper*. New York: Van Nostrand Reingold Company, 1971. (Translation of Japanese edition, Tokyo: Orion Press, 1967.)

[Resch 71] Ron Resch. "Yellow Folded Cones: Kissing (1969–1970)." Available at http://www.ronresch.com/gallery/yellow-folded-cones-kissing, 1971.

[Sternberg 09] Saadya Sternberg. "Curves and Flats." In *Origami⁴: Fourth International Meeting of Origami Science, Mathematics, and Education*, edited by Robert J. Lang, pp. 9–20. Wellesley, MA: A K Peters, 2009.

[Wertheim 04] Margret Wertheim. "Cones, Curves, Shells, Towers: He Made Paper Jump to Life." *The New York Times*, June 22, 2004.

[Wingler 69] Hans M. Wingler. *Bauhaus: Weimar, Dessau, Berlin, Chicago*. Cambridge, MA: The MIT Press, 1969.

Simulation of Nonzero Gaussian Curvature in Origami by Curved-Crease Couplets

Cheng Chit Leong

1   Introduction

A flat piece of paper is malleable. We may define its extrinsic geometry by bending and folding, without stretching, tearing or deforming its surface. Its intrinsic geometry does not change. An intrinsic property is an inherent property of the paper. For example, a straight line drawn on a paper is intrinsically straight, although the line in 3D space is extrinsically not straight when you bend the paper.

Surfaces, such as those on a box, a pyramid, a cylinder or a cone can be modeled accurately by folding a piece of paper. These surfaces are termed "developable" and can be generated by the motion of straight lines. A developable surface that can be flattened on a plane without distortion is a ruled surface. Every developable surface is a ruled surface, but not every ruled surface is a developable surface [Wolchonok 59].

We can simulate a nondevelopable surface by folding a piece of paper with curved creases or a combination of curved and straight creases. Several papers on the application of curved creases in simulating curved surfaces have been published, which generally cover geometrical models [Sternberg 06, Koschitz et al. 08, Demaine and Demaine 98]. In animal models, curved-crease folding, when applied, is often a complement to flat folding.

The focus of this paper on curved-crease folding is a specific couplet consisting of an intrinsic straight crease and an intrinsic curved crease.

53

The paper discusses the simulation of curved surfaces, which are not developable, using the couplet as the fundamental folding unit for designing and folding both geometrical and animal models.

2 Geometry of a Curved Surface

In 1827, Gauss published an analysis of surface curvature [Gauss 65]. To quantify the curvature at a point p on a surface, he introduced a quantity $k(p)$. If $k(p) < 0$ then the immediate neighborhood of p resembles a saddle; if $k(p) > 0$, then it bends in the same way in all directions, like a sphere; and if $k(p) = 0$, the immediate neighborhood of p is intrinsically flat, although it is bending in one direction [Needham 97, p. 274].

In mathematics, a surface with zero Gaussian curvature, $k(p) = 0$, is a developable surface that can be flattened onto a plane without distortion, i.e., without "stretching" or "compressing." Surfaces represented by $k(p) < 0$ and $k(p) > 0$ are not developable.

All three categories of Gaussian surface curvature are found on a typical Greek vase. The vase on the left in Figure 1 is the common Greek vase, and the model on the right is the geometrical representation of its surface. Consider the triangles T_1, T_2, and T_3 drawn on different parts of the model. The lines tracing the triangles are geodesic. These are lines drawn on the surface with the shortest distance between the two points. Clearly, the sum of the angles of triangle T_1 is greater than 2π, and the sum of the angles of T_3 is 2π, and the sum of the angles of T_2 is less than 2π. We assume that at T_3 the surface is, or is a very close approximation to, that of a cone.

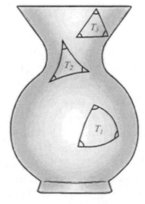

Figure 1. Typical Greek vase (left). Geometrical model of the Greek vase showing the three categories of Gaussian curvature (right).

We demonstrate how the three categories of Gaussian surface curvature of the Greek vase can be simulated by folding.

3   Representation of Curved Surfaces by Paper Folding

This paper discusses the use of the curved-crease couplet as the basis for generating nonzero Gaussian curvature in the design of origami geometric and animal models. Design, not mathematical exactness, is the main criterion.

Art is chiefly a visual concept. The origami model of an object we fold is only a representation of the actual object. From a practical point of view, the more we try to fold the model to conform to the reality or mathematical equivalent of the real object, the more exact we must be in both technique and measurement. As an art form, reality or accuracy is not or need not be specifically sought as an objective.

How then do we represent a surface with nonzero Gaussian curvature in origami based on the couplet, consisting of a straight mountain crease and a curved valley crease (or a straight valley crease and a curved mountain crease)? Based on what the eye sees and the mind interprets, Figure 2 presents two representations.

Figure 2(a) shows a fold of a nonzero Gaussian curvature formed by a series of couplets. The curved straight creases are visible. In the origami model, the mind could interpret them as an "unbroken" surface, traced by an envelope of the external creases, which the eye sees. The opposite parity curved creases of the couplets are buried and are hidden. The eye does not see these creases.

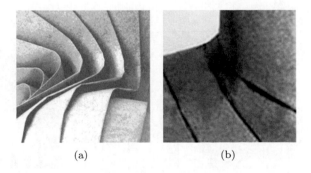

(a)   (b)

Figure 2. (a) Curved surface simulated by the envelope of curved straight creases of curved-crease couplets. (b) Curved surface simulated by curved creases of the couplets.

In Figure 2(b), a nonzero Gaussian curvature is traced by the curved creases. This is the opposite of the envelope of the curved straight creases, which are on the other side of the closed surface and are hidden. The couplet has a straight valley and a curved mountain fold. This is the Gaussian curvature $k(p) < 0$, which resembles a saddle.

4   Ruled Surfaces

In geometry, there are ruled surfaces that are nondevelopable. Several ruled surfaces are cited in *The Art of Three-Dimensional Design* [Wolchonok 59, pp. 5–12]. An example of a nondevelopable ruled surface given in the book is the hyperbolic paraboloid. The motion of a straight line touching two straight lines and parallel to a plane surface generates this surface. The hyperbolic paraboloid is, in fact, a doubly ruled surface for which the Gaussian curvature is $k(p) < 0$.

Figure 3 presents a hyperbolic paraboloid [Wolchonok 59], with the two sets of straight lines generating the surface shown as well as the crease pattern (CP) and a folded model. For the hyperbolic paraboloid, we have $m = l \cos \frac{\theta}{2}$, where θ is the angle between the left and the right edges in Figure 3(a), and m is the length of the shortest distance between these edges. The overlap is $n = \frac{l-m}{2N}$, where N is the number of segments. Therefore, $k = m + 2Nn$.

For each of the curved creases, we therefore have three reference points. We should be able to find a sinusoidal line joining the points for the curved crease and three points for the top and bottom curves. (The reader should see that more points can be obtained by varying θ for the other lines.)

Mathematically, the analysis is not strictly rigorous, as it does not account for the slightly twisted segments. However, that is not a major factor,

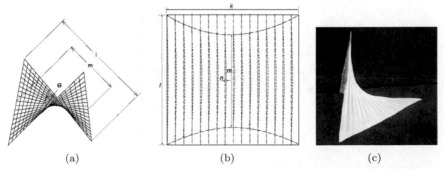

(a)   (b)   (c)

Figure 3. Hyperbolic paraboloid: (a) doubly ruled surface, (b) crease pattern, and (c) paper model.

Figure 4. Paper models of ruled surfaces of a hypoid (left) and conoid (right).

as paper is malleable and having more segments would minimize the twisting. As mentioned earlier, strict mathematical accuracy is not an absolute criterion in design.

The paper "Polyhedral Sculptures with Hyperbolic Paraboloids" presents a popular method for folding the hyperbolic paraboloid by square pleating [Demaine et al. 99]. Again, the segments are slightly twisted, but because paper is malleable, it adjusts for some surface bending.

Other examples of nondevelopable ruled surfaces are the hypoid and the conoid [Wolchonok 59], as shown in Figure 4. A similar analysis can be used to obtain the CPs to fold these surfaces.

5   Radial Formed Rotational Symmetric Models

Radial formed rotational symmetric models appear to be favorites among origami artists using the couplet or flange as the basis for designing geometric origami models. Among the creative folders are Philip Chapman-Bell, Robert Lang, Jun Mitani, and Chris Palmer [Chapman-Bell 08, Lang 08, Mitani 10, Palmer 03]. Representative models are shown in Figure 5.

The horizontal cross-section of a radial formed rotational symmetric model is shown in Figure 6. The model has an outer envelope formed by the curved straight creases and an inner envelope formed by the inner curved creases.

Lang has a complete Mathematica notebook for the generation of couplet-based rotational models [Lang 09]. The starting point of this program is to define the vertical cross section of the vessel or pot to be folded. It then will trace the CP of the object. Conceptually, opposite pairs of the intrinsic

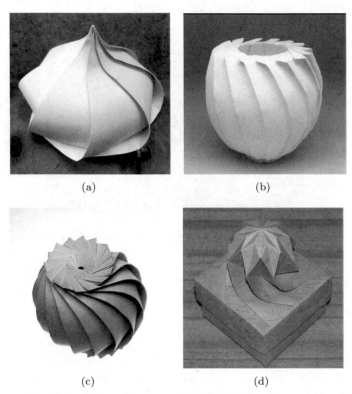

Figure 5. (a) "Onion" by Chapman-Bell, (b) "Rim Pot 15" by Lang, (c) "20101006" by Mitani, and (d) "Origami Box" by Palmer.

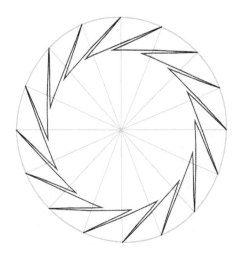

Figure 6. Horizontal cross section of radial formed rotational symmetric models.

curved creases trace the vertical cross section of the vessel. This inner envelope of curved creases represents the actual form of the object and is generally hidden. The curved straight creases trace the outer envelope, which the eye sees and the mind would interpret as the vessel's surface.

Mitani presented a similar method of simulating the surface of the actual vessel traced by the rotation of the inner intrinsic curved creases of the folded model [Mitani 09]. Again, the outer envelope is that of the curved straight creases. The surface representing the actual vessel is the inner envelope.

It is possible to fold a radial formed rotational symmetrical model such that the envelope of the outer creases and not the inner envelope of the curved creases represents the surface of the vessel. The Greek vase, whose surface curvature exhibits the Gaussian curvatures of $k(p) < 0$, $k(p) > 0$, and $k(p) = 0$, provides an excellent object for folding such a model.

5.1 Folding Radial Formed Rotational Symmetric Models from a Crease Pattern

Before the age of computers, geometric drawing for generating segments of a curved surface on a metal sheet was the common method used. The segments were cut and welded together to form the curved surface [Dickason 67]. Figure 7 provides the CP generated from an engineering drawing for folding the Greek vase, and Figure 8 shows two views of the folded model. The model (15 cm in height) was folded from circular elephant hide paper (110 gsm (grams per square meter)) with a radius of 20 cm.

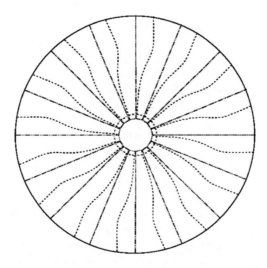

Figure 7. Crease pattern of the Greek vase.

Figure 8. Two views of a folded model of the Greek vase.

The CP was enlarged to about 18 cm radius, which provided paper at the rim to fold into the vase. The CP was scored onto the paper, which was then folded.

5.2 Folding Radial Formed Rotational Symmetric Models without Crease Patterns

If mathematical accuracy is not an important criterion, an experienced folder may be able to fold a model without a crease pattern. This paper presents two examples: a wine glass and a baseball cap. To fold such models, much judgment will be needed. Depending on the skill of the folder, it is possible to obtain a model that is a fairly close approximation of the object to be represented. Further refinements can be made in subsequent models. An idea of how the couplets define the curvature of the surface (such as in the Greek vase) will help.

Folding a wine glass. To select a paper large enough to fold the wine glass, half the length of the paper must be greater than a line running on the surface from the bottom center point to the lip of the mouth. The model was created using circular elephant hide (110 gsm) with a radius of 35 cm. Radial straight creases were made first, ensuring there would be sufficient number of couplets (64 in this case) to allow for the folding of the stem at the bottom of the cup bowl. The overlap of the couplet at the critical levels was calculated for the lower stem, the upper stem, the cup, and the mouth. These markers then were duplicated for the other couplets. The curved creases would be lines joining these critical markers. Due to the

Figure 9. Two views of a folded wine glass.

large number of couplets, the overlaps needed to be folded as accurately as possible, so that 4 couplets would give one-eighth of the circumference of the cup, 8 couplets would give one-quarter of the circumference, 16 couplets half the circumference, etc. The folded model is shown in Figure 9.

A CP for folding subsequent models based on the folded model can be drawn based on the markers. Scoring with a sharp pointer, a laser scriber, or a cutting plotter (e.g., Craft ROBO) will facilitate folding.

Folding a baseball cap. Rectangular paper is required for the baseball cap, to allow for the brim of the cap. A life-size cap may be folded from B2 size paper. Again, elephant hide paper (110 gsm) was used. The semispherical top occupied a square close to one end of the paper. The button located at

Figure 10. Baseball cap.

the center of this square was formed by a 16-gonal twist. The base line of the cap measures the circumference of the head. Knowing the circumference of the head of the wearer allows calculation of the amount of overlap of each of the couplets of this semispherical top. Extra paper was tucked in and the brim shaped by flat folding. Figure 10 shows the finished model. Again, the CP or folding diagrams can be prepared for folding subsequent models based on the folded model.

6   Cylindrical Formed Rotational Symmetric Models

For completeness of the paper, consideration will be given to the class of cylindrical formed rotational symmetric models. The CPs of such models are relatively easy to define. Two examples, both designed by me, are shown—the sphere in Figure 11(a) and the cap in Figure 11(b), along with the CPs.

For the sphere CP, the width of the paper is half the circumference of the sphere and the length its circumference. To provide proper locking of both ends, allowance is made for the extension of the length by one segment. The curved crease of each segment is sinusoidal, extending from the midpoint on one side to the top and bottom of the other side; they form the longitudes of the sphere. It is easy to calculate the overlap of the couplet at each

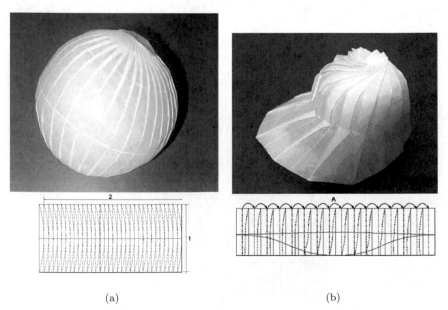

(a)   (b)

Figure 11. Folded models and CPs for (a) a sphere and (b) a cap.

latitude point. At latitude θ the overlap is $(nR(1 - \sin\theta))/2N$, where R is the radius and N is the number of couplets. The envelope of the inner curved creases defines the surface of the sphere; the couplets at top and bottom tend to flatten at these points. With more couplets, the circular buttons at the top and at the bottom reduce in size.

For the cap, the CP of the bowl is half a sphere shown at the top part of the rectangle. The rest gives the brim of the cap and the tucked-in part.

7   Applications of Couplets to Folding Animal Models

Curved creases have been and are used for folding animal models. Many of the models by Akira Yoshizawa and Eric Joisel, for example, incorporate a high proportion of curved creases in their creations [Yoshizawa 99, Joisel 09]. Most models start off with flat folding, and the incorporation of curved creases comes later. Several models, such as Joisel's masks, are composed primarily of curved creases.

All surfaces, including those of real life objects and animals, exhibit Gaussian curvature. Gaussian curvatures of $k(p) < 0$ and $k(p) > 0$ are not developable, but the basis of the straight and curved crease couplet can be used to capture the nonzero Gaussian surface curvature.

Design of animal origami models using the couplets requires imagination and experimentation. Folding requires judgmental skill. It is possible to define certain markers, but the aesthetics of the final model will depend on how well the creases are formed and positioned.

Figure 12 shows a face model in which I used curved crease couplets extensively [Leong 01]. I used tracing paper (vellum) to fold the face and many of the couplets are apparent in the photo (Figure 12(a)). The number of couplets is minimized, and these couplets are strategically located and folded so that a 3D surface of a face is formed. In designing the face, I took much care on the positioning and folding of the couplets. The diagram in Figure 12(b) shows how the nose and cheeks are formed with the couplets; the couplet on either side has two cusps. Such a couplet may be considered a combination of two or more couplets. The couplet at the top of the nose shapes the nose and the eyes. Popping in the cusps provides character to the eyes. Four couplets form the chin, with two below the lower lip and two below the chin, simulating the Gaussian curvature $k(p) > 0$. The mouth is formed by sinking in a valley between the upper and lower lips. A row of couplets at the top partitions the hair from the forehead.

The seal is also another of my models that involves mainly the application of curved crease couplets [Leong 00]. The body of the model of the seal in Figure 13(a) is shaped by several couplets. The diagram in Figure 13(b) shows how this is formed. It allows for the capture of the pose of the seal.

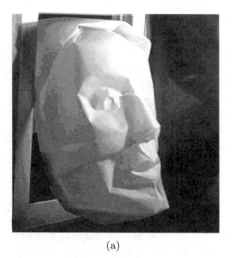

(a)

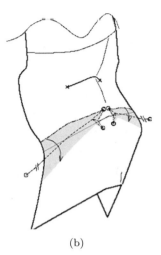

(b)

Figure 12. (a) Photograph of a face folded using couplets; (b) detail of the folding of the nose and cheeks.

Here the mountain curved creases, defining the surface curvature, are exposed. The surface is closed, with simulation of the Gaussian curvature $k(p) < 0$. Except for some features of the face and other parts of the body, the rest of the features of the seal are generally flat folded.

Figure 14 shows my more complex model of a lion head and its CP, consisting mainly of couplets. The model may be folded by scoring the CP

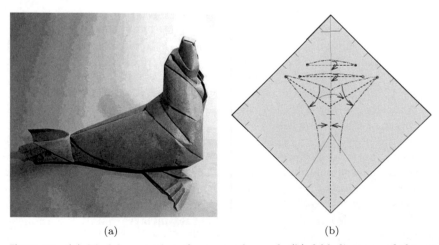

(a)   (b)

Figure 13. (a) Model capturing the pose of a seal; (b) fold diagram of the seal body.

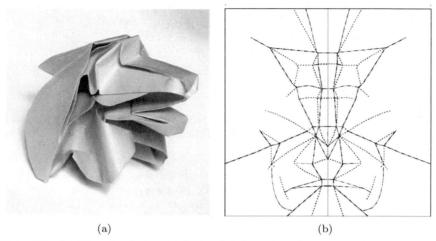

(a)   (b)

Figure 14. (a) Lion head, in which the face has clean features and projects the lion's majesty. (b) CP of a lion head; the sharp creases are the thicker lines and the soft creases are the thinner lines.

on the paper. Recommended paper size is a square of 35 cm with paper thickness of 120 gsm. Alternatively, as was done with the model in the picture, one can print the CP on thin photo copy paper, stick it on the paper to be folded by gluing the corners with rubber glue, and fold the two together, with removal of the photo copy paper and glue after folding, followed by molding the model into shape.

The lion head was designed by using the couplets like brush strokes of a painting. I started with the couplets on either side of the nose between the eyes. I added two more couplets to form the length of the nose. The design of the nose above the mouth was done with two adjacent couplets, bringing two points of the paper on either side together. More couplets strategically positioned were folded and collapsed to form the mouth. Forming the chin came next. Other parts of the head followed. The last step was molding the model into shape.

8 Conclusion

The application of the straight mountain and curved valley (or curved mountain and straight valley) crease couplet in the design and folding of origami models, both geometric and animal models, is still in its infancy and has not been widely explored. This paper provides a rational basis for its application in simulating surface curvature in the folding of a wide range of origami models: ruled surfaces, radial formed rotational symmetric models, cylindrical formed rotational symmetric models, and animal models. A balance between mathematics and aesthetics is achieved by approaching the creation of the models from a design perspective.

Bibliography

[Chapman-Bell 08] Philip Chapman-Bell. *Onion.* Available at http://www.flickr. com/photos/oschene/sets/72157601506905133/, 2008.

[Demaine et al. 99] Erik D. Demaine, Martin L. Demaine, and Anna Lubiw. "Polyhedral Sculptures with Hyperbolic Paraboloids." In *Proceedings of the 2nd Annual Conference of BRIDGES: Mathematical Connections in Art, Music, and Science (BRIDGES '99)*, pp. 91–100. Winfield, KS: BRIDGES Society, 1999.

[Demaine and Demaine 98] Erik D. Demaine and Martin L. Demaine. *History of Curved Origami Sculpture.* Available at http://erikdemaine.org/curved/ history/, 1998.

[Dickason 67] Alfred Dickason. *The Geometry of Sheet Metal Work.* Essex, UK: Longman Group Limited, 1967.

[Gauss 27] Carl F. Gauss. *General Investigation of Curved Surfaces.* New York: Raven Press, 1965. (Originally published October 8, 1827.)

[Joisel 09] Eric Joisel. *BOS Booklet 69:3D Masks and Busts.* London: British Origami Society, 2009.

[Koschitz et al. 08] Duks Koschitz, Erik D. Demaine, and Martin L. Demaine. "Curved Crease Origami." In *Abstracts from Advances in Architectural Geometry (AAG 2008), Vienna, Austria, September 13–16, 2008*, pp. 29–32. Vienna: RFR and Waagner-Biro Stahlbau AG, 2008. (Available at http://www.architecturalgeometry.at/aag08/.)

[Lang 08] Robert J. Lang. *Rim Pot 15.* Available at http://www.langorigami. com/art/gallery/gallery.php4?name=rimpot15, 2008.

[Lang 09] Robert J. Lang. *Origami Flanged Pots.* Available at http:// demonstrations.wolfram.com/OrigamiFlangedPots/, 2009.

[Leong 00] Cheng Chit Leong. "Seal." In *British Origami Society Magazine* 204 (October 2000), 10–14.

[Leong 01] Cheng Chit Leong. "Face." In *Annual Collection 2001*, edited by Mark Kirschbaum, p. 219. New York: OrigamiUSA, 2001.

[Mitani 09] Jun Mitani. "A Design Method for 3D Origami Based on Rotational Sweep." *Computer-Aided Design and Applications* 6:1 (2009), 69–70.

[Mitani 10] Jun Mitani. "20101006." Available at http://www.flickr.com/photos/ jun_mitani/5056579868/, October 6, 2010.

[Needham 97] Tristan Needham. *Visual Complex Analysis*. New York: Oxford University Press, 1997.

[Palmer 03] Chris   Palmer.   *Origami   Box*.   Available   at   http://www. bridgesmathart.org/nexus-mirror/bridges03-conf-review/index.html, 2003.

[Sternberg 06] Saadya Sternberg. "Curves and Flats." In *Origami⁴: Fourth International Meeting of Origami Science, Mathematics, and Education*, edited by Robert J. Lang, pp. 9–20. Wellesley, MA: A K Peters, 2009.

[Wolchonok 59] Louis Wolchonok. *The Art of Three-Dimensional Design*. London: Dover Publication, Inc., 1959.

[Yoshizawa 99] Akira Yoshizawa. *Akira Yoshizawa: ORIGAMI*. Tokyo: Asahi Shimbun, 1999.

Compression and Rotational Limitations of Curved Corrugations

Christine E. Edison

1 Introduction

Curved corrugations caught my interest two years ago. It started with a piece I did called "Degrees of Freedom in Blue," shown in Figure 1. During that time, I was playing with computer-aided origami crease patterns and corrugations, heavily influenced by Polly Verity [Verity 10] and Ray Schamp [Schamp 10]. It only made sense to combine corrugations and curves; this led to the creation of "Degrees of Freedom in Blue." The piece had peculiar properties of rotation and compression, and as I played with similar pieces, I found that changing the curves substantially changed the ability of the pieces to move in varying directions.

2 Method

I wanted to understand what was happening mathematically and if there were any specific rules to these types of behaviors. Talking to many individuals who have much more of a background in mathematics and/or engineering than I, I was unable to find the information on the limitations of compression and/or rotation of curved corrugations that I wanted. I then undertook a systematic test of the general observations I had made. This paper reflects results of tests with sine curves that had differing amplitudes,

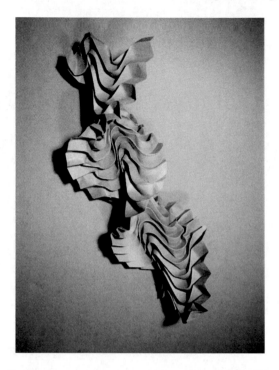

Figure 1. "Degrees of Freedom in Blue." (See Color Plate II.)

numbers of periods, and distances between sine waves. The corrugations were printed from a CAD (computer-aided design) file I created, on Canson paper (Mi-Teintes line of charcoal and pastel paper), scored, and then pleated in alternating directional folds(i.e., mountain then valley). This is akin to accordion pleating paper; the only difference is that it is along a sine curve rather than a straight line.

The equation for the sine curve is $y = a\sin(bx + c) + d$, where a is amplitude, b is used to find the period $(2\pi|b|)$, c is used to find the phase shift, and d is the vertical shift. For all the tests $c = d = 0$ and $b = 1$. The distance between each curve placed on the y-axis is denoted α. The standard period is 2π, and the period is not changed in any test; what is changed is how many periods there are, which is denoted as Pd. The measurements taken for each corrugation were length (prior to scoring, folding, and compression/rotation), width (prior to scoring, folding, and compression/rotation), compressed length, compressed width, compressed depth, and degrees of rotation (denoted DOR). Rotation is measured in degrees and is measured by rotating the corrugation around a central axis; all other measurements are in centimeters. When I refer to the rigidity of a tessellation, I refer to how little a corrugation rotates around the

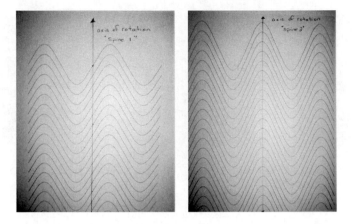

Figure 2. Axes of rotation: spine 1 on left and spine 2 on right.

central axis and how little it compresses for both length and width, without rotation around the central axis. Density refers to how closely the curves pack together. There are two different axes of rotation, referred to as "spine 1" and "spine 2," as shown in Figures 2 and 3. In the first three tests, G, DB, and B, rotate around the first spine; in the fourth and fifth tests, T and P, rotate around the second spine.

The results have multiple limitations: The scoring was done by hand, which introduces human error, and measurements were taken with an attempt to keep the structural integrity of the curves while reaching the true

Figure 3. Rotated tests: P test with $a = 0.25$ (top left) and T test with $a = 0.25$ (top right) both with spine 2; G test with $a = 0.25$ (bottom). The picture does not show how much the T test is rotated internally.

Figure 4. G tests: clockwise from upper right, $a = 0.25$, $a = 0.5$, $a = 1$, $a = 1.5$, $a = 2$, $a = 2.5$.

Figure 5. DB tests: clockwise from upper right, $a = 0.25$, $a = 0.5$, $a = 1$, $a = 1.5$, $a = 2$, $a = 2.5$.

compression limit without collapsing the paper. The curves were graphed in a CAD program to the tenths place value, as allowed within the constraints of the program. The rotational limitations may be the most problematic as they are measured only in multiples of $22.5°$, due to my inability to be more precise. When rotations went past $360°$, the measurements became more challenging, as the edges were hard to identify clearly. In certain cases, not all tests could be made at all amplitudes as the limitation of the paper and the scoring made collapsing structurally impossible. In two cases, all I ended up with was a crumpled mess. The final problem was humidity. Measurements were taken over many days, and Chicago's climate, well known for its summer humidity, vacillated widely, which may have affected the measurements.

Each of the five tests had six corrugations for six different amplitudes, $a = 0.25, 0.5, 1, 1.5, 2, 2.5$. It was not possible to do the collapses in two cases, as the limitation of the paper and my skills created a wad of paper that was not measurable. The G test (Figure 4) had the following parameters: spine 1, $\alpha = 1$ cm (α is the distance between the sinusoidal curves), Pd = 2. Tests DB and B all had the same parameters except $\alpha = 2$ cm for the DB test and $\alpha = 0.5$ cm for the B test. The initial flat sheet had the same length and width for all three tests, which had a total of 18 sinusoidal corrugations. Figure 5 shows the DB tests, and Figure 6 shows the B tests. The initials on the tests, G, DB, B, T, and P, merely identify the tests by constrained variable. The variables that changed per test are listed below:

G: Spine 1, $\alpha = 1$ cm, Pd = 2
DB: Spine 1, $\alpha = 2$ cm, Pd = 2
B: Spine 1, $\alpha = 0.5$ cm, Pd = 2
T: Spine 2, $\alpha = 1$ cm, Pd = 1.5
P: Spine 2, $\alpha = 1$ cm, Pd = 2.5

Figure 6. B tests: clockwise from upper right, $a = 0.25$, $a = 0.5$, $a = 1$, $a = 1.5$.

Figure 7. T tests: clockwise from upper right, $a = 0.25$, $a = 0.5$, $a = 1$, $a = 1.5$, $a = 2$, $a = 2.5$.

Figure 8. P tests: clockwise from upper right, $a = 0.25$, $a = 0.5$, $a = 1$, $a = 1.5$, $a = 2$, $a = 2.5$.

The T tests and P tests (Figures 7 and 8) utilized a different axis of rotation, spine 2. The parameters were $\alpha = 1$ cm, Pd = 1.5, $a = 0.25, 0.5,$ 1, 1.5, 2, 2.5 for test T; and $\alpha = 1$ cm, Pd = 2.5, $a = 0.25, 0.5, 1, 1.5, 2, 2.5$ for test P. Test P was added after the fact, when such a strong difference in degrees of rotation was found between the first three tests and the T test, to identify if it was merely the shorter width that allowed for the greater rotational freedom.

3   Discussion of Results

Figures 9 and 10 are graphs that shows all the results of the G tests with the x values being the amplitude and the y values being the compression

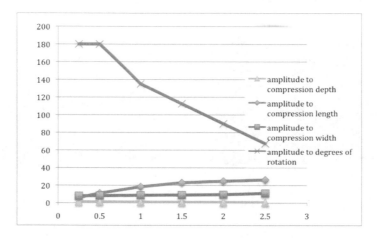

Figure 9. Comparison of various attributes for G tests: spine 1, $\alpha = 1$, Pd $= 2$.

in centimeters of depth, length, and width and the degrees of rotation or how much the corrugation was able to be rotated around its central axis.

All the tests showed similar relationships. The amplitude has a positive correlation to the final compressed length and width of the corrugation. This can be seen visually as the two scatterplots for amplitude to compression length (diamonds) and width (squares) are going up to the right. The relationship of amplitude to width and depth is difficult to see in Figure 9,

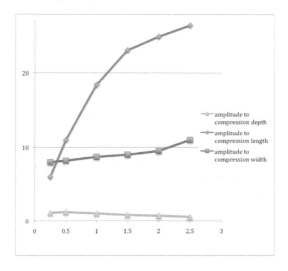

Figure 10. Comparison of various attributes for G tests: spine 1, $\alpha = 1$, Pd $= 2$ (from Figure 9 with expanded vertical scale).

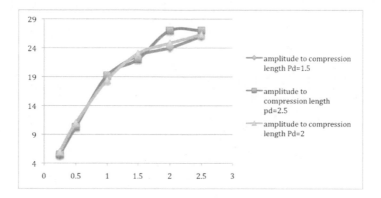

Figure 11. Comparison of G, T, and P tests: spine 1 for G and spine 2 for P and T, $\alpha = 1$ for all three. The periods are Pd = 2, 1.5, and 2.5, respectively.

so Figure 10 has the degrees of rotation removed and a change in the scale to reveal the relationships. The higher the amplitude, the less I was able to compress the corrugation. The difference in the length and width of the corrugation in the lower left of Figure 10 is less than that in the upper right. The only variable that changed was the amplitude of the sinusoidal curve; the higher the amplitude, the more rigid was the sheet. This was true across all tests.

A negative relationship was observed between amplitude and the number of degrees that the corrugation could be rotated around the spine (x's) and amplitude to depth (triangles). There is a very noticeable relationship between amplitude and degree of rotation. It was a negative correlation, which meant the higher the amplitude, the less the corrugation was able to rotate, with less torsion possible. Depth also had a negative relationship; as the corrugation could not compress, the depth did not increase. These relationships held true for both spines and all values of α.

Figure 11 shows the results for the three sets of tests G, T, and P, when comparing amplitude (x values) to compression length (y values). It shows that the relationship of amplitude to compression was similar when comparing different periods. Across each set of tests, the lower the amplitude, the more compression that could occur; this applied to both compressed width and length (not shown in Figure 11). The points lined up against each other consistently for length, width, and depth. From a visual standpoint, the less pronounced the hills and valleys of a curve, the more a corrugation could be pushed together. At $a = 0.25$ it was almost possible to completely bring the sinusoidal curves together.

All plots show the same trend when looking at amplitude (x values) to degrees of rotation (y values), shown in Figure 12, a measure of how much one can twist the piece around a spine (a human analogy is rotating your

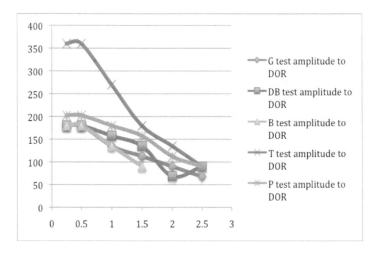

Figure 12. Comparison of G, DB, B, T, and P tests.

upper body when sitting—usually about 90 degrees). There was a negative correlation between amplitude and DOR. The higher the amplitude, the less rotation the corrugation exhibited. The lower the amplitude, the higher the DOR. There was a significant difference in rotation for the T test. This may in part be due to the width being less than the width of the other tests, as it was a test with a period of 1.5. There are issues in measuring DOR by hand, as I used only multiples of 22.5 degrees to identify the amount of rotation. It also appears that the lower the amplitude, the more degrees of rotation, although this may have lower limits, since in each set of tests the amplitudes of 0.25 and 0.5 both had the same degrees of rotation. There are both upper and lower limits to the rotations. Eventually, no compression can occur because the sinusoidal curve approaches a straight line as amplitude decreases. The structures are more rigid when the amplitude is higher. The corrugations with the highest amplitudes had almost no rotation.

The "spine" along which the rotation occurs for the T and P tests is different than for the G, DB, or B tests. The rotation occurs at the midpoint between the maximum and minimum y values of the sine curve (Figure 2, left); this happens whenever the number of periods is whole. In the fourth test the "spine" of the rotation is shifted to the maximum/minimum of the sine curve (Figure 2, right). This shift allowed for a much greater rotation. Since the width was less in the T tests of the paper before corrugation, I could not determine whether the extra paper width was what limited the rotation in tests G, DB, or B. To probe this question, the P test was added. Three of the tests with different periods are shown in Figure 13.

Figure 13. Corrugations before rotation around the spine: P test with $a = 0.25$ (top left) and T test with $a = 0.25$ (top right) both with spine 2; G test with $a = 0.25$ (bottom).

The P test was checking rotation when 2.5 periods were compressed and rotated. In each case, when comparing the two periods to the 2.5 tests, the DOR was significantly larger when the "spine" was along the maximum/minimum y value of the sine curve, except for $a = 2.5$. Figure 14 shows the DOR of the two spine 2 tests as compared to the DOR of all the amplitudes of the other three tests combined. Again, the x values are the amplitudes and the y values are the degrees of rotation around the spines.

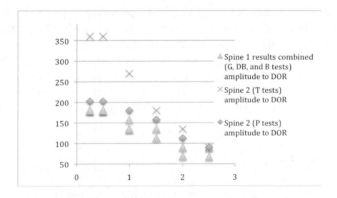

Figure 14. Figure 14. Comparison of G, DB, B, T, and P tests.

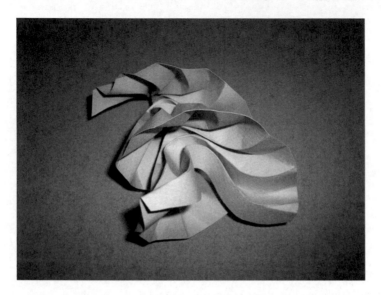

Figure 15. T test underside.

The largest differences are at the smallest amplitude $a = 0.25$. Even with the extra width, the P tests had much higher DOR than the DB, B, and G tests. The varying rotations can be seen in Figure 3, although the T test corrugation is deceptive, as some of the rotation is hidden internally as the piece had wrapped around itself (see Figure 15).

As the amplitude increases, it appears that the minimum distance between curves approaches zero. That is the minimum distance between the sine curves before corrugating the piece. The depth of the compressed corrugation also appears to be directly related to the minimum distance, although this relationship needs further exploration.

The density of the compressed curves increases as the amplitude decreases; in other words, the curves can compress closer and closer to each other, becoming more like an accordion pleated paper. It seems that the higher the density, the stronger the strength of the compressed corrugation, although this was not measured. While only anecdotal, placing more than 150 pounds on top of one of the corrugations (with $a = 0.25$) did not crush it, which indicates that the corrugation had impressive structural integrity.

4   Conclusions

Although some general trends were identified, this series of explorations opens up more questions than it answers, such as the following:

1. What other curves would work or would not work with corrugations?

2. Do the curves need to be parallel, or does it work when the curves are shifted?

3. How do varying curves relate to one another when corrugated?

4. Which properties of curved corrugations are the same with straight-line corrugations, and which are different?

5. What are the actual strength relationships between the various compressed corrugations?

Using a cutting plotter (e.g., Craft ROBO) or laser cutter to create tests would offer more precision and remove one of the variables that can affect results. There are programs, such as those described by Schenk and Guest and Klett and Drechsler, that could precisely measure the exact compression that occurs and identify when the structural breakdown of the corrugations is occurring [Schenk and Guest 11, Klett and Drechsler 11]. I printed out a total of 48 more corrugations than were created due to time constraints, and many more tests are needed to reach a more extensive understanding of compression and rotational limitations and the possibilities of curved corrugations both for aesthetic and functional purposes.

Bibliography

[Klett and Drechsler 11] Yves Klett and Klaus Drechsler. "Designing Technical Tessellations." In *Origami⁵: Fifth International Meeting of Origami Science, Mathematics, and Education*, edited by Patsy Wang-Iverson, Robert J. Lang, and Mark Yim, pp. 305–322. Boca Raton, FL: A K Peters/CRC Press, 2011.

[Schamp 10] Ray Schamp. *Ray's Origami*. Available at http://fold.oclock.am/, accessed November 28, 2010.

[Schenk and Guest 11] Mark Schenk and Simon D. Guest. "Origami Folding: A Structural Engineering Approach." In *Origami⁵: Fifth International Meeting of Origami Science, Mathematics, and Education*, edited by Patsy Wang-Iverson, Robert J. Lang, and Mark Yim, pp. 291–303. Boca Raton, FL: A K Peters/CRC Press, 2011.

[Verity 10] Polly Verity. *Polyscene*. Available at http://www.polyscene.com, accessed November 28, 2010.

Polygon Symmetry Systems

Andrew Hudson

1 Introduction

Geometric origami has long been limited by the constructibility of its structures. Models based on square and hexagonal grid symmetries are commonly known, and some interesting work has been done with octagons and the occasional pentagon, but very little else. More recently, Alex Bateman and Robert Lang have used computer-aided design algorithms to create geometric works based on unusual or irregular polygons [Bateman 10, Lang 09] by specifying a desired folded structure, then engineering a crease pattern to fit their constraints, which is certainly an interesting avenue of research. However, I have used a different approach to get around the problem of constructibility, and demonstrate here the consequences of this method in my own work.

2 Polygon Construction

Polygon construction has been one of the classic problems in the field of computational origami. By tackling this problem, we open up a relatively unexplored region of geometric origami—that of designing and folding models from nonsquare polygons.

Several paper-folding mathematicians have explored this problem before—most notably, Robert Geretschläger, whose book *Geometric Origami*

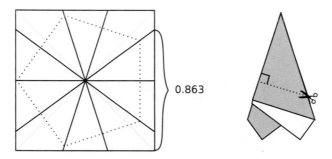

Figure 1. Fold-and-cut algorithm.

[Geretschläger 08] is the most complete treatment of axiomatic origami I have ever seen; he uses polygon constructions to demonstrate its axiomatic construction methods. Generally, exact polygons are constructed using translations of existing straightedge and compass constructions. From a mathematical standpoint, these are very interesting. In practice, however, they have several flaws: they seem arbitrary and are therefore hard to memorize, their folding sequences are often difficult to perform accurately, and they often produce a large number of extra creases that cannot be used later in the model.

My first attempt at solving the problem of polygon construction, which I developed with Ben Parker [Parker 10], was a simple fold-and-cut problem, using Robert Lang's ReferenceFinder [Lang 10a] and some rudimentary trigonometry to construct a sort of n-sided waterbomb base from creases radiating outward from the center of the square, going through the vertices and midpoints of the desired polygon. See Figure 1 for a construction of a pentagon using this algorithm.

However, because it relied on arbitrary approximations and a folding sequence that tended to maximize error, this algorithm was neither useful nor mathematically significant, so I started looking for other ways to construct polygons. Having seen rigorously proven methods for dividing lengths into an arbitrary number of equal parts [Lang 10b], and other unified methods for tackling categories of constructions, I wondered whether it would be possible to devise a unified algorithm for polygons as well; this led me to look at polygon constructions I already knew from square and hexagonal geometry and to try to extrapolate from there.

In this paper, I present the algorithm I developed. It turned out to be easy to implement, and relatively unified, although there are still small differences between the way odd and even polygons are constructed, due to the symmetry of the starting rectangle. Furthermore, the principles that underlie this method of construction allow for the referencing of complex geometric structures in a wide variety of symmetries.

3 Unified Algorithm for Polygon Construction

Regular polygons have a number of common elements that are well defined. They have a center point, n identical edges, and the same number of vertices; by connecting these vertices, we obtain a set of diagonals, which are all rotations by angles that are multiples of a common divisor, $180°/n$. These grid-like symmetrical structures allow us to construct and manipulate the polygons themselves, with aesthetically pleasing results in a surprising variety of forms.

Each polygon and associated symmetry system has a set of angles and proportions peculiar to it, defined by its diagonals and triangulation. For example, angle multiples of $36°$ and length ratios of $1 : \phi$ (that is, the Golden Ratio) are native to the pentagon, whereas the square has angle multiples of $45°$ and proportions of $1 : \sqrt{2}$. These determine the structure of diagonal grids and other constructible shapes within the polygon. Defining a folding algorithm in terms of these symmetrical elements allows it to be generalized to other polygons, which gives us a better understanding of its structure, and gives us a wider variety of visual effects for a model. These symmetry systems can be used to exploit the commonalities of different polygons and create powerful tools for designing geometric origami.

Constructing the regular polygons themselves using a symmetrical approach turns out to be relatively straightforward. When inscribing a regular polygon in a circle, we can show that each of the angles formed by adjacent diagonals of the polygon meeting at a vertex are equal. By applying this procedure to two opposite vertices at once, as shown in Figure 2, we can reference all the points of the polygon; by constructing between two

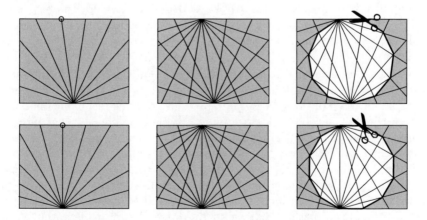

Figure 2. Polygon construction algorithms: maximum odd-sided polygon from a rectangle (top); maximum even-sided polygon from a rectangle (bottom).

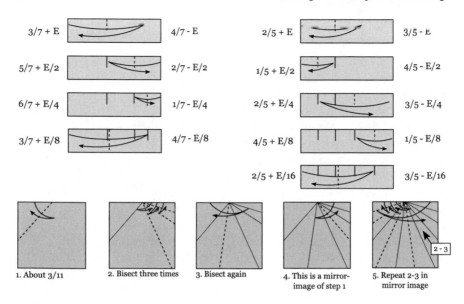

Figure 3. Linear Fujimoto approximation (top); angular Fujimoto approximation (bottom).

parallel lines, which can be found in any rectangle, two such points are readily produced while keeping the polygon contained inside the space. This approach generalizes to squares as well; however, the process is more involved. The sides of the square impose additional constraints because most polygons have diagonals wider than their height, and so for odd-sided polygons, these constraints are used to define the remaining points of the polygon, instead of the opposite vertex. But for even-sided polygons obtained from a square, and for any polygon obtained from a rectangle, the only obstacle is to find a functional n-section for a straight angle that works well enough.

Finding an exact method of angular n-section that works for any integer n would be quite a difficult task, but for all practical purposes, a good approximation will do. Shuzo Fujimoto developed a technique that used a repeating sequence of folds to divide an angle or line segment into any number of equal parts. Fujimoto's algorithm [Fujimoto and Nishikawa 82], shown in Figure 3 to divide into fifths and sevenths, is only an approximation, but due to its recursive nature, the error is reduced by half with each step, and soon the error becomes insignificant in comparison to human error and the thickness of the paper [Hull 06]. This is similar to a more generalized method discovered by James Brunton [Brunton 73] for finding a fraction using its binary expansion; for each 0, the left edge is folded inwards, and for each 1, the right edge is folded in. The repeating pattern that results is similar to the results from Fujimoto's method. Both division methods can be applied to angle n-sections as well, so they are a good choice for this step. I prefer Fujimoto's way of thinking about the method because it doesn't require the conversion to binary, but either one can be used in this situation.

This method of construction gives us a polygon with creases only along its diagonals, i.e., only within the symmetry of the polygon. Using the polygon and its diagonals, we can start to work with the polygon and define algorithms to fold. Many of these involve twist folds and related structures; I focus on twist-fold-based constructions for the remainder of the article.

4   Polygonal Grids and Their Properties

Using the diagonals of the polygon, we can start to construct other creases within the symmetry and form an aperiodic grid with a finite number of component widths. In polygonal grids, each crease is constructed parallel to a side of the polygon. The reference points for the creases come from the intersections of the polygon's diagonals. These grids have some unusual properties because of this; for example, consider the pentagonal grid, as shown in Figure 4. Pentagonal grids are somewhat self-similar and may have some fractal properties; this self-similarity extends to other polygons as well. Each iteration of the grid contains sets of parallel creases that alternate between large (L) and small (S) in a pattern defined by the substitution system $\{L \rightarrow LS, S \rightarrow L\}$. Note that these are the same rules as those that generate the Fibonacci sequence; at higher resolutions, the ratio of L:S approaches the golden ratio, $1 : \phi$. Furthermore, because of this property and the pentagonal symmetry, we can say that the pentagonal grid is identical to a subset of the Ammann bars of a Penrose tiling [Lord 91]. Similar grids can be created on any polygon, and indeed square and hexagonal grids can be constructed using this process; however, larger grids become increasingly more difficult to utilize, because there are so many creases necessary per iteration.

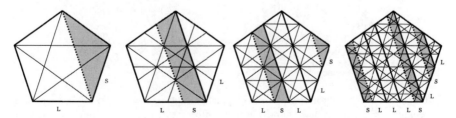

Figure 4. Pentagonal grid iterations.

Other grids can be analyzed using substitution systems, however, the mechanics are much more complex because there are more than two proportions at any given resolution. Heptagonal grids have three components at any given grid resolution and can be generated by the rules {L → LMS, M → LM, S → L}. Much more care must be taken when constructing heptagonal grids than with pentagons. There are far more intersections available; however, some of these are not usable in the immediate next generation of grid lines, if one is working only within the substitution system.

5   Symmetrical Twist-Fold Constructions

Polygonal grids create the framework for some very interesting origami; because of the embedded symmetrical elements, it is possible to construct twist-fold patterns that are similar to those commonly found in the origami tessellations genre [Gjerde 09].

When working with a polygon, twist folds preserve the radial symmetry of the original polygon, so, aesthetically speaking, they are a good choice for polygon-based structures. These twist-fold structures can then be mod-

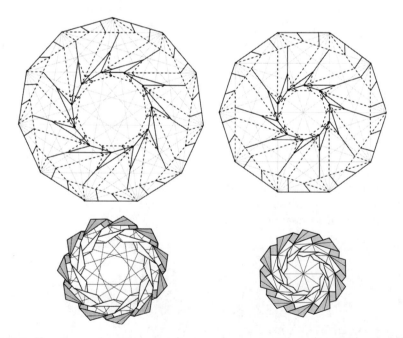

Figure 5. Hurricane algorithm: hendecagonal crease pattern (top left), decagonal crease pattern (top right); hendecagonal folded form (bottom left); decagonal folded form (bottom right).

ified through the use of swivels, sinks, pleats, layer reordering, and other conventional origami operations. Through these modifications, we can produce surprisingly complex structures, such as the two models in Figure 5. Here, a twist fold with some layers rearranged is used to create a color change along the outer regions of the polygon, then another twist is put on top of this referenced by the third diagonals, and that twist is then pursed to create a final model, which is not recognizable as a simple twist fold. Each fold is referenced the same way regardless of which polygon is being used (e.g., the first color change always uses the outer set of diagonals, the primary twist fold uses the third set of diagonals, etc.). In this case, I named the generalized folding algorithm the *Hurricane* algorithm, naming the algorithm itself instead of the individual instances that result from this algorithm's application to different polygons. Two of these instances are shown in Figure 5.

6   Generalizing Folding Algorithms to Different Polygons

Generalizing a folding algorithm to a different polygon involves several considerations. First, each substep and operation must be defined in terms of the symmetry system. Second, these steps are geometrically analyzed to make sure that they represent the correct aspect of the symmetry; most of the time, a fold can be referenced several ways, but often only one or two of these will work for a generalized algorithm. Sometimes more than one construction method works out, so the designer must decide which method to use. Finally, the algorithm is analyzed to figure out on what range of polygons it will work. Often a complex algorithm will not work on lower-order polygons; for example, the "Hurricane" folding algorithm, examples of which are shown in Figure 5, uses the third diagonal in from the side as a reference mark, so it will not work on octagons and lower-order polygons, which do not have third diagonals.

The aesthetic effects of generalizing an algorithm to a different polygon obviously vary depending on the algorithm, but in general: there is more emphasis on the radial nature of the piece with higher-order polygons; the symmetries involved are more obvious with lower-order polygons; and with higher-order polygons, the "details" formed by the edge of the paper get smaller, whereas the open space at the center of the polygon gets larger. Visually, this has the effect of opening and closing an iris aperture, such as that in a camera lens. Switching to a different polygon can help balance out the different elements of the algorithm and can give the designer a few more options for modification before the model is finalized.

7 Conclusion

These techniques open up some very interesting tools for the geometric designer, and I hope that they will be used by other designers. Efficient polygon construction has always been a problem, and this contribution may have opened the door to further explorations with polygons and their symmetry systems by other designers.

Bibliography

[Bateman 10] Alex Bateman. "Tess: Origami Tessellation Software." Available at http://www.papermosaics.co.uk/software.html, 2010.

[Brunton 73] James Brunton. "Mathematical Exercises in Paper Folding." *Mathematics in School*, Longmans for the Mathematical Association, 2:4 (July 1973), 25.

[Fujimoto and Nishikawa 82] Shuzo Fujimoto and M. Nishiwaki. *Sojo Suru Origami Asobi Eno Shotai (Invitation to Creative Origami Playing)*. Tokyo: Asahi Culture Centre, 1982.

[Geretschläger 08] Robert Geretschläger. *Geometric Origami*. Shipley, UK: Arbelos, 2008.

[Gjerde 09] Eric Gjerde. *Origami Tessellations: Awe-Inspiring Geometric Designs*. Wellesley, MA: A K Peters, 2009.

[Hull 06] Thomas Hull. *Project Origami: Activities for Exploring Mathematics*. Wellesley, MA: A K Peters, 2006.

[Lang 09] Robert J. Lang. "9-Fold Star Woven Tessellation." Available at http://www.flickr.com/photos/langorigami/4052747535/, 2009.

[Lang 10a] Robert J. Lang. "ReferenceFinder." Available at http://www.langorigami.com/science/reffinder/reffinder.php4, accessed November 12, 2010.

[Lang 10b] Robert J. Lang. "Origami and Geometric Constructions." Available at www.langorigami.com/science/hha/origami_constructions.pdf, accessed December 3, 2010.

[Lord 91] Eric A. Lord. "Quasicrystals and Penrose Patterns." *Current Science* 61:5 (10 Sept 1991), 313.

[Parker 10] Benjamin Parker. "brdparker's photostream." Available at http://www.flickr.com/photos/brdparker/, accessed December 11, 2010.

New Collaboration on Modular Origami and LED

Miyuki Kawamura and Hiroyuki Moriwaki

1 Introduction

We have developed modular origami models containing light-emitting diodes (LEDs) inside. We use a high-power LED chip that is extremely bright. The chip is a small square with an edge length of 5 mm, allowing placement of it in a narrow space. Unlike light bulbs, it does not crack; the LED is powered by low energy from a few dry cells, thus its temperature is lower than a common electrical lamp.

The low temperature and lightweight features of an LED are very good for use with paper models. We can use many kinds of paper, even very thin and delicate sheets, without worrying about fire hazards. At the same time, light from the high-power LED is very bright, and we can see this brightness even outside under full sunlight, allowing us also to use very thick paper for LED origami lamps. This brightness is the most powerful reason to use high-power LED chips for origami models. Light from an LED chip will work as a kind of "x-ray," irradiated from of the interior of origami models. It can penetrate through more than 15 sheets of TANT paper, a type of Japanese construction paper. High-power LEDs give us a new way to show the hidden structure of origami models without taking the model apart.

2 AKARI-ORIGAMI: Some Modular Works

In the origami world, there are many works combining origami and light. We can see many lampshade works of origami at exhibitions, in books,

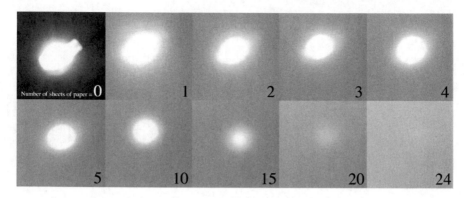

Figure 1. Brightness of high-power LED light: numbers indicate the number of sheets of TANT paper.

and on the Internet. Tessellations are good examples of such collaborations. For example, there are beautiful works by Joel Cooper [Cooper 06a, Cooper 06b] and Sipho Mabona [Mabona 06]. And also there is a nice origami lamp images collection by Eric Gjerde [Gjerde 11]. At the Fifth International Meeting of Origami in Science, Mathematics and Education (5OSME), Tomoko Fuse presented a recent creation, an origami twist lamp, and Jun Mitani exhibited an origami lamp.

Some works show shadows made by the internal structure of the model. For example, Kunio Suzuki researched shadow shapes of internal patterns of two-dimensional origami snowflakes made from a sheet of hexagonal paper [Suzuki 97]. Suzuki used a fluorescent lamp to show the shadow. That is an example of work that positively uses light to show a shadow pattern.

We use a high-power LED light inside a three-dimensional model. The properties of the LED chip are as follows:

Model number:	LP-5060H196W-S
Color of light:	Warm Tone (2700 k)
Forward voltage:	3.2 V (typ)
Forward current:	60 mA
Luminous intensity:	3190–4150 mcd
Irradiation angle:	100 degrees
Cost:	About 1–2 dollars per one LED chip

When the LED chip is turned on inside the origami model, brightness on the surface depends on the number of sheets of layered paper in the model (see Figure 1).

For the model in Figure 1, we used Japanese construction paper, TANT. At first the LED chip is naked (0 sheets). Then some sheets of TANT paper are placed over the chip one by one. We can see clearly the LED

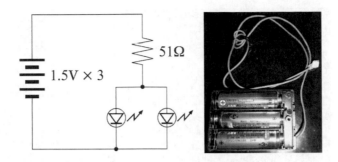

Figure 2. The circuit of the Moriwaki module.

light through 20 sheets of TANT paper. In the case of the modular model, some sheets of paper are overlapped based on geometrical patterns with each other inside the model. We can see the patterns as a shadowgraph on top of the model's surface by using the high-power LED.

The irradiation angle of our LED chip is about 100 degrees. Moriwaki combined the two high-power LED chips and made one chip that irradiates light in almost all directions. This improved LED chip requires 120mA of current, but it is too bright for our eyes. So, one resistor is built into the circuit as shown in Figure 2, cutting the current in half.

This circuit has one battery module, and it needs three AA dry cells to emit the LED light. The LED chip of the Moriwaki module is about 5 mm square, so the chip can be placed inside an origami model through the narrow slit between papers. It does not need another structure to hold the LED chip. The battery module remains outside the origami model. This battery module is a small black box 7 cm in width, 4.5 cm in depth, and 1.5 cm in height. This small box can be used as a stand for the model.

Kawamura designed some modular models for this improved LED chip. We named these works *AKARI-ORIGAMI* [Kawamura 10]. The Japanese word *akari* means light and/or lamp. In the 1950s, Isamu Noguchi, the famous artist, named his lamp works "AKARI," which became familiar outside of Japan.

Figure 3 is an example of AKARI-ORIGAMI. The title of the model is "Checkered Lantern" [Kawamura 10]. On the left side, the image is of the model with the LED turned off, and on the right side is the model with the LED turned on. The checkered lantern model is made of six modules, as shown in Figure 4. These diagrams are a modified version for AKARI-ORIGAMI. (For the original model, make mountain folds at two corners in the first step.) There are six squares on the model's face, and one square is divided into four small squares by the difference in the number of sheets of paper. The bright areas are made of one layer, and the dark areas are made of three layers. The checkered lantern modules and model are very

Figure 3. "Checkered Lantern" with LED off (left) and on (right).

simple to construct, making the model a good choice for introducing people to the world of AKARI-ORIGAMI.

Figure 5 is another example of AKARI-ORIGAMI. The title of this model is "Earthen" [Kawamura 10]. This model is made of six modules, each of which has four smooth triangular surfaces. The shadow of the overlapped paper appears on the surface of this model, which is composed of 24 flat triangles. There are seven different densities of shadow due to the differences in the number of layers.

More examples are shown in Figure 6. These models are older works by Kawamura not designed originally for AKARI-ORIGAMI, but they have very good shadow patterns.

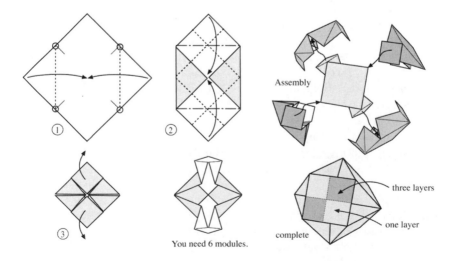

Figure 4. "Checkered Lantern" folding diagrams.

Figure 5. "Earthen."

"Wedge" [Kawamura 03] is made from two types of modules, with the finished model composed of 20 modules. The overlapped areas are very thick, with three-dimensional shapes inside, so the shadow outlines are not very clear.

"Dogwood" [Kawamura 09a] is made of 12 modules. The eight large triangular holes on the face of model may make it unsuitable as a desktop lampshade due to the brightness of the light.

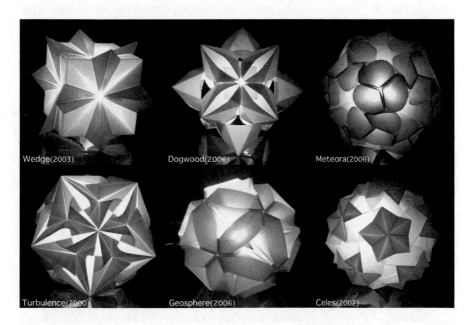

Figure 6. Models by Kawamura: "Wedge," "Dogwood," "Meteora," "Turbulence," "Geosphere," and "Celes." (See Color Plate III.)

"Meteora" [Kawamura 07a] is made of 30 modules. The shadow pattern, which is made by overlapped paper, is not distributed along the surface but seems to be standing up along the vertical directions from the center of the model.

"Turbulence" [Kawamura 01, Kawamura 10] is made of 30 modules. With almost all layers laid down on the surface, the shadow pattern appears very clear. This model serves as a good application of AKARI-ORIGAMI, but assembly of this model is difficult.

"Geosphere" [Kawamura 07b] is made of 30 modules. This model has 20 large triangular windows. We can see the center of the model easily, so this also is unsuitable as a desktop lampshade.

"Celes" [Kawamura 02, Kawamura 10] is made of 30 long rectangles. The "bridge" [Kawamura 00, Kawamura 09b] area of the module is made from one sheet of paper, and these areas are strongly lit. We can see 12 stars clearly on the surface of the model.

3   Workshop

Kawamura has conducted a number of AKARI-ORIGAMI workshops. On June 27, 2010, we presented a workshop in Japan's Saga prefecture in the Kyushu area. There were 26 workshop participants, ranging in age from 11 years old to 85 years old (Figure 7). The AKARI-ORIGAMI kit for this workshop included one Moriwaki module and 18 sheets of white square TANT paper. Participants made one or two "Checkered Lantern" models

Figure 7. AKARI-ORIGAMI Workshop in Japan, June 27, 2010.

Figure 8. AKARI-ORIGAMI Workshop at 5OSME, July 17, 2010.

in one and a half hours. The participants were not origami beginners, so they had no problem making the models. They just inserted the LED chip into the "Checkered Lantern" model, turned it on, and they saw the shadow pattern on the surface of the model, even under the light of the room's fluorescent lamps. Due to the light from the LED, the structure of the connecting parts of the model can be observed directly.

Another workshop was presented on July 17, 2010, to 14 participants at the 5OSME PLUS! convention in Singapore (Figure 8). They enjoyed the workshop very much and were surprised that the light from the LED chips was so bright.

4   Future of AKARI-ORIGAMI

Usually, modular folders understand the structure of modules, and they can follow the instructions for assembly. However, it is difficult to make a clear image of the inside pattern of the completed model. High-power LED is an excellent tool to show the hidden structure of the interior of modular works. The structure appears as a shadow pattern on the model's surface. Now we can see these patterns directly from the outside of the models. A high-power LED will produce not only beautiful lamps but also some interesting viewpoints about the origami world.

A book entitled *AKARI-ORIGAMI* was published in November 2010 by the Japanese publisher Gakken [Kawamura 10] and is shown in Figure 9 (left). The book includes one LED lamp kit designed by Moriwaki, which consists of an electric wiring base, an LED chip, and a battery case (see

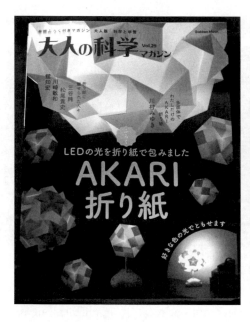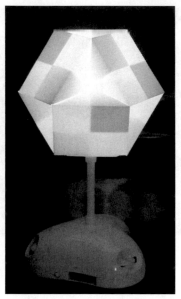

Figure 9. The book *AKARI-ORIGAMI* (left); "Checkered Lantern" made from the book's kit (right).

Figure 9 (right)). The book also includes enough white TANT paper for construction of one "Checkered Lantern," one "Celes," and a third model of one's own choice or design.

We plan to have future exhibitions and workshops. We hope that people will enjoy AKARI-ORIGAMI and new discoveries in the origami world.

Bibliography

[Cooper 06a] Joel Cooper. "Fujimoto Lampshade." *flickr.* Available at http://www.flickr.com/photos/origamijoel/87054668/in/set-72057594048880232/, 2006.

[Cooper 06b] Joel Cooper. "Lampshade." *flickr.* Available at http://www.flickr.com/photos/origamijoel/97312151/in/set-72057594048880232/, 2006.

[Gjerde 11] Eric Gjerde. "Archive for the 'Lighting' Category." *Origami Tessellations.* Available at http://www.origamitessellations.com/category/lighting/, 2011.

[Kawamura 00] Miyuki Kawamura. "Quick Snap." *Origami Tanteidan Magazine* 60 (2000), 11–12.

[Kawamura 01] Miyuki Kawamura. "Turbulence" In *BOS Convention 2001 Autumn*, edited by British Origami Society, pp. 51–52. London: British Origami Society, 2001.

[Kawamura 02] Miyuki Kawamura. "Celes." In *Origami Tanteidan Convention Book 8*, edited by Japan Origami Academic Society, pp. 148–150. Tokyo: Japan Origami Academic Society, 2002.

[Kawamura 03] Miyuki Kawamura. "Wedge." In *Origami Tanteidan Convention Book 9*, edited by Japan Origami Academic Society, pp. 26–28. Tokyo: Japan Origami Academic Society, 2003.

[Kawamura 07a] Miyuki Kawamura. "Meteora." *Origami Tanteidan Magazine* 102 (2007), 4–5.

[Kawamura 07b] Miyuki Kawamura. "Geosphere" *Origami Tanteidan Magazine* 101 (2007), 4–7.

[Kawamura 09a] Miyuki Kawamura. "Dogwood." In *Origami Tanteidan Kansai Convention Book 10*, edited by Kansai Tomonokai, pp. 8–9. Kobe: Kansai Tomonokai, 2009.

[Kawamura 09b] Miyuki Kawamura. "The Celes Family of Modular Origami." In *Origami⁴: Fourth International Meeting of Origami Science, Mathematics, and Education*, edited by Robert J. Lang, pp. 21–30. Wellesley, MA: A K Peters, 2009.

[Kawamura 10] Miyuki Kawamura, Hiroyuki Moriwaki, and Atuhiko Isiguro. *AKARI-ORIGAMI*, Otona no Kagaku 29. Tokyo: Gakken Education Publishing Co., 2010.

[Mabona 2006] Sipho Mabona. "S-PHERE." *flickr*. Available at http://www.flickr.com/photos/sipmab/190630015/, 2006.

[Suzuki 97] Kunio Suzuki. "Creative Origami 'Snow Crystals': Some New Approaches to Geometric Origami" In *Origami Science and Art: Proceedings of the Second International Meeting of Origami Science and Scientific Origami*, edited by K. Miura, pp. 361–378. Shiga, Japan: Seian University of Art and Design, 1997.

Using the Snapology Technique to Teach Convex Polyhedra

Faye Goldman

1 Introduction

There are many ways of making polyhedra using origami. John Montroll specializes in polyhedra made with a single sheet of square paper [Montroll 02, Montroll 04, Montroll 09]. Tomoko Fuse [Fuse 90] describes face units and edge units. The most famous unit was created by Mitsonobu Sonobe [Mukerji 07]. Usually the Sonobe unit is used as an edge unit; however, the unit can also be used as a face unit to make a cube (six units). With additional creases, objects have been made with the number of units ranging from three (Toshie's Jewel) [Kasahara 87] to thousands.

If a modular technique were being used, it was usually left to the folder to figure out which part of the polyhedron the unit represented. It is only in the past several years that authors have begun to define the part of the polyhedron the unit is making, but many books are available on the subject (e.g., [Kawamura 01, Simon 99, Cundy 61, Franco 99, Holdern 91, Wenninger 71]).

This paper demonstrates a technique for folding polyhedra from ribbons, which Heinz Strobl, using strips of paper, calls *Snapology* because of the way the units snap together. Snapology uses edge modules (hinges) to connect face units (scaffolds). The focus will be on a simple form of Snapology, called *Special Snapology*, which works for many polyhedra. The paper then describes a generalization of the technique, called *General Snapology*, which is useful for a wider range of polyhedra.

2   Polyhedra Review

The Snapology technique is used to build polyhedra. Therefore, a brief
review of the terms used to define polyhedra and the classification of polyhedra might be helpful.

Special Snapology is limited to convex polyhedra. There are no dimples
or depressions on the surface of a convex polyhedron. Additionally, a line
connecting any two points on the surface of a convex polyhedron will be
on the surface or within the interior.

The name *polyhedron* comes from the Greek *poly* for "many" and *hedron*
for "face." Thus, a polyhedron has many faces. The plural of polyhedron is
polyhedra, but current usage also allows polyhedrons. Polyhedra consist of

- vertices: corner points;

- edges: connecting lines between two vertices;

- faces: flat polygonal-shaped surfaces that are bounded by the edges.

The dihedral angle, or face angle, is the interior angle formed by two
adjacent polygons forming the surface of the polyhedron.

The major families of convex polyhedra are Platonic and Archimedean
solids. Several resources contain pictures and more detailed discussions of
these [Coxeter 97, Cromwell 97, Sutton 02, Montroll 09, Weisstein 11a,
Weisstein 11b, Weisstein 11c]. Other convex polyhedra include Catalan
solids, prisms, antiprisms, Johnson solids, pyramids, and dipyramids.

2.1   Platonic Solids

Platonic solids are the most uniform of all the polyhedra and are comprised
of the same regular polygon and identical vertices. A regular polygon has
equal-length edges and all of the angles between adjacent sides are the
same. There are five Platonic solids, identified by the number of faces:

- The *tetrahedron* has 4 triangular faces.

- The *cube* or *hexahedron* has 6 square faces.

- The *octahedron* has 8 triangular faces.

- The *dodecahedron* has 12 pentagonal faces.

- The *icosahedron* has 20 triangular faces.

2.2 Archimedean Solids

The faces of the *Archimedean solids* are also regular polygons, but each Archimedean solid consists of two or more different polygons. These are also known as semiregular polyhedra. Each Archimedean solid has vertices that are identical. There are 13 Archimedean solids: truncated tetrahedron, truncated octahedron, truncated cube, truncated icosahedron, truncated dodecahedron, cuboctahedron, icosidodecahedron, small rhombicuboctahedron, small rhombicosidodecahedron, truncated cuboctahedron, great rhombicosidodecahedron, snub cube, and snub dodecahedron.

3   Snapology Technique

Heinz Strobl invented Snapology at the end of the 1990s to use up the ends of strips of ticker tape he employed in the development of *Knotology* [Strobl 10]. The Knotology technique creates pentagons in strips of paper by creating a flat knot. Beautiful three-dimensional models can be created [Strobl 06].

The Snapology construction creates prisms with rectangular sides on top of each of the faces of the polyhedron being constructed. We detail Special Snapology here. Special Snapology is a subset of General Snapology. Special Snapology is limited to the use of equal-width scaffold and hinge modules and a four-unit hinge. In General Snapology, some of the basic rules of Special Snapology are relaxed.

3.1   Terminology

The Special Snapology technique uses two types of strips, referred to as the scaffold and the hinge. The *scaffold* module makes the inside structure. It outlines each face of the polygon with a doubled wall of ribbon or paper. The length of each scaffold module is $2N$, where N is the number of edges of the polygonal face. For a triangular face, we use a strip six units long; for a square, we use eight; etc. The *hinge* module connects two pieces of scaffold modules, i.e., two adjacent faces. In Special Snapology the hinge module is always four units long. It wraps around the scaffold module and "snaps" into place.

Initially, the two strips should be of equal width. With experience, different width strips may be used. Models made with scaffold modules much thinner than hinge modules will result in a more compact model, and scaffold modules much thicker than hinge modules will result in a spikier model.

Table 1 contains a complete list of Platonic and Archimedean solids and the numbers and lengths of scaffold and hinge modules needed to

	Vertex description	# of edges	# of vertices	Inside scaffolding: # strips of length x	Outside hinges: # strips of length 4	# of squares inside or outside	Total squares	Approx dihedral angle	180-dihedral angle
Platonic									
1 Tetrahedron	(3,3,3)	6	4	4 of 6	6	24	48	71	109
2 Octahedron	(3,3,3,3)	12	6	8 of 6	12	48	96	110	70
3 Hexahedron (cube)	(4,4,4)	12	8	6 of 8	12	48	96	90	90
4 Icosahedron	(3,3,3,3,3)	30	12	20 of 6	30	120	240	138	42
5 Dodecahedron	(5,5,5)	30	20	12 of 10	30	120	240	117	63
Archimedean									
6 Truncated tetrahedron	(3,6,6)	18	12	4 of 6 / 4 of 12	18	72	144		
							hex-hex	71	109
							hex-tri	110	70
7 Truncated octahedron	(4,6,6)	36	24	6 of 8 / 8 of 12	36	144	288		
							sq-hex	125	55
							hex-hex	109	71
8 Truncated cube (hexahedron)	(3,8,8)	36	24	8 of 6 / 6 of 16	36	144	288		
							Oct-Tri	126	54
							Oct-Oct	90	90
9 Truncated icosahedron	(5,6,6)	90	60	12 of 10 / 20 of 12	90	360	720		
							hex-hex	138	42
							hex-pent	143	37
10 Truncated dodecahedron	(3,10,10)	90	60	20 of 6 / 12 of 20	90	360	720		
							dec-dec	117	63
							dec-tri	143	37
11 Cuboctahedron	(3,4,3,4)	24	12	8 of 6 / 6 of 8	24	96	192	125	55
12 Icosidodecahedron	(3,5,3,5)	60	30	20 of 6 / 12 of 10	60	240	480	143	37
13 (Small) Rhombicuboctahedron	(3,4,4,4)	48	24	8 of 6 / 18 of 8	48	192	384		
							sq-sq	135	45
							sq-tri	144.5	35.5
14 Small Rhombicosidodecahedron	(3,4,5,4)	120	60	20 of 6 / 30 of 8 / 12 of 10	120	480	960		
							pent-sq	148	32
							tri-sq	159	21
15 Truncated cuboctahedron/ Great rhombicuboctahedron	(4,6,8)	72	48	12 of 8 / 8 of 12 / 6 of 16	72	288	576		
							oct-sq	135	45
							oct-hex	125	55
							hex-sq	145	35
16 Great rhombicosidodecahedron/ Truncated Icosidodecahedron	(4,6,10)	180	120	30 of 8 / 20 of 12 / 12 of 20	180	720	1440		
							hex-dec	142	38
							sq-dec	149	31
17 Snub cube	(3,3,3,3,4)	60	24	32 of 6 / 6 of 8	60	240	480		
							sq-tri	143	37
							tri-tri	153	27
18 Snub dodecahedron	(3,3,3,3,5)	150	60	80 of 6 / 12 of 10	150	600	1200		
							pent-tri	152	28
							tri-tri	164	16

Table 1. Number, strip lengths, and dihedral angles for Platonic and Archimedean solids using Heinz Strobl's Snapology technique.

construct them. There is also a column for the dihedral angles formed between adjacent faces of a solid. This information is useful to determine how stable the construction will be, based on the dihedral angles of the underlying polyhedron.

Figure 1. Finished witch's ladder.

3.2   Assembling the Basic Icosahedron

This section explains the assembly of a Snapology icosahedron. Even though an icosahedron is relatively complex, it offers a good starting point because the angles between the walls of adjacent prisms are acute. Acute angles between adjacent prisms lead to greater stability in the finished model, and the large number of faces have large dihedral angles (138°). The angle between the walls of the prisms is 180° minus the dihedral angle of the underlying polygons, resulting in an acute angle of 42°, which is smaller than those found in the other Platonic polyhedra. The larger the angle between the walls of the prisms, the less stable is the construction.

Using two different colors of paper or ribbon differentiates between the scaffold module and the hinge module. For the icosahedron, 120 squares are used for the inside (scaffolding) color and 120 squares for the outside (hinge) color (see Table 1). For three-quarter inch ribbon, two strips of slightly over 45″ are needed for each color (120 squares at 3/4″ per square = 90″, which is divided in half for ease of work).

There are several ways to make the pre-creases necessary to create the scaffold and hinge modules. A quick and simple way is to fold two strips alternately over each other. This is called *muizentrapje* (mouse's staircase) in Holland, and in Germany it is a *hexentrap* (witch's staircase). Heinz Strobl has heard English and American people call it a "witch's ladder" [Versnick 00] (Figure 1).

It is not necessary for the scaffolds and hinges to have the same width. However, it is important that if ribbons of different widths are being used, one ribbon of each width must be used to make the witch's ladder. If ribbons of different width are used, the pre-creases become rectangles, and the total length of each ribbon is based on the width of the other ribbon.

The creation of a witch's ladder is begun by making a fold at the beginning of each strip approximately one width from the end. The ribbons are linked at right angles, and the strips are alternately folded over each other. The back and forth crossing of the ribbons is continued until all the ribbon is used. Two ladders are made using one of each color ribbon in each ladder. Figure 2 shows detailed pictures of the making of a witch's ladder. Precisely made creases make it easier to assemble the model. Once

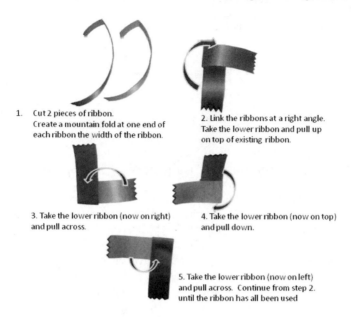

1. Cut 2 pieces of ribbon.
Create a mountain fold at one end of
each ribbon the width of the ribbon.

2. Link the ribbons at a right angle.
Take the lower ribbon and pull up
on top of existing ribbon.

3. Take the lower ribbon (now on right)
and pull across.

4. Take the lower ribbon (now on top)
and pull down.

5. Take the lower ribbon (now on left)
and pull across. Continue from step 2.
until the ribbon has all been used

Figure 2. Making of witch's ladder.

the creases are made, the ladder is unfolded, and a decision needs to be made as to which color will be the scaffold (inside) and which will be the hinge (outside).

For the scaffold units, the ribbon is cut into 20 pieces each of six unit lengths. These units form the 20 faces of the icosahedron. Using the good (or pretty) side, the zigzag ribbon folds of the witch's ladder are all turned into mountain folds.

The hinge units are cut into 30 pieces each of four unit lengths. One hinge unit is needed for each edge, and each hinge unit connects two "faces." Mountain folds are made on the good side of the ribbon.

To begin building the model:

- Two triangles of scaffold modules are created by wrapping each strip around itself to form a triangle with double-sided walls.

- They are joined with a hinge unit as shown in Figure 3:

  - The edge of the triangle is covered with the loose outside raw edge first to keep the triangle from unraveling.

  - The hinge unit comes up from the bottom, into the triangle over the top and "snaps" into place in the "V" between the two triangles.

Figure 3. Two linked triangles.

The following steps are continued until the model is completed. Detailed pictures of the creation of the first ring of five triangles are shown in Figure 4.

- A *combo* module consists of a scaffold module and an attached hinge (see Figure 4, Step 2)

- Combo modules are added to form rings of five (always added so the hinge is at the bottom or inside of the model).

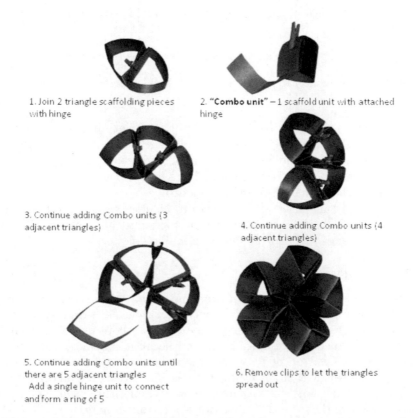

1. Join 2 triangle scaffolding pieces with hinge

2. **"Combo unit"** – 1 scaffold unit with attached hinge

3. Continue adding Combo units (3 adjacent triangles)

4. Continue adding Combo units (4 adjacent triangles)

5. Continue adding Combo units until there are 5 adjacent triangles
   Add a single hinge unit to connect and form a ring of 5

6. Remove clips to let the triangles spread out

Figure 4. Joining scaffold and hinge modules to create a ring of five triangles.

Figure 5. Finished icosahedron.

- The last hinge in each group of five scaffolds connects two existing triangles.

- Making rings of five as soon as possible helps to stabilize the model.

- The process continues until all 20 triangles of the icosahedron are built.

The last triangle may be difficult to put into place because there is so little room to work. It may be easier to add the last (and hardest) triangle by adding the hinge modules to the model first, and then building the triangle by wrapping the ribbon around the loose hinge modules.

Figure 5 shows the completed icosahedron.

Hint: Miniature clothespins can help to hold the pieces together. Once a ring of five triangles is made, the clothespins must be removed. The five connected triangles will spread out to look like a flower.

3.3 Other Platonic Solids

Figure 6 shows a completed tetrahedron and cube, which have larger angles between the walls of the built-up prisms than the icosahedron (Figure 5). The four-unit hinge module will not stay closed without using glue to hold it in place. This unstable hinge is created because Snapology is based on building prisms on top of the polygons underneath the solid being constructed. The angle that the prisms form is 180° minus the dihedral angle of the underlying adjacent polygons. For example, the angle the prisms make outside of the cube is 90°, and for the tetrahedron it is $\sim 109°$. Contrast this with the $\sim 42°$ angle made with the icosahedron. Dihedral angles

Figure 6. Tetrahedron and cube.

of greater than 135° make the most stable Snapology structures. A dihedral angle of more than 135° will make the angle between the Snapology prisms less than 45°. A modified hinge module is discussed in the next section.

4 Brief Introduction to General Snapology

Special Snapology uses a four-unit hinge module. The scaffold and hinge modules are also the same width. Special Snapology borrows the six-unit hinge from General Snapology because not all Platonic and Archimedean solids can be made with the four-unit hinge without necessitating the use of glue.

According to Heinz Strobl [Strobl 10], General Snapology grew out of Special Snapology. Expansions from Special to General Snapology include the following:

- Different widths are used for the scaffold and hinge modules.

- Different lengths are used for the polygon sides. Special Snapology is restricted to polygons of equal edge lengths but not simply regular polygons (rhombi are allowed in Special Snapology).

- Hinge modules may be placed on both ends of the polygonal scaffold (torus).

The distinctions between Special and General Snapology are still evolving.

Thus far, I have made all of the Platonic and Archimedean solids, some Archimedean duals (Catalan solids), two ovoid (egg) shapes, bracelets (cylinder), and a torus (Figure 7). The torus is a nonconvex model. It is impossible to make a torus with all the hinge units on the same side of the model. The innermost ring of the torus is made by connecting the triangles

Figure 7. Torus.

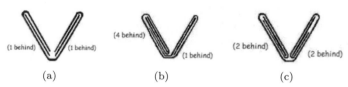

(1 behind)   (1 behind)   (4 behind)   (2 behind)
                                             (1 behind)   (2 behind)

(a)   (b)   (c)

Figure 8. Cross section of the (a) original four-unit hinge, (b) modified six-unit hinge for easier assembly, and (c) modified six-unit hinge that is harder to assemble.

in groups of seven. Within each group of seven, some of the hinge modules are made on the opposite side. This allows for the negative curvature necessary in a torus.

Some of the polyhedra made with the Snapology technique don't hold together as well as others due to the unstable hinge modules. For example, the hinge modules on the tetrahedron and cube (see Figure 6) are particularly unstable and will not stay in place without glue.

To overcome the need for glue, a new hinge unit is needed. This strip of six units is borrowed from Heinz Strobl's General Snapology [Strobl 10]. One additional unit is added at the beginning and another at the end of the four-unit hinge. These extra units are wrapped behind and underneath the scaffold. Figure 8 provides cut-away views of the new hinge module.

5   Observations

In any Snapology structure made using the four-unit hinge, the number of squares needed for all the scaffolds (if equal width ribbon is used) is equal to the number of squares needed for all the hinges. This piece of insight allowed for easy calculation of the length of ribbon needed for a particular polyhedron. The proof for this follows.

It is given that in Special Snapology each polygon of the scaffold needs length $2N$, where N is the number of edges for each polygon, and each hinge unit uses four squares.

At each edge of the polyhedron are squares of scaffold from each adjacent polygon. There are two polygons that meet at each edge. Thus, a total of four scaffold squares make up each edge.

There is one hinge module at each edge of the polyhedron. Each hinge module has four squares. Thus, there are an equal number of squares from the scaffold hinge modules.

6 Conclusions

This paper describes the basic concepts of the Special Snapology technique developed by Heinz Strobl through the building of an icosahedron. Special Snapology with the use of glue can be used to make all convex polyhedra. Heinz Strobl has been working on additions to Special Snapology that allow for more complex structures. He calls this General Snapology. Using six-unit hinge modules rather than four-unit hinge modules allows for completion of polyhedra with smaller dihedral angles, resulting in the prisms built on top of the faces that can hold together without gluing the hinges. Placing the hinge unit on either side allows for more complex creations, as in the torus (Figure 7).

Although creating these models takes time and concentration, the end result is wonderful. Snapology is a unique way to get students engaged in the abstract idea of polyhedra and interacting with it in a way that results in very concrete and beautiful objects.

Acknowledgment. I thank Brian Kolins for help with the figures.

Bibliography

[Coxeter 97] H. S. M. Coxeter. *Regular Polytopes*. New York: Dover, 1997.

[Cromwell 97] Peter R. Cromwell. *Polyhedra*. Cambridge, UK: Cambridge University Press, 1997.

[Cundy 61] H. M. Cundy and A. P. Rollett. *Mathematical Models*. Oxford, UK: Oxford University Press, 1961.

[Franco 99] Betsy Franco. *Unfolding Mathematics with Unit Origami*. Emeryville, CA: Key Curriculum Press, 1999.

[Fuse 90] Tomoko Fuse. *Unit Origami: Multidimensional Transformations*. New York: Japan Publications, 1990.

[Holdern 91] Alan Holden. *Shapes, Space and Symmetry*. New York: Dover, 1991.

[Kasahara 87] Kunihiko Kasahara and Toshie Takahama. *Origami for the Connoisseur*. New York: Japan Publications, 1987.

[Kawamura 01] Miyuki Kawamura. *Polyhedron Origami for Beginners*. Tokyo: Japan Publications, 2001.

[Montroll 02] John Montroll. *A Plethora of Polyhedra in Origami*. Mineola, NY: Dover, 2002.

[Montroll 04] John Montroll. *A Constellation of Origami Polyhedra*. New York: Dover, 2004.

[Montroll 09] John Montroll. *Origami Polyhedra Design*. Wellesley, MA: A K Peters, Ltd., 2009.

[Mukerji 07] Meenakshi Mukerji. *Marvelous Modular Origami*. Wellesley, MA: A K Peters, Ltd., 2007.

[Simon 99] Lewis Simon, Bennett Arnstein, and Rona Gurkewitz. *Modular Origami Polyhedra*. New York: Dover, 1999.

[Strobl 06] Heinz Strobl. "My Strip Tease Gallery." *Knotology*. Available at http://www.knotology.eu/index_en.html, 2006.

[Strobl 10] Heinz Strobl. Personal communications, 2010.

[Sutton 02] Daud Sutton. *Platonic and Archimedean Solids*. New York: Walker and Company, 2002.

[Weisstein 11a] Eric W. Weisstein. "Archimedean Solid." *MathWorld—A Wolfram Web Research*. Available at http://Mathworld.wolfram.com/ArchimedeanSolid.html, 2011.

[Weisstein 11b] Eric W. Weisstein. "Johnson Solid." *MathWorld—A Wolfram Web Research*. Available at http://Mathworld.wolfram.com/JohnsonSolid.html, 2011.

[Weisstein 11c] Eric W. Weisstein. "Polyhedron." *MathWorld—A Wolfram Web Research*. Available at http://Mathworld.wolfram.com/Polyhedron.html, 2011.

[Wenninger 71] Magnus J. Wenninger. *Polyhedron Models*. Cambridge, UK: Cambridge University Press, 1971.

[Versnick 00] Paula Versnick. "The Kotology of Heinz Strobl." *Paula's Orihouse*. Available at http://www.orihouse.com/knotology.html, 2000.

A Systematic Approach to Twirl Design

Krystyna Burczyk and Wojciech Burczyk

1 Introduction

The idea of twisted spirals used for joining modules into a modular origami model was proposed by Herman van Goubergen about ten years ago [Goubergen 00] as an example of the paper tension technique. His curler unit is a waterbomb with all four flaps twisted into spirals (Figure 1). It is used as a vertex module to build a limited number of models based on polyhedra with all vertices of degree four.

Goubergen's idea [Goubergen 00] was a starting point for our research on how such a technique can be used to create origami models. We have considered the following problems:

- Is a square the only paper shape that produces a well-working twirl module?

- What shape of a flap is suitable for a spiral?

- What parameters describe a twirl module and a twirl model, and how can we change them?

- Can we mix spirals and other techniques to join modules together?

Hundreds of origami models called "Twirls" resulted from these investigations and have been published [Burczyk 03a, Burczyk 03b, Burczyk 08,

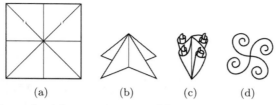

Figure 1. Curler unit: (a) crease pattern; (b) a waterbomb base; (c) the finished module; (d) a schematic representation.

Burczyk and Burczyk 09a, Burczyk and Burczyk 09b, Burczyk and Burczyk, to appear]. The books contain diagrams and detailed descriptions of many models illustrated in this paper as well as some models outside the scope of this paper.

The sequence of results presented here is not chronological. We recognized some structural features of twirls just after designing a model, and others were identified later after examining and comparing the structures of independently designed models.

Our research has focused on finding features and parameters describing twirl models. Some parameters are discrete, and others are continuous. Such parameters may be used to classify and describe existing models as well as to provide guidance for creating new twirl models.

2   The First Attempt: Small (Change) Is Beautiful

A small modification of a standard module was our first attempt to create new twirl models. We sunk a waterbomb and noticed that a small change of a base module produced a completely different appearance of a model: spiky, round, and like pasta (Figure 2(a)–(c)).

The center point of a module does not play any role in joining the modules. We divided a module into two areas: flaps twisted into spirals (structural area), essential for joining modules together, and a central point (decorative area) that may be folded in any way to create interesting visual effects. As a finished model consists of several modules, any change in the decorative area is duplicated many times, like in a kaleidoscope, yielding dramatic changes in the finished model.

A change in the spiral's direction is another simple modification that produces an astonishing visual effect (Figure 2(d)). That small modification illustrates important features of twirl models:

- A small change of a base module repeated many times results in a significant change of the visual appearance of the finished model (kaleidoscope effect).

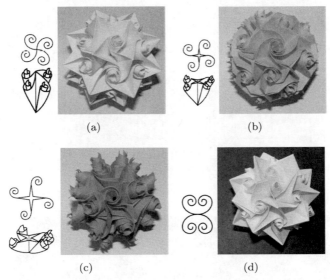

(a)   (b)

(c)   (d)

Figure 2. Results of a small change of a base module: (a) a standard module [Goubergen 00]; (b) shallow sink [Burczyk 03a]; (c) deep sink [Burczyk 03a]; (d) a twirl differently twisted [Burczyk 03b].

- A base module has two different regions: flaps that are twisted into spirals (structural area), and the remaining part of a module that may be folded as one wishes (decorative area).

3   A Square Is a Rectangle: Metamorphosis

We next asked how the shape of a starting sheet of paper might affect the end result. A rectangle is one of the interesting possibilities. The structure of a module made from a rectangle is not very different from the structure of a module based on a square. The recipe is simple: extend the center part (a line) of a waterbomb base, so the line becomes a rectangle (Figure 3). A center rectangle forms a linear decorative area between halves of the waterbomb base. Such development opens new possibilities. In the case of a linear decorative area, there may be more than one peak to sink, with more combinations of different sizes of sink at particular peaks (Figure 3). The size of a sink may be increased to create a gradually changing series of models (Figure 4).

A linear module leaves more space for decorative folding. Repeating a pattern is another technique used in the case of a linear module. The "Metamorphosis IV" series (Figure 5) is an example of this technique.

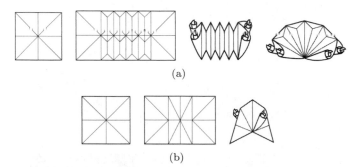

(a)

(b)

Figure 3. Crease patterns of twirl modules based on rectangles: (a) peaks and fan; (b) spidron.[1]

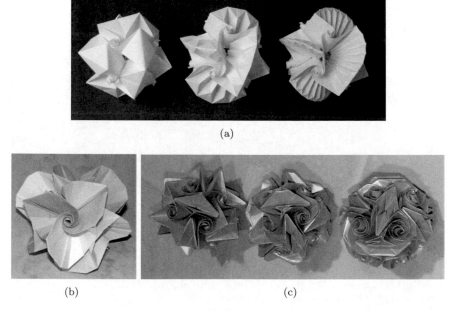

(a)

(b)   (c)

Figure 4. Models from rectangles: (a) "Metamorphosis I"; (b) model inspired by spidron; (c) "Metamorphosis II."

That phase of our research produced a new shape of the decorative area and more possibilities of folding for the linear decorative area. A series of models may be created by a gradual change of a parameter or by repeating a motif.

---

[1] The term *spidron* was coined in the 1970s by Dániel Erdély to describe a spiral-like geometric figure composed of triangles [Erdély 00].

Figure 5. "Metamorphosis IV," an example of the pattern repetition technique: an increasing number of peaks produces a series of models.

4 Planar Decorative Area: From Octahedron to Cube, Mosaic Twirls

To construct a decorative area we may not only separate halves of a waterbomb base, but also move every flap outward (Figure 6(a)). The resulting module has a square decorative area. The size of this area may be changed gradually, as shown in the case of the "From Octahedron to Cube" series (Figure 6(b)).

Any polygonal shape of the decorative area may be obtained in a similar way. A polygonal decorative area offers lots of space for creativity and standard origami techniques. In general, any origami folding may be performed in the central part of the paper (the part that becomes the decorative area of a module) as long as the outer part is saved for flaps (the structural part of a module).

Tessellations produce especially good results (Figure 7(e)–(g)), but other techniques work as well, such as the decorative area filled with crease pat-

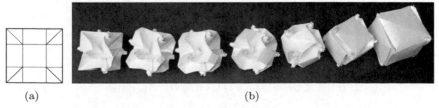

(a)   (b)

Figure 6. Planar decorative area: (a) basic form of square decorative area; (b) "From Octahedron to Cube" series [Goubergen 00, Burczyk 09].

Figure 7. Examples of polygonal decorative areas: (a–c) triangles and rectangles; (d) decorative area filled with creased pattern borrowed from UFO2 [Mitchell 06]; (e–g) Mosaic twirls, simple tessellation used for the decorative area; (h–i) folding sticking out from the plane.

terns (Figure 7(d)). Decorative folding can also stick out to 3D. Figure 7(h) and (i) show geometric examples of such folding, but one can adapt figurative folding as well.

Results of that phase of our research revealed that the decorative area may be a polygon, gradual change of a continuous parameter (size of the decorative area) gives new possibilities, and a planar decorative area offers space for utilization of advanced origami techniques.

5   Shapes and Lengths: Different Spirals

A waterbomb folded from a square has flaps in the shape of a right isosceles triangle. But what happens if we change the shape of the flaps? First, the

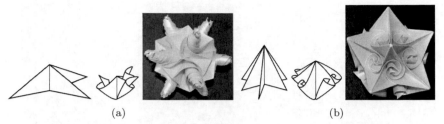

Figure 8. Different lengths of spirals: (a) long flaps; (b) short flaps.

flaps may be longer or shorter. Short flaps produce vortices sunk inside a model, and long flaps produce vortices spiking outside a model. Folding a waterbomb from an equilateral triangle or from a pentagon is the easiest way to change the length of flaps (Figure 8).

Second, we changed the shape of the flaps (Figure 9). A flap can be twisted not only from a triangle, but from a rectangle too. A rectangle flap twisted into a spiral gives a tube-like appearance (Figure 10(b)), whereas a triangle usually gives flower-like appearance.

The lengths of the flaps and the spirals twisted from these flaps limit the structures of the polyhedra that are possible to assemble. Unlike flap-and-pocket systems, spirals are tolerant in terms of distance and angle between adjacent modules. That feature enables the assembly of many different polyhedral structures from the same module. But in extreme cases, the standard design of a twirl module is not sufficiently tolerant.

When a twirl module is used as a vertex module in a polyhedron, flaps must be long enough to reach the center of a face to hook flaps coming from other vertices of the face. In the case of the Platonic solids, there is no problem. All faces around a vertex are the same polygons, and the distance from a vertex to the center of a face is the same for all the faces around the vertex (Figure 11(a)). In the case of the Archimedean solids, there are different faces around a vertex, and distances from a vertex to the center of different faces vary. In the case of the rhombic cuboctahedron (Figure 11(b)), the ratio of the longest to the shortest distance is

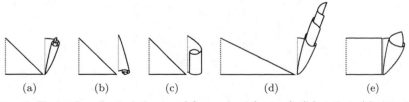

Figure 9. Examples of spiral shapes: (a) standard (a cone); (b) rocket; (c) cylinder; (d) screw; (e) petal.

Figure 10. Shape and length effects: (a) two different triangles; (b) rectangles; (c) a cube by Wojtek; (d) hidden spirals from triangles; (e) different triangles.

$d/d' \approx 1.22$, and it fits the tolerance of spirals. The same spirals may be used around a module. In the case of the truncated dodecahedron (Figure 11(c)), the ratio is $d/d' \approx 2.8$, and a special design of a module is necessary. We need three flaps—two long and one short. An isosceles 120° triangle is a good starting point for such a module.

How can we produce a module that fits our needs? There are two basic approaches. First, a folding sequence that produces a base with the required number and length of flaps may be prepared (Figure 12). Second,

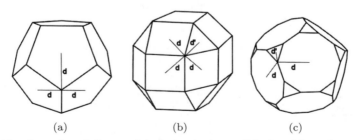

Figure 11. Lengths of flaps: (a) dodecahedron; (b) rhombic cuboctahedron; (c) truncated dodecahedron.

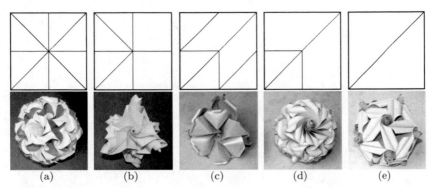

Figure 12. Different sizes and shapes of flaps generated by crease patterns: (a) "Square Rose" (from a preliminary base); (b) "Minimum Rose"; (c) petals; (d) "Triangle Rose"; (e) an edge "Butterfly" module.

the starting shape of the paper may be changed to fit our needs (Figures 13, 14, and 15).

Results of that phase of our research revealed the following:

- Short flaps produce a vertex sunk in a model, and long flaps produce a spiky vertex sticking out of a model.

- Triangle flaps produce a pointy vertex, and rectangular flaps produce tubular or leaf-like structures.

- The flexibility of spiral joining makes it possible to assemble most polyhedra structures from modules with the same size flaps; however, some extreme cases of models require different lengths of flaps.

- Different polygons may be folded into a base, producing a collection of flaps that can be twisted.

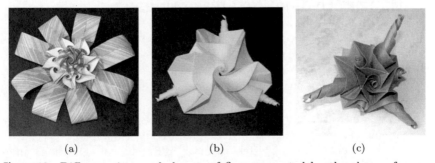

Figure 13. Different sizes and shapes of flaps generated by the shape of paper: (a) rectangle; (b) trapezoid; (c) deltoid.

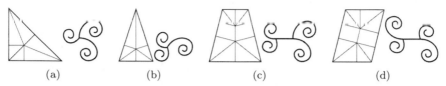

Figure 14. Examples of shapes of sheets of paper: (a) right triangle; (b) isosceles triangle; (c) trapezoid; (d) parallelogram.

(a)   (b)   (c)

Figure 15. Different shapes of modules from rectangles: (a) "Bends"; (b) "Union Monument"; (c) "White Corncockles (Agrostemma)."

- A required number of flaps, or length of flaps, may be obtained from different starting shapes of paper or different folding sequences.

- Different shapes and lengths of spirals may be mixed in the same module; however, spirals in the same vortex should have the same size to join the modules firmly.

6 Minimal Folding: No Crease Origami

So far, we have discussed modules with many creases and modifications that resulted in more creases. But how many creases do we really need to make a twirl model? We have investigated this question over the last year, and we have found a surprising answer leading to a family of appealing models.

The answer to this question is that actually we do not need any creases. The unfolded square, rectangle, or triangle already has flaps in place. And flaps may be located and twisted in many different configurations (Figure 16). Discovery of this fact opened the way to several new models where no crease, or only one crease is used to make a module (Figure 17).

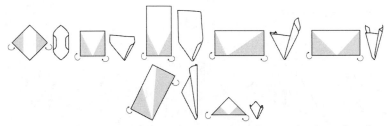

Figure 16. Examples of unfolded square, rectangular, and triangular sheets of paper as twirl modules (shadows mark flap area).

Results of that phase of our research showed that

- an unfolded sheet of paper may be divided into areas corresponding to flaps and decorative area to produce a twirl module;

- there are many different arrangements of divisions of unfolded sheet of paper;

- minimal folding modules are edge modules.

Figure 17. Examples of minimal folding: (a) squares, no creases; (b–c) rectangles, no creases; (d) squares and rectangles, no creases.

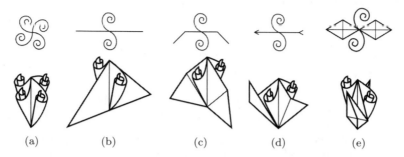

(a)   (b)   (c)   (d)   (e)

Figure 18. Examples of modules based on different types of spirals: (a) SSSS [Goubergen 00]; (b) SLSL [Burczyk 09a]; (c) SFSF [Burczyk 09a]; (d) SOSK [Burczyk 10]; (e) SPSP [Burczyk 10] (where S = spiral, L = line, F = fold, O = hook, K = pocket, P = petal).

7 Consensus Building: Spirals Work Together with Flaps-and-Pockets, Macro-modules

So far, we have discussed modules where all flaps are twisted into spirals. What about the other types of joining widely used for modular models? We replaced a number of spirals by flap-and-pocket joining (Figure 18). New base modules create new, interesting models. Such modules are especially useful for models based on a macro-modular technique.

A macro-modular approach previously has been used to build large polyhedral structures from Sonobe modules [Kasahara and Takahama 87], to arrange large constructions [Mitchell 99], and to simplify the assembly diagram [Goubergen 00]. We use a macro-modular approach as a systemic tool to create kusudama models [Burczyk 08] (Figure 19). The flap-and-pocket system is used to form macro-modules corresponding to polygons

Figure 19. Macro-modular twirl models. (See Color Plate IV.)

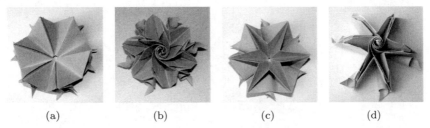

(a) (b) (c) (d)

Figure 20. Two configurations of SFSF and SSFF modules in macro-modules: (a) a flower (SSFF); (b) a flower (SFSF); (c) a star (SSFF); (d) a star (SFSF) (where S = spiral, F = fold).

(hinges created by a single crease offer enough flexibility to join modules into a polygon shape). Spirals make a joint that is flexible in terms of both distance and angle between macro-modules. Thus, macro-modules may form an arbitrary polyhedron structure.

Macro-modules provide additional parameters to our model. First there are base modules that may be arranged in two ways in a macro-module [Burczyk and Burczyk 09b]. Such configurations are called star and flower (Figure 20).

Depending on the base module type, a macro-module may have one or two outward spirals at each vertex of a polygon. In the case of a single outward spiral, there are two different methods to join macro-modules together [Burczyk and Burczyk 09a]. In the case of double outward spirals, there are four different methods to join macro-modules together [Burczyk and Burczyk 09b] (Figure 21). The same methods of joining are also applicable to base modules when they have a longitudinal direction (all modules have linear and polygonal decorative area).

In the case of macro-modules, spirals usually are twisted clockwise. When macro-modules are joined vertex to edge, they turn a little against each other. They may turn clockwise or counterclockwise resulting in two

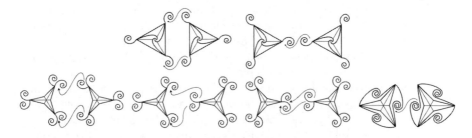

Figure 21. Two ways of joining single outward spiral modules (top), and four ways of joining double outward spiral modules (bottom).

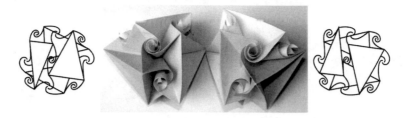

Figure 22. Example of chirality.

different chiral configurations of a model (Figure 22). Usually there is no obvious visual difference between these configurations, but one of them looks nicer. We do not recognize chirality as a regular construction parameter, but such behavior of the model must be observed in the case of a series of similar models

Results of that phase of our research showed that

- spirals may be mixed with flap-and-pocket joining;

- a mixed system of joining is especially effective when a model is build from macro-modules;

- different types of spirals and flaps, different configurations of macro-modules, and different ways of joining macro-modules and chirality enhance the variety of twirl models.

8   Conclusion

We have described different paths of twirl development. Let us summarize the features leading to understand the diversity of the world of twirls.

A large number of independent parameters generate a huge variety of models. The system of parameters acts as a generator of models, as most of the attributes of the configurations produce actual models. Moreover, a new design in the scope of an attribute leads to a full class of new models. Even a small modification of a base module produces a significant difference in a final model (kaleidoscope effect), and the final effect is hard to predict (one must make a model to see the effect).

A spiral offers a flexible connection in terms of distance and angle, which makes such a connection a universal joint. Such a joint is very stable. These two features, flexibility and stability, are keys to the variety of twirls.

As a base module is divided into a structural area and decorative area, the design process may be separated into two relatively independent parts, which opens the field for creativity and amazing models.

The most important thing in twirl designing is a set of parameters describing twirl models. Such a set converts a random collection of models into an organized and structured system. We have discussed in this paper 11 parameters related to the geometric structure of a model, including the assembly process; number, shape, length, and position of spirals; and decoration motif. These parameters are relatively independent of each other. Arbitrary selection from possible values of parameters usually corresponds to an actual origami model. Moreover, a change in a single parameter usually results in a significant change of the final model. Numerous independent parameters and sensitivity of the final model to a parameter change produce a huge variety of appealing models. Some of the configurations that are easy to describe may be hard to fold, however, as the thickness and tension of paper play a role. The parameterization creates a framework for building new models of twirls and still leaves a lot of space for creativity.

Twirl models look appealing due to their flower-like (in most cases) appearance and their regular geometric structure. Last, but not least, these models are relatively simple to fold.

Copyright notice. All models were designed by Krystyna Burczyk, except where otherwise stated. All models were folded by Krystyna Burczyk.

Bibliography

[Burczyk 03a] Krystyna Burczyk. *Kręciołki (Twirls)*. Zabierzów: self-published, 2003.

[Burczyk 03b] Krystyna Burczyk. *Kręciołki kręcone inaczej (Twirls Differently Twisted)*. Zabierzów: self-published, 2003.

[Burczyk 08] Krystyna Burczyk. *Kręciołkowe kusudamy 1 (Twirl Kusudamas 1)*. Zabierzów: self-published, 2008.

[Burczyk 09] Krystyna Burczyk. "Pozdrowienia z Polski (Greetings from Poland)." *The Polish Origami Association Bulletin* 7 (2009), 84–85.

[Burczyk and Burczyk 09a] Krystyna Burczyk and Wojciech Burczyk. *Kręciołkowe kusudamy 2 (Twirl Kusudamas 2)*. Zabierzów: self-published, 2009.

[Burczyk and Burczyk 09b] Krystyna Burczyk and Wojciech Burczyk. *Kręciołkowe kusudamy 3 (Twirl Kusudamas 3)*. Zabierzów: self-published, 2009.

[Burczyk and Burczyk, to appear] Krystyna Burczyk and Wojciech Burczyk. *Kręciołkowe kusudamy 4 (Twirl Kusudamas 4)*. Zabierzów: self-published, to appear.

[Erdély 00] Dániel Erdély. "Spidron System: A Flexible Space-Filling Structure."
Symmetry: Culture and Science 11:1–4 (2000), 307–316.

[Goubergen 00] Herman van Goubergen. "Curler Unit." *British Origami* 205
(2000), 17–18.

[Kasahara and Takahama 87] Kunihiko Kasahara and Toshie Takahama.
Origami for the Connoisseur. Tokyo: Japan Publications, 1987.

[Mitchell 99] David Mitchell. *Building with Butterflies: An Introduction to the
Art of Macro-modular Origami Sculpture*. Kendal, UK: Water Trade, 1999.

[Mitchell 06] David Mitchell. "UFO2." Workshop at the XVIII Internationales
Treffen von Origami Deutschland, Dresden, Germany, April 22, 2006.

Oribotics: The Future Unfolds

Matthew Gardiner

1   Introduction

Oribotics [Gardiner 09] is a field of research concerned with the aesthetic, biomechanic, and morphological connections among nature, origami, and robotics. In my current research, the focus is on the actuation of fold-programmed materials such as paper and synthetic fabrics. The design of the crease pattern, the precise arrangement of mountain and valley folds, and the way they fold and unfold directly inform mechanical design. There-fore, a key area of current research is focused on the discovery of patterns that have complex expressions that can be actuated repeatedly. This re-search has resulted in some artistic exhibitions; the largest to date was at the 2010 Ars Electronica Festival in Linz, Austria (Figure 1).

2   Industrial Evolution of Oribotics

I began building robots with LEGO Mindstorms in 2003, and since 2010 I have been making my own customized snap-lock polymer parts with 3D design and printing. This evolutionary process took seven years and was realized through the creation of five generations of Oribots. In the pa-per I presented at the Fourth International Meeting of Origami Science, Mathematics, and Education (4OSME), I discussed the 2004 and 2005

Figure 1. Oribotics at the 2010 Ars Electronica Festival. (See Color Plate V.)

generations [Gardiner 09]. In this paper I discuss the 2007 and 2010 generations. Figure 2 shows a linear evolution of oribotic skeletons from 2005 to 2010. Each progression has retained the five-petalled symmetry and hand mechanism; the sculptural form became more complex with production techniques.

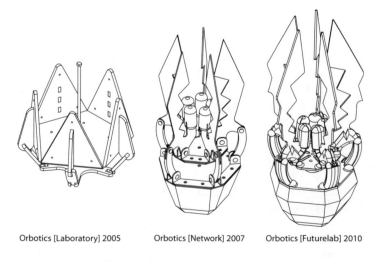

Orbotics [Laboratory] 2005   Orbotics [Network] 2007   Orbotics [Futurelab] 2010

Figure 2. Evolution of the species.

In 2007, I received an Arts Victoria Arts Innovation grant. My partners for this project were Etheira Technologies (a.k.a. Ray Gardiner) for electronics, and Design Sense, an industrial design company based in the northern suburbs of Melbourne. I had worked with Joe Iacono at Design Sense prior to 2007 on an origami-inspired folding set design. Design Sense brought a refined level of design to my artwork by advising on material choice, finishes, and production techniques.

3   Crease Pattern and Mechanical Design

Artistically, my work deals with the subject of interconnectivity. The following story describes the interconnections discovered during the development of the crease pattern and mechanical design.

After 4OSME, I returned to Australia inspired by the potential of the crease pattern design for the heart stent presented by Zhong You [You and Kuribayashi 09]. I began experimenting with various expanding crease patterns and looked for ways in which I might actuate them. I was seeking a mechanical way to actuate each individual crease, a long-term goal, and I performed a range of experiments with pneumatics. The results became somewhat complicated, and so I looked for a simpler, more elegant way. The inspiration for the mechanical design eventually came from reading a paper titled "The Geometry of Unfolding Tree Leaves" [Kobayashi et al. 98]. While the topic of the paper addressed my interests exactly, it was the intuitive illustrations by Biruta Kresling that revealed the natural mechanism. Kresling's illustrations instantly inspired a clarity of understanding. During a fleeting trip to Paris in August 2007, I took time to meet with Kresling, and during a five-hour interview, showed her my recent works, including one pattern that mixed the leaf mechanism with the waterbomb pattern. Kresling explained how she was a nexus for the intercultural, intercontinental transportation of the pattern known by her as "pineapple" (*ananas* in German), and by the Western world as the waterbomb pattern. During one of Kresling's bionics classes at Supinfocom in France, a 19-year-old French student, N. Maillard, discovered the pattern with support and guidance from Kresling. Kresling shared the pattern with Emeritus Koryo Miura, who explained its mathematical properties. Kresling then introduced the pattern and its properties to Kaori Kuribayashi [Kuribayashi 04], thus providing inspiration for Kuribayashi's discovery of an application in a medical stent design, the very same stent presented by You at 4OSME [You and Kuribayashi 09]. The interconnections had come full circle, and I had discovered the history of the ways in which the pattern traveled the world over before it became my inspiration.

Figure 3. Oribotics [de] outcome of artistic residency in Schöppingen, Germany.

After the establishment of the leaf unit, my mechanical design followed the path of the 2005 bots, a fivefold symmetrical arrangement of a simple lever system to open and close the pattern. The design was so successful that I kept it in the 2010 version and focused on the industrial design. The key changes were materials and production methods. The 2007 bots used laser-cut and folded aluminium, along with cutting-plotter scored and folded plastic paper membranes. The plastic paper is very strong and is impossible to tear, as it lacks a fiber structure, but the scoring weakened the paper along the crease lines and at the fold intersections. Within three years of exhibition, the paper developed holes due to stress at the intersections. The paper also acted like a spring, meaning that closing the bot required more force because the folds stored kinetic energy.

During a 2008 residency in Schöppingen, Germany (see Figure 3), I didn't have a particular project; instead, I used the time for my own professional development. I taught myself 3D modeling, including the simulation of simple mechanics and 3D printing (rapid prototyping) through a series of simple experiments. This foundation was further developed during 2009, when I was commissioned by the arts company Arena Theatre to create an oribotic installation concept for a children's theatre work called House of Dreaming. The House of Dreaming Oribots became the precursors to the refined design for Oribotics [Futurelab]. A second aspect of the House of Dreaming development was to find a faster, stronger, lighter, folded membrane that is easier to produce. The answer lay in fabric as well as in pleating.

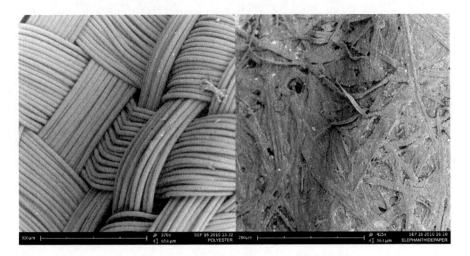

Figure 4. Scanning electron microscope comparison of creases in polyester fabric (left) and elephant-hide paper (right).

4   Paper versus Fabric

Fabric has proven to be far superior to paper for oribotic applications. Because of its fibrous structure, paper is prone to failure after repeated folding actions. Anyone who has owned and folded an old map (except a Miura-ori map, of course) knows that holes develop in the corners and along the crease lines, and eventually the paper will tear. A solution is to make holes at the corners to minimize stress, which is a necessary choice for metal or rigid applications, as in Kuribayashi's stent design [Kuribayashi 04, You and Kuribayashi 09]. For my work, however, I didn't consider holes to be an aesthetically pleasing option, so between 2007 and 2010 I sought a new solution, which I found in the world of fabrics.

Figure 4 shows the surface of two materials: polyester fabric and elephant-hide (*Elephantenhaut*) paper. The polyester is a standard dressmaking fabric, of which there are many kinds, compositions, and weaves that vary from manufacturer to manufacturer. The fabric was selected for its translucency and plain weave without a noticeable pattern. Looking at both images in Figure 4 together, the material difference is obvious: the polyester fabric is made of thousands of unbroken strands woven in an orderly manner, whereas the paper is fibrous and organic in structure, with the fibers bound together with pulp. On the surface of each material is a diagonal deformation, a crease, from approximately top right to bottom left. The crease in the polyester was made through application of heat during shaping (see Section 4.1). The paper crease was made with laser etching, then folded by hand. The crease in the polyester fabric is a plastic deformation with no broken strands; the paper crease is also visible as a deformation, but signs of fibers delaminating along the fold are evidence of material distress. The winner in practice is the polyester fabric, and these images help to show why: the polyester is composed of ordered woven flexible plastic fibers that do not break during the folding process.

4.1   Folding Polyester Fabric

Traditional techniques should be examined very closely, as they are often simple and therefore robust, and oftentimes they are useful in a modern context. Pleaters are the professionals who put pleats in dresses, and they use a centuries-old technique. In 2009 I met Matthew Bennett, a pleater from Specialty Pleaters in Melbourne, and I instantly understood the technique his company employed: one takes a folded pattern in paper, makes two copies, places the fabric between the two sheets, re-collapses the sandwich of paper and fabric very carefully to avoid creep and gathering of fabric inside the paper, clamps or holds the pattern tightly, and then heats it. Pleaters have a special steam oven that can hold many garments at once, but for my work I use a domestic oven with a bowl of water at the bottom. It is rather crude, but it works and is readily accessible in the domestic environment. I cook polyester at 170°C for 15–20 minutes and then allow the "cooked" product to cool completely before unfolding.

4.2   Paper Choice

Pleaters use a range of weights of Kraft paper, between 120 and 200 gsm (grams per square meter), because of its long fibers and long life. I conducted some experiments to test the process of machine scoring in the process of fabric pleating, and to test the best weight of paper. Unfortunately, since the global financial crisis of 2009, worldwide paper production has been limited by economic factors, and I could not acquire many weights in large sheet sizes, as most stock comes in rolls and would require de-curling before scoring. I worked with a local Melbourne company, and had some Kraft paper machine scored using their flatbed CNC machine, which has two main tools, a knife and a scoring tool. I tried the weights 90 gsm and 170 gsm. I then folded the scored paper and conducted my first pleatings of polyester fabric. The results were promising, but a combination of the paper weight of the 170-gsm paper, and the thickness of the scoring tool on 170-gsm paper caused soft corners in the pleated fabric. The 90-gsm paper was too flimsy to be of practical use in repetitions of the pleating process, as it was corrupted quickly with additional creases. Later in the year, inside the Ars Electronica Fablab, I conducted experiments with laser engraving of elephant-hide paper, known in origami circles because of its strength, and found it to be ideal. I use an 110-gsm sheet lightly etched by

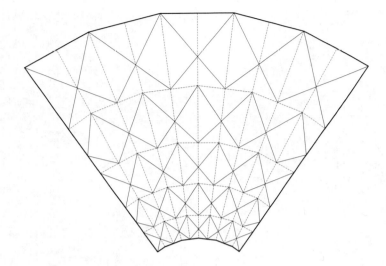

Figure 5. WB75 crease pattern.

laser. Its lifespan as a useful mold is somewhat decreased by the etching, but the results are crisp, clean folds in the fabric. I attribute the success to the strength of the paper and the thinness of the laser etch (0.3 mm). The paper molds do not need to be retired until after every 50 or so cycles.

4.3 Crease Pattern Design

Figure 5 is my variation on the ananas/waterbomb corrugation. The main variant is that the pattern uses radially arranged creases. The pattern was modified using a visual algorithm created in Adobe Illustrator. I took a unit of the waterbomb and, selecting the bottom and center points, scaled them by varying amounts. I then copied the pattern and rotated it to the next position, building horizontal rows, then copied the row, and scaled the copy by the same amount, and rotated it into place. The process required some pattern tweaking at the intersections of the rows. A variety of scale factors were test folded, and I eventually made an aesthetic decision and settled on 75 percent as the scale factor; hence, the pattern is known as waterbomb 75 (WB75) (Figure 5).

5 Interaction Design

Oribotics [Network] (Figure 6) was first exhibited at the Melbourne International Art Festival [Gardiner 09]. It was, in my opinion, a sculpturally polished artwork, but the complex details of the interaction were a mystery

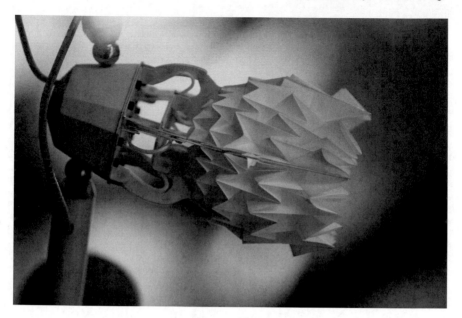

Figure 6. Oribotics [Network].

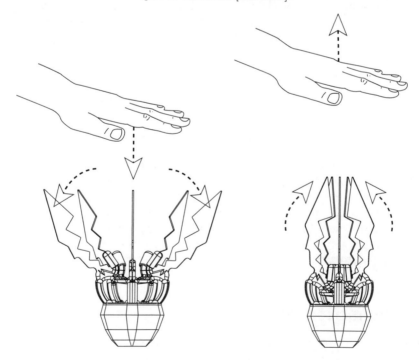

Figure 7. Micro interaction with a human hand.

Figure 8. Detail of the blossom head from Oribotics [Futurelab]. (See Color Plate VI.)

to much of the audience. The idea was to create "food" for the oribots that would effect the movement of the oribotic blossom and the expression of color. The experience was intended to be like that of watering a plant, with cumulative results achieved over time.

The immediate interaction left much to be desired, so one of the first tasks I undertook during my artistic residency at the Ars Electronica Futurelab was to find a meaningful but very simple interaction design for the work. I found inspiration by reflecting on my memory of the audience's physical interactions with the artwork. I remembered that people intuitively placed their hand in front of the bot, in the hope of getting a physical reaction (Figure 7).

My intention with Oribotics [Futurelab] (Figure 8) was to reveal, in a simple way, the interconnectivity of the folded pattern through micro and macro interactions.

In an oribotic pattern, actuating a single fold causes every other fold to move; each fold is mechanically interconnected. Micro interactions occur with sensors; inside each bot a proximity sensor measures objects in front of its mouth. As an object (a human hand) approaches, the oribot blossom opens, causing 1050 folds to actuate in the bot. Macro interactions (Figure 9) occur via the network and software; each micro interaction is broadcast to every other oribot in the installation, causing a ripple ef-

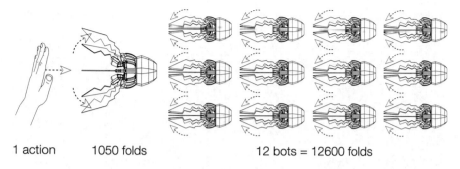

1 action   1050 folds   12 bots = 12600 folds

Figure 9. Micro interaction with macro effect.

fect of movement over 52,500 folds across the entire installation, creating a stunningly complex moving image.

6   Conclusion and Further Work

The development of oribotics since 2007 has been a process of refining, perfecting and building for longevity rather than reinvention. The transition to using fabrics for the folded membrane has been the most significant advance in longevity. The membrane can fold and unfold repeatedly without causing damage to itself. The artistic residency at Ars Electronica was my longest period of sustained research and subsequent production, resulting in exhibitions at the Ars Electronica Festival in Linz, Austria; Tokyo Design Touch at Tokyo Midtown, Japan; and Innovators 3 at Linden Gallery in Melbourne, Australia. These exhibitions increased the exposure and the profile of my work to an international level.

Oribotics is a field of research that evolves with the technological proficiency of the artist toward the idea of self-folding systems. Self-folding is a topic being researched by several labs (e.g., [Hawkes et al. 10]). In my future work, as with my past work, I am thinking about the full cycle of folding systems, including the phase of unfolding, and I think the problem of designing a system that folds and unfolds with bidirectional fold capabilities is a great design challenge, one that will require some dynamic cross-disciplinary partnerships.

Acknowledgments. The development of oribotics has been successful only by the support, funding, and generous collaboration of the Australia Council for the Arts, Arts Victoria, The City of Port Phillip, The Rupert Bunny Foundation, Ars Electronica, Novamedia, Ray Gardiner, My Trinh Gardiner, Josh Gardiner, Aphids, Design Sense, Melbourne International Arts Festival, and Federation Square.

For further information, exhibition photographs, and video, please see http://www.oribotics.net.

Bibliography

[Gardiner 09] Matthew Gardiner. "A Brief History of Oribotics." In *Origami⁴ : Fourth International Meeting of Origami Science, Mathematics, and Education*, edited by Robert J. Lang, pp. 51–60. Wellesley, MA: A K Peters, Ltd., 2009.

[Hawkes et al. 10] Elliot Hawkes, B. Nadia Benbernou, Hiroto Tanaka, Sangbae Kim, Erik D. Demaine, Daniella Rus, and Robert J. Wood. "Programmable Matter by Folding." *Proceedings of the National Academy of Sciences* 107:28 (2010), 12441–12445.

[Kobayashi et al. 98] Hidetoshi Kobayashi, Biruta Kresling, and Julian F. V. Vincent. "The Geometry of Unfolding Tree Leaves." *Proceedings of the Royal Society* 265 (1998), 147–154.

[Kuribayashi 04] Kaori Kuribayashi. "A Novel Foldable Stent Graft." Ph.D. thesis, Univesity of Oxford, Oxford, UK, 2004.

[You and Kuribayashi 09] Zhong You and Kaori Kuribayashi. "Expandable Tubes with Negative Poisson's Ratio and Their Application in Medicine." In *Origami⁴ : Fourth International Meeting of Origami Science, Mathematics, and Education*, edited by Robert J. Lang, pp. 117–127. Wellesley, MA: A K Peters, Ltd., 2009.

Part II

Origami in Education

Origametria and the van Hiele Theory of Teaching Geometry

Miri Golan

1   Introduction

The geometry curriculum implemented by the Israeli Ministry of Education follows closely the theories of the van Hieles [Fuys et al. 84, van Hiele 86, Burger and Shaughnessy 86]. In this paper, we show how the practices developed for the Origametria program can provide the solid foundation students need to prepare them to move into higher levels of abstraction in geometry.

Since 1992, the Israeli Origami Center (IOC) has trained teachers to teach the Origametria program in schools, and kindergarten Origametria in preschool. In 2010, after several years of close scrutiny, the Israeli Ministry of Education gave formal approval to the IOC to train preschool teachers for the kindergarten Origametria program. During the 2009–2010 academic year, the program operated nationwide in 35 primary schools (approximately 2.5% of all primary schools), half of which are Arab. The Origametria kindergarten and primary school programs are continually evolving.

The Origametria program teaches topics of curriculum geometry through the use of origami models. Unlike almost all other programs that teach geometry with origami, the models taught are not geometric subjects such as boxes, cubes, pyramids, modulars, and the like, but animals and action toys, which the students find fun and motivating. Also, the geometry of the final model is rarely, if ever analyzed. Instead, the geometry of the paper during the folding of the model is analyzed for its geometric content.

The wide experience of the IOC in running year-long programs in schools of different faiths, abilities, class numbers, and educational systems for almost two decades has enabled it to develop a distinctive style of classroom teaching which it considers integral to the growth and success of Origametria. Several schools have participated in the program for many years, citing that it helps the students attain better grades in the national TIMMS mathematics tests.

2   The van Hiele Theory of Geometric Teaching

The van Hiele theory was developed in the 1950s by two Dutch mathematics teachers, Pierre and Dina van Hiele [Fuys et al. 84]. The theory attempts to explain how students learn geometry and why many have difficulty with higher-level cognitive processes, especially when they are expected to give geometric proofs.

According to this theory, the development of the mathematical thought process, especially geometry, can be divided into five levels:

- level 0: visualization,

- level 1: analysis,

- level 2: abstraction,

- level 3: deduction,

- level 4: rigor.

Note that the levels are sometimes described as running from level 1 to level 5, creating some confusion as to which level is being discussed. There are also alternative names given to each level. A useful primer on the van Hiele theory can be found in [Burger and Shaughnessy 86].

3   Origami and the Van Hiele Theory

After the IOC established the Origametria program in schools, many parallels were found with the van Hiele method of teaching geometry. In Israel, many students learn geometry at the van Hiele Deductive Level (level 3) in middle and high schools, before they have established their knowledge at the earlier levels. These students are required to formulate proofs when they still cannot identify a side or an angle, cannot find a polygon within a polygon, or do not know basic geometric definitions.

The Origametria programs teach geometry to kindergarten and early primary school students first at the visual level (level 0) then later at the analysis level (level 1). There is no sudden jump from one level to another, but a gradual shift of emphasis from one level to the next. The Origametria program uses the first three levels of the van Hiele model, although most of the teaching focuses on levels 0 and 1. Following is an overview of the van Hiele levels and their comparison with Origametria:

- van Hiele level 0: Visualization

  - *van Hiele:* "At this level, students learn the names of many geometric terms and forms. They can identify geometric forms and understand the differences between them."

  - *Origametria:* In kindergarten Origametria and in early primary school, students are exposed to basic geometric terms and forms such as side, vertex, square, rectangle, and triangle. In every lesson, Origametria makes repeated reference to these terms and forms, so that a basic understanding is reinforced many times.

- van Hiele level 1: Analysis

  - *van Hiele:* "At this level, students can identify and analyze characteristics of geometric forms."

  - *Origametria:* In the process of folding a model, a geometric subject (such as an isosceles triangle or a line of symmetry) appears many times in different guises, and its character is discussed each time by the students. By this cumulative analysis of different examples, the characteristics of a geometric subject are learned.

- van Hiele level 2: Abstraction

  - *van Hiele:* "Students can understand the relationships and differences between polygons, and understand the importance of accurate definitions. Students can identify sets of shapes and their subsets (e.g., why all rectangles are in the family of parallelograms)."

  - *Origametria:* While folding a model, students test the characteristics of a geometric subject in different contexts, learning to separate and define similar-looking forms, such as scalene and isosceles triangles or parallelograms and rhombuses.

4   Time of Learning

The students understand what is being taught when it is within their time of learning, that is, when they are mature enough to understand the sub-

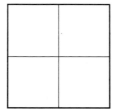

Figure 1. Find the number of squares in the figure.

ject. The class structure in Origametria generates an ambience that enables the student to arrive at the required time. This ambience is characterized by a constant analysis of the paper throughout the process of folding a model. The teacher helps the students to investigate selected geometric subjects with the students.

This process will enable each student to explore and learn the geometric subject at his/her own maturity and pace. The use of folding, identification of shapes, and investigation during the folding process will enable learning while actively and gradually building the knowledge at each student's pace. Even if at first the student only partially understands a concept, the repetitive investigation process while the paper is being folded will enable the student to gradually learn, but without feeling that he or she does not understand.

A good example of how students learn by analysis is shown in Figure 1, which asks how many squares the student can identify. At first, the students will say four; only later will they identify the fifth. A student within the time of learning will identify five squares, whereas a student who is not within this time will not see the fifth square. This latter student will learn from the answers of his or her peers and the teacher's explanations. The next time a similar question is posed, the student will immediately search and find the additional square. It is possible to extend this exercise, and ask how many quadrangles can be identified.

5   Gradually Building Knowledge and Concepts

Geometric terms are introduced to the students first using visual descriptions such as "side" or "vertex." This is the van Hiele level 0, where definitions are not taught, but are learned at an intuitive level of understanding. Later, at the van Hiele level 1, definitions are given, and the students' understanding will move from unconscious intuition to conscious knowledge. Origametria enables the same term to be used in many different circumstances at many different steps during the folding of a model, and from model to model. Thus, an intuitive understanding of a definition is built up and then confirmed when a definition is given.

One example is the diagonal of a polygon. A student can understand the definition of a diagonal of a polygon only after learning and understanding what the nonadjacent vertices are.[1] Thus, a student can grasp what a diagonal is only after having understood what the adjacent and nonadjacent vertices of a diagonal are. However, in doing origami, the student has already heard that term, at least for squares.

6 Using Origametria to Eliminate Misconceptions

During Origametria lessons in kindergarten and primary school, the students repeatedly experience creating and identifying polygons. This experience occurs in every lesson, enables the students to accumulate knowledge based on increasingly accurate intuition, and assists in eliminating misconceptions.

One example of a misconception is in the identification of a square. A familiar case is the one in which students identify and recognize a square in its familiar vertical-horizontal orientation, but when it is rotated 45 degrees, students no longer recognize it as a square but as a diamond. This occurs also with other polygons, such as isosceles triangles or right-angle triangles, where any rotation away from symmetry on the page, or from a polygon with a horizontal base, can lead to misidentification.

In Origametria lessons, the paper is folded into different polygons. The students investigate and learn to identify and define the polygons in different orientations during the natural rotation of the paper throughout the process of folding. This process enables them to avoid these misconceptions in their later studies.

7 Origametria and van Hiele: An Example from the Classroom

Below is an example of how an origami model—in this instance, the traditional Chinese duck—can be taught by the van Hiele–Origametria method (Figures 2–16). Depending on the level of the students, the questions asked are either at the visual level (van Hiele level 0) or at the analytic level (van Hiele level 1). The example is particularly appropriate to be taught in grades 1–3, although it may also be taught in other grades at the discretion of the teacher.

---

[1] A definition of a diagonal within a polygon is "a line segment linking two nonadjacent vertices" [Page 09].

Visual level   What polygons can you find?
What triangles can you identify?
What kind of angles do you find
in a square?

Analytic level   What is the sum of the angles in
a square?

Figure 2. Step 1.

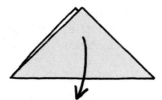

Visual level   What is the polygon?
What triangles can you identify?
What are the angles?

Analytic level   What is the sum of the angles in
a triangle?

Figure 3. Step 2.

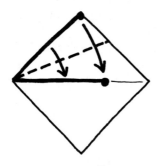

The paper is folded without analysis. In Origametria, not every folding step is examined for its geometric content. However, if this model is being used
to teach bisections, this step would be discussed.

Figure 4. Step 3.

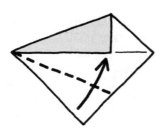

Visual level   What polygons can you find?
What triangles can you identify?
What kind of angles can you
find?

Analytic level   What are the angles of the
quadrilateral created after fold-
ing the step?

Figure 5. Step 4.

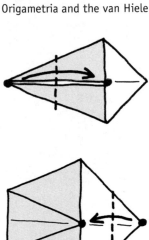

Visual level   What polygons can you find?
What triangles can you identify?
What kind of angles do you see in the polygons?

Analytic level   What are the angles in each corner?

Figure 6. Step 5.

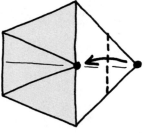

Visual level   What polygons can you find?
What triangles can you identify?
What kind of angles do you see in the pentagon?

Analytic level   What is the total number of degrees in a pentagon?

Figure 7. Step 6.

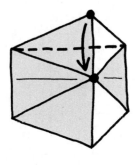

No questions.

Figure 8. Step 7.

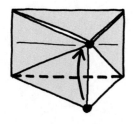

No questions.

Figure 9. Step 8.

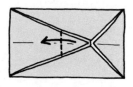

Visual level   What polygons can you find?
What triangles can you identify?
What kind of angles do you see?

Analytic level   Show that the total number of degrees where the four triangles meet is 360°.

Figure 10. Step 9.

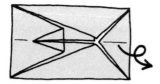

No questions.

Figure 11. Step 10.

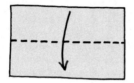

Visual level   How many rectangles can you find?

Figure 12. Step 11.

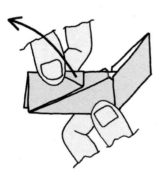

No questions.

Figure 13. Step 12.

No questions.

Figure 14. Step 13.

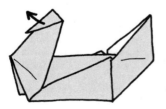

No questions.

Figure 15. Step 14.

No questions.

Figure 16. Step 15.

The students are not told what model they are folding. This approach helps to abstract the paper so that a sharp protruding point can be identified as an isosceles triangle and not as a "leg," or whatever. It also frees a child's imagination to complete a model and to name it. The van Hieles also used play in the classroom, allowing children to play imaginatively with tangrams.

8   Conclusion: The Benefits of Using Origametria in the van Hiele System

It can be seen that Origametria supports the van Hiele method of teaching geometry, particularly at levels 0 and 1 and can offer benefits over traditional methods of teaching geometry. The van Hieles identified the lack of teaching of levels 0 and 1 as the main reason for poor performances in middle and high schools at levels 3 and 4. The focus of Origametria is on levels 0 and 1, helping to give students a strong foundation of geometric knowledge for performing successfully later at higher levels.

Making origami models in each lesson keeps the students' motivation to learn very high. Origami puts fun and fascination into learning topics that would otherwise be too dry and abstract for many children to enjoy learning. The van Hieles accepted that fun and creativity in a lesson motivated children to learn.

Using this approach, Origametria has helped less able children to improve. The constant repetition of topics in levels 0 and 1 helps all children to learn and to enter middle school better able to learn at higher levels of van Hiele. Further, there are many anecdotal reports from IOC teachers of children with learning difficulties or behavioral problems enjoying origami, succeeding in folding a model and thus, being motivated to learn more. For these children, the acquisition of geometric knowledge is incidental, but occurs nonetheless.

The main contribution of Origametria to the van Hiele theory at levels 0 and 1 is to help students better recognize and define terms and shapes fundamental to an understanding of geometry at higher levels. This ability is achieved by the constant rotation, turning over, manipulation and folding of the paper through a multiplicity of shapes, so that terms and shapes are identified many times, but each time in a unique context. This theme and variation approach to teaching strengthens each student's flexibility in thinking, ability to recognize and define, and at higher van Hiele levels, to deduce and extrapolate.

Origami is also suited for mathematical investigation at levels 3 and 4, though the IOC does not currently teach the Origametria program by the van Hiele method at these levels. Once Origametria is firmly established at the lower levels, we will build upon these foundations and expand into the higher levels.

Acknowledgments. This article was translated from the original Hebrew by Boaz Shuval and edited by Paul Jackson.

Bibliography

[Burger and Shaughnessy 86] William F. Burger and J. Michael Shaughnessy. "Characterizing the van Hiele Levels of Development in Geometry." *Journal for Research in Mathematics Education* 17 (1986), 31–48.

[Fuys et al. 84] David Fuys, Dorothy Geddes, and Rosamond Tischler (eds.). "English Translation of Selected Writings of Dina van Hiele-Geldof and Pierre M. van Hiele." Technical report, U.S. Department of Education, Washington, DC, 1984. (Available at *Education Resources Information Center (ERIC)*, http://www.eric.ed.gov/ERICWebPortal/detail?accno=ED287697.)

[van Hiele 86] P. van Hiele. *Structure and Insight: A Theory of Mathematics Education.* New York: Academic Press, 1986.

[Page 09] John Page. "Diagonals of a Polygon." *Math Open Reference.* Available at http://www.mathopenref.com/polygondiagonal.html, 2009.

Student Teachers Introduce Origami in Kindergarten and Primary Schools: Froebel Revisited

Maria Lluïsa Fiol, Neus Dasquens, and Montserrat Prat

1 Introduction

This paper describes our ten-year journey to foster creativity and imagination through the use of origami in the teaching of geometry. Over the years in our geometry classes in the Faculty of Education at the Universitat Autònoma de Barcelona (UAB), we became aware that many people seem to suffer from math phobia and believe they cannot do mathematics. We often notice that some students have serious difficulties with mathematics, but in other fields are intelligent and productive. Fortunately, resources exist for working with these individuals to overcome their math anxiety [Adams 93; Caine and Caine 97, Chapter 5]. Moreover, specifically in the case of geometry teaching and learning, it is striking that even though human perception is so powerful from the first months of life (or even days) in recognizing and identifying shapes [Mehler and Dupoux 94], so many errors are made in identifying and classifying simple geometrical figures.

These observations forced us to consider how to structure and organize our curricular offerings, specifically in three areas: (1) methods of teaching mathematics, first level of kindergarten teachers; (2) methods of teaching geometry, third level of primary school education teachers; and (3) Practica III and IV, third level of kindergarten teachers in public schools. At first, we began making polygons with paper and doing some drawings related to numeric and algebraic calculations, using a ruler and a compass.

The paper-folding activities were a great success, and we decided to progressively introduce paper birds, boxes, figures, polyhedra, and modular pieces. The student teachers have access to four compilations or internal dossiers on mosaics and puzzles, symmetry, 2D versus 3D, and papiroflexy (origami). In class, on the very first day, a paper-folding activity is introduced.

Our experiences over the past ten years have given us the chance to receive and review the feedback on all our projects and procedures. We observed that our student teachers enjoyed working with paper, cardboard, and scissors. Thus, it seemed a good idea to extend our work with these materials. We suggested to student teachers at the diverse kindergarten and some primary levels that they use manipulative exercises in their teaching of mathematics when they entered their field practicum. This decision involved three steps: (1) the learning of traditional origami basics; (2) transmitting, step by step, the skills to these future teachers by presenting one or two paper figures per lesson, in two lessons per week; and (3) helping these future professionals understand origami as a methodological didactic option for use in the schools where they would do their practicum or in other educational sites such as clubs or holiday camps.

2   Objectives

As time went by, the objectives of our work with future teachers have become clearer:

- to integrate the learning of geometry by creating shapes and identifying them;

- to immerse the students in the possibilities of educational, scientific, aesthetic, and creative work with paper models (*papiroles*, as we say in Catalan) adapted to the different education levels;

- to build knowledge connecting the geometric world with language and imagination;

- to enhance the joy of working with one's hands, using inexpensive material and developing imagination.

The literature in such directions provided a source of support for our project.

3   Review of the Literature

The review of literature has focused on three main areas: the psychology of learning; those specific aspects of learning in young children especially as they relate to psychomotor activities such as origami; and the aspects of brain function and its intersection with psychomotor skills in the consideration of creativity and imagination. The well-known research findings of notables [Pestalozzi 15, Froebel 89, Decroly 86, Erikson 65, Piaget 62, Wertheimer 91, Vygotsky and Luria 07, Rogers 96] serve as the foundation for our decision-making. Essentially, the total history and works of Friedrich Froebel's (1782–1852) ideas and notions of implementation accompanied the whole of our efforts. Froebel attended Pestalozzi's school in 1808 to 1810. While accepting Pestalozzi's basic ideas on permissive school atmosphere and the object lesson, Froebel moved from a total emphasis on nature to one involving a spiritual mechanism in early education. His main psychological ideas hovered around free self-expression, creativity, social participation, and motor expression. In 1837, close to the 400th anniversary of the discovery of Gutenberg's movable type, he opened the first kindergarten in Germany. He then introduced paper folding into his kindergarten as one of the children's occupations. There are three categories to Froebelian folding: folds of truth, folds of life, and folds of beauty. In the first category, the folds of truth were those that helped children discover the elementary principles of Euclidian geometry. The second area, the folds of life, included those basic, traditional folds that we still use in modern origami models. The third area, folds of beauty, made up the greater part of Froebelian folding, and were intended to inculcate in the children a sense of artistic beauty and creativity.

These folds began with the "blintz" form and then moved on to squares, hexagons, and octagons. These areas all existed for a while, and then because the creativity of the third area did not penetrate the extension of the first two, it became the task and mission of others to carry Froebel's initial impetus in this direction to other countries around the world. In the late 1800s and early 1900s, we see the establishment of kindergartens in Florence (1871), London (1873), Tokyo (1876), St. Petersburg (1897), and Boston (1859, by Elizabeth Palmer Peabody). In 1855, Charles Dickens visited a London kindergarten and wrote, "by cutting paper, patterns are produced in the Infant Garden, that would often, then the work of very little hands, be received in schools of design with acclamation."

The big surprise for us has been that making paper folds has taken us into a world of educational experience of ideas and events that we had not recognized before even though we were professionally engaged as teacher educators. This new inspiration has been a big motivation in the study presented here.

We cannot pursue paper folding to any further degree in an observational way without realizing the complexity of the brain and its relation to the use of the hand, and then the creativity employed as it calls upon the human imagination. Every teacher comes into contact with such notions in the everyday classroom. Certainly, the brain is a complex organ, which generates many questions and inferences in observing actual occurring events. Bishop talks of "two complementary modes of thinking: one logical, one-dimensional, language-oriented and the other more dimensional, visual and more intuitive" [Bishop 84]. This, while a very simple scheme, can get as complex as we wish, always from a deep respect for the complexity of the human brain, whether it is the child's or the adult's [Deglin 77, Watzlawick 94, Caine and Caine 97, Rubia 00, Blakemore and Frith 05].

Paper folding demands accuracy and fine motor skills. The learning process has been focused on the spoken and written words and calculations, leaving aside characteristics associated with a global way of thinking. Wilson insists on the fact that we should reconsider our tandem teaching and learning and start focusing on the integration of hand and brain [Wilson 02]. To defend his idea, Wilson declares,

> When a personal desire drives us to create something with our hands, a process begins to work hard to give, as a result, a heavy emotional burden. It seems that people change when they are melded with movement, thought and sensibility.

Pestalozzi, who wrote a brief, but blunt statement, reached a similar conclusion: "head, heart, and hands" [Pestalozzi 15].

Consequently, the human hand, a biomechanical miracle generated some four million years ago as a result of biological adaptation, plays a crucial role in the learning process, but has not yet been valued sufficiently [Wilson 02]. According to Wilson, Bell (a Darwinian contemporary and a Scottish surgeon [Bell 40]) and Froebel [Cuéllar Pérez 05], we can consider that an intelligent use of the hand, along with the language instinct, is a basic stimulus in the development of the mind, which is activated at birth. Figure 1 shows four-year-old hands working with paper.

We have mentioned two other aspects of our interests beyond those of physical entities. We now seek an operational definition of creativity and imagination. We would want our teaching efforts to involve both. For a long time in educational history there has been a reluctance to consider the manipulation of objects as a strategy to stimulate thought or to solve problems. Traditionally, manipulation has been identified with the head and brain, but not with the hands, and with logical thinking and reason, but not with imagination.

We still seek a definition of what we understand as creativity and imagination. For the time being, and following Wild's [Wild 03] recommendations: "Right now we are alert about how children focus on their work

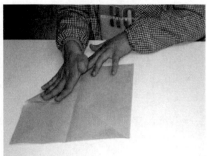

Figure 1. Four-year-old hands.

when they feel motivated, when they feel free to produce, go their way, experiment" [Foucault 73, Johnson 91, Fernández and Peralta 98, Root-Bernstein and Root-Bernstein 99, Fiol 02, Costa 08]. A child, often in the actual folding process, will see that certain images emerge from an otherwise planned model, purely from that child's unique background of prior individual experience. Can we operationally call this "imagination"? Is the individual who is able to engage in this behavior showing a type of "creativity"? To answer such an observation in the classroom, our purpose for this investigation emerged: To observe and record interactions of trained student teachers and young children in classrooms where mathematics and geometry are goals of learning in the use of origami methods to effect this end. We now consider what questions we would ask as we moved ahead with the investigation.

4 Problem Statements

To move ahead, we had to consider what data-collection methods would be necessary and sufficient to attain the data that were needed. The best way was to formulate the following questions:

- What type of curricular inclusions would prepare our student teachers to enter classrooms of young children and teach concepts of math and geometry appropriate for the ages of 2–5 years?

- What would be the sequence of such content within the programming of the UAB classes? In mathematics, geometry, origami?

- What would be the expectations of the teaching faculty?

- What would be the required skills and knowledge acquisitions of the student teachers before their specific practica?

- What interactions between teaching faculty at UAB and students would aid the generation of these?

- In the classrooms, what would be the form and substance of teaching plans at the different levels?

- What type of interactions would be meaningful between teacher educators and student teachers when in the field?

- What observations were made with children's responses and behavior, both as a group and then individually?

- Which of all of these were expected, unexpected, and/or unique?

- Of the observations, which would answer some of our concerns with hand and brain involvement, affect, enjoyment, skill development, and acquisition of curricular-required notions in mathematics, and specifically geometry?

- If necessary, how could these data change our plans for future interactions?

- How could/should we fill in the gaps with our methods for future study?

With these inquiries in mind, we collected the data, as discussed in the next section.

5 Data Collection, Analysis, and Discussion

Our initial attempts were supported by the notion that whatever we would do or hope to accomplish would have to be based on cogent and intensive focused education of our student teachers. Obviously, our students must learn paper folding in order to transmit its aspects to their own future students. Curiously enough, before we realized that we were doing actual paper folding, we began working on making polygons from rectangles and circles with our primary education students before moving on to the college students studying Early Childhood Education. We noticed there was an overall general interest in such activities. One of us (M. L. Fiol) started then to introduce, in parallel, work with a compass and ruler and polygons made with paper [Donovan and Johnson 75]. Fiol noticed that the students were enjoying these activities, and somebody began to talk about papiroflexy. Although not Fiol's original purpose, she nevertheless felt curious and began to focus her interest on paper models and choosing some for classroom activities [Clemente 01, de la Peña 01, Palacios 02, Riglos 03, Delgado et al. 04].

We realized that such was a method that allowed us to use words in their "natural context," as later we confirmed this notion with a group of three-year-old children. A week after this population folded a diagonal crease in a square, they were using the word "diagonal" naturally. We then began to introduce more paper-folding models in the continuing study of geometric forms. At the same time, we structured a curricular format for keeping records of student learning, achievement, and teacher educators' feedback. This involved keeping a portfolio by each student teacher, and then a planned concurrent evaluation setup and meeting with their educator mentors. In the global portfolio, the student teachers collected and put in order all the paper work done, such as drawings, photographs, and folded paper models.

Throughout the project, it has been interesting to see the resources created by the different student teachers. They had been asked only to make the figure they knew from the classes in order to memorize it. This figure was to be on a page of the portfolio in such a way that it could be unstuck for purposes of reproduction for their own future young students. The figure was always glued and photographed only if the result was three-dimensional or a modular type of origami. Spontaneously, many student teachers developed different strategies to improve "the storing of the figure." They listed the different images they made, in a step-by-step way that followed the process of folding, with the appropriate international signs for each move, or they even invented new directions and signs that followed the suggested procedure. We took notice of how each student teacher was often prone to make up his or her own method to remember the folding sequence.

The classroom climate was affected by the total learning situation. When a new paper figure was introduced, student teachers got involved in the activity and exhibited an active attitude toward learning. They exchanged opinions, asked questions, and shared their doubts. Although the international signs were used in the class, they were not officially taught until the last sessions. Learning was done by imitation, which allowed the teacher educator to focus on the important topics of measurement estimation and specific actions to remember the fold, because it could be difficult or specific. The university students, organized in small tables of four to six students, questioned insistently, exchanged opinions, asked for more information, repeated the steps when necessary, and then finally evaluated the figure. It would seem that they wanted to do a good job. The resulting figure acted as a mediator between instructor and student. A strong emotional tie seems to have been created between the finished product and

Figure 2. University student practicing paper folding.

the creator. The student teachers were eager to apply what they learned as they went out into the field and became instructors themselves. Over the past ten years, most of these former students (now classroom teachers) became collaborators with us, the educators, and the new student teachers in learning. Figure 2 shows student-teacher hands working during a lesson with paper. Table 1 provides a summary of the courses and the organization of the work between the university and the practicum schools.

Because of the positive feedback from the university students and their practicum experiences, gained and growing since 2003 and 2004, the Paper Project School was initiated. Fourteen kindergarten classrooms have hosted our student teachers. They have suggested paper-folding activities related to geometry and origami with groups of two- to five-year-old children. The goal has always been to introduce the children to paper folding and, at the same time, give new ideas to teachers. The initial meetings between the professor and the internship student determined just what activities were important for children at the young ages of two to five years. We tried to offer them enjoyable and motivating proposals that allow the children to work in an atmosphere with a high level of freedom. At the same time, the adult could also be observing the results of the activities. Actually we were aware that children of the same age in Japan were doing similar activities. So we asked ourselves, "If Japanese children can do it, why don't we try it with our kids?" But which models must we introduce first? How can we do it? Which is an appropriate methodology? We found some answers to our questions from the literature [Bueno et al. 06, Fiol et al. 10]. Also, constant feedback from our new teachers in the field has allowed us to modify our methods and make them more specific.

Subject and college specialty	Students	Number of models completed	Material	Evaluation	Feedback
Didàctica de la Matemàtica (methods of teaching mathematics), first level, kindergarten teachers	More than 200	More than 20	Self-learning portfolio (30 pages)	Four questions in an exam, one of which consists of the realization of a paper figure.	Nonmandatory interview. Student and teacher exchange their opinions about the portfolio and the exam; 8–10 minutes per student.
Didàctica de la Geometria (methods of teaching geometry), second level, primary school teachers	More than 900	More than 30	Self-learning portfolio (70 pages)	Five questions in an exam, one of which consists on the realization of a paper figure.	Mandatory interview. Student and teacher exchange their opinions about the portfolio and the exam; 10–15 minutes per student.
Practicum III (internship experiences), third level, kindergarten teachers	More than 30	10–20	School attendance (100 hours minimum)	Every student prepares an internship memory before the end of January (school and class notes and pedagogic information)	Eight interviews during the first university term (September–February).
Practicum IV (internship experiences), third level, kindergarten teachers	More than 30	The university students work with children at schools. They organize the activities, offering 2–8 models, depending on the children's ages.	School attendance (100 hours minimum) preparing one intervention with the children every week	Every student prepares an internship memory before the end of June (diary about every activity and pedagogic information)	Fifteen interviews during the second university term (February–July).

Table 1. General view: organizing the work between the university and the practicum schools.

Figure 3. A glass on the table. (See Color Plate VII.)

Figure 4. A jumper in a pool. (See Color Plate VIII.)

We realized, first of all, that children need to feel the basic material. So we started with movement activities (happenings), interactions with different types of paper, manual explorations, and recognition of the different textures. Next, we moved on to creative activities such as composing a picture with different shapes and pieces of paper. The next phase involved the encouragement of the first free folds. The children were invited to handle the paper and fold it freely into a figure. Finally, they were asked to imagine what the figure was and name it. Once the children gained confidence in the use and handling of paper, they were able to move on to more complex issues in folding, such as following an order of folds.

Figures 3 and 4 show two children's creations: they participated in a lesson where they were free to fold the paper as much as they wanted. Finally they were invited to give a name to the figure obtained.

An initial motivation was the introduction of meaningful mathematical concepts to the children. We found that just getting the children into the activity involved three different strategies that, once implemented, could move on to our further goals of making these fit into a meaningful mathematics and geometry curriculum for our children in the schools: (1) imitating the steps that another person explains; (2) folding a piece of paper and then imagining what it might be (giving a name to the figure and explaining it to the teacher); and, (3) creating what the child imagines (we found that only the most persistent children reach this plateau.)

We have had the opportunity to change our methodology because of the initial successful implementation of our plans. Specifically, the area of the adult's attitude when in contact with the children during the classroom activities engaged our focus. The teacher should initiate an activity by creating a calm atmosphere that promotes concentration. This climate in the classroom assures that the children are ready for the activity. Moreover, the teacher should encourage the children to experiment with their ideas, and should accompany them in their efforts, never leaving them on their own, being very positive, and respecting every one of the child's actions. Giving the children enough time to find inspiration and making the figure that each one wishes helps them share with the adult their creations. It also gives the teacher the opportunity for greater understanding of where each student is by listening to each child's explanations, intuitions, and ideas.

Finally, the teacher must respect the individual products of the students as a strategy to enhance the innate creativity in every child. We understand that creativity now, in this frame of reference, is a process through which children express their imagination, ideas, and emotions, and thereby obtain a high level of satisfaction: "Creativity is defined by the grade of satisfaction that the child obtains from his or her creations more than by how much the results get close to the guidelines established by other people" [de Artola and Hueso 06].

As we approach a definition of creativity supported by the observations made in the initial stages and growth of our project, we subscribe to Malaguzzi's notion that the adult understands the child as a being who is complex in his or her own individuality [Malaguzzi 96]. The teacher must stimulate individuality, fighting against the homogenization of students in institutional schools. Furthermore, the child who faces the challenge of learning is usually thrilled with the idea of producing an object with his or her own hands. And, as we have realized by observing children and their actions day by day in the schools, this excitement, mixed with the pleasure that paper-handling provides, is translated along with the experience gained over time into (1) a self-demanding attitude toward the production itself, (2) a growing tolerance toward researching other ways, and (3) a greater intuitive capability.

6   Conclusion and Next Steps

Our data collection thus far has been concentrated on the first part of our project mission: finding a way to get the student teacher and the professor to interact to develop a method that would be appropriate for young children in learning mathematics and geometry by, among others, the ingenious use of papiroflexy. In the search and accomplishment of this, we may have come upon some working definitions of creativity and imagination.

We shall continue in this direction as well as move forward to determine whether such methods aid in the learning of age-relevant mathematics and geometric concepts, or just what brain and motor skills working in unison can do or achieve. We still have a long way to go. We anticipate that in the next 36 months, when Neus Dasquens will be in charge of a P3 class with 18 students, where tutors will be our researchers M. L. Fiol or Montserrat Prat, working in unison with Dasquens in charge of a class of three-year-olds, that we shall be able to answer the remaining questions of our initial effort. The hope is to be there, make observations, conduct interviews and perhaps focus groups, photograph, record actual workstations and language use, and use constant comparative research methods that will enable us to come up with some themes that pervade all of our efforts in the areas where we yet must present our findings.

Acknowledgment. We thank Ryda Rose for helping us to reorganize and rewrite this paper.

Bibliography

[Adams 01] James L. Adams. *Conceptual Blockbusting: A Guide to Better Ideas*, Third Edition. Cambridge, MA: Perseus Publishing, 2001. (First eidtion published in 1974; Spanish translation available: *Guía y juegos para superar bloqueos mentales.* Barcelona: Gedisa, 1993.)

[Bell 40] Charles Bell. *The Hand, Its Mechanism and Vital Endowments as Evincing Design (Treatise IV).* New York: Harper & Brothers, 1840.

[Bishop 84] Alan Bishop. "The Mathematics Teacher and New Developments." *Euclides* 2 (1984), 15–20.

[Blakemore and Frith 05] Sarah-Jayne Blakemore and Uta Frith. *The Learning Brain: Lessons for Education*, 2005. (Spanish translation available: *Cómo aprende el cerebro. Las claves para la educación.* Barcelona: Ariel, 2008.)

[Bueno et al. 06] Marisa Bueno, Maria Lluïsa Fiol, and Laura Soler. "Somnis de paper" ("Paper Dreams"). *Perspectiva Escolar* 306 (2006), 68–76.

[Caine and Caine 97] Renata Nummela Caine and Geoffrey Caine. *Education on the Edge of Possibility.* Alexandria, VA: Association for Supervision and Curriculum Development, 1997.

[Clemente 01] Eduardo Clemente. *Papiroflexia.* Barcelona: Plaza & Janés, 2001.

[Costa 08] Joan Costa. *La forma de las ideas. Cómo piensa la mente. Estrategias de la imaginación creativa (The Shape of Ideas. How the Mind Thinks. Strategies of Creative Imagination).* Barcelona: Costa Punto Com Editor, 2008.

[Cuéllar Pérez 05] Hortensia Cuéllar Pérez. *Froebel, la educación del hombre (Froebel, the Education of Man).* Madrid: Trillas, 2005.

[de Artola and Hueso 06] Teresa de Artola González and María Antonia Hueso. *Cómo desarrollar la creatividad en los niños (How to Develop Children's Creativity)*. Buenos Aires: Astor Juvenil Palabra, 2006.

[Decroly 86] Ovide Decroly. *El Juego educativo : iniciación a la actividad intelectual y motriz*. Madrid: Morata, 1986.

[de la Peña 01] Jesús de la Peña. *Matemáticas y papiroflexia*. Madrid: Associación Española de la Papiroflexia, 2001.

[Delgado et al. 04] Laura Delgado, Soledad Zapatero, and Maria Lluïsa Fiol. "2D versus 3D: La papiroflèxia, un recurs didàctic" ("2D versus 3D: Origami, a Didactic Resource"). *Perspectiva Escolar* 284 (2004), 59–65.

[Deglin 77] Vadim Lvovich Deglin. "Nuestros dos cerebros" ("Our Two Brains"). *Infancia y Aprendizaje* 7 (1977), 37–53.

[Donovan and Johnson 75] Richard Donovan and Albert Johnson. *Matemáticas más fáciles con manualidades de papel (Easy Mathematics with Paper Crafts)*. Barcelona: Distein, 1975.

[Erikson 65] Erik H. Erikson. *Childhood and Society*. London: Penguin Books, 1965.

[Fernández and Peralta 98] Rosa Fernández and Felisa Peralta. "Estudio de tres modelos de creatividad: criterios para la identificación de la producción creativa." *Faísca: revista de altas capacidades* 7 (1998), 67–85.

[Fiol 02] Maria Lluïsa Fiol. "Duendes en el desván: tanteo, intuición, creatividad" ("Elves in the Attic: Testing, Intuition, and Creativity"). In *Aportaciones de la Didàctica de la Matemática a diferentes perfiles profesionales*, edited by M. Carmen Penalba, German Torregrosa, and Julia Valls, 21–42. Alicante, Spain: Universidad de Alicante, 2002.

[Fiol et al. 10] Maria Lluïsa Fiol, Neus Dasquens, and Montserrat Prat. "La papiroflèxia a l'escola: imaginació, emoció i geometría." Girona: Congrés de Didàctiques Generals, 2010.

[Foucault 73] Michel Foucault. *Ceci n'est pas une pipe*. Montpellier: Fata Morgana, 1973. (Spanish translation available: *Esto no es una pipa*. Barcelona: Anagrama, 1981.)

[Froebel 89] Friedrich Froebel. *La educación del hombre i el jardín de infancia (The Education of Man and Kindergarten)*. Barcelona: Eumo, 1989.

[Johnson 91] Mark Johnson. *The Body and the Mind: The Bodily Basis of Meaning, Imagination and Reason*. Chicago: University of Chicago Press, 1987. (Spanish translation available: *El cuerpo y la mente. Fundamentos corporales del significado, la imaginación y la razón*. Madrid: Debate, 1991.)

[Malaguzzi 96] Loris Malaguzzi. "Malaguzzi i Educació Infantil a Reggio Emilia" ("Malaguzzi and Kindergarten in Reggio Emilia"). Special issue, *Temes d'infància: educar de 0 a 6 anys* 25 (1996).

[Mehler and Dupoux 94] Jacques Mehler and Emmanuel Dupoux. *Nacer sabiendo. Introducción al desarrollo cognitiva del hombre.* Madrid: Alianza, 1994.

[Palacios 02] Vicente Palacios. *Papiroflexia.* Barcelona: Miquel A. Salvatella, 2002.

[Pestalozzi 15] Johann Heinrich Pestalozzi. *How Gertrude Teaches Her Children: An Attempt to Help Mothers to Teach Their Children and an Account of the Method,* translated by Lucy E. Holland and Francis C. Turner, edited by Ebenezer Cooke. London: G. Allen and Unwin, Ltd., 1915.

[Piaget 62] Jean Piaget. *Play, Dreams, and Imitation in Childhood.* New York: Norton and Company, 1962.

[Riglos 03] Grupo Riglos. *El libro de las pajaritas de papel (The Book of Paper Folds),* Second Edition. Madrid: Alianza Editorial, 2003.

[Rogers 96] Carl Rogers. *Libertad y creatividad en la educación (Freedon and Creativity in Education).* Barcelona: Paidós, 1996.

[Root-Bernstein and Root-Bernstein 99] Robert Root-Bernstein and Michèle Root-Bernstein. *Sparks of Genius: The Thirteen Thinking Tools of the World's Most Creative People.* Boston: Houghton Mifflin, 1999. (Spanish translation available: *El secreto de la creatividad.* Barcelona: Kairós, 2000.

[Rubia 00] Francisco J. Rubia. *El cerebro nos engaña (The Brain Cheats on Us).* Madrid: Temas de Hoy, 2000.

[Vygotski and Luria 07] Lev S. Vygotski and Aleksandr R. Luria. *El Instrumento y el signo en el desarrollo del niño.* San Sebastián de los Reyes, Spain: Fundación Infancia y Aprendizaje, 2007.

[Watzlawick 94] Paul Watzlawick. *El lenguaje del cerebro (The Language of the Brain).* Barcelona: Herder, 1994.

[Wertheimer 91] Max Wertheimer. *El pensamiento productivo (Productive Thinking).* Barcelona: Paidós, 1991.

[Wild 03] Rebeca Wild. *Calidad de vida. Educación y respeto para el crecimiento interior de niños y adolescentes.* Barcelona: Herder Editorial, 2003.

[Wilson 02] Frank R. Wilson. *The Hand: How Its Use Shapes the Brain, Language and Human Culture.* New York: Pantheon Books, 1998. (Spanish translation available: *La mano. De cómo su uso configura el cerebro, el lenguaje y la cultura humana.* Barcelona, Colección Metatemas, Tusquet, 2002.

Narratives of Success: Teaching Origami in Low-Income Urban Communities

Christine E. Edison

1 Introduction

This paper offers some anecdotal accounts of the positive, and potentially life-changing, effects of origami on students in low-income urban communities. The origami journey began seven years ago when I was teaching seventh grade at a private school. Teachers were required to have a club on Fridays and we were given a budget of $10 for the school year. Since drama and drawing were already taken, I thought paper was cheap and suggested an origami club. I picked up an origami book [Gross 04] and a ream of paper and off I went.

From that private school I moved to Chicago Public Schools to teach high school mathematics. The high school, on the south side of Chicago, was a school within a school that enrolled students who had aged out[1] of grade school that focused on students getting their eighth grade diploma. The population was approximately 95% African-American and 5% Hispanic, and 98% were from low-income homes. For the girls, that often meant they had been pregnant, were in unstable homes where they rarely went to school, and/or had spent time in jail. For the boys, it usually

---

[1] These students technically were not promoted. They were earning their eighth grade diploma at the same time they were taking high school classes. The program was put in place because the Chicago schools do not enroll students after a certain age (15 years old) in elementary school out of safety concerns for the younger students.

meant they had been in jail and/or had an unstable home life or no home at all. I taught geometry and saw students for 90-minute periods five days a week—a long time for any subject, much less for students who had limited attention span and were not engaged with school. These students simply had not been acculturated to the norms and expectations of middle-class society; they were often in a perpetual state of emotional and physical crisis, and education was perceived to have little value in their immediate situations. I found origami to be a tool of engagement, through which students who typically did not perform well in school found application and understanding.

Over the years, I have met several people who also use origami in similar environments to impact positively the lives of children and adults. This paper is a collection of first-person accounts of people who teach origami in the inner city and/or low-income urban environments, beginning with my own accounts of using origami in mathematics classes.

All names of students and adults have been changed to protect their privacy.

2 Origami in Mathematics

The following accounts are from three different schools. The experiences are predominantly with African-American and Hispanic students in low-income urban schools in Chicago, Illinois.

2.1 Englewood, Illinois

My first day teaching as a fully certified teacher in Chicago public schools (CPS) was in January, as replacement for a teacher who left midyear (in the United States, the school year begins in August or September). My students were considered tenth graders (15–16 years old), but they ranged in age from 16 to 18 years. On that first day, I decided to introduce these students to origami through the construction of a spinning octahedron [Gross 04, p. 35]. Since I had previously taught the model to children as young as four years of age, I thought it would be a fairly accessible model for students with limited dexterity. The model has a great finished look, and the spinning aspect always pleased the students I had taught previously. The students constructed a skeletal octahedron and found its surface area and volume. We all enjoyed the experience; students taught me a new term, "raw," and were excited, and I was hooked. Importantly, this origami lesson met the required curriculum standards. We used a specific curriculum, but every time we utilized origami I found student engagement went up, often with 100% participation, which was a rare occurrence in most classes. My students were also surprised that they were allowed to take their octahedrons home.

Students in Englewood are poor in a way the majority of Americans are not used to or even aware of. It is in many respects an area apart. Even living in the same city for years, I never realized the depth of poverty that my fellow citizens experience. In our city the news media had often ignored an entire community; guns in school and arrests were common, and I did home visits where the only visible thing in the house was a crack pipe or a stained mattress. Finding out they could use paper to make something that was fun and moved was an eye-opener to these students.

2.2   Achievement Academy in Englewood

While student academic success is a primary concern for me as a teacher, there is a social benefit to school that in some respects is more important. Students from difficult homes often find themselves unable to connect with school and teachers. Many of them are dealing with emotional burdens that are incompatible with functioning in a mainstream school setting. At the Achievement Academy, many of the boys were in and out of jail. One student, Darnell, was released from jail in the middle of the year. He was angry and non-responsive; I was scared he was going to snap in the classroom, and I did not want to end up in the hospital.

Winter break was coming, and I decided to teach a wreath, a simple modular [Pederson 99] that students tend to like (Figure 1). I explained the activity and started handing out the paper. Darnell took the paper

Figure 1. Wreath designed by Mette Pederson and folded by Christine Edison.

and didn't brush it to the floor, which was his usual behavior. We went through the steps of how to make the module, and I drew instructions on the board. Darnell completed the activity quickly and asked for more paper. I was in complete shock, as this was the first time he spoke directly to me. I gave him the paper and complimented him on his finished model. Darnell told me he knew all about origami, as his grandma had taught him a bird. I offered to show him another model or let him help other students. Darnell didn't want to work with the other students, but he took the book [Pederson 99] and independently made several simple models.

In many classrooms, communication is a normal and expected part of the class, but appropriate, understandable communication is not something an urban schoolteacher can take for granted. The level of tension in the room decreased after that wreath activity; Darnell would make an effort to participate, and he began addressing me by my name. It also allowed for dialogue about his family and grandmother, which helped solidify the connection between us.

2.3   Origami and Student Self-Confidence

In my inclusion geometry class, Jose was very unwilling to even write down a problem and was disruptive during instruction. At the high-school level, the ability to do algebraic manipulation is required and is a significant focus of the curriculum. Before Thanksgiving break, I included an activity that combined building Tom Hull's dodecahedrons [Hull 06] with a small business experience (Figure 2). Students had to find the area, estimate what they needed for supplies, and calculate what they would charge for time and materials if making them to sell.

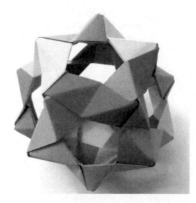

Figure 2. Skeletal dodecahedron designed by Tom Hull and folded by Christine Edison.

Since I have started teaching, the one recurring anecdotal observation I have made is that the student with the strongest spatial skills in the class has always been a student with special needs. Jose did not need me to explain construction; he could figure out the construction just by looking at the model. He was able to orally explain the mathematics necessary to solve the equations. Jose was allowed to go to each group of students and help them construct their dodecahedron. Every single school day for the rest of the school year, Jose asked to do origami. One day I asked him why we should do origami every day. Jose replied it was because it was the first time he had been smarter than anybody else in anything. It also allowed students to see Jose in a different light and to gain new respect for him, which helped to integrate him into the community and enabled him to work with others. Jose still struggled, but he was more willing to try harder in class and he still had a reference that he was good at something; in fact, he was not just good, but the best in the class.

I do not use as much origami in the classroom now that I teach Algebra II, but I still use it on X-days (no classes) and field days, and similar results still emerge. For the last three years, Jose consistently has come after school and learned new models, and he has been pleased to discover his abilities. He is willing and able to learn and teach others, even getting service learning hours for it. He graduates this year, and I will miss our discussions and lunchtime origami while he played Yu-gi-oh with friends.

3   Origami in Art Class

I was an art teacher in Chicago for years, and I believe strongly in the power of origami to change students for the better. Students learn so much about sequence and transformation. They take a flat piece of paper and make it into a recognizable object. I have used origami in tutoring programs and I have seen that it has powers and gives students focus and teaches them to use their hands. Origami teaches art, mathematics, and fine-motor skills. I have used origami at every school where I have taught, although the power of origami has not often been recognized by principals and the school administration.

3.1   Learning to Follow Instructions with Origami

Through origami, students who couldn't follow instructions learned how to sequence events, the power of transformation, and that anything can be created from this two-dimensional sheet of paper. When students will not or cannot seem to learn in the traditional classroom and they learn origami, the teachers then learn about the students' potentials.

I have taught in many places, and I really believe that origami should be an integral part of any art program. One student who stood out was Terrance. Terrance was a fighter, a terror. I was at the school for only a year, but we did origami and I taught the students how to make a star. Terrance loved origami, and he could make the star in no time at all. Terrance wsa often sent to me because no one else could deal with him. He always did well in my class, and he would make origami for everyone. Several years after I left that school, I ran into some teachers who were still at the school, and I asked about the art program and the students. Terrance was still there, and the teacher noted he still would beat anybody up, but he would be holding an origami star he made in one hand and would give it to anyone, even the kid he just beat up.

4   Origami in Elementary School

This section contains a first-person account from another teacher who teaches English to sixth graders in a school that is predominantly Latino with a highly mobile population. More than 60% of her students are learning English as a second language.

4.1   Learning English through Origami

The school is considered a horrible school, but it is not. The staff members work hard, and the students are great kids, but many of them don't finish the school year; things are different for them than for students in a middle class school. The parents take them back to Mexico for long periods of time, and they just don't get enough learning time. The students also struggle with English and even with proper Spanish. The girls miss a lot of school to take care of their brothers and sisters.

In the second half of the school year last year, my students were not doing well. They and I were struggling with the curriculum, which is scripted and not enjoyable at all. I love crafting and do a lot of crafting with my 11-year-old daughter. She wanted to make a crane, so we went online and learned to make one, and we had a super time. I decided that I needed to do some origami in class.

I bought some paper, and in class we learned to make a simple crane. I asked my students to write instructions in English, and then they had to write stories about the crane. Students who were really struggling were paired up with better writers, and they worked on writing the stories together. Finally, everyone switched stories with their reading buddy and then read them to each other. We also focused on the skill set of the week, and I made sure we checked the stories for the proper use of commas and tenses that week. My students loved the activity. I noticed that the quiet kids became more talkative. One of the little girls came in the next day with 20 cranes, one even made out of a candy wrapper in Spanish, which I told her was great recycling. Every week, we made a new origami friend and wrote about it.

Some of my students became avid writers; they wrote more in that part of the year than they ever had in a whole school year. The students were very enthusiastic when I bought some pretty scrap-booking paper with hearts all over it; students were able to make cranes, flowers, hearts, frogs, and boxes that they gave to the eighth graders when they graduated. One of the boys, who was absent for several weeks when visiting relatives in Mexico, came back with an origami dog, car, and bug with a story written for each piece. My students' scores were better than those of the other teachers' classes for the same grade level, and they also were behaving better. When the coach for our curriculum arrived, though, we had to stop, but I still taught a new origami model each week for the last month of school. I am looking forward to working with the coach to integrate origami into the curriculum for the entire next school year.

Another thing that was also really gratifying was how students who wouldn't focus for ten minutes could practice making models over and over again.

4.2 Origami in a Summer Program

I also used origami in a summer program I did with the kids at the park near the school. My daughter, Sarah, attended the one day a week when I taught origami. I used many of the models I had taught during the year, and Sarah would help. One of the girls, Rosy, a small seven-year-old, was really quiet and had almost no understanding of English; she didn't play with anyone and cried a lot. She wouldn't play games and would physically run away and hide if we tried to get her involved. Rosy always hid behind Sarah and played with her hair. Sarah got her involved in doing origami with us and always was extra patient with her. Rosy loved making things for Sarah; even when Sarah wasn't there, she wanted to fold. Since she would not participate in games, I brought paper for her to fold. She always folded two things: a flower and a dog. This went on for six weeks, and on the last day I met the woman with whom Rosy was staying. She thanked Sarah and me because Rosy previously had never lasted through an entire summer program. This woman had been taking care of Rosy for three years while Rosy's mother was in jail. Rosy ran over and hugged Sarah before she left and gave her a pile of flowers made from newspaper she had cut into squares at home.

5   Discussion

As a teacher, I would prefer to base my teaching choices on clearly proven methodologies that have been tested in effectively designed and executed studies, but we do not live in that world. It is difficult to isolate variables, and quantifying aspects of our humanity and intellect is tenuous science at best—so teaching is in large part trial and error. I have listened to brilliant men and women who have contributed to a variety of disciplines through the study of, or application of, origami. When I teach, it is rarely to students who are statistically likely to go to university and complete their master's degrees or doctorates. I find origami is a tool that can teach, but it also can help a student find his or her voice. It is a means of communication and connection. Children who have never succeeded can be good, even if only for a day, with paper folding. Some students can see the Platonic solid they folded and synthesize an understanding of polyhedra they otherwise would not be able to do in the traditional classroom setting. And some students find a voice through a folded piece of paper. The transformative aspect or origami is not limited to the paper—it also has the potential to transform people in ways that are difficult to quantify. I do not use origami every day or even every month, but I still find it one of the most surprising activities to do with my students, and I hope long term it is incorporated into our grade schools and high schools with specific curricular goals so that all children will get the chance to experience the benefits that it has to offer.

Bibliography

[Gross 04] Gay Merrill Gross. *Origami: The Art of Paper Folding*. San Diego, CA: Laurel Glen Publishing, 2004.

[Hull 06] Thomas C. Hull. *Project Origami: Activities for Exploring Mathematics*. Natick, MA: A K Peters, 2006.

[Pederson 99] Mette Pederson. "Ring 1 Unit." Available at http://www.metteunits.com/, 1999.

Origami and Spatial Thinking
of College-Age Students

Norma Boakes

1 Introduction

Spatial ability is viewed as one of the prime indicators of intelligence [Mann 05, West 97]. Renowned individuals such as Picasso, Einstein, and Edison all had this spatial gift [Mann 05]. Examining how intelligence was developed and strengthened, Piaget [Piaget and Inhelder 56] and Gardner [Gardner 93] identified spatial thinking as a key to cognition and learning. Piaget and Inhelder broke down the very nature of knowledge into three distinct types: mathematical-logical, physical, and social [Piaget and Inhelder 56]. Exploration of spatial concepts was seen as an important element in the development of these intelligences. Seeking to categorize the different types of intelligences, Gardner identified nine types, of which one is spatial intelligence [Gardner 93].

Individuals possess and utilize spatial intelligence every day. Spatial skills are used when directions are given to a local shopping center. Putting up a tent when camping or assembling a new toy during the holidays utilizes this skill to visually manipulate two- and three-dimensional objects in space [Olson 84]. Beyond visualization, spatial ability also "concerns the locations of objects, their shapes, their relations to each other, and the paths they take as they move" [Newcombe 10].

We attend school to build and develop the skills and knowledge we need to be successful in life. One of these skills is spatial thinking. Spatial skills are seen in many subject areas, especially mathematics, science, and the arts. Perhaps we are constructing a ramp for a physics experiment or using geometric shapes to tessellate a plane. From elementary age through college and beyond, we use, develop, and strengthen this mental skill. In fact, studies over the past few decades have shown that spatial thinking can be specifically targeted through specialized activities [Newcombe 10]. Results from the Trends in International Mathematics and Science Study (TIMSS) in 2007 [Martin et al. 08] have documented the poor performance of US students in geometry and measurement relative to other areas assessed: number sense, algebra, and data and statistics. Clearly, there is a need for better instruction to improve students' spatial and logical thinking.

In *Origami4*, I presented a study on how origami was used to teach mathematics to a group of middle school students. In this study, origami was shown to be a potential tool for developing spatial skills [Boakes 09]. Although this approach is far from unique, I found through extensive literature searches that there exist limited data to substantiate the claims that origami builds spatial skills [Boakes 06]. My study was an attempt to begin the process of developing quantitative evidence on what impact origami truly has on students' abilities.

This paper presents another study targeting the impact of the use of origami on a person's spatial skills. At the time of my 2006 study, I was focused on middle school curriculum and how origami infused in instruction could benefit middle school students. My current work shifts focus to college-age students. Just as spatial skills are malleable at the middle school level, the same is true for college undergraduates. Studies have shown that practice in spatial skills improves college students' performance on visual test items and has even helped educational outcomes in science, technology, or mathematics (STEM) degrees, where many of the courses necessitate spatial acuity [Sorby 05, Terlecki et al. 08]. With the flexibility of a small liberal arts college and a strong general studies program, I designed a course for non-mathematics majors based on my research called "The Art and Math of Origami" [Boakes 10]. As the title suggests, origami was the centerpiece of the course. The work presented here describes the course content, the students who have taken it over the past three spring semesters, and the impact of this course on college students' spatial abilities.

2 Design and Purpose

The purpose of this study is to build upon the investigation I conducted with a group of middle school students in 2006 [Boakes 06, Boakes 09]. The study for this paper involved college-age students taking a semester long course, "GNM 2257—The Art and Math of Origami," designed from the premise of my 2006 research. The purpose remains the same, with the investigation centering on what impact origami instruction has on students' spatial abilities. Since the nature of the course was different from the original mathematics setting in the middle school classrooms, the focus was narrowed slightly to look solely at students' spatial abilities after regular exposure to the practice of origami. Although mathematics is a part of the course, the direct mathematical content comprises only about 20% of the course. Spatial skills play a direct role in mathematical abilities, so it was this focus that drove the second study.

2.1 Methodology

The design of this study is similar to the one conducted in 2006 [Boakes 09]. The advantage here was a more prolonged, concentrated exposure to origami instruction and practice. To maintain consistency in determining spatial abilities, the three spatial tests used were the same as in the 2006 study. Further, the same pre- and posttest experimental design was used. Due to scheduling restrictions, however, it was not possible to have a control and experimental group as I did previously. Instead, I gathered data from three consecutive spring semesters using all students enrolled in the origami course. Data analysis methods were mainly quantitative in nature. A paired-sample t-test, or repeated measures test, was used to explore how participants' spatial test scores changed over time, from pre- to posttest. In addition, an analysis of covariance (ANCOVA) [Pallant 07] was used. In this statistical test, the difference between groups can be detected while controlling for some mitigating factor. In this case, it was important to know how spatial experiences outside of the course influenced participants' spatial test scores. At the same time, it was essential to consider that pre-existing differences in actual spatial ability could influence this relationship. The ANCOVA allowed me to control for this possibility. In the 2006 study [Boakes 09], there was no test that analyzed how natural spatial ability influenced performance. This element is new to this study. It is the overall intent here to delve further into spatial abilities and their connection to the practice of origami through these analyses.

2.2 Sample

The sample used for this study consisted of undergraduate college students at Richard Stockton College in Pomona, New Jersey. Each of these students selected this elective general studies course, GNM 2257 (described in Section 2.5). The four-credit course took place during a typical college semester (approximately four months). One section of this course was offered every spring. Enrollment was first come, first serve, with a limit of 25 students. Due to the popularity of the course (filling usually the first

	Low spatial experience	Mid-to-high spatial experience
STEM amjors	37% (7)	63% (12)
Non-STEM majors	52% (28)	48% (26)

Table 1. Comparison of spatial experience and academic major of participants.

day of registration), the makeup of the student sample was mainly juniors and seniors, who have the advantage of registering first. To date, a total of 75 students have taken this course, with 25 students each term between Spring 2008 and Spring 2010. Among those, 45 students were female and 30 male. In terms of ethnicity, the majority of the sample was Caucasian (79%) along with 4–5% African American, Asian, Hispanic, or other. Because the course is open to all majors, the students had a wide variety of academic backgrounds, ranging from computer science to criminal justice. Students majoring in STEM areas such as mathematics or biology may have some bearing on spatial ability testing, so the number of STEM versus non-STEM students was calculated. Of the 75 students, 19 students reported majors in the areas of science, technology, or mathematics.

Another key identifying factor for the sample was students' spatial experiences prior to taking this course. (Activities such as video gaming or construction could potentially influence spatial skills.) To identify students' perceived levels of spatial experience, a set of six everyday types of tasks (all requiring spatial skills) were listed on a survey administered on the first day of the course with a Likert-style scale from 1 (little to no activity) to 5 (daily). Students rated each of these items, then the values were added together to create a "spatial experience" indicator. The values reported ranged from 3 to 20 points with a maximum of 30 possible points for the 6 combined items. In an effort to create two equal groups, students with total points 0–9 were deemed "low spatial experience," having little to no exposure to spatial activities in their daily lives. Students with "mid-to-high spatial experience" ranged 10–20, claiming anywhere from monthly to daily exposure to spatially related tasks. Among the 75 participants, 35 of them were identified with a low level of spatial experience prior to taking the course. Of these 35 students, 28 identified themselves as non-STEM majors (see Table 1). The remaining 7 students reported that they were STEM majors but with little exposure to spatial-related tasks beyond their academics.

2.3   Instruments

Four main instruments were utilized for this study: a survey and three pre-assessments of students' current spatial skills. The first instrument was a survey given at the start of the semester. This survey gathered basic demographic information as well as academic background and exposure to inherently spatial tasks. (See the Appendix for an excerpt from this part of the survey.) The survey blended both qualitative and quantitative items. Of those collected, specific questions were targeted for inclusion in this study. Demographics allowed for a basic understanding of the participant makeup. The question related to major field of study allowed for identification of STEM versus non-STEM students. In addition, six questions were designed to quantify to what extent students were exposed to activities that are known to build spatial skills. Students rated each spatially related task listed with a 1–5 response (1 indicating limited exposure and 5 denoting daily exposure). As noted in the discussion of participants above, this survey allowed for identification of students who would likely be strong or limited in spatial skills due to practices outside of their college experiences.

The remaining three instruments were selected to quantify students' spatial abilities. The same instruments were used in the 2006 study [Boakes 09]. After extensive review of past studies [Boakes 06], I concluded that the assessment that best matched the definition of spatial skills was a set of three subtests taken from the *Kit of Factor-Referenced Cognitive Tests* [Ekstrom et al. 76]. The card rotation test was linked to the ability to mentally manipulate two-dimensional figures. The test consisted of two parts, each with 80 responses. For the 2006 study, only one part of the test was used due to the age of participants. In this study, both parts were used, for a total of 160 points possible on the test (based on the number of problems correct). The paper-folding test involved mentally manipulating a square sheet of paper, imagining punching a hole in a specific location on the paper, and then identifying which unfolded sheet of paper would match the hole-design created. As the name suggests, it is mental paper folding involving both two- and three-dimensional visualization skills. There were two parts to the test, each containing ten questions. Both parts of the test were used with a maximum cumulative score of 20. The third and final test, the surface development test, involved visualizing the process of matching a 2D figure (its mathematical "net") to its constructed 3D solid object. In each part of the test, six figures were given. Each of these figures had five responses, making a total of 30 possible points in each part. All three cognitive tests not only matched the spatial skills targeted but also were appropriate for the age group of this study. Reliability was found to be reasonably high [Fleishman and Dusek 71].

The spatial tests were given at two distinct times over the semester-long course. The pretest was given prior to any exposure to formal origami instruction. The second set of tests, the posttests, were given at the completion of the course, prior to the final exam. All procedures set forth by the cognitive tests were followed, including prescribed time limits and parts to the tests. This differs from the previous study in that only one of the two parts of each spatial test was used due to the young age of participants. With a college-age sample, I was able to use the full, original form of the tests. (For more details, see the full report of the 2006 study [Boakes 09].)

2.4 Treatment

In the case of this study, the treatment is the course itself, GNM 2257 ("The Art and Math of Origami"). The course acronym, GNM, stands for General Natural Sciences and Mathematics and is one of five general studies areas that are part of the college's general studies program. All students' undergraduate degree coursework comprises at least 25% from the general studies areas, which include arts and humanities, integration and synthesis, interdisciplinary studies, social science, and natural mathematics and science. GNM courses seek to share the nature and processes of science through distinctive links to science and/or math.

This course used paper folding as a vehicle for teaching about mathematics, culture, and history. Over the span of four months, students attended class for two hours twice a week. During this time, they practiced, discussed, and learned origami models. At a minimum, students were folding at least four models per session. In the beginning, they were given direct instruction on how to read and interpret origami diagrams and printed instructions. As students built their comprehension of the specialized notation and terminology, attention shifted to the origami models' connections to other fields. Students learned a variety of types of origami, including unit, traditional, and action, and using material other than the traditional square sheet of paper (money, for example). They learned about connections to various cultures and history through readings from *Complete Origami* [Kenneway 87] and class discussions. An additional feature was the inclusion of "free exploration days," when class time was dedicated to paper folding. The only requirement on these days was that the students had to fold at least one model.

In addition to receiving folding instruction during class, students were asked to complete a number of assignments as part of their coursework. One assignment required students to fold a model and then compose a related essay based on an assigned theme (such as the significance often attached to specific origami models such as the crane). In addition, students were asked to keep a journal of their weekly readings, discussions, and thoughts as they progressed through the course. The largest assignment for students was the creation of their own origami collection based on specified guidelines. A minimum of five origami models were required for this collection. Each model had an assigned significance, including: one related to mathematics, one made of an unique material, one that challenged the student's folding abilities, one model of their choice, and one referred to as the "final project." The final project required not only the model but that the student also script, prepare, and digitally record an origami instructional video.

Origami was the major activity of this course. Through both in- and out-of-class experiences, the students completed many origami models, with some students using packs of 400 sheets of paper in only a few weeks. As the instructor of this course, I provided all origami instruction during class. Eventually, students became self-guided with intervention by a classmate or me only when help was needed. Models were used throughout the course to illustrate concepts or highlight discussion topics. With one of the topics being mathematics, about 20% of the course was spent exploring and describing the mathematics of paper folding. Not all the time was spent directly discussing mathematics concepts, the thought being that the act of paper folding would serve as a treatment for building spatial skills.

3   Results

I used several methods to analyze the data. Quantitative data consisted mainly of pre- and posttest results on the spatial assessments. Additional data were collected by converting qualitative data from the survey given at the start of the course. STEM and non-STEM majors were coded to allow for tracking students who may have come into the course with previous spatial ability. The spatial experience quantitative Likert-scale questions in the survey were collapsed and coded into *low* and *mid-to-high* spatial ability categories. Both indicators of spatial abilities allowed for an analysis of variance to explore how spatial experiences outside of the course might influence test results.

The original sample size for this study consisted of 25 students in each course over three consecutive spring terms. Data gathered in this time period were combined to provide a larger sample size of 75 total students. With only 25 students each term, the large sample allowed for more reliable generalization of results. Further, the tests conducted were voluntary. For those participants who opted out of one or more of the tests, the data points were eliminated from the analysis. Thus, the value of the sample varies from 69 to 70 subjects, depending on the analysis.

3.1   Comparing Pretest and Posttest Spatial Skills

The main question in this study was how a semester-long college origami course might influence participants' spatial ability. To begin the process of analysis, a paired-sample t-test was used. The categorical variable in this case was the time when each test was taken, one prior to the beginning

Test	n	Pretest	Posttest	Gain	Significance
Card rotation	68	120.03	130.31	10.28	$p < .0005$
Paper folding	70	11.66	13.21	1.56	$p < .0005$
Surface development	70	39.19	44.96	5.77	$p < .0005$

Table 2. Paired sample t-test results on spatial abilities.

course instruction and one after instruction was completed. The continuous, dependent variables were the results on the completed spatial tests.

A paired-sample t-test was run on the scores of each of the three spatial tests. In all three cases, there was a significant increase in spatial ability scores between the pretest and posttest. Table 2 shows the means and standard deviations for the pre- and posttests in each of three spatial areas. In each case, the p value was less than .0005. For the card rotation test, the mean increase between tests was 10.28 with a 95% confidence interval ranging from 4.13 to 5.31. In the case of the paper-folding test, the mean increase was 1.56 with a 95% interval ranging from 1.02 to 2.10. A mean increase of 5.77 was found for the final of three spatial tests, surface development. The 95% confidence interval ranged from 3.71 to 7.83. Beyond these results, the η^2 calculated for all three tests indicated a large effect size (card rotation, .20; paper folding, .33; surface development, .31). Overall, the differences between pretest and posttest were substantial.

3.2   Influence of Spatial Experiences on Results

Another question this research sought to answer was whether the amount of spatial experience students reported beyond school had any bearing on their test results. To answer this question, a one-way between-groups ANCOVA [Pallant 07] was conducted to compare the impact of the origami course on two types of students (low and mid-to-high spatial experience). The independent variable was the level of spatial background the student reported (low or mid-to-high level of experience), and the dependent variable consisted of the posttest scores on the spatial test administered after the course was completed. Participants' scores on the pretest, prior to the course instruction, was used as the covariate to control for preexisting differences in abilities. In cases where a participant did not complete one or more of the spatial tests, that participant's data set was eliminated from the analysis.

Preliminary checks were conducted to ensure that there was no violation of the assumptions of normality, linearity, homogeneity of variances, homogeneity of regression slopes, and reliable measurement of the covariate. In one case, the surface development test, the value of the Levene's test of equality of variances [Pallant 07] indicated that the assumption of

		STEM	Non-STEM	Low spatial experience	Mid-to-high spatial experience
Spatial test	n	19	49	34	34
Card rotation — Pretest	Mean	134.42	114.45	111.97	128.09
	SD	20.51	25.66	26.17	23.12
Card rotation — Posttest	Mean	145.58	124.39	124.62	136
	Adjusted mean	137.26	127.62	129.69	131.00
	SD	15.25	25.07	26.78	21.06
Paper folding* — Pretest	Mean	13.47	10.98	10.71	12.65
	SD	2.53	3.69	3.83	3.03
Paper folding* — Posttest	Mean	15.37	12.39	12.56	13.88
	Adjusted mean	14.14	12.86	13.28	13.17
	SD	2.39	3.31	3.54	3.04
Surface development — Pretest	Mean	47.58	35.43	37.47	40.18
	SD	8.92	14.37	14.28	14.05
Surface development — Posttest	Mean	51.58	42.00	41.76	47.59
	SD	8.80	14.47	16.30	10.10
*Significance shown on ANCOVA					

Table 3. Descriptive statistics by grouping variable and spatial test.

the equality of variance was violated. The nonparametric alternative, the Kruskal-Wallis test [Pallant 07] was then utilized to compare the posttest surface development scores.

Two of the three tests were analyzed using the ANCOVA. Results are shared in Table 3. For the card rotation test, the analysis of the data controlling for pretest scores revealed no significant difference between the two levels of spatial experience on the posttest scores, $F(1,64) = .17$, $p = .68$, partial $\eta^2 = .003$. There was a strong relationship between the pre- and posttest scores, however, as indicated by a partial η^2 value of .42. This translates into 42% of the variance of the dependent variable (posttest) being due to the covariate (pretest).

For the paper-folding test the same analysis was run. The ANCOVA again revealed no significance difference between the two levels of spatial experience, $F(1,65) = .04$, $p = .83$, partial $\eta^2 = .001$. However, a strong tie between pre- and posttest scores was found with 60% of the variance (partial $\eta^2 = .63$) in paper-folding posttests explained by the pretest results.

As noted, the ANCOVA could not be used for the surface development test because Levene's test revealed unequal variances. Using the nonparametric Kruskal-Wallis test as an alternative, no significant difference in scores among spatial ability levels was revealed (Group 1, $n = 33$: low experience; Group 2, $n = 36$: mid-to-high experience), $\chi^2(1, n = 69) = 1.61$, $p = .21$. The average-to-high spatially experienced students recorded a higher median score on the posttest (Md = 37.93) than the lower spatially experienced group (Md = 31.80).

3.3   Influence of Fields of Study on Results

Beyond grouping by the spatial experience self-reported by participants, students were also grouped based on their chosen majors. Students were identified as STEM or non-STEM, with STEM referring to any major in mathematics or the sciences. With the logic that STEM students would have more exposure to mathematics and science, both of which have a distinct connection to spatial skills [Newcombe 10], a second analysis using an ANCOVA would further explore the significance spatial experiences play in performance on these measures of spatial ability. Thus, an ANCOVA was run again to analyze how students' academic backgrounds may influence scores. All variables remained the same except that the covariate became the STEM or non-STEM designation of participants. Preliminary checks were conducted to ensure there was no violation of the assumptions associated with ANCOVA.

The ANCOVA on the card rotation and surface development tests revealed results similar to the previous ANCOVAs run as shown in Table 3. No significance on posttest scores was found when the pretests were taken into account. STEM students did not perform significantly better than their non-STEM counterparts (card rotation: $F(1,65) = 3.42$, $p = .07$, partial $\eta^2 = .05$; surface development: $F(1,65) = .001$, $p = .98$, partial $\eta^2 = .000$). Although this was the case, there was a strong relationship again between the pre- and posttest scores (card rotation: partial $\eta^2 = .38$; surface development: partial $\eta^2 = .61$).

In the paper-folding test, the ANCOVA did reveal a significant difference, with STEM students performing better than non-STEM students ($F(1,65) = 4.14$, $p = .03$, partial $\eta^2 = .07$). In addition, there continued to be a strong relationship between the pre- and posttests, with 57% of the variance in the posttest performance explained by the pretest performance.

4   Conclusion

During the past three spring terms, the Art and Math of Origami course was offered at Richard Stockton College in New Jersey. As professor of the course, I taught 75 students about the ancient art of paper folding. I gathered data each term in the hopes of quantifying how regular and repeated exposure to the art impacts students' spatial skills. The results are revealing and promising. Consider first the simplest of analysis, the paired-sample t-tests. Statistical significance from pretest to posttest was high on all three spatial skills tested. Now, one must be careful here. It cannot be said for certain that it is purely the origami instruction that resulted in such gains. A different, non–origami-related class taken the same term similarly might improve a student's spatial skills. The students might get their hands on a copy of the spatial tests ahead of time and study it. Their jobs outside of school might provide them with exposure to highly spatial tasks. Did these factors influence scores? In terms of the test, I can safely say that is doubtful. The test is secured and not easily obtained. However, the existing spatial skills of students may have some influence on scores.

To control for students' existing spatial skills, the next two analyses were completed. In both cases, students' spatial backgrounds were accounted for by using an ANCOVA test. This allowed me to determine whether outside exposure to spatial thinking might influence scores. Another advantage was the ability to control for preexisting differences in abilities and to safeguard results from being tainted by a nonrandom sample of students. The first set discussed dealt with students' perceived spatial skill levels. For both the paper-folding and card rotation tests, results showed the level of spatial ability to have no significant influence on posttest scores. The partial η^2 values did show that the variance in the posttest scores was at least partially explained by the pretest results. The nonparametric alternative to the ANCOVA analysis for the surface development test was similar. Perceived spatial practices outside of the course did not have a major impact on posttest scores.

The same set of ANCOVAs was run for STEM versus non-STEM students. In this case, the surface development and card rotation tests indicated no significant differences among posttest scores when pretest scores were taken into account. Interestingly, in one case, the paper-folding test, a significant difference was found: STEM and non-STEM students were significantly different in their performance, with STEM students doing much better on the posttest. Unique to this test is the fact that it is so similar to the actual act of doing origami (hence the test's name, paper folding). Although the sample of students involved, particularly those with a STEM concentration, was small, it supports an earlier finding that noted spatially specific tasks are beneficial for STEM students [Sorby 05].

Beyond the statistics themselves, it is helpful to look more generally at the results. Table 2 provides an overview of the three spatial tests and the scores associated with them. Scores were consistently higher on posttests than on pretests. The students differed in spatial experience levels, but both groups showed improvement. Even with the means adjusted for pretest differences, students performed better on posttests. One trend that contrasts spatial versus nonspatial students can been seen when focusing on how the posttest mean changes when it is adjusted for pretest results (as part of the ANCOVA completed). In both the card rotation and paper-folding tests, those with less spatial experience had greater means when pretest values were taken into account. Although purely speculative, it may be that nonspatial students benefit more from repeated and regular exposure to origami instruction.

When looking at STEM and non-STEM students, similar results were found. Regardless of students' exposure to coursework in the fields of math, science, and technology, students showed improvement in their spatial ability from pretest to posttest even when adjusted posttest means were considered. There was one case, for the paper-folding tests, in which STEM students showed significantly more improvement than their non-STEM counterparts. However, both groups showed a gain in their skills on all three spatial tests.

An element that also should be considered is the design of the study as a whole. With only one group of students to test, there is no ability to compare this particular method of instruction over another. In my past research, I was able to compare traditional instruction versus instruction that had origami lessons infused within it [Boakes 09]. Thus, it cannot be said how the origami activity from this course compares with other methods that may improve spatial ability, such as constructing mathematical models or working with computer software. Nonetheless, the data do show that origami has a positive impact on spatial skills. Future research with a variation on the design structure will have to be conducted to determine how the practice of origami might compare to another method of developing spatial skills.

This study was designed to extend my research on the impact of origami on students' spatial skills. In this case, the data offer further support for the use of origami as a tool for developing spatial skills. Regular and repeated exposure to origami benefited both the spatially strong and weak college-age students. As we know, spatial skills are useful in many avenues in life and serve as strong indicators of intelligence. Thus, origami provides one of many ways to help develop and strengthen spatial ability, even at the college level.

One must also consider the connection spatial skills have to mathematics. With spatial ability seen as a core element of geometry, origami is again shown to have promise in the development of mathematical skills. As the TIMSS data have shown, US students continue to perform poorly in the area of geometry [Martin et al. 08]. Perhaps the use of origami can help improve the learning of geometry in the mathematics classroom. It is my hope that my research sparks and encourages others to join me in trying to determine the mathematical and other benefits origami may provide the learner.

Appendix: Excerpt from the Student Survey

Name _____   **GNM 2257- Background Survey**

This survey is intended to gather some basic information relative to this course. It will not impact your grade in any way and is meant simply for the instructor to prepare for the variety of backgrounds likely in a general studies course.

1. Circle the appropriate choice:
    a. Gender   *male*   *female*
    b. Ethnicity   *African American   Asian Caucasian   Hispanic   other*

2. How many credits have you currently accumulated at Stockton? _____

3. What is your current major? _____

4. Have you ever folded Origami models before? _____

5. If yes to #4, rate how experienced you'd say you are on a scale of 1(very little) to 5 (very well versed) _____

6. Indicate whether you have participated in any of the following activities. If so, rate how often you perform them.

				Not often at all	Only a few times	Monthly	Weekly	Daily
a.	Video/computer games	Yes	No					
b.	Construction	Yes	No					
c.	Sports (formal or informal)	Yes	No					
d.	Drawing, painting, or other art form	Yes	No					
e.	Fix cars, bikes, motorcycles, etc.	Yes	No					
f.	Do an activity that requires the use of visualization of objects in 2- or 3-D Activity: _____	Yes	No					

Bibliography

[Boakes 06] Norma Boakes. "The Effects of Origami Lessons on Students' Spatial Visualization Skills and Achievement Levels in a Seventh-Grade Mathematics Classroom." EdD dissertation, Temple University, Philadelphia, PA, 2006.

[Boakes 09] Norma Boakes. "The Impact of Origami-Mathematics Lessons on Achievement and Spatial Ability of Middle-School Students." In *Origami⁴: Fourth International Meeting of Origami Science, Mathematics, and Education*, edited by Robert J. Lang, pp. 471–481. Wellesley, MA: A K Peters, 2009.

[Boakes 10] Norma Boakes. "GNM 2257 Art and Math of Origami." *Richard Stockton College of New Jersey*. Available at http://stockton.edu, accessed June 29, 2010.

[Ekstrom et al. 76] Ruth Ekstrom, J. French, Harry Harman, and Diran Derman. *Kit of Factor-Referenced Cognitive Tests*. Princeton, NJ: Educational Testing Service, 1976.

[Fleishman and Dusek 71] Joseph Fleishman and Ralph Dusek. "Reliability and Learning Factors Associated with Cognitive Tests." *Psychological Reports* 29 (1971), 523–530.

[Gardner 93] Howard Gardner. *Multiple Intelligences*. New York: Basic Books, 1993.

[Kenneway 87] Eric Kenneway. *Complete Origami*. New York: St. Martin's Griffin, 1987.

[Mann 05] Rebecca Mann. "The Identification of Gifted Students with Spatial Strengths: An Exploratory Study." PhD dissertation, University of Connecticut, Storrs, CT, 2005.

[Martin et al. 08] Michael Martin, Ina Mullis, and Pierre Foy. *TIMSS 2007 International Mathematics Report: Findings from IEA's Trends in International Mathematics and Science Study at the Fourth and Eighth Grades*. Boston: TIMSS & PIRLS International Study Center, 2008.

[Newcombe 10] Nora Newcombe. "Picture This: Increasing Math and Science Learning by Improving Spatial Thinking." *American Educator* 34:2 (2010), 29–35.

[Olson 84] Meredith Olson. "What Do You Mean by Spatial?" *Roeper Review* 6 (1984), 240–244.

[Pallant 07] Julie Pallant. *SPSS Survival Manual*, Third Edition. New York: Open University Press, 2007.

[Piaget and Inhelder 56] Jean Piaget and Barbel Inhelder. *The Child's Conception of Space*. London: Routledge & Kegan Paul, Ltd., 1956.

[Sorby 05] Sheryl Sorby. "Assessment of a New and Improved Course for the Development of 3-D Spatial Skills." *Engineering Design Graphics Journal* 69:3 (2005), 6–13.

[Terlecki et al. 08] Melissa Terlecki, Nora Newcombe, and Michelle Little. "Durable and Generalized Effects of Spatial Experience on Mental Rotation: Gender Differences in Growth Patterns." *Applied Cognitive Psychology* 22:7 (2008), 996–1013.

[West 97] Thomas West. *In the Mind's Eye: Visual Thinkers, Gifted People with Learning Difficulties, Computer Images, and the Ironies of Creativity.* Buffalo, NY: Prometheus Books, 1997.

Close Observation and Reverse Engineering of Origami Models

James Morrow and Charlene Morrow

1 Introduction

Suppose that you encountered the paper object shown in Figure 1—perhaps on the street, on a bar counter, in a classroom on a desk, ... And, because it was really cool and beautiful, you wanted to make one of your own, exactly the same—or maybe a little bit different. If you hadn't encountered anything like it before, what would you do? Perhaps ask some friends, do some googling, examine the object? But, suppose you couldn't find anything or anyone to help. Would you want to start to take it apart to see how it was made?

This paper describes ways to facilitate independent reverse engineering of such a modular origami piece (starting with something a bit simpler than the piece shown in Figure 1!) in a classroom environment. The paper is written for an audience interested in teaching problem-solving and learning strategies by guiding students through experiences with origami. The observation, exploration, and reverse engineering experience is designed to strengthen students' learning and problem-solving strategies and, more generally, to foster creativity. As teachers of mathematics, we seek to help young people become, in the words of the mathematician Keith Devlin, "innovative mathematical thinkers" [Devlin 10]. This paper is a preliminary report on an approach that shows much promise based on the student responses we have gotten to our methods; we will be instituting assessment procedures in future implementations.

Figure 1. A 120-PHiZZ unit Bucky Ball by Tom Hull [Hull 06] folded by Charlene Morrow. (See Color Plate IX.)

2   Rationale/Goals

The construction of origami models has been used in the service of education in many ways. It has been used to build an understanding of aspects of geometry, ranging from learning names of shapes to making geometric constructions of angles to exploring the vertex-edge-face structure of polyhedra and other polytopes [Gurkewitz and Arnstein 95, Burczyk and Burczyk 02, Kawamura 02, Maekawa 07] It has also been used as a setting in which algebraic expressions and equations are constructed and solved [Franco 99, Tubis and Mills 06]. It is used to develop such personal skills and attitudes as manual dexterity, persistence, and self-confidence. Origami construction can provide experiences of concrete models of abstract ideas and a rich source of problems and questions in higher-level mathematics that can be used by teachers of mathematics at many educational levels [Alperin 00, belcastro and Hull 02, Lang 02, Morrow 09, Dacorogna et al. 10]. This latter use is perhaps not thought of as an educational use, but it can be very useful to illustrate the interplay of concrete models of mathematical concepts and mathematical modeling of the real world. It can also be a wonderful way for teachers of mathematics to reconnect to their learning of mathematics, both for the joy of learning and to recall the frustration that inevitably accompanies learning mathematics.

What this paper illustrates is a strategy for working with students so that they can construct an origami model by reverse engineering an existing model. The strategy involves encouraging them to use experimentation, communication, and observation skills, rather than just giving them folding instructions. In doing so, students develop general skills and habits for (1) connecting conclusions to information and data, (2) communicating effectively, (3) solving problems, and (4) being more creative. For other uses of origami to stimulate creativity, see the work by Pope and Lam [Pope and Lam 09].

Critical to reverse engineering is what we call "close observation." Close observation is a term that perhaps originates in art history as a necessary skill of the discipline. It has more recently been used in the environment of an art museum by medical schools to develop clinical observation skills [Naghshineh et al. 08]. Much of the importance of close observation comes from the fact that it forces one to become more aware of just what it is that is observed, in contrast to what the observer "makes of" and concludes from it. Although such a distinction is perhaps impossible in principle, close observation forces one to be more aware of what one sees and to actually see more and see more accurately. Observation, close or not, is the starting point for any analytical reasoning.

What we mean here by *close observation* in the context of reverse engineering modular origami models is observation coupled with an organized and systematic touching, verbalizing of properties and qualities of the piece, communicating about it, taking it apart, and noting how units fit together.

When we use close observation as part of a mathematics class, our primary goal is to enable students to use structured observation as a powerful problem-solving and learning tool. As teachers of both university and school-level students, we have heard students many times express the desire, "Just tell me how to do it!" Often, if one can just get started on a problem, one can finish it easily. But, one usually isn't sure about where to start a problem if it isn't just an exercise. Thus, the popularity of the teacher-given "hint." Close observation is a way of starting to work on any problem. We have seen students, prompted we think by their experience with close observation, begin to ask probing questions, make conjectures, and try a multitude of strategies when unsure about how to start solving a particular problem or when stuck at any other point while trying to solve a problem.

Our belief is that this sort of careful and active observation also helps students take a creative approach to problem solving, learning, and construction of new origami models. Our hope, in terms of origami, is that the experience will help some students design their own modular pieces. Beyond our hopes for students to design origami, we hope that they become more confident and independent learners in other contexts.

3 Origins and Context

We developed the practice of reverse engineering and close observation for a long-running summer mathematics program for young women in secondary school, and we have used it with teaching staff, students in that program, and other students of widely varying ages. Our systematic use of reverse engineering modular origami began when we used it as a centerpiece of staff team-building in preparation for the summer program. Even though nobody actually completed the complex origami piece chosen, the experience was exciting, challenging, frustrating, and stimulating; it was wonderful as a team-building exercise, too. It was frustrating in a *good* way because part of the objective of the exercise was to feel some of the things that our students would feel in the program. Since that initial use in team building, we have used only origami pieces that are challenging but doable.

We were led to close observation by the desire to develop *problem-posing* skills among students, in the belief that *posing* problems, rather than *solving* problems, is perhaps the most fun and quintessential part of being a mathematician. Upon putting students into an environment where they are shown objects, patterns, and processes and asked, "What questions do you have?" and "What do you wonder about?" and "What do you think is true?" we realized that a significant amount of time was needed to be spent actually carefully *observing* the object/process. Without such observation time, there was not so much to wonder about and question. So, we structured class time to allow, and require, that a significant amount of time be spent observing, organizing observations, and discussing those observations. We then included close observation as a more structured aspect of reverse engineering. Subsequently, we discovered that medical students are now being asked to do such close observations as part of their clinical training, often utilizing paintings and sculptures in an art museum [Naghshineh et al. 08]. In the example of clinical training, close observation in an environment outside the clinic helps make the point of the value of taking more care in looking, rather than just a brief look until what is seen can be put into a category that the clinician knows about. Close observation in the context of mathematical problem solving is analogous to the in-depth interview and observation of a patient in clinical practice.

Since our initial forays into reverse engineering and close observation with origami, we have used variations on the practice with many age groups and situations, which we describe in more detail in Section 7.

4   An Approach to Close Observation and Reverse Engineering

This section provides a detailed description of the sequential steps for carrying out the reverse engineering of a unit.

4.1   Close Observation of the Modular Origami Object

We start by presenting students with a modular origami piece that they haven't seen before. After giving them a brief chance to take a look at the piece, we tell them that their eventual goal is to construct the object on their own, starting with squares of paper, but that they will begin reverse engineering by closely inspecting the object. Sometimes students work alone, but more often, they work in pairs or in a larger group, which fosters more communication and explicit verbalization of what is observed. Students working alone each get a model to work and play with; otherwise, each pair or larger group gets a single model. In what follows, for simplicity, we assume that the students are working in pairs.

We encourage students to hold and move the object, make notes about it and communicate what they see, in writing and orally, to a partner. We often ask them to make a rough sketch of anything they think might be especially important. We also encourage them to form conjectures about the piece, list its properties, and ask questions about it. In describing properties of the object, the need for vocabulary that is more technical than that used in ordinary English often comes up. We encourage sticking to ordinary English unless they are sure of what a technical term means. Later, we discuss technical terminology and its role in science and mathematics. We suggest that observation of color, as well as shape, be made. Students make a written record of their observations, organizing them in some way that seems reasonable to them. The listing of properties of a piece takes on additional importance when students go beyond reverse engineering to invent a new piece; i.e., to re-engineer the piece.

4.2   Deconstruction of the Piece

Once the entire modular piece has been observed and the observations recorded and discussed, the delicate process of taking the piece apart begins, with the tension of wanting and not wanting to take it apart. Some students fear being unable to put it back together; others are impulsive, plowing ahead with no thought about putting it back together. We use this tension to encourage patience and notetaking of what they see as they carry out the disassembly. Of course, some who have very good visual memory and good observation skills may not need to make written notes for the purpose of remembering, but we encourage the written notes as a way of communicating to others what they see. Before taking on understanding how the unit is folded, we generally ask that they put the piece back together, and then take it apart once more.

4.3   Close Observation of the Unit

The process of describing what they see begins anew with the units of the piece. We ask students to observe and describe the color patterns and symmetry of the unit as well as how one unit fits into another. We provide some scaffolding of thinking and terminology at this point by rephrasing, sometimes using terms such as *tabs* and *pockets*.

4.4   Unfolding the Unit

Then unfolding of the unit begins, with an objective of being able to replicate the unit starting with an uncreased piece of paper. The objective also is to identify points and lines on the unfolded unit that have a simple way of being folded or way of being used to make other creases. We stress that the idea is not just copying the crease pattern by "eye-balling" it, but, rather, to make precise folds by using landmarks, such as folding a line segment in half or folding a line that joins two points or folding one point onto another point. In this way, one should be able to fold the unit without using the original as a template.

4.5   Re-assembly

The fun gets serious as folding begins: How do you actually get exactly the right crease pattern and fold so the units are copied? There are many potential difficulties. Hints may be given and techniques shown, depending on the experience of the student and the difficulty of the piece. Ideally, the students work independently, but this happens only if the piece is chosen to fit the students' experience very well. It often isn't possible to tell the folding order just from observation, and in this case we give considerable guidance and hints.

4.6   Reflections and Revisions: Re-engineering

We now assume that the reverse engineering has been successful, and the students have been able to construct an exact copy of the original piece given to them. Or, perhaps some students have made conscious variations. In either case, we next propose that each group or individual try to construct a variation (or a larger variation than they have so far done) on the model they have. We encourage them to go beyond reverse engineering an exact copy to *re-engineering* a different, but related, modular piece—or maybe not even modular. Some may have ideas for variation. Others may have no idea, in which case we have some prompts to stimulate thinking. Our prompts come primarily from the book on the art of problem posing [Brown and Walter 90]. The idea is to reflect on things one has observed and think about the properties of the finished piece, or think about the construction process, the folding of the unit, the shape or color of the units, or anything else about constructing the modular piece, and then ask, "What if some one thing, or more than one, changed?" "What if not?" "Could we still form some object of interest?" Once a "what if not" idea is identified, the students try to carry out some alternative to "what is."

5   An Example

We now give a brief description of a close observation and reverse engineering experience we had with one group of 18 high school students, all of whom had, at most, very little paper folding experience.

The students were participants in the four-week summer program that we direct and teach. Prior to their first close observation and reverse engineering (CORE) experience, we had conducted several sessions in which they worked in pairs to solve some "word problems," i.e., written problems involving real-world objects. In those sessions, the students had been given no instructions for solving the problems other than to talk about how they were going about solving each problem and to draw a picture to represent their solution process. They already had experienced the frustration—which we expect—of not being given instructions for getting solutions.

For their first CORE experience, we presented each pair with the modular origami polyhedron with six faces shown in Figure 2. In talking to students as they conducted their observation of the piece, we noted that there was much talk about the colors present, the number of colors, the pattern of colors, and the idea that the piece is a polyhedron, but not so much discussion on the number of faces, edges, or vertices, nor how many edges

Figure 2. The modular piece to reverse engineer.

Figure 3. One unit of the model in Figure 2.

and faces met at a vertex. We asked the whole class to share observations and then to move on to disassembling it, again asking for observations, now focusing on how the pieces were joined. This time, in circulating about the room, as students talked about how one part of a unit fit into another part of another unit, we introduced the idea of *tab* and *pocket* to give a little more structure to their observations. We always find it difficult to decide when to introduce terminology, in this case choosing to err on the side of more rather than less imposition of language because the specific terms in this situation have the same meaning colloquially as they do technically. For this object, the disassembly itself is fairly straightforward, which we believe is important for the first CORE.

After a brief discussion in which students shared their thoughts about the process, we asked them to put the pieces back together. Most students completed that task easily; for the others, we asked them to refer to their notes and thoughts about how they had taken the pieces apart and what they had noticed about how one piece fit into another. We also made some observations about how to identify congruence of tab and pocket by fitting a tab of one unit onto the outside of a pocket before inserting the tab into the pocket.

Once everyone had done the reassembly, we asked the students to take it apart again, and then focus on observing the unit (see Figure 3) to note its characteristics—another close observation.

After sharing characteristics that they observed about the unit, the students unfolded the unit; we asked that they record any obvious order to the unfolding they might notice. On sharing what they recorded, they noted that there seemed to be several different orders to unfolding and wondered whether there were better and worse orders to follow in the folding to come. Figure 4 shows the crease pattern on the entire square, which we asked students to describe in words and to sketch.

Figure 4. The unit completely unfolded.

We then asked the whole class whether there were any creases that they could make immediately with their "blank" square. Several students observed the two diagonals and two "book folds" (not their terminology). Students generally referred to the book fold as "folding to the middle," to which we asked, "*What* is folded to the middle?" Our goal was to encourage some precision of language, though not necessarily so much precision as to avoid *any* ambiguity—and not conventional mathematical or origami terminology. We suggested that the students construct the diagonal folds and book folds, and then unfold. Some then observed the "rows of little squares" lined up along the diagonal in the model unit, and then one student noticed and shared the sequence of alternating mountain and valley folds heading off on a diagonal in one half of the square. Once these folds had been observed, by looking at the crease pattern of the unfolded unit, it was relatively easy for everyone to make all the remaining folds. They then folded the other two units and, with some help, completed the piece successfully.

All of the students were very excited to have made the complete modular piece—especially because the modular piece had seemed rather complicated when first viewed. Many were surprised that they actually could do it. They shared their methods of folding, and, again, they were surprised to see that there were several different ways of making the necessary folds. There was a whole-class discussion about the several different methods students used to fold the units. Students talked about folding that was easier, simpler, "cool," or surprising; we guided the discussion toward explaining what *made* one method seem easier than another or cool, maintaining a focus on observation and conclusions made from observation.

We would like to have continued with "what if not?" where we ask students to look at characteristics of the units, colors, the folding process, etc., and consider what would happen if it did *not* have some characteristic.

They would then imagine some alternative and carry out an experiment in which they see what would happen if they used such an alternative. For example, one could use something as simple as different colors, or start folding color side up (on all or some of the units), or use more or fewer than three units. However, on consideration of the time constraints of the program, we decided, with regret, that we needed to forgo this final stage.

6 Close Observation and Reverse Engineering in Learning and Problem Solving

We believe that the habits of observing closely, actively, and carefully are of extreme importance to lifelong learning. Being a good learner and problem solver is aided immensely by reading information carefully, making notes, organizing information, asking questions, and drawing sketches to represent information. However, many classroom practices and societal pressures are not conducive to taking the time needed to make observations in such a time-consuming way. In many mathematics classes, we know that speed and learning "tricks" are often stressed. We have often heard students praise a teacher who made things *easy* for them. So, CORE is counter to most of many students' prior educational experience. Despite such cultural pressures, developing such skills and applying them in the engaging process of reverse engineering an origami piece has proved to be very stimulating to most people with whom we have worked. They seemed quite motivated to engage in such an exploration—at least once they got over the idea that we should just tell them how to do it!

The method described above, in which we first engage students in the replication of a modular origami piece and follow with re-engineering, is well suited to developing creativity in mathematics. We believe that there is great potential to the "What if not" aspect, and we will try it in our college geometry classes. Using CORE with re-engineering challenges the "only one way to do it" idea that is the take-home message of many mathematics classes. Our own experience of doing the things we ask of our students has been one of stimulating our own creativity as well as appreciating the artistry of origami.

One caveat to the above statements about our experiences is that nearly all of our participants have been people who *chose* to participate in the activity. Exceptions include the very youngest, who were in a Montessori school, some students in a middle school class, and the students in our summer program, which focused on mathematics. So, although not all of our participants had chosen an origami activity, they all had made some commitment to learning and to the larger educational goals.

...liked making the hexagon box	...forced me to concentrate more	...liked all of the stuff we made
...liked that I could fold boxes that will be awesome as gift boxes!	...got confused and I liked figuring out that that flap goes into that wall!	...projects were challenging but fun
...helped us make beautiful origami!	...liked that we made so many things that we could take home	...liked making boxes
...liked the boxes that stand on each other	...liked getting instructions	...liked taking on the different challenges of folding
I made some very hard masterpieces because I never saw it coming!	I loved all the origami	

Table 1. Student responses to evaluation question about their origami experience from the 2009 SEARCH program (14 of the 18 students responded to this question).

7 Our Observations of CORE

Our CORE experiences have included a group of 15 pre- and elementary school children, another group of 5 students aged 8–11 along with two teachers, an after-school origami club consisting of 8 high school students and a teacher, a group of 18 young women participating in the program for high school students that we mentioned above, and a group of 10 of that program's adult staff members, whom we will call students as well. Everyone became engaged in the reverse engineering aspect of CORE and all made observations during the process, most in a somewhat less-structured way, by way of speaking, rather than writing as in the scenario envisioned above. At least two teachers were present and interacting and/or observing each of the sessions.

In these sessions we observed the following behavior of students engaging in CORE activities:

1. All students were active in the process of observing and in communicating their observations.

2. With the exception of some of the adult group members, all students completed the modular piece.

3. More than 90% of the students expressed satisfaction and pride in what they accomplished.

Our most formal evaluation of the effects of these processes comes from the 2009 summer program. We asked students for feedback on the origami sessions, which included styles of teaching other than CORE. None of the 18 students had any negative feedback. Fourteen of them were positive about their experiences (Table 1). Of the remaining four individuals, two expressed positive feelings and the other two negative feelings about origami in speaking to staff during the program. Such feedback provides some documentation of CORE's potential, admittedly minimal, but encouraging at this stage. We are much more encouraged by what we saw and heard from students as they did origami. We also asked students about our much more general goals of becoming confident, persistent, and independent learners. We have many detailed positive responses to that query, but they are in response to the program as a whole, which included several activities other than origami. Thus, we do not include them in this preliminary report. No formal evaluation was done for the other CORE experiences, which were all one-time sessions.

8 Challenges, a Question, and Next Steps

We have identified three challenges to using CORE:

1. Taking a significant amount of time to record, organize, and discuss their observations is frustrating to many people, and it is a bit out of step with "faster is better" cultures.

2. The first reverse engineering session should be one in which everyone is successful in completing the given piece; it is fine for people to be frustrated *while* going through the process, but we do not want to turn people off from trying something difficult, nor do we want to jeopardize future working relationships with anyone.

3. For some, engaging in the close observation, particularly if they don't come from a culture of close observation, is counter to their intuition and their experience of "how things should work." Furthermore, some students ask, "What am I supposed to observe?" It is important not to stall out at this point.

To address each of the challenges, we propose the use of short, simple reverse engineering projects initially. We, of course, need to choose as compelling an origami piece as we can, but fortunately there are many good choices! By choosing initially compelling pieces with fewer steps, we can guide students gradually into being more willing to take more time, tolerate frustration, and learn to take time and care with observations. In general, the more students find the piece to be something they would really like to make, the more they will have the patience for a lengthy observation. The example of Section 5 illustrates the point that even in this very simple modular piece there is great potential for students to feel initial nervousness, be surprised that they can actually do the re-engineering, and finally, feel a sense of accomplishment. In contexts where more origami and more reverse engineering are being done, the challenge level should increase, at least for some students, beyond that of the initial encounter.

Concerning the question of the students who ask, "What am I supposed to observe?" the teacher can reiterate that students can look at shape, color, number of faces, etc., while not giving specifics; continue to encourage them to keep trying; and note that we are not taking away any of their learning strategies, but asking them to try something new that they will be free to drop if they don't find it useful or practical. In all of our uses of CORE, in which students are allowed time to think and encouraged to persist, all eventually made some useful observations.

Constructing origami objects provides good and natural incentive for developing patience and persistence due to the wide appeal of colorful and artful origami. Not everyone is interested in learning to make origami pieces, nor even finds origami interesting, and we are not suggesting that origami has universal appeal. What we are saying is that using origami expands greatly both the number of students who become engaged by a mathematical investigation and the number of modes through which they can learn mathematics.

9   Conclusion

We have been asked about CORE, "*Who* is this kind of stuff for?" We would like to say, "Well, everyone"—in part because the question seems to presume a slotting into categories with which we feel uncomfortable, and in part because our experience has been so overwhelmingly well received. However, we'll interpret the question to mean, "Are there people this won't work with—and if so, who are they and what do you do about that?" The honest, but misleading, answer to this variation on the original question is that no student to this point has expressed anything but enthusiasm for the activity. What this response ignores is that every participant has had an interest in origami, openness to origami, or a commitment to the larger endeavor. For example, at least one adult had anxieties about doing origami but suppressed that, because he believed, as we do, that going through the process would be very helpful to his teaching. Going back to the original question, we'll say, "You can't just pull people off the street and expect them to get much out of the CORE experience. However, if they have committed to an educational experience and the expectations and goals are explained up front, we believe that most anyone will benefit from the experience." That said, we should note that the experience requires a bit of manual dexterity. But we can help out there, by providing a couple extra hands. Of course, such exceptions would require accommodations for many activities other than the CORE activities.

Most important to making good use of CORE is to make a good determination of the level of challenge needed for each student. To do so requires us to talk with, carefully listen to, and observe each student. In a classroom, there is always a diversity of abilities, skills, ways of learning, and knowledge. Some students have difficulty with persisting in a challenging task, others tend to be impatient with an observation process, some work quickly, and others work more slowly. As teachers, we need to make the process engaging and appropriately challenging for all; as a practical matter, this means developing for CORE a variety of models that address the diversity of skills, quickness, and patience of our students.

Bibliography

[Alperin 00] Roger Alperin. "A Mathematical Theory of Origami Constructions and Numbers." *New York Journal of Mathematics* 6 (2000), 119–133.

[belcastro and Hull 02] sarah-marie belcastro and Thomas Hull. "A Mathematical Model for Non-flat Origami." In *Origami³: Proceedings of the Third International Meeting of Origami Science, Mathematics, and Education*, edited by Thomas Hull, pp. 39–51. Natick, MA: A K Peters, Ltd., 2002.

[Brown and Walter 90] Stephen Brown and Marion Walter. *The Art of Problem Posing*, Second Edition. Hillsdale, NJ: Lawrence Erlbaum Assoc., 1990.

[Burczyk and Burczyk 02] Krystyna Burczyk and Wojciech Burczyk. "Exploring the Possibilities of a Module." In *Origami³: Proceedings of the Third International Meeting of Origami Science, Mathematics, and Education*, edited by Thomas Hull, pp. 257–267. Natick, MA: A K Peters, 2002.

[Dacorogna et al. 10] Bernard Dacorogna, Paolo Marcellini, and Emanuele Paolini. "Origami and Partial Differential Equations." *Notices of the American Mathematical Society* 57 (2010), 598–606.

[Devlin 10] Keith Devlin. "Wanted: Innovative Mathematical Thinking." *Devlin's Angle*. Available at http://www.maa.org/devlin/devlin_07_10.html, July 2010.

[Franco 99] Betsy Franco. *Unfolding Mathematics with Unit Origami*. Emeryville, CA: Key Curriculum Press, 1999.

[Gurkewitz and Arnstein 95] Rona Gurkewitz and Bennett Arnstein. *3-D Geometric Origami Modular Polyhedra*. New York: Dover Publications, 1995.

[Hull 06] Thomas Hull. *Project Origami: Activities for Exploring Mathematics*. Wellesley, MA: A K Peters, Ltd., 2006.

[Kawamura 02] Miyuki Kawamura. "Origami with Trigonometric Functions." In *Origami³: Proceedings of the Third International Meeting of Origami Science, Mathematics, and Education*, edited by Thomas Hull, pp. 169–178. Natick, MA: A K Peters, Ltd., 2002.

[Lang 02] Robert Lang. "Polypolyhedra in Origami." In *Origami^3*, edited by Thomas Hull, pp. 153-167. Natick, MA: A K Peters, Ltd., 2002.

[Maekawa 07] Jun Maekawa. *Genuine Origami*. Tokyo: Japan Publications Trading Company, Ltd., 2007.

[Morrow 09] Charlene Morrow. "How Many Ways Can You Edge-Color a Cube?" In *Origami⁴: Fourth International Meeting of Origami Science, Mathematics, and Education*, edited by Robert J. Lang, pp. 517–524. Wellesley, MA: A K Peters, Ltd., 2009.

[Naghshineh et al. 08] Sheila Naghshineh, Janet P. Hafler, Alexa R. Miller, Maria A. Blanco, Stuart R. Lipsitz, Rachel P. Dubroff, Shahram Khoshbin, and Joel T. Katz. "Formal Art Observation Training Improves Medical Students' Visual Diagnostic Skills." *Journal of General Internal Medicine* 23:7 (July 2008), 991–997.

[Pope and Lam 09] Sue Pope and Tung Ken Lam. "Using Origami to Promote Problem Solving, Creativity, and Communication in Mathematics Education." In *Origami⁴: Fourth International Meeting of Origami Science, Mathematics, and Education*, edited by Robert J. Lang, pp. 517–524. Wellesley, MA: A K Peters, Ltd., 2009.

[Tubis and Mills 06] Arnold Tubis and Crystal Mills. *Unfolding Mathematics with Origami Boxes*. Emeryville, CA: Key Curriculum Press, 2006.

Origami and Learning Mathematics

Sue Pope and Tung Ken Lam

1 Introduction

Many countries are modifying their school curriculum to include a greater emphasis on problem-solving, creativity, and developing robust knowledge and technical skills (e.g., England, Wales, Northern Ireland, Scotland, Ireland, South Korea, Japan, the Netherlands, Norway, and New Zealand). The International Review of Curriculum and Assessment Frameworks contains a summary of more than 20 national curricula and their priorities [INCA 10].

Practical and purposeful collaborative activity is an essential component of the implementation of such curricula, in line with the theories of social constructivism [Wood 98]. In England, video case studies [QCDA 09] designed to support teachers implementing the National Curriculum include using origami with 13-year-olds. In these case studies, the students worked together in groups and the lesson was led by older students.

Origami is an accessible activity that helps develop fine motor skills for young children (see Froebel's work on kindergarten education [Heewart 92]) and is a source of problem solving and creativity across the entire age range (e.g., [Frigerio 02, Haga 02, Cornelius and Tubis 09, Pope and Lam 09, Hull 06]). These examples show how origami can support understanding of essential geometric principles such as symmetry, similarity, and congruence. It can also support fundamental mathematical notions of division and ratio and provide contexts for developing sophisticated mathematical arguments (e.g., [Leroux and Santos 09]).

Many teachers find that folding paper is a motivating context for learning (e.g., [Golan and Jackson 09, Pope and Lam 09, Boakes 09, Wollring 03]). They find that students will persevere to achieve success and will want to share that success. These outcomes resonate with our experience of working with hundreds of students in mathematics classrooms and extra-curricular workshops. We have found that encouraging collaboration and discussion among students of all ages can help to develop confidence and understanding alongside success.

However, several problems may have to be overcome on the way. In this paper we identify some of the challenges that arise when using origami as a context for learning and how these challenges might be used to enable the development of mathematical understanding. We also suggest some ways that origami can be used when teaching mathematical concepts that are notoriously difficult to understand.

2   Using Origami in Mathematics Lessons

Origami offers a potentially rich starting point for learning mathematics. In this section, we discuss possible approaches and common practical problems based on our extensive experience with a wide range of learners.

2.1   Learning to Fold

Many teachers are unsure about how to use origami in mathematics lessons. We believe that a range of approaches will be most likely to be effective for the majority of learners. This approach contrasts with that of Golan and Jackson, who propose a fixed set of principles in their method of *Origametria teaching* [Golan and Jackson 09, pp. 461–462]:

1. The way it is taught—no criticism of folding.

2. We never check the accuracy of folding.

3. Our teachers never touch a student's model—teacher repeats the demonstration with their own paper.

4. The choice of model—teacher explains the folding procedure while the students follow.

5. Positive reinforcement.

6. The model is never named while being folded.

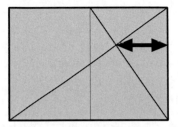

Figure 1. A fraction challenge: "The marked length is what fraction of the rectangle's long edge?"

(We call principle 4 *demonstrability*.) Some of these principles resonate with our own approach: e.g., avoid criticizing a learner's folding, use positive reinforcement, and make careful selection of models.

When using origami for teaching mathematics, our criterion for the choice of model is rarely based on demonstrability but rather on the model's mathematical potential. Similarly, we do not always follow the principle of never naming a model while it is being folded. The *surprise* of the result can be enjoyable, but we regularly adopt the approach advocated by Wollring, who recommends providing a pair of completed models that learners are challenged to recreate by working collaboratively [Wollring 03]. In our experience with students and teachers, this approach fosters creativity and cooperation. We also offer challenges, such as that described by Hull [Hull 06], in which students (typically, 14-year-old and above) are invited to fold an equilateral triangle from a rectangle, or having made some folds, they are challenged to prove the apparent properties, as in Figures 1 and 2.

We regularly check the accuracy of folding, but sometimes we assist learners or recommend they seek peer assistance when they request help. One strategy we find particularly valuable is encouraging "those who can" to assist "those having difficulties." This is particularly important in work-

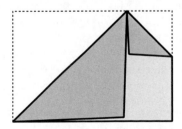

Figure 2. A challenge starting from a 1:$\sqrt{2}$ rectangle: "Prove that this figure is a kite. What is the perimeter (as a multiple of the shortest side)?"

chop sessions in which the participants have a limited time to achieve success.

We believe that it is not necessary for teachers to adopt a rigid set of principles to make effective use of origami in their teaching of mathematics.

2.2   Common Folding Problems

The seven problems described here are not restricted to children; they arise with a range of learners, including teachers and university students.

1. *Soft folding:* Most geometrical folds require sharp creases made with a finger or thumbnail.

2. *Inaccurate folding:* This is generally caused by failing to line up edges precisely. For example, when folding a diagonal of a square, some learners try to make a fold through two corners, even though it is much easier to join opposite corners together and make the crease by sweeping the paper flat away from the joined corners.

3. *Difficulty manipulating paper:* Beginners benefit from folding on a hard surface because of the extra support provided for the paper. Positioning the paper is important. For example, to fold a square into a 2-by-1 rectangle, it's easier to "book fold" with the "spine" closest to the body, not farthest away.

4. *Halving:* Many learners can halve an edge, but halving an angle often presents particular challenges, maybe because they do not appreciate that an angle is "at a point" and to bisect the angle the edges that meet at the point must be brought together through it.

5. *Reverse folds:* Beginners often find it difficult to make reverse folds, partly because their original crease is too soft. Folding back and forth will help to reinforce the crease and achieve success.

6. *Mirror image modules:* Many modular models, such as the Sonobe unit, need to be folded in a consistent way to ensure they will fit together [Kasahara and Takahama 87]. It is very common for learners to end up with a mixed set of right- and left-handed units that won't fit together. The shuriken star [Petty 01] is invaluable for exploring this difficulty, as it is made of two units that *are* mirror images.

7. *Visualization and understanding instructions:* We encourage students to "help your neighbor" and "if you can do it, help someone else" so that the responsibility for success is shared throughout the group.

Some problems can be compounded by poor choice of paper—too small, too big, too soft, too thick, etc. When adopting Wollring's approach [Wollring 03], it is helpful to use paper that is identical to that used in the sample models. A larger piece of paper needs to be used for demonstration, but it needs to be the same type as that used by the learners (e.g., only duo paper should be used when students are using duo paper (and vice versa)). Even if the model will work with any sized rectangle (e.g., the magazine box [Petty 98]), the paper for demonstration should be geometrically similar to that used by the learners.

Our suggestions are for basic folding techniques, which if absent can inhibit learner engagement in the *mathematical* activity. A number of videos available at the British Origami Society website may help address some of the problems discussed [BOS 10].

3 Using Origami to Teach Conceptually Demanding Mathematics

In this section we explore how origami can be used when teaching conceptually demanding mathematics, namely, angles, polygons (especially triangles), symmetry, and fractions. We finish this section by discussing the use of origami as a stimulus and motivator for the important mathematical process skills of reasoning and proof.

3.1 Angles

Learners at all levels, including teachers, can look at an angle that is 60°and say that it is 45°. We believe that a lack of varied and practical experience leads them to expect that folding rectangles and squares inevitably leads to angles of 45°. While learners from the age of six can easily halve lengths and shapes, they often struggle to halve angles. Despite being introduced to angles as a measure of turn from early primary school, their conceptual understanding of angles is weak, and they do not appreciate that it is "at a point" and the edges through the point need to be brought together through that point in order to halve the angle.

The practical activity of moving through a full turn, half turn, and quarter turn is rarely associated with a classic diagram of an angle (i.e., a point with two rays). Young children are likely to fold squares in half as an introduction to fractions and to discover there are two ways in which this can be done, but the geometric properties of these halves are rarely discussed (Figure 3). This omission is typical of classroom practice in England, where the teaching of different topic areas frequently is not linked and there is relatively little emphasis on teaching for understanding [Ofsted 08].

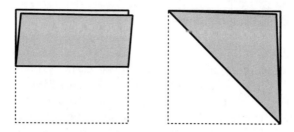

Figure 3. Exploiting folds for geometric learning: "Fold a square in half; what are the geometric properties?"

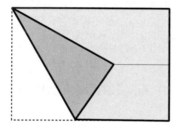

Figure 4. Developing understanding of halving angles: "How do you halve the obtuse angle?" (Figure 6 shows the result.)

Using models that require the bisection of angles and drawing attention to how this has been done, including unfolding the paper to notice the location of the creases, can help to address this difficulty. Folding an equilateral triangle is a particularly fruitful activity in this regard [Hull 06]. As well as folding 60°(not 45°), bisecting the obtuse angle, which most learners will recommend as a means of obtaining a further 60° angle, makes it possible to reinforce what halving an angle means (Figure 4).

3.2 Polygons

Many learners have almost no practical experience with manipulating shapes and using the associated mathematical vocabulary correctly. It is not uncommon for a child to insist that a rotated square is a diamond (Figure 5). This response may, in part, be due to the different meanings associated with words used in mathematics compared with colloquial use [Pimm 87]. "The properties of any geometric shape are those features which remain invariant for that shape" [Hopkins et al. 04, p. 162]. The naming convention for polygons is based on their distinguishing properties. A square has distinct geometric properties: it is a four-sided regular polygon with equal lengths and equal angles. On the other hand, a diamond is a rhombus made of two equilateral triangles. A rhombus has four equal sides and equal opposite

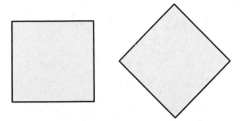

Figure 5. A question to stimulate discussion about the invariant properties of shapes: "Is this a square and another square—or is it a diamond?"

angles. A diamond is a member of the polyiamonds family [O'Beirne 61], made from a number of congruent equilateral triangles joined edge to edge.

Similarly, learners will regularly assert that a square is not a rectangle. They do not appreciate that rectangle means "right angles" and therefore both squares and oblongs are included.

Some learners do not appreciate the essentially "double-barreled" naming of triangles—thinking that there are only four types of triangles: equilateral, isosceles, scalene, and right-angled. They don't understand that triangles have names associated with lengths (equilateral, isosceles, and scalene) and names associated with the greatest angle (acute-angled, right-angled, and obtuse-angled).

Not all nine possible triangles exist on a plane (see Table 1). For example, equilateral triangles are acute-angled, and right-angled and obtuse-angled equilateral triangles are not possible on a plane. However, a thought experiment on a sphere shows that this is simply a restriction of the plane: imagine you're at the equator, go directly to the pole (either will do), turn right, go back to the equator, turn right, go back to where you started—assuming Earth is spherical, you have just traveled around a right-angled equilateral triangle!

Many learners think that all polygons are convex and some even believe they are all regular; when they fold a shape with an interior angle that is reflex, they may not even appreciate that they have a polygon. The folding

	Equilateral	Isosceles	Scalene
Acute	1 only	Y	Y
Right-angled	N	1 only	Y
Obtuse-angled	N	Y	Y

Table 1. Classification of triangles on the plane by angle and length: N means that no combination exists (e.g., right-angled equilateral), and Y means that many versions of the combination are possible (e.g., acute-angled scalene triangle).

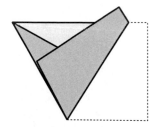

Figure 6. A concave pentagon—a rare species of pentagon in schools.

of an equilateral triangle is useful again in addressing this issue, as the penultimate shape, produced by bisection of the obtuse angle, is a concave pentagon (Figure 6).

Encouraging the correct use of mathematical language and emphasizing geometric properties when discussing origami models can be very helpful in avoiding these misconceptions.

Origami models of a range of geometric objects in both 2D and 3D provide valuable starting points for exploration, both of the geometric properties and why they arise.

3.3 Symmetry

In our experience, many learners do not appreciate that folding a shape with bilateral symmetry into matching halves gives a mirror line (see, e.g., Figure 3). Origami provides a practical experience that helps to reinforce the concept of symmetry.

Any fold using bilateral symmetry automatically generates a line of symmetry, albeit possibly localized. Investigating the number of different ways a regular polygon can be folded in half allows learners to discover for themselves that an n-sided regular polygon has n lines of symmetry, whereas other polygons might have no lines of symmetry at all (e.g., a parallelogram or a scalene triangle). Unfolding can help to draw attention to and aid understanding of this important geometric property. As many origami models are symmetric, either reflective or rotational, origami provides an excellent practical context through which to develop conceptual understanding of symmetry.

Folding many origami models requires working symmetrically. Generating identical units to make a modular origami model, for example, making a cube with Sonobe units [Kasahara and Takahama 87], means learners quickly discover the importance of ensuring they are producing identical units if their units won't fit together, even when well folded. The shuriken (Ninja star) is a great model for teaching symmetry, as it is made of two units that are mirror images [Pope and Lam 09].

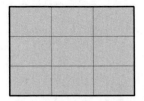

Figure 7. A rectangle folded into a 3-by-3 grid is a practical context for fractions.

3.4 Fractions

Fractions are a notoriously difficult topic for children to learn. Despite improvements in performance in international comparisons and national tests, it appears that English children's understanding of fractions has changed little since the 1970s [Hodgen et al. 10]. The question "What fraction is this?" arises naturally in many paper-folding contexts.

Folding any rectangle into a 4-by-4 grid allows exploration of halves, quarters, eighths, and sixteenths, equivalence, and many other important relationships. It is possible to begin arithmetic of fractions using such a simple resource.

Folding a regular hexagon from an equilateral triangle [Pope and Lam 09] gives easy access to thirds and ninths, as does folding a square into a 3-by-3 grid [Haga 02]. In the practical context of origami, it is relatively easy to establish that one way of understanding fractions is through appreciating the relationship between part and whole and which is which (Figure 7).

3.5 Reasoning and Proof

The role of proof in mathematics has been important since the time of the ancient Greeks [Davis and Hersh 81], and requires logical deduction to reach a conclusion based on a set of starting principles. Origami can provide a valuable starting point for developing mathematical reasoning and proofs. For example, one way of folding the diagonal of any oblong is to join opposite corners and pinch the ends of the crease. Joining the pinches will give the diagonal of the most oblate rectangle (Figure 8). Challenging

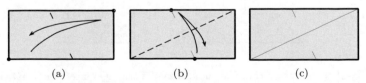

(a)   (b)   (c)

Figure 8. A motivator for proof: folding the diagonal of a rectangle exactly by "pinching" landmarks.

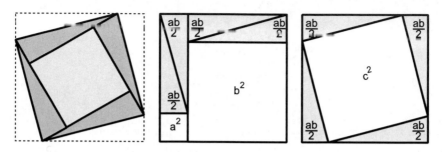

Figure 9. Visual demonstration of the Pythagorean theorem: $c^2 = a^2 + b^2$

learners to explore the question "Why does this folding sequence work?" requires understanding of geometric properties and the construction of a logical and convincing argument.

Using a paper cup with undergraduates [Frigerio 02] and Origamics [Haga 02] can provide accessible starting points for rigorous mathematical activity. The two examples mentioned in Section 2 (Figures 1 and 2) are accessible to many 14-year-old students and lead naturally to proof. There are many paper-folding exercises that illustrate important mathematical results, such as the Pythagorean theorem (Figure 9) or the angle sum of a triangle (Figure 10) and the area of a triangle [Row et al. 66].

The folding sequence is merely a demonstration, and while it may be convincing to the student, and even to a friend, it will not "convince an enemy," using John Mason's hierarchy of mathematical argumentation [Mason et al. 82]. The experience of folding the artefact can provide insights into possible starting points for the development of a rigorous mathematical proof, which *will* "convince an enemy."

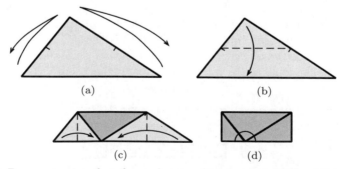

Figure 10. Demonstration that the angle sum of a triangle is 180° and that the area of the triangle is double half the base length multiplied by half the perpendicular height.

4   Conclusions

Careful choice of origami models such as those mentioned in this paper can enable the development of robust mathematical conceptual development. Working with an origami model provides opportunities to use mathematical vocabulary correctly in a context that is meaningful and purposeful. Whether demonstrating a particular fold or asking learners to determine their own folding sequence, correct mathematical language can and should be encouraged at all times.

Unfolding models and examining the geometric properties of the creased paper can help to develop understanding about symmetry and fractions. Asking "Why does this folding sequence work?" can lead naturally to proof through the development of mathematical reasoning.

In addition, origami models of geometric objects can provide a practical starting point for exploring mathematics.

Bibliography

[Boakes 09] Norma J. Boakes. "The Impact of Origami-Mathematics Lessons on Achievement and Spatial Ability of Middle-School Students." In *Origami⁴: Fourth International Meeting of Origami Science, Mathematics, and Education*, edited by Robert J. Lang, pp. 471–482. Wellesley, MA: A K Peters, 2009.

[BOS 10] BOS.   "British Origami Society."   Available at http://www.britishorigami.info/, 2010.

[Cornelius and Tubis 09] V'ann Cornelius and Arnold Tubis. "On the Effective Use of Origami in the Mathematics Classroom." In *Origami⁴: Fourth International Meeting of Origami Science, Mathematics, and Education*, edited by Robert J. Lang, pp. 507–515. Wellesley, MA: A K Peters, 2009.

[Davis and Hersh 81] Philip J. Davis and Reuben Hersh. *The Mathematical Experience*. Brighton, UK: Harvester Press, 1981.

[Frigerio 02] Emma Frigerio. "In Praise of the Papercup: Mathematics and Origami at the University." In *Origami³: Third International Meeting of Origami Science, Mathematics, and Education*, edited by Thomas Hull, pp. 291–298. Natick, MA: A K Peters, 2002.

[Golan and Jackson 09] Miri Golan and Paul Jackson. "Origametria: A Program to Teach Geometry and to Develop Learning Skills Using the Art of Origami." In *Origami⁴: Fourth International Meeting of Origami Science, Mathematics, and Education*, edited by Robert J. Lang, pp. 459–469. Wellesley, MA: A K Peters, 2009.

[Haga 02] Kazuo Haga. "Fold Paper and Enjoy Math: Origamics." In *Origami³: Third International Meeting of Origami Science, Mathematics, and Education*, edited by Thomas Hull, pp. 307–328. Natick, MA: A K Peters, 2002.

[Heewart 92] Eleonore Heewart. "Course in Paper Folding—One of Froebel's Occupations for Children." In *COET'91: Proceedings of the First International
Conference on Origami in Education and Therapy*, edited by John Smith,
pp. 101–153. Birmingham, UK: British Origami Society, 1992. (Originally
published 1895.).

[Hodgen et al. 10] Jeremy Hodgen, Dietmar Kuechemann, Margaret Brown, and
Robert Coe. "Lower Secondary School Students' Knowledge of Fractions."
Research in Mathematics Education 12:1 (2010), 75.

[Hopkins et al. 04] Christine Hopkins, Sue Pope, and Sandy Pepperell. *Understanding Primary Mathematics*. London: David Fulton Publishers, 2004.

[Hull 06] Thomas Hull. *Project Origami: Activities for Exploring Mathematics*.
Natick, MA: A K Peters, 2006.

[INCA 10] INCA. "International Review of Curriculum and Assessment frameworks (INCA)." Available at http://www.inca.org.uk/, 2010.

[Kasahara and Takahama 87] Kunihiko Kasahara and Toshie Takahama.
Origami for the Connoisseur. Tokyo: Japan Publications, 1987.

[Leroux and Santos 09] Helen Leroux and Sara Santos. "Unfolding Geometry."
Research in Mathematics Education 216 (November 2009), 16–19.

[Mason et al. 82] John Mason, Leone Burton, and Kaye Stacey. *Thinking Mathematically*. London: Addison-Wesley, 1982.

[O'Beirne 61] T. H. O'Beirne. "Pentominoes and Hexiamonds." *New Scientist*
12 (November 1961), 379–380.

[Ofsted 08] Ofsted. "Mathematics: Understanding the Score." Available at http://www.ofsted.gov.uk/Ofsted-home/Publications-and
-research/Browse-all-by/Documents-by-type/Thematic-reports/
Mathematics-understanding-the-score, 2008.

[Petty 98] David Petty. "Model of the Month: Dec '98—Multibox." Available
at http://www.davidpetty.me.uk/mom/mom4.htm, 1998.

[Petty 01] David Petty. "Model of the Month: Dec '01—Four Point Star." Available at http://www.davidpetty.me.uk/mom/mom40.htm, 2001.

[Pimm 87] David Pimm. *Speaking Mathematically: Communication in Mathematics Classrooms*. London: Routledge & Kegan Paul, 1987.

[Pope and Lam 09] Sue Pope and Tung Ken Lam. "Using Origami to Promote
Problem-Solving, Creativity, and Communication in Mathematics Education." In *Origami⁴: Fourth International Meeting of Origami Science, Mathematics, and Education*, edited by Robert J. Lang, pp. 517–524. Wellesley,
MA: A K Peters, 2009.

[QCDA 09] QCDA. "The Gherkin Shapes Up." Available at http://curriculum
.qcda.gov.uk/key-stages-3-and-4/case_studies/casestudieslibrary/
case-studies/Gherkin_shapes_up.aspx, 2009.

[Row et al. 66] T. Sundara Row, Wooster Woodruff Beman, and David Eugene
Smith. *Geometric Exercises in Paper Folding*. New York: Dover Publications, 1966.

[Wollring 03] Bernd Wollring. "Working Environments for the Geometry of Paper Folding in the Primary Grades." In *Proceedings of the International Symposium Elementary Mathematics Teaching (SEMT'03)*, edited by J. Novotna. Prague: Charles University, 2003.

[Wood 98] David Wood. *How Children Think and Learn: The Social Contexts of Cognitive Development*, Second edition. Oxford, UK: Blackwell, 1998.

Hands-On Geometry with Origami

Michael J. Winckler, Kathrin D. Wolf,
and Hans-Georg Bock

1 Introduction

The project outlined in this paper describes our first steps to use origami in classroom work. Under the auspices of the MINTmachen! school project of Heidelberg University, we developed a series of learning units for grade 8 students to rediscover and relearn geometry by folding origami figures.

In the next section we motivate the development of the project from a constellation of four different factors of influence that made the use of origami a reasonable and interesting choice for a teacher's thesis. Section 3 outlines some of the teaching lessons we developed and highlights the interplay between competences that eighth grade students should acquire and topics we addressed.

Section 4 sheds light on the evaluation of the project and the lessons we learned. Since the project was an isolated test carried out with a class of gifted students in mathematics, no comparison with a control group could be established for the evaluation. However, some cautious conclusions are still possible. We conclude with an outlook from the first origami project we conducted to possible next investigations that we plan.

2 New Trends in Teaching

2.1 Teacher Education in Germany

Our first reason to investigate possibilities to develop teaching units in a new fashion is rooted in the German education system—mainly in the changes that were introduced after the first shock that arose from Germany's poor PISA (Programme for International Student Assessment) results [OECD 00].

Teacher education in Germany is carried out in special teacher training colleges except for those individuals who aim to work at a *Gymnasium.* The *Gymnasium* is the only one of three middle-/high-school types that leads students to higher education at the university level. Therefore, many German states, which are responsible for all decisions regarding education, established *Gymnasium* teacher education at *comprehensive universities* instead of teacher training colleges.

Although this decision is considered to be positive, it leaves university faculties, which have a strong focus on researcher training, with the obligation to organize teacher education as well. In practice, in the State of Baden-Württemberg (to which our university belongs), teacher education in *content* is done along the regular bachelor/master courses, while the education on *didactics* (pedagogy) is shifted to teacher seminars. The downside of this divided approach is that students come into contact with school classes only very late in their education.

During the seminar, students have some *practicum* at nearby schools, but the majority of didactical education is trained on the job during their first practical year at school—after the completion of education in their subject area.

On top of that, large changes in the curriculum in reaction to very bad reviews from PISA demanded changes in the didactical concepts of teaching as well. As teacher training on the job usually occurs through state training programs, distributing these new influences to the everyday life at school is a long and slow process.

Observation 1. *Gymnasium* teachers are educated by the universities. This education is largely restricted to subject area, not didactics. The methods for training on the job react rather slowly to trends and changes.

2.2 The MINTmachen! Project

MINTmachen! (http://www.mintmachen.de) was founded in 2005 to host educational projects at primary and secondary school levels conducted by university researchers in mathematics and computer science. The main goal of MINTmachen! is to introduce new educational concepts into everyday school life.

Examples for such concepts range from rent-a-prof—a program to rent a university professor for a few hours to visit a school or conduct a seminar at a classroom—at robotics and mathematics labs organized at the university and led by researchers and senior students, to the introduction of teaching modules into regular education at the local classrooms.

The workforce to conduct the projects is recruited from student teachers (see Section 2.1): many students seek a chance to improve on their didactical skills as early as possible. When they develop a project for MINTmachen!, they usually get the opportunity immediately to carry out the first prototype of the project at one of our contact schools.

Observation 2. Student teachers find the MINTmachen! project an ideal environment to develop and test their ideas for new concepts and innovations in standard school education settings.

2.3 Research-Driven Projects

While the MINTmachen! project is largely centered around education and didactics, the Interdisciplinary Center for Scientific Computing (IWR), with which the project leaders are affiliated, is a research institute. Its main topic, scientific computing, is the interdisciplinary research field of mathematical and computational methods in scientific applications.

For this reason, the educational projects conducted within the MINTmachen! framework focus on a practical approach to teaching mathematics over a purely theoretical design. The hands-on aspect of using some topic that can be touched, used, or handled by the students is a valuable asset when teaching an abstract subject such as mathematics.

One way to teach mathematical principles to high-school students is by using origami, which is the focus of this paper. It brings together the leading principles of IWR with the needs to improve mathematics education along the lines of the PISA guidelines. The details of this approach are fleshed out in Section 3, but the main observation is the link between mathematical theory (in this case, principles of geometry and axiomatic structures) and applied use of such concepts (geometry in origami graphs and axioms of origami folding).

Observation 3. The combination of theoretical concepts and application areas is a key concept for interesting and authentic education projects.

2.4 Teacher Exam Thesis

The regulations of the state of Baden-Württemberg require all teachers to write an exam thesis in the framework of regular university research. One goal of this exam thesis is to prove that the student has understood the principles and has sufficient knowledge of the field of his or her studies to conduct research.

Putting this thesis to good use, several researchers at Heidelberg University propose research topics in the field of developing educational methods for topics of high-school (*Gymnasium*) mathematics. The aims of such projects are divided into two parts: One is to investigate the current state of knowledge on a topic taught in classrooms, and the other asks for fresh ideas to convey the concepts that form the backbone of this topic.

This setting is a win-win situation for both the supervisor and the student teacher: while the supervisor can issue a closed subject that is often not part of the student's main research to a student teacher, the student receives a topic that is useful for his or her further career as a teacher and helps the student to get first-hand experience in developing new teaching topics.

Observation 4. The teacher exam thesis provides an ideal opportunity for both supervisor and student to carry out educational/didactical research.

2.5 Putting It All Together

The four key observations from the previous sections fall into place when we put them together:

1. as student teachers trained at universities should have practical experience as early as possible,

2. the MINTmachen! project offers an ideal setting to do so,

3. allowing the use of hands-on applications to make theoretical methods easier to experience, and

4. the teacher exam thesis is used to develop this bridge between theory and practice using modern didactical concepts.

Figure 1. Four factors contributing to the origami project.

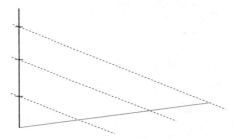

Figure 2. Trisecting a line by the ruler-and-compass method.

3 Design of the Coursework

3.1 Overall Goal

The goal of the project described in this paper is to teach a geometry technique to eighth-grade students using origami. Besides the training of basic principles of ruler-and-compass constructions, a further aim of the project is to convey to students the idea of an axiomatic structure, which is often used in mathematical theory.

A detailed description of the coursework can be found in K. Wolf's teacher thesis [Wolf 09]. The three examples in this paper were selected to illustrate key goals of the didactical concepts formulated in that thesis:

- Haga's theorem is used as a practical example to teach in a very compact way the value of a single technique in geometry (in this case, the use of similar triangles) as a tool for proofs.

- The angle trisection problem leads to the interpretation of origami crease patterns as an equivalent to geometric constructions.

- Finally, the axiomatics of origami are contrasted with the axiomatics of ruler-and-compass constructions to show a meta-concept in mathematical theory.

3.2 Example 1: Haga's Theorem

The teaching unit starts with a review of the trisection of any given line segment using a ruler and compass. The principal technique of this trisection is a topic in sixth grade. The method of marking three arbitrary equal sections on an auxiliary line (black) first and using parallels to transmit this construction to the line in question (gray) implicitly uses similar triangles (Figure 2).

As this is also the key idea to prove Haga's theorem (for a reference, see, e.g., [Kasahara 88, pp. 76–77]), the first free exercise after this review is

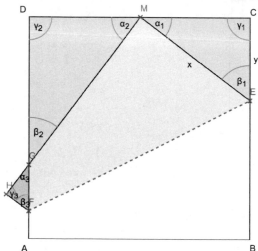

Figure 3. Similar triangles—the main learning goal in the figure of Haga's construction.

to find similar triangles in the geometric construction, which can be either colored or cut out from paper.

In the second part of the teaching unit, we apply Haga's theorem to the same question of dividing a line into thirds. While it can also be used for the more general case of constructing $1/n$th of a section, the specific case proves to be an ideal introduction to the geometry of origami figures. The decisive folding—after bisecting the top edge—is illustrated in Figure 3. We use *this figure alone* to instruct eighth-grade *Gymnasium* students to prove that point G in fact divides line AD with the ratio of $1:2$.

The key idea of the proof is again an observation of similar triangles. This time, the triangles are a bit more difficult to spot, because they do not have a common corner and sides. For the triangles found and identified in group work, the size of each with respect to the others has to be established.

This step is best carried out with a fixed side length: we use 8 units of length. From this first information we show that $|MC| = 4$, and using the Pythagorean theorem, that $y = 3$ and $x = 5$. Using the fact that $\triangle ECM$ and $\triangle MDG$ are similar and that $|MD| = 4$, one arrives at $|DG| = 16/3$ and hence $|GA| = 8/3$, which proves the point.

At the conclusion of the teaching unit, the teacher collects the information on the size of each of the triangles from all groups. Recalculation on the blackboard enables the teacher to make sure that all students have identical values and therefore can understand the proof. A simple GFS subject would be to repeat this proof at a more abstract level for an arbitrary side length a.

Remark 1. Gleichwertige Feststellung von Schülerleistungen (GFS) is used in the German school system to indicate a student presentation (oral with additional handout) that is counted toward the final grade. All teaching units should give ample opportunity for students of various levels to present a GFS.

3.3 Example 2: Trisecting the Angle

The unit on Haga's theorem mainly relies on the geometric figure resulting from the folding itself; this second example makes the transition from the folded figure to crease lines.

Therefore the angle trisection problem places the students in the situation to interpret a *series* of foldings and their resulting crease pattern. This approach shifts the focus in the direction of origami constructions (similar to ruler-and-compass constructions), where the students find a direct equivalence between crease lines/crease intersections and pencil lines/line intersections.

From the worksheet excerpt in Figure 4, we can see that several folding steps lead to the final result. The definition of each folding step (e.g., the first step folds the paper in half) is needed to finally arrive at a proof that indeed the angle given by the line AP is trisected by this process (for a proof using classroom techniques see, e.g., [Henn 03] or [Hull 06]).

The subtle shift from Haga's construction (Figure 3) to the crease pattern in Figure 4 is usually not noticed by the students. However, it pays to indicate the difference, namely that in the proof of Haga's process, the visible triangles used to complete the chain of arguments vanish as soon as the folding is undone. Such lines can be made permanent by adding a

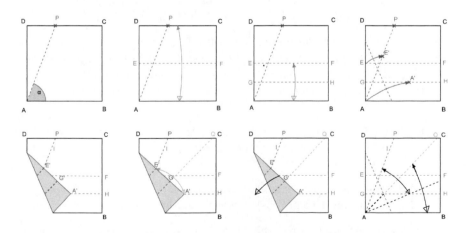

Figure 4. Worksheet: Construction of angle trisection

crease while the figure is still folded or by linking marked points such as E and M by a crease.

From our experience, a series of two observations make this teaching unit something special. First, each student can construct his or her own angle to be trisected. This leads to the unusual situation that students can have (slightly) different crease patterns, which makes it more complicated to compare results. As a consequence, constructions have to be explained using *relative descriptions* that operate more on the identification of lines depending on their relative position in the pattern and not on their absolute position on the paper.

We used a classroom discussion on this topic to make the students aware of the fact. The course of this discussion directly leads to the third example: the axiomatic view on geometry and origami.

3.4   Example 3: Axiomatics in Geometry and Origami

The first two examples were concrete constructions, mainly to hone the students' skills in constructing figures and deriving proofs; this third lesson sheds light on the axiomatics of origami. Much in the same way as the basic axioms of classical geometry define the set of geometrically constructible points, a set of axioms for allowed foldings defines the set of origami foldable points.

We start out with the classic axioms as defined by Humiaki Huzita and the added seventh axiom by Koshiro Hatori (for a proof of completeness of this set of axioms, see, e.g., Lang [Lang 03]). Some of these axioms have a one-to-one correspondence with similar geometric axioms. A direct example is

Origami: Through any two points there exists a fold. $\leftrightarrow$ Geometry: Through any two points there exists a line.

Because the circle has no immediate equivalent in origami, however, it is too involved for eight-grade students to prove that the set of origami axioms is in fact more powerful than the ruler-and-compass constructions.

To explain to the students that these axioms form the core of origami, we do two things:

- After introducing the first axiom we try to find more by asking the students to play with the paper and come up with possible new isolated (axiomatic) construction steps.

- We present instruction candidates and ask the students to investigate (usually in pairs or groups) if these instructions are in fact axioms.

One example for the latter is the candidate "Given two pairs of points in the plane, A, B and C, D, the fold that brings A to C and B to D is well defined." As this is obviously not possible if the distances between the pairs are not equal, in a further investigation we add this constraint.

By playing with the axioms, students begin to understand how fundamental they are for the mathematical model under investigation. This is a good starting point to either address the capabilities of the two models, geometry and origami, and point out their differences (see [Schweizer Fernsehen 09]) or to investigate a system in which axioms were changed (e.g., looking at non-Euclidean geometry by changing the parallel axiom). These, again, could be topics for a GFS alongside the regular school lessons.

3.5   Remarks on the Educational Standards

With the changes in the German school system introduced as a reaction to the PISA results, many states, including Baden-Württemberg, changed their curricula from teaching *knowledge* to acquiring *skills and competences*. In this new framework, more emphasis is placed on *how* students are able to work with knowledge, not merely *what* knowledge they acquire.

Origami is especially suited to shape some of these competences. Two examples illustrate this point—especially in the context of mathematics, which is often understood to be knowledge alone without consideration for competences.

Reasoning. The art of presenting a chain of arguments that finally leads to an uncontested proof of an assumption is a skill much needed by true mathematicians, but seldom a focus in schools. Many hours in mathematics courses are spent on practicing calculation skills, which can be taught in a mechanical, algorithmic fashion. Reasoning, on the other hand, asks for the skillful and imaginative combination of arguments from several sources.

The net woven through this act cannot be constructed in a straightforward way, but is usually built by intuition and experience. The two proofs we presented are ideal examples of this work. With the trisection proof, students learn to combine all the information at hand to finally reach the conclusion they are seeking—an excellent test case through which they can be guided following the origami folding steps one after the other.

Modeling. Modeling the world with science is another skill that students should develop throughout their school careers. Mathematics itself, although the language of all models, is never understood as such in school classes. When we ask students what they think is typical for mathematics, "models" is never an answer: The most obvious answers are "numbers," "calculations," and "abstract symbols."

Origami has the power to bridge this gap. The foldings we did in the exercises are hands-on models for the mathematics behind the processes. Furthermore, building paper models in two and three dimensions,

Figure 5. Platonic solid with modular origami—a classroom project used in MINT-machen!.

as exemplified by the platonic solid models we build with larger classes (see Figure 5) are touchable counterparts to the pseudo-three-dimensional drawings usually encountered in textbooks on this topic. As Flachsmeyer [Flachsmeyer 09] puts it: "With origami, an experimental side of mathematics opens up!"

4   Evaluation and Lessons Learned

As outlined at the beginning, we tested this project with a class of mathematically gifted students. The installation of such a class does not happen regularly in each school year, and the topics taught in these special mathematics classes vary with the teacher teaching the course. Combining the fact that the class is in addition to the regular math education and that the students in the class have a high affinity with mathematics anyway, a meaningful comparison to a control group was impossible in the project.

To evaluate the approach to teach geometry using origami, we therefore relied on the students' evaluations on the usefulness of the approach and the general applicability. We state here the results we obtained on some specific questions. Generally, we used the German grading system (1 = very good, 6 = deficient) as a scale for the students to evaluate the respective question.

- *Did you like the topic of the course?* The answer to this question was undisputed: five 1s and eight 2s gave our topic selection a very good mark (average: 1.62).

- *Did you like to work with origami?* Here, the answer was even more positive: seven 1s and six 2s (average 1.46) told us that origami was very popular in our course—maybe more than the geometry review.

- *How was the complexity level of the course? (1 = very difficult, 6 = very easy)* Three students rated the complexity with 2 (difficult), eight students 3 (somewhat difficult), one student 4 (somewhat easy), and only one student 5 (easy). Overall, then, we avoided the extremes and created a somewhat challenging environment.

- *Is the teaching unit suited for general classroom education?* The answer again was very positive: eight 1s, two 2s, one 3, and one 4 is a very encouraging response, but still two students had doubts about the general usefulness of this kind of course (average: 1.46; one invalid mark).

From the responses to these and several other questions, we drew some conclusions regarding the project. First, the overall approach to teaching origami can be considered a success. The students in our class were all highly motivated to follow the exercises, even though geometry itself can sometimes be a dry subject. The motivation engendered by the hands-on work folding origami can be used to improve the attitude of the students toward mathematics as a subject.

Another lesson learned was that just replacing geometry subunits with equivalent origami exercises is only partially successful. In cooperation with the teacher at our partner school, we relied on some of the previous coursework for the general layout of the course. This was a viable approach given the severe time constraints of a teacher's thesis (six months from start to end, including preparation, holding the course, and writing the thesis), but the most interesting moments during the course were those when the students were playing around with the paper, exploring the different approaches that origami offers over using a pencil on paper. We all agreed that these exploratory units should be increased for the next instance of the course.

Finally, the axiomatic approach to origami and geometry was not very well received by the students. When asked for the most interesting parts of the project, only one student mentioned the axiomatic (and another student named this the most boring part of the course). Other lessons were evaluated much more favorably, such as the angle trisection (3 pros, no cons) or the construction of a parabola as an envelope (4 pros, 2 cons; not presented in this paper). Since the axiomatic is one of the learning goals for mathematics in eighth grade, we have to analyze the project and possibly search for contact points to the students' experience to motivate this complex topic.

Future Plans

From the lessons learned, we conclude the following major development plans for future projects in this area.

The project needs further application, possibly to several classes, to enable a thorough and comparable evaluation. Having a control group for the project for comparison is also a must. The construction of specific problems to be included into class tests to assess the influence of the proposed method on the learning effect is also important.

Contentwise, a reformulation of the teaching units will involve more time for free folding and experimental, open learning environments. This approach, encouraged by the guidelines for high-school development (e.g., by the workgroup Application-Oriented Mathematics [Höger 10] of the school council in Karlsruhe, Germany), is nevertheless viewed suspiciously by many teachers, mainly due to the feeling of an "uncontrolled learning situation."

An extension that we have in mind is to construct a "long night of mathematics and origami" in spring 2011 as an open offer to all our partner schools. Judging from previous experiences with similar events, such an activity invites many interested people (from students to teachers and parents) to experience experimental educational settings. It also would give us a wider range of feedback to provide insight into projects with different age groups.

Acknowledgments. The authors wish to thank the MINTmachen! school project of Heidelberg University for their support. Special thanks go to Ms. Johanna Brandt, mathematics teacher of St. Raphael Gymnasium, Heidelberg, who supervised the teaching unit in her classroom.

Bibliography

[Flachsmeyer 09] Jürgen Flachsmeyer. "Mathematische Belange des Origami (Mathematical Issues in Origami)." *Mathematische Semesterberichte (Kategorie Mathematik in der Lehre)* 56:2 (2009), 201–214.

[Henn 03] Hans-Werner Henn. "Origamics—Papierfalten mit mathematischem Spürsinn (Origami with Mathematical Serendipity)." *Die Neue Schulpraxis* 6/7 (2003), 49–63.

[Höger 10] Christoph Höger. "Anwendungsorientierte Mathematik (Application-Oriented Mathematics)." Available at http://www.anwendungsorientiert.de, 2010.

[Hull 06] Thomas Hull. *Project Origami: Activities for Exploring Mathematics.* Wellesley, MA: A K Peters, Ltd., 2006.

[Kasahara 88] Kunihiko Kasahara. *Origami Omnibus*. Tokyo: Japan Publications, 1988.

[Lang 03] Robert J. Lang. "Origami and Geometric Constructions." Available at http://www.langorigami.com/, 2003.

[OECD 00] OECD. "Programme for International Student Assessment." Available at http://www.pisa.oecd.org/, 2000.

[Schweizer Fernsehen 09] Schweizer Fernsehen. "Origami löst unlösbare Probleme (*Origami Solves Unsolvable Problmes*)." *Einstein*, video, 2009.

[Wolf 09] Kathrin D. Wolf. "Mathematik und Origami in Forschung und Lehre (Mathematics and Origami in Research and Teaching)." Master's thesis, Heidelberg University, Heidelberg, Germany, 2009.

My Favorite Origamics Lessons on the Volume of Solids

Shi-Pui Kwan

1 Introduction

Origamics is a term introduced by Kazuo Haga to describe the branch of study connecting origami and mathematics [Haga 08]. In this paper, I share with readers some of my favorite lecture ideas on using origami in guiding student teachers to learn and teach high school geometry. The main issue of this paper is not the detailed exploratory procedures, but the structure of these processes. I describe five closely related explorations, building mathematical complexity with each exploration, and ending in a set of pedagogical thought questions. They form the background framework on which my lesson design is built.

I agree with George Polya that mathematics education is an art [Polya 81]. As teachers, we all know that lesson design has to meet the needs of the learners and to adjust to changes of time and environment. For further elaboration of my thoughts, I append some of my teaching notes for Exploration 1 as an example.

2 Exploration 1: Origami Masu Cubic Box

Our series of exploratory activities starts with a masu box. A traditional masu box is a half-cube with base length twice its height. Maintaining

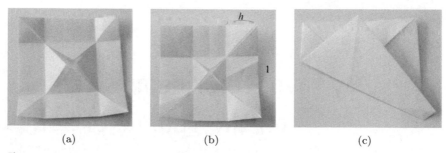

(a)   (b)   (c)

Figure 1. Masu box: (a) normal crease pattern, (b) $h = 1/3$, and (c) Haga's method.

the familiar folding instructions for the masu box, how can we get a cube? Instead of dividing the sheet into the "normal" masu pattern, as shown in Figure 1(a), we crease it into three equal parts; the $h = 1/3$ pattern (Figure 1(b)) will produce the cubic box. Many books on origami suggest the "zigzag" method (e.g., [Beech 03]) in constructing the cube. Nevertheless, not only is that just an estimation, the method can hardly be generalized to divide a length into $1/n$ ($n \in \mathbb{Z}^+$) of its original length. With this motivation, Haga's theorem [Haga 08] is introduced (Figure 1(c)).

Thought questions.

- Haga's theorem yields $2/3$ from $1/2$ (Figure 1(c)). Halving $2/3$, we get $1/3$. How can we generalize this theorem?

- Given a sheet of paper, what is the maximum volume of the masu box so produced?

- Why is the cube not the solution expected?

- Compare this problem with the famous problem in two dimensions on finding the maximum area among rectangles with the same perimeter; what are the similarities and differences?

Referring to the volume maximization problem above, this is a good chance to introduce the Cauchy mean theorem [Weisstein 10] to senior secondary students; for those at the junior level, we can use the Microsoft Excel worksheet (Figure 2(b)) to do the numerical computations.

For students who have learned differential calculus, here is an alternate discussion. Suppose the length of the square in Figure 1(b) is 1 and the

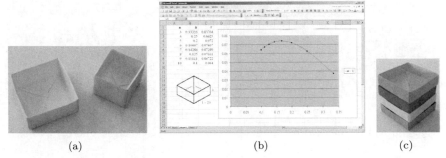

Figure 2. Volume of a masu box: (a) Two boxes with $h = 1/4$ and $h = 1/3$. (b) Spreadsheet and graph of volumes. (c) Four $h = 1/6$ boxes stack to form a cube.

height of the box is h. Then, we have

$$V_{\text{masu box}} = h(1 - 2h)^s$$
$$= h - 4h^2 + 4h^3$$

$$\frac{dV_{\text{masu box}}}{dh} = 1 - 8h + 12h^2$$
$$= (1 - 2h)(1 - 6h)$$

$$\left[\frac{dV_{\text{masu box}}}{dh}\right]_{h-0} \Rightarrow h = \frac{1}{6}.$$

Thought questions.

- Why is $h = 1/6$ a maximum?

- Why is $h = 1/2$ rejected? (The first root $h = 1/2$ is mathematically correct but not foldable. It yields a local minimum.)

We can tackle the problem differently, depending on the prior knowledge of the learners.

3   Exploration 2: Origami CK-Octahedron

Let us proceed with the volume computation of an antiprism, which is more demanding than that of a cube. Figure 3 shows folded models of trigonal antiprisms. We pick one bounded by six identical isosceles right-angled triangular lateral surfaces (lengths l and $(\sqrt{2})l$) and two equilateral triangles (lengths $(\sqrt{2})l$) situated at the top and bottom. For ease of communication, we name this particular type of trigonal antiprism a *CK-octahedron*

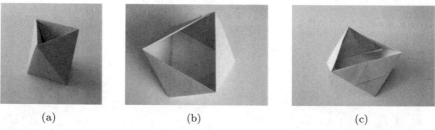

(a)   (b)   (c)

Figure 3. Antiprisms: (a) by Miyuki Kawamura, (b) by Verrill, and (c) by Fuse.

with lateral edge length l, or simply a CK(l) solid (sometimes just CK if the idea is clear), and we denote its volume by $V_{\mathrm{CK}(l)}$. A CK is actually a dissected portion from a cube (C stands for cube and K stands for cut), but in my experience, students at this stage can hardly recognize this fact. We then fold a CK according to Helena A. Verrill [Verrill 05] (Figure 3(b)) or Tomoko Fuse [Fuse 90], who calls it a triangular-pattern belt unit (Figure 3(c)).

Consider CK(1). Since we do not have a volume formula for a CK, students familiar with calculus naturally think of expressing the cross-sectional area A in terms of the height, h, so $V_{\mathrm{CK}(1)}$ can be obtained from $\int_0^h A(h)dh$. The idea is fine, but how can one find $A(h)$?

Let us unfold the model. Draw the intersecting line of a cross section with the CK at a distance x along the slant edge from its base vertex (Figure 4(a)). It is easier for us to express various lengths in terms of x.

In Figure 5(a), $A(x) = \triangle XYZ - 3\triangle XAB$, therefore

$$
A(x) = \frac{1}{2}\left(\left(\sqrt{2}(1-x)+2\sqrt{2}x\right)^2 - 3\left(\sqrt{2}x\right)^2\right)\sin 60°
$$
$$
= \frac{1}{2}\left(\left(\sqrt{2}+2\sqrt{2}x\right)^2 - 6x^2\right)\sin 60°
$$

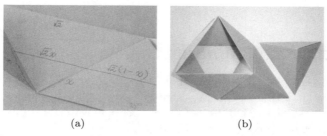

(a)   (b)

Figure 4. CK-octahedron: (a) its various lengths and (b) compared with a triangular pyramid.

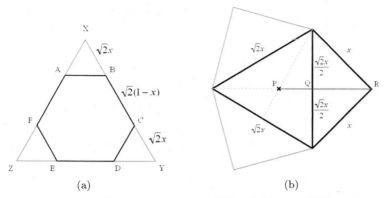

Figure 5. (a) Cross section of CK and (b) net of DB.

$$= \frac{\sqrt{3}}{4}\left(2 + 4x - 4x^2\right)$$

$$= \frac{\sqrt{3}}{2}\left(1 + 2x - 2x^2\right).$$

To establish the relationship between x and h, we examine closely a small triangular pyramid by the side of the CK(1) (Figure 4(b)). Let us name it a *DB-tetrahedron* (D stands for double and B stands for Bienao [Shen et al. 99]) of lateral edge length 1 or DB(1). (We discuss DB at greater length in the last two explorations.) Both CK(1) and DB(1) have the same base, the same lateral faces and the same height. Figure 5(b) shows the net of a DB. Using the properties of a centroid, we have

$$h = \sqrt{QR^2 - PQ^2}$$

$$h^2 = \left(\sqrt{x^2 - \left(\frac{\sqrt{2}x}{2}\right)^2}\right)^2 - \left(\frac{1}{3}\sqrt{\left(\sqrt{2}x\right)^2 - \left(\frac{\sqrt{2}x}{2}\right)^2}\right)^2$$

$$= \left(x^2 - \frac{1}{2}x^2\right) - \left(\frac{1}{3}\sqrt{\frac{3}{2}x^2}\right)^2$$

$$= \frac{1}{2}x^2 - \frac{1}{6}x^2$$

$$\therefore h = \sqrt{\frac{1}{2}x^2 - \frac{1}{6}x^2}$$

$$= \frac{1}{\sqrt{3}}x.$$

When $x = 0$, $h = 0$; when $x = 1$, $h = 1/\sqrt{3}$; when $x = 1/2$, $h = 1/2\sqrt{3}$ and $dh/dx = 1/\sqrt{3}$. Hence,

$$V_{CK(1)} = \int_0^{\frac{1}{\sqrt{3}}} A(h)\,dh$$

$$= 2 \int_0^{\frac{1}{2\sqrt{3}}} A(h)\,dh \qquad \text{(by symmetry)}$$

$$= 2 \int_0^{\frac{1}{2}} \frac{\sqrt{3}}{2} \left(1 + 2x - 2x^2\right) \frac{1}{\sqrt{3}}\,dx$$

$$= \int_0^{\frac{1}{2}} \left(1 + 2x - 2x^2\right) dx$$

$$= \left[x + x^2 - \frac{2x^3}{3} \right]_0^{\frac{1}{2}}$$

$$= \frac{1}{2} + \frac{1}{4} - \frac{1}{12}$$

$$= \frac{2}{3}.$$

Thought questions.

- What shape is the cross section at different heights?

- What are the variants and the invariants of these sections?

- Are there still other ways to find the volume?

4   Exploration 3: The Building Block of CK and KC

Are there simpler ways to evaluate $V_{CK(1)}$? A hint is to dissect it into two identical solids with a volume formula known to us. Yes, a rectangular pyramid, as shown in Figure 6(a), will do. Take the rectangle as the base (width 1 and length $\sqrt{2}$). Its area is $\sqrt{2}$ and its height is $(\sqrt{2})/2$. So the volume of this pyramid can be calculated easily:

$$V_{\text{rectangular pyramid}} = \frac{1}{3}(\sqrt{2}) \left(\frac{\sqrt{2}}{2} \right) = \frac{1}{3}.$$

Its crease pattern is shown in Figure 6(b). Combining two such pyramids with their rectangular bases side by side, we have the CK(1) (Figure 6(c)). In other words, a CK can be dissected into two identical rectangular pyramids. Therefore,

$$V_{CK(1)} = 2(V_{\text{rectangular pyramid}}) = \frac{2}{3}.$$

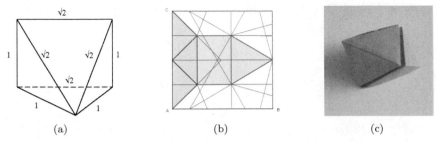

Figure 6. (a) The pyramid, (b) its crease pattern, and (c) the CK-octahedron.

If we rotate one of the pyramids by a straight angle, the two rectangular faces overlap again (Figure 7(a)). Two of the lateral surfaces merge to form a square, and the solid so formed is a heptahedron. We follow our notation above and call it a KC-heptahedron (The letters C and K are interchanged). A KC is also a dissected portion of a cube. It is easier for students to discover this fact if the square is taken as the base (Figure 7(b)).

Thought questions.

- Can you tell why they are named CK and KC?

- CK and KC form a solid twin. What can you say about $V_{CK(l)}$ and $V_{KC(l)}$? How about their surface areas? Can you name another pair of solids with such a relationship?

- KC is a dissected portion of a very familiar solid. Can you tell what it is? Take out your cubic masu box and put it next to the KC (Figure 7(c)). Examine them carefully. What have you discovered?

- How is CK related to this familiar solid?

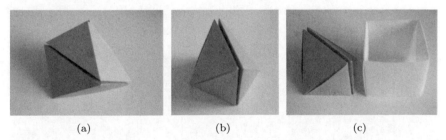

Figure 7. (a) KC-heptahedron, (b) rotated so that the square side is the base, and (c) comparison to a cube.

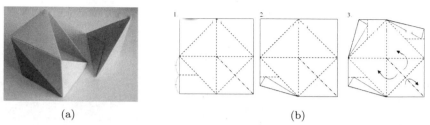

(a)   (b)

Figure 8. (a) DB-tetrahedron "cut" from the cube. (b) Instructions for folding DB.

5   Exploration 4: Origami KC-Heptahedron

Observe that either a CK or a KC is a cube with two identical DBs cut off (Figure 8(a)). Let us follow the instructions in Figure 8(b) and construct the DBs so as to examine the volume of both CK and KC carefully. Thus,

$$V_{\mathrm{CK}(1)} = V_{\mathrm{KC}(1)} = V_{\mathrm{cube}(1)} - 2V_{\mathrm{DB}(1)} = 1^3 - 2\left(\frac{1}{3}\right)\left(\frac{1}{2}\right)(1) = \frac{2}{3}.$$

Take a closer look at the KC. Two more such tetrahedrons can be dissected from it. What is left behind? Yes, a KC is, in fact, formed by two DBs and a regular tetrahedron, as shown in Figure 9. So let us construct the origami model for a regular tetrahedron designed by Haga [Kinsey and Moore 02] and the square base (Figure 10) for it. Now we have an alternate method to construct a KC model.

Thought questions.

- How did you previously find the volume of a regular tetrahedron with edge $\sqrt{2}$? What alternate method do you have now?

- Compare our ways of finding volume in three dimensions with that of area in two dimensions. What are the similarities and differences?

Figure 9. Structure of a KC-heptahedron.

Figure 10. The square base.

6 Exploration 5: Tessellating Solids

All the methods discussed thus far derive from either the idea of calculus or the volume formula for a solid. Can we get V_{KC} without using these concepts? How did the great minds in the past derive the volume formulae? We are going to follow the footsteps of Liu Hui, an ancient Chinese mathematician, and imitate his yangma[1] method [Shen et al. 99] in determining the volume ratio $V_{\mathrm{KC}} : V_{\mathrm{DB}}$ by considering their dissection portions in a cube of unit length.

Liu's approach is best illustrated in Figure 11 [Nelson 00, p. 111]. He called it "proof without words." It looks like the famous Sierpinski triangle. Suppose a triangular pizza is divided into four smaller self-similar identical parts, two are given to A and one is given to B, leaving a part behind (Figure 11(a)). If we repeat the division algorithm in this remaining part again (Figure 11(b)), and this process continues (Figure 11(c)), what is the ratio of the total obtained by A to that by B? The answer is 2 : 1.

Mathematically, A is $1 - (1/4 + (1/4)^2 + (1/4)^3 + \cdots) = 1 - 1/3 = 2/3$ of the pizza.

Returning to the cube dissection, we have $V_{\mathrm{cube}(1)} = V_{\mathrm{KC}(1)} + 2V_{\mathrm{DB}(1)}$. It is more commonly known, however, that $V_{\mathrm{cube}(x)} = 8V_{\mathrm{cube}(x/2)}$. That is why we say that the volume of a cube is eight times as large as one with a side half of its side. Can we tessellate eight $\mathrm{KC}(x/2)$s to form a larger $\mathrm{KC}(x)$? Can we do the same with the triangular pyramids DBs?

This 8 : 1 volume ratio is clearly observed in the case of a cube by tessellating eight congruent cubes to form a bigger one with linear dimension doubled. Nevertheless, the method does not work for solids in general.

---

[1]Yangma is a rectangular pyramid whose vertex is above one corner of its base.

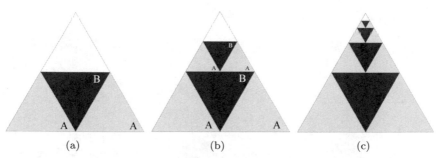

(a)   (b)   (c)

Figure 11. Proof without words.

What if we release the constraint by allowing the use of both $KC(x/2)$ and $DB(x/2)$ to do the tessellation? Can we do it this time? In Figure 12(a) we see that one $KC(x/2)$ and four $DB(x/2)$s tessellate a $DB(x)$. And in Figures 12(b) and 12(c), six $KC(x/2)$s and eight $DB(x/2)$s tessellate a $KC(x)$. Expressing these findings in mathematical terms, we set up the following pair of equations:

$$V_{DB(x)} = V_{KC(x/2)} + 4V_{DB(x/2)},$$
$$V_{KC(x)} = 6V_{KC(x/2)} + 8V_{DB(x/2)}.$$

Substituting $V_{cube(x/2)} = V_{KC(x/2)} + 2V_{DB(x/2)}$, we get

$$V_{DB(x)} = V_{cube(x/2)} + 2V_{DB(x/2)},$$
$$V_{KC(x)} = 4V_{cube(x/2)} + 2V_{KC(x/2)}.$$

Since one KC and two DBs of the same order give a cube, we multiply the first equation by 2 to get

$$2V_{DB(x)} = 2V_{cube(x/2)} + 4V_{DB(x/2)},$$
$$V_{KC(x)} = 4V_{cube(x/2)} + 2V_{KC(x/2)}.$$

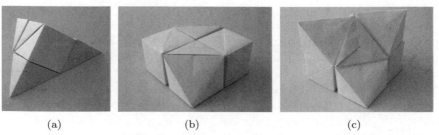

(a)   (b)   (c)

Figure 12. Tessellating solids: (a) $DB(x) = KC(x/2) + 4DB(x/2)$, (b–c) $KC(x) = 6KC(x/2) + 8DB(x/2)$.

Summing the two equations, we get

$$2V_{\text{DB}(x)} + V_{\text{KC}(x)} = 6V_{\text{cube}(x/2)} + 4V_{\text{DB}(x/2)} + 2V_{\text{KC}(x/2)},$$

which simplifies to
$$V_{\text{cube}(x)} = 8V_{\text{cube}(x/2)}.$$

By considering the origins of the eight smaller cubes, we see that

$$V_{\text{cube}(x)} = (4V_{\text{cube}(x/2)} + 2V_{\text{cube}(x/2)}) + 2V_{\text{cube}(x/2)}.$$
$$\quad\uparrow \text{ from KC alone} \quad \uparrow \text{ from DB alone} \quad \uparrow \text{ from both}$$

Let us interpret this result by geometrical dissection. A cube of side x is dissected into 8 smaller cubes half of its length (if the 2 smaller cubes form 1 part, then here are 4 parts). Four of them are designated to KC (2 parts) and two to DB (1 part), leaving two smaller cubes (1 part) in a hybrid of KC and DB in the same ratio. We can repeat the dissection and the designation process again and again in this manner, with both KC and DB diminishing in volume. Recalling the triangle dissection in Figure 11, can you tell what will be the eventual ratio of KC : DB?

We have
$$V_{\text{KC}(x)} : V_{\text{DB}(x)} = 2 : 1.$$

Therefore,

$$V_{\text{KC}(x)} = \frac{2}{3}V_{\text{cube}(x)}$$
$$= \frac{2}{3}x^3.$$

Thought questions.

- Although the yangma method involves only mathematics learned at the junior secondary level, its mastery demands a strong conceptual foundation and sensitive spatial awareness. Can you see the fundamental concepts of limits and infinite series hidden behind this "proof without words"?

- The volume ratio between similar solids converts $V_{\text{DB}(x)} = V_{\text{KC}(x/2)} + 4V_{\text{DB}(x/2)}$ to $8V_{\text{DB}(x/2)} = V_{\text{KC}(x/2)} + 4V_{\text{DB}(x/2)}$ and hence $V_{\text{KC}(x/2)} = 4V_{\text{DB}(x/2)}$.
  Have you found another method in finding $V_{\text{KC}(x)}$?

- The volume ratio 8 : 1 for two cubes with length ratio 2 : 1 can be illustrated easily by dissection, but it is not apparent for solids in general. Small CKs and KCs can be tessellated to give the larger ones. What other solids have such a nice property?

- What extended explorations can be conducted? Interested readers may refer to the article "The van Hiele Phases of Learning in Studying Cube Dissection" [Kwan and Cheung 09].

7   Origami, Science, Mathematics, and Education

Besides volume computations, mathematics also plays an important role in origami designs. For example, to match the dissected solids CK, KC, DB, and the regular tetrahedron and fit them into one single box (Figure 13), a careful selection of paper size is crucial. It is left for the reader to examine these solids closely and do a little mathematics.

With the rapid advancement of technologies, many java applets (Figure 14) can assist in the learning of geometry. The WisWeb applets [Freudenthal Institute 08] are helpful tools for teachers.

Polydron [Polyhedron 10] and Gigo [Genius Toy 10] products can also be used to construct the geometrical solids mentioned above. However, sheets of paper are readily available for students, and they are far less expensive. Not only do folded models help develop visual perception, their artistic forms/shapes and their underlying thoughtful designs enrich the learning process.

The whole exploratory sequence has been developed gradually in the past two years. Bit by bit, I merge these ideas into my geometry lectures with revisions and modifications based on successive classroom experience. The fine-tuning process is lengthy but unforgettable. I still recall the wonderful moment my colleague Ka-Luen Cheung and I discovered the interesting properties of the CK and KC solid twin and the persistent joyous and attentive facial expressions of the student teachers from various groups and cohorts in carrying out the above folding and tessellation activities. Many breakthroughs actually have been initiated through their inquiries.

With the development and maturation of these exploratory ideas, the next step is to integrate them into actual classroom practices. One difficulty I have encountered is that these explorations are distributed across various

Figure 13. CK, KC, and DBs in cubic boxes.

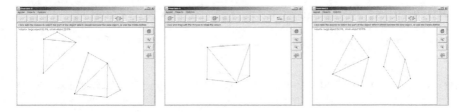

Figure 14. Wisweb applets.

levels and topics in the formal mathematics curriculum, which makes it unlikely to be able to complete all the trials with one selected target class in a short period of time. Recently, I tried out the explorations in a class of secondary 2 (grade 8) students. The teaching plan in the appendix has been modified to meet the standards of students.

8   Conclusion

Figure 15 is excerpted from a student's journal on the completion of Exploration 1. This student mentions that previously she was weak in learning three-dimensional solids and could hardly conjure up mental images of their shapes. But now, she states, it is easier for her to learn geometry in this manner.

The students' words are encouraging in my pilot studies. It is my wish that with the introduction and promotion of these teaching ideas to pre-service/in-service secondary teachers, these classroom explorations can be merged into our curriculum study in the near future.

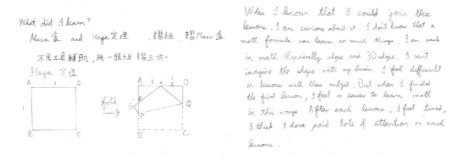

Figure 15. A student's journal on folding cubic boxes.

Appendix: My Teaching Notes for Exploration 1

All the origamics activities in this exploration start with a square sheet of side x units long.

1. How do you fold a masu box? What solid is obtained? What are its dimensions and volume?

   - Teacher demonstrates to students how to fold a masu box. (Folding instructions: Marc Kirschenbaum (2005) from http://www.origami-use.org/files/masu.pdf.)
   - Ask students to find the solutions.
     The masu box is a half cube.
     Length of masu box $= x/2(\sqrt{2})$
     Width of masu box $= x/2(\sqrt{2})$
     Height of masu box $= x/4(\sqrt{2})$
     Volume of masu box $= x^3/32(\sqrt{2})$

2. How do you modify the folding instructions to obtain a cubic box?

   - Teacher introduces to class how step 3 (Marc Kirschenbaum's instructions) is modified by applying Haga's first theorem.

     If AP is $(1/2)AB$ then BQ is $(2/3)AB$.
     M is the midpoint of BQ.
     Therefore BM is $(1/3)AB$.

   - Guide students to prove Haga's first theorem. Hints: congruence and similarity.

3. How do you extend Haga's theorem to the case of 1/3? And how do you generalize it to $1/n$?

   - Teacher extends the case 1/2 to 1/3 in class.

     If AP is $(1/3)AB$ then BQ is $(2/4)AB$.
     M is the midpoint of BQ.
     Therefore BM is $(1/4)AB$.

   - Guide students to follow the same line of thought and generalize it to $1/n$.
     If AP is $(1/n)AB$ then BQ is $(2/(n+1))AB$.
     M is the midpoint of BQ.
     Therefore BM is $(1/(n+1))AB$.
     This is my web page constructed for illustration purposes: http://home.ied.edu.hk/~spkwan/HagaThm.html.

- Discuss recurrence relationship.

4. Given a sheet of paper, what is the maximum volume of the masu box produced?

  - Teacher reviews with students a similar two-dimensional problem: Given a rectangle of constant perimeter the maximum area is that of a square.

  - Ask students to form small groups, discuss, make conjecture, and devise experiment to find out the solution of our masu box problem.

  - The cube is usually the most frequent guess. Students are then guided to construct masu boxes with different heights and make measurements to find the answer. (See Figure 2(a).) This can be carried out in a cooperative manner.

  - Teacher collects data from groups and holds a discussion with the whole class.

5. How do you prove the empirical finding in a rigorous manner?

  - Teacher may take different approaches toward the problem, depending on the attainment levels of students.
    - *Lower level:* Instead of doing direct measurements and calculations, use a spreadsheet (e.g., Microsoft Excel worksheets) to carry out these tedious tasks and plot curve to get the maximum value. (See Figure 2(b).)
    - *Intermediate level:* Use differential calculus to determine the maximum volume.
    - *Higher level:* Recall the Cauchy mean theorem and apply it in this situation:
      (a) For each height, construct four identical masu boxes.
      (b) Why four? ($4h + l + w = 4(1/n)(x/\sqrt{2}) + 2(x/\sqrt{2} - (2/n)(x/\sqrt{2}))$ is independent of n.)
      (c) Pile them up in a column.
      (d) The pile that forms a cube attains the maximum volume. (See Figure 2(c).)

  - The maximum volume $= x^3/27(\sqrt{2})$ when $n = 6$ (i.e., when height is $1/6$ of length).

  - Will the solution be the same if n takes real values in $\mathbb{R}^+$?

Acknowledgments. I would like to take the opportunity here to express my heartfelt thanks to my colleague, Cheung K. L., for his guidance and support throughout the last two years in the development of these mathematics investigations, and to my student teachers, for I have learned a lot from the frontier classroom experiences with them. I particularly thank Ng K. Y., Chung S. S., Wong H. C., and Lo M. S. for their encouraging feedback.

Bibliography

[Beech 03] Rick Beech. *Origami Handbook: The Classic Art of Paperfolding in Step-by-Step Contemporary Projects.* London: Hermes House, 2003.

[Freudenthal Institute 08] Freudenthal Institute. *WisWeb Applets.* Available at http://www.fi.uu.nl/wisweb/applets/mainframe_en.html, 2008.

[Fuse 90] Tomoko Fuse. *Unit Origami: Multidimensional Transformations.* Tokyo: Japan Publications, 1990.

[Genius Toy 10] Genius Toy Taiwan Co. *Gigo.* Available at http://www.gigo.com. tw/index_en.php, accessed September 12, 2010.

[Haga 08] Kazuo Haga. *Origamics: Mathematical Explorations through Paper Folding.* Hackensack, NJ: World Scientific, 2008.

[Kinsey and Moore 02] L. Christine Kinsey and Teresa E. Moore. *Symmetry, Shape, and Space: An Introduction to Mathematics through Geometry.* Emeryville, CA: Key College Publishing, 2002.

[Kwan and Cheung 09] Shi-Pui Kwan and Ka-Luen Cheung. "The van Hiele Phases of Learning in Studying Cube Dissection." In *Proceedings of the 10th International Conference: Models in Developing Mathematics Education,* pp. 358–363. The Mathematics Education into the 21st Century Project, 2009.

[Nelson 00] Roger B. Nelson. *Proofs without Words II: More Exercises in Visual Thinking.* Washington, DC: The Mathematical Association of America, 2000.

[Polya 81] George Polya. *Mathematical Discovery: On Understanding, Learning, and Teaching Problem Solving.* New York: John Wiley & Sons, 1981.

[Polyhedron 10] Polyhedron, Ltd. *Polydron.* Available at http://www.polydron. co.uk/, accessed September 12, 2010.

[Shen et al. 99] Shen Kangsheung, John N. Crossley, and Anthony W. -C. Lun (eds.). *The Nine Chapters on the Mathematical Art: Companion and Commentary.* Oxford, UK: Oxford University Press, 1999.

[Verrill 05] Helena A. Verrill. *Pieces for Simple Dissection Problems.* Available at http://www.math.lsu.edu/~verrill/origami/, 2005.

[Weisstein 10] Eric W. Weisstein. "Cauchy's Formula." *MathWorld—A Wolfram Web Resource.* Available at http://mathworld.wolfram.com/CauchysFormula.html, 2010.

Part III

Origami Science, Engineering, and Technology

Rigid-Foldable Thick Origami

Tomohiro Tachi

1 Introduction

Rigid-foldable origami (or *rigid origami*) is a piecewise linear origami that is continuously transformable along its folds without deformation by bending or folding of any facet. Therefore, rigid origami can realize a deployment mechanism using stiff panels and hinges, which has advantages for various engineering purposes, especially for designs of kinetic architecture.

In a mathematical context, origami is commonly regarded as an ideal zero-thickness surface. However, this is no longer true when we physically implement an origami mechanism. In particular, when we utilize the stiffness of panels for large-scale kinetic structures, it is necessary to consider a mechanism that accommodates thick panels. For example, in the design of architectural space, we need structures composed of thick panels or composite three-dimensional structures that have finite (nonzero) volume in order to cope with gravity, bear loads, and to insulate heat, radiation, sound, etc.

Thick panel origami with symmetric degree-4 vertices using shifted axes have been proposed [Hoberman 88, Trautz and Künstler 09]. However, no method that enables the thickening of arbitrarily designed rigid origami has been proposed; such a freeform rigid origami can be obtained as a triangular mesh origami or as a generalized rigid-foldable quadrilateral mesh origami [Tachi 09a]. This paper proposes a novel geometric method for implementing a general rigid-foldable origami as a structure composed of tapered or nontapered thick, constant-thickness plates and hinges without changing the mechanical behavior from that of the ideal rigid origami. Since we can easily obtain a valid pattern for a given rigid-origami mechanism, the method can contribute to improving the designability of rigid-foldable structures.

2 Problem Description

In this section, we review the problem of thickening origami, and show some existing approaches that tackle this problem. The simplest thick rigid origami structure is a door hinge, which is a thick interpretation of a single line fold. In this case, the rotational axis is located on the valley side of the fold line. Here we call this type of approach *axis-shift* because the axis is shifted to the valley side of the thick panel. Axis-shift can also convert a corrugated surface without interior vertices, such as the repeating mountain and valley pattern used in a folding screen composed of thick plates. This type of structure can fold and unfold completely from 0 to π. However, the axis-shift method is usually not successful for typical rigid origami mechanisms that have interior vertices. This problem is described next.

2.1 Rigid Origami without Thickness

First, we illustrate the kinematics of ideal folding, i.e., nonthick, rigid origami. The configuration of rigid origami is represented by the folding angles of its fold lines, which are constrained around interior vertices. This constraint can be represented as the identity of the rotational matrix, as studied by Kawasaki [Kawasaki 97], belcastro and Hull [belcastro and Hull 02], Balkcom [Balkcom 02], Watanabe and Kawaguchi [Watanabe and Kawaguchi 09], and Tachi [Tachi 09b]. This approach produces three degrees of constraint for each interior vertex that fundamentally correspond to the rotations in the x-, y-, and z-directions of facets around the point of intersection of incident fold lines. As a result, a rigid origami produces a kinetic motion where the fold lines fold simultaneously. Since the number of vertices, facets, and edges are related by the Euler characteristic of the surface, which is 1, the degrees of freedom (DOF) of the overall system is limited. Specifically, a model has at most $N_0 - 3$ degrees of freedom (assuming that all facets are triangulated), where N_0 is the number of vertices on the boundary of the surface.

Notably, in the case of quadrilateral-mesh-based origami such as Miura-ori, the number of fold lines is smaller than the number of constraints. This produces either an overconstrained structure without kinetic motion or a

1-DOF kinetic structure with redundant constraints. Tachi investigated the condition for a quadrilateral mesh origami to have kinetic motion to allow freeform generalization of the Miura-ori [Tachi 09a].

In the context of utilizing the kinetic behavior of general origami, the axis-shift approach has a problem: for origami in which every edge folds simultaneously, the typical kinetic behavior of origami is produced by the interior vertices that constrain the folding motion. In the case of thick origami that is implemented with axis-shift, an interior vertex generally produces six constraints throughout the transformation (three rotations and three translations) since the fold lines are no longer concurrent. This normally produces an overconstrained system in which no continuous motion can be achieved. Even if we succeeded in designing a consistent pattern for a finite number of states, this would at best produce a multistable structure without rigid-foldabilty.

2.2   Existing Methods

A few approaches have been proposed to solve the problem of thickening.

Symmetric Miura-ori vertex. Hoberman [Hoberman 88] designed a degree-4 vertex by thick panels that connects shifted axes of rotation using plates with two levels of thickness. This gives a structure that enables a 1-DOF folding motion between completely unfolded and folded states represented by rotation angle (0 to π) (Figure 1). The structure can be applied to designing Miura-ori or Miura-ori-based cylindrical surfaces.

The most significant limitation of this structure is that it cannot be applied to nonsymmetric or non-flat-foldable vertices. A variational design of flat-foldable degree-4 vertices thickened with this approach forms a bistable structure in which the connectivity breaks unless it is completely unfolded or folded. In fact, the application is limited to *only* the symmetric vertex (i.e., a vertex of Miura-ori), which enables only one parameter variation.

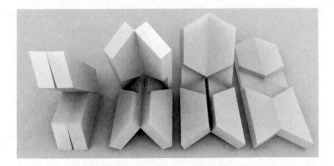

Figure 1. The folding motion of a thickened symmetric degree-4 vertex.

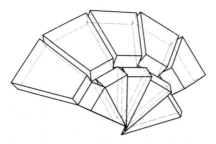

Figure 2. An example of slidable hinges, where the sliding value is accumulated at the hinges on the right.

Another notable limitation of this approach is that it cannot allow multiple overlaps of plates. In the case where alternately adjacent facets, (those sharing the same adjacent facet) overlap in the folded state, the panel of the shared facet is separated into two, as the half-thickness volume of the overlapped part is removed in this approach.

Slidable hinges. An implementation method by slidable hinges has been proposed by Trautz and Künstler [Trautz and Künstler 09]. This method adds extra degrees of freedom by allowing the fold lines to be slid along the rotational axes. The number of variables is doubled by such slidable hinges to compensate for the doubled number of constraints around each vertex. Trautz and Künstler have shown thick panel kinetic structures with symmetric degree-4 vertices that can be folded to an angle of $\pi - \delta$, where δ relates to the amount of slide. Since the sliding amount of an edge is shared by adjacent vertices, the behavior is determined globally for a general case, although the global behavior of slidable hinges structures have not been fully analyzed.

In fact, this global behavior can be a critical problem for some patterns. We can easily show an example for which this approach fails. In Figure 2, the sliding value is accumulated at one of the edges, which will produce separation or intersection of volumes. Therefore, slidable hinges do not allow direct interpretation of general origami structures.

3 Proposed Method

3.1 Tapered Panels

To enable the construction of a generalized rigid-foldable structure with thick panels, we propose kinetic structures that precisely follow the motion of ideal rigid origami with zero thickness (Figure 3(b)) by locating the rotational axes to lie exactly on the edges of ideal origami. This has a great

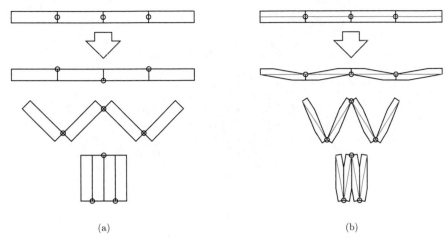

Figure 3. Two approaches for enabling thick panel origami: (a) axis-shift, and (b) the proposed method based on trimming by bisecting planes. The gray path represents the ideal origami without thickness.

advantage over previous axis-shift approaches (Figure 3(a)) for which the folding motion is only approximated by the kinematics of ideal origami. With tapered panels, the correspondence between ideal and rigid motion is exact.

The procedure for creating the thick panels is as follows. First, a zero-thickness ideal origami in the developed state is thickened by offsetting the surface by a constant distance in two directions; in this state, the solids of adjacent facets collide when the origami tries to fold. Then the solid of each facet is trimmed by the bisecting planes of the dihedral angles between adjacent facets in order to avoid the collision of volumes (Figure 4). The shape of the solid changes according to the desired range of folding angles of the edges. By first assuming the maximum and minimum folding angles that the thick origami can fold, we can obtain a solid that works entirely within that range. However, since half of the volume of the solid becomes zero when the maximum folding angle of an edge is equal to π, we cannot completely fold the ideal model to be flat. Therefore, we limit the maximum folding angle to $\pi - \delta$ for some value δ. Here, the structure follows the kinetic motion of rigid origami without thickness because all fold lines are located on the center of the panel (i.e., on the ideal origami surface).

The upper bound of each folding angle $\pi - \delta$ is determined by the thickness of the panels. If we project a solid facet onto a plane, an edge on the top facet is an offset of the original edge by the distance of $t \cot \frac{\delta}{2}$, where t is the half thickness of the panel and the maximum folding angle of the edge is given by $\pi - \delta$ (Figure 5). The intersection of adjacent offset edges

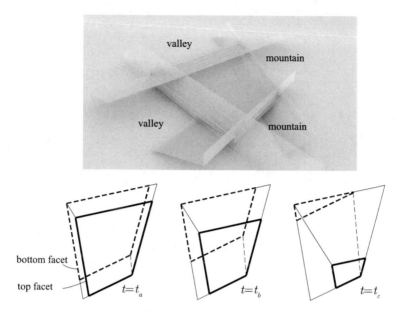

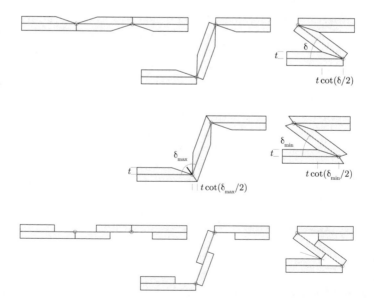

Figure 4. Trimming the volume by bisecting planes of dihedral angles between adjacent facets (top). The top view of the panel for different thicknesses $t_a < t_b < t_c$ (bottom).

Figure 5. Several variations of thickening: Tapered volume (top), tapered volumes that limit the unfolding motion (middle), and constant thickness (bottom).

determines the corresponding corner points on the offset volume (Figure 4). This is equivalent to calculating the weighted skeleton of the polyhedron. This process generally stops when the area of the top surface becomes zero; in the implementation, we stop when two offset corners are merged into one. Therefore, the amount of possible offset is limited by the size of the panel, and thus the dihedral angle δ is related to the thickness of the panel as $\tan \frac{\delta}{2} \propto t$. So, if we try to thicken the panel, the packaging efficiency of the structure is lowered.

This relation is generally nonlinear due to the nonlinear global folding motion, but we can understand the basic behavior by observing the thickening of a simple pleat model, where the depth in the stacked direction is approximately proportional to $\sin \delta/2$, and thus to the thickness t when δ is very small (in a close-to-flat-folded state).

3.2 Limiting the Unfolded State

Figure 5 (middle) shows a variation of this method. Instead of trimming the volume, we can add extra volume to the side of the panels to restrict the unfolding motion. For example, if we offset the edge by $-t \cot \delta_{\max}/2$ (negative weight), the unfolding of the fold line is limited to $\pi - \delta_{\max}$. This technique is useful for avoiding the singular configuration of the completely unfolded model (since unfolded origami in general can infinitesimally fold in the wrong direction(s); see [Watanabe and Kawaguchi 09]) and/or for creating a mechanism that deploys and fixes itself in a specified three-dimensional state.

3.3 Constant Thickness Panels

If the thickness-to-width ratio for each panel is small enough compared to incident minimum dihedral angles, so the top and bottom facets share a significant amount of area in a top view, the tapered solid can be substituted by two-ply constant thickness panels (Figure 5, bottom). A structure with constant thickness panels can be easily manufactured via a simple 2-axis cutting machine. This significantly simplifies the cutting procedure, although it produces holes at the corners of panels. Figure 6 shows the folding motion of an example model with constant thickness panels.

3.4 Global Collision

In our proposed method, we have assumed that the collision between thick panels occurs only along a fold line. Even though this approximation works for many models, this is not true in a general sense because there can occur global collisions between nonadjacent thick panels that did not arise in the ideal (zero-thickness) case. In order to avoid global collision between

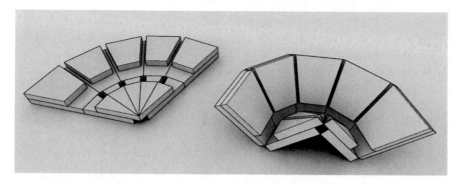

Figure 6. A model with constant thickness panels. Notice the difference from the slidable hinge method shown in Figure 2.

nonadjacent panels, we can naturally extend the proposed method: calculate the bisecting plane for each pair of intersecting facets and cut out the volume of panels along this plane, or rather, along the swept plane to allow continuous motion.

3.5 Characteristics

Since any fold line cannot fold up completely to π, we cannot produce a folding mechanism with two separable motions (singular motion), such as exists in a vertex with four $\pi/2$ corners. This is a disadvantage of our method, since the axis-shift method applied to a symmetric Miura-ori vertex can produce singular motion with four $\pi/2$ corners. Therefore, our method is most suitable for producing mechanisms with simultaneous folding motions.

4 Application for Designs

The proposed thickening method has been implemented as a parametric design system using *Grasshopper* [McNeel 11] and VC# script. This successfully yielded a rigid-foldable structure with thickness producing a mechanism identical to the ideal rigid origami (Figures 7 and 8). The connection part can be realized as embedded mechanical hinges whose rotational axes are located exactly on the ideal edges. Also, non-mechanical hinges can be constructed by sandwiching a strong fabric or film between the two panels, since the rotational axes are located on the center plane.

A realized example design of constant-thickness rigid origami composed of quadrilateral panels is shown in Figure 9 [Tachi 10]. This 2.5 m × 2.5 m square model is manufactured from two layers of double-walled cardboards

Figure 7. A quadrilateral-mesh hypar model with thick tapered panels.

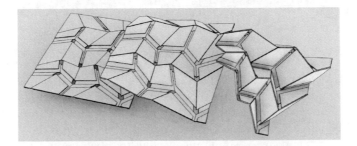

Figure 8. Volume substituted by two constant-thickness panels.

Figure 9. An example design of rigid foldable origami materialized with cloth and cardboard. (See Color Plate XII.)

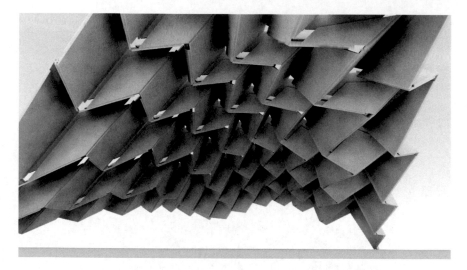

Figure 10. An image of architectural-scaled rigid origami.

(each of which is 10 mm thick) sandwiching a piece of cloth. Because of its 1-DOF mechanism, a simultaneous motion that counterbalances the weight has been produced. This enabled a smooth and dynamic motion by lightly pushing the rim of the structure. A prospective design possibility is applying the method for kinetic architectures (Figure 10).

5   Conclusion

This paper has presented a novel method for enabling a rigid-foldable origami structure with thick panels while preserving the kinetic behavior of an ideal origami surface. The method trims the intersecting material between the panels and produces a kinetic mechanism that folds between predefined minimum and maximum folding angles. The maximum folding angle $\pi - \delta$ and the thickness of panels t are related by $\tan \delta / 2 \propto t$. Our method successfully produced rigid origami designs applicable for human-scale structures.

Acknowledgments. This work was supported by Grant-in-Aid for JSPS Research Fellow by the Japan Society for the Promotion of Science. I would also like to acknowledge NTT InterCommunication Center (ICC) for realizing the construction of the 2.5-m rigid origami model mentioned in Section 4 [Tachi 10].

Bibliography

[Balkcom 02] Devin Balkcom. "Robotic Origami Folding." Ph.D. thesis, Carnegie Mellon University, Pittsburg, PA, 2002.

[belcastro and Hull 02] sarah-marie belcastro and Thomas Hull. "A Mathematical Model for Non-Flat Origami." In *Origami³: Proceedings of the Third International Meeting of Origami Science, Mathematics, and Education*, edited by Thomas Hull, pp. 39–51. Natick, MA: A K Peters, 2002.

[Hoberman 88] Charles Hoberman. "Reversibly Expandable Three-Dimensional Structure." US Patent 4,780,344, 1988.

[Kawasaki 97] Toshikazu Kawasaki. "$R(\gamma) = I$." In *Origami Science and Art: Proceedings of the Second International Meeting of Origami Science and Scientific Origami*, edited by Koryo Miura, pp. 31–40. Shiga, Japan: Seian University of Art and Design, 1997.

[McNeel 11] McNeel. "Grasshopper—Generative Modeling for Rhino." Available at http://grasshopper.rhino3d.com/, 2011.

[Tachi 09a] Tomohiro Tachi. "Generalization of Rigid-Foldable Quadrilateral-Mesh Origami." *Journal of the International Association for Shell and Spatial Structures* 50:3 (2009), 173–179.

[Tachi 09b] Tomohiro Tachi. "Simulation of Rigid Origami." In *Origami⁴: Fourth International Meeting of Origami Science, Mathematics, and Education*, edited by Robert J. Lang, pp. 175–187. Wellesley, MA: A K Peters, 2009.

[Tachi 10] Tomohiro Tachi. "Architectural Origami." In "Exploration in Possible Spaces" Exhibition, NTT InterCommunication Center, January 16–February 28, 2010. (See http://www.ntticc.or.jp/Archive/2010/Exploration_in_Possible_Spaces/.).

[Trautz and Künstler 09] Martin Trautz and Arne Künstler. "Deployable Folded Plate Structures: Folding Patterns Based on 4-Fold-Mechanism Using Stiff Plates." In *Proceedings of IASS 2009 Symposium*, pp. 2306–2317. Madrid, 2009.

[Watanabe and Kawaguchi 09] Naohiko Watanabe and Ken-ichi Kawaguchi. "The Method for Judging Rigid Foldability." In *Origami⁴: Fourth International Meeting of Origami Science, Mathematics, and Education*, edited by Robert J. Lang, pp. 165–174. Wellesley, MA: A K Peters, 2009.

Folding a Patterned Cylinder by Rigid Origami

Kunfeng Wang and Yan Chen

1 Introduction

Rigid origami is a branch of origami for which foldability is kept even when its facets and crease lines are replaced by rigid panels and hinges. Therefore, the folding process of the rigid origami pattern is, in fact, the motion of mechanisms.

In general, several crease (hill or valley) lines meet at a single point called a *vertex*. If the facets and crease lines are replaced by the rigid panels and hinges (also called *revolute joints* in kinematics) at each vertex, the panels and hinges are always same in number. They also form a single closed loop in a spherical linkage, as all the hinges meet at the same point. The degrees of freedom (DoF) for this linkage (also called *mobility* in kinematics) is

$$m = n - 3, \tag{1}$$

where m is the degrees of freedom and n is the number of links or joints, which is equal to the number of crease lines meeting at one vertex. Thus a spherical linkage with four links has only one DoF. This shows that the necessary condition of foldability is that there must be at least four crease lines in one vertex [Miura 89a]. When there are six crease lines in one vertex, as in many folding patterns, the corresponding mechanism has three degrees of freedom. It should be pointed out that Equation (1) shows the degrees of freedom of a mechanism with a single vertex. However, in most origami patterns, there are many more than one vertex. The calculation of degrees of freedom becomes more complicated because the corresponding mechanism has multiple loops [Gogu 05]. Therefore, it is difficult to detect whether an origami folding is rigid and even more difficult to calculate the degrees of freedom of the whole pattern. Two mathematical models have been proposed to study rigid origami "Gaussian Curvature" by Huffman [Huffman 76] and Miura [Miura 89a], which gave the properties of a rigid origami pattern, and a matrix model [belcastro and Hull 02], which derived the necessary condition of the rigid origami patterns with either single or multiple vertices. Recently, Watanabe and Kawaguchi used both the diagram method and numerical method for judging rigid foldability deriving only the necessary condition [Watanabe and Kawaguchi 09]. "Miura-ori" [Miura 89b] is an example of a famous pattern of rigid-foldability. It has recently been generalized to a family of rigid foldable origami with quadrilateral mesh [Tachi 09a].

In contrast, a typical profile of an origami pattern is the cylindrical shape. The cylindrical-shaped patterns that satisfy the condition that, when folded from one piece of flat paper longitudinally into a flat configuration, the paper does not break have the characteristic called *flat-foldability*. Examples include the triangulated foldable cylinder proposed by Professor C. R. Calladine during an investigation of the mechanics of biological structures [Guest and Pellegrino 94] and the "Yoshimura Pattern" ("Diamond Pattern") derived from the buckling modes of thin circular cylindrical shells under axial loads [Lu and Yu 03], which has been used for beverage cans [Han et al. 04] and the timber structure [Buri and Weinand 08]. More are listed in [Nojima 02]. Some other examples show the folding of the cylinder in the radial direction, such as a pineapple folding pattern using triangular units [Kresling 95] that has been applied for a foldable stent graft [Kuribayashi 04]. The only pattern that has one DoF rigid origami, which folds the cylindrical structure in the longitude direction, was proposed by Tachi [Tachi 09b].

In this paper, we focus on the design of several origami patterns to fold one piece of flat paper into a closed patterned cylinder while remaining rigid with only one DoF. Further, the patterned cylinder formed in this way is a static structure without further flat foldability. The process of folding a flat sheet into a closed patterned cylinder with internal mobility can be used for the fabrication of energy-absorbing structures with smart materials. The layout of the paper is as follows. First, the spherical $4R$ linkage and its one DoF assembly is introduced in Section 2. Section 3 gives the corresponding rigid origami pattern all with one DoF. The conclusion and discussions in Section 4 end the paper.

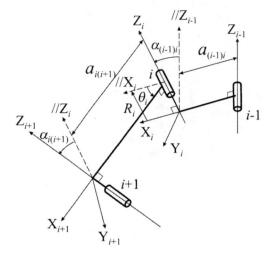

Figure 1. Coordinate systems, parameters, and variables for two adjacent links connected by revolute joints.

2 Kinematics of Spherical 4R Linkage and Its One DoF Assembly

Kinematics studies the geometric properties of the motion of mechanisms. The position of one link relative to another in a kinematic chain is defined mathematically by a coordinate transformation between reference frames attached to each body. The link is rigid, so this transformation must preserve the distances measured between points. This is called a *rigid transformation*, which consists simply of rotations ($\mathbf{Q}_{i(i+1)}$) and translations ($\mathbf{q}_{i(i+1)}$). Figure 1 shows the important geometric features of two binary links $(i-1)i$ and $i(i+1)$, which are connected by the revolute joint i. The coordinate system on the links and joints are set up so that Z_i is the axis of revolute joint i; X_i is the axis commonly normal to Z_{i-1} and Z_i, positively from joint $(i-1)$ to joint i; and Y_i is the third axis following the right-hand rule.

According to the coordinate system, the geometric parameters of the links are defined as follows: the *length* of link $(i-1)i$, $a_{(i-1)i}$, is the distance between axes Z_{i-1} and Z_i; the *twist* of link $(i-1)i$, $\alpha_{(i-1)i}$, is the angle of rotation from axes Z_{i-1} to Z_i positively about axis X_i; and the *offset* of joint i, R_i, is the distance from link $(i-1)i$ to link $i(i+1)$ positively about Z_i. The kinematic parameter of the revolution of joints, the *revolute variable* of the linkage, θ_i, is the angle of rotation from X_{i-1} to X_i positively about Z_i.

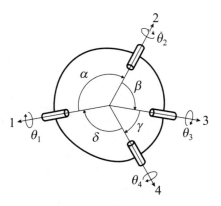

Figure 2. A general spherical $4R$ linkage.

Therefore, the homogeneous transformation between the system of link $(i-1)i$ and the system of link in Figure 1 is

$$
\mathbf{T}_{i(i+1)} = [\mathbf{Q}_{i(i+1)}, \mathbf{q}_{i(i+1)}]
$$
$$
= \begin{bmatrix}
C\theta_i & S\theta_i & 0 & -a_{i(i+1)} \\
-C\alpha_{i(i+1)}S\theta_i & C\alpha_{i(i+1)}C\theta_i & S\alpha_{i(i+1)} & -R_i S\alpha_{i(i+1)} \\
S\alpha_{i(i+1)}S\theta_i & -S\alpha_{i(i+1)}C\theta_i & C\alpha_{i(i+1)} & -R_i C\alpha_{i(i+1)} \\
0 & 0 & 0 & 1
\end{bmatrix},
$$
$$
\tag{2}
$$

where C stands for cosine and S for sine.

Denavit and Hartenberg pointed out that, for a simple close-loop in a linkage, the product of all the transform matrices equals the unit matrix [Denavit and Hartenberg 55]. So the *loop closure equation* becomes

$$
\mathbf{T}_{n1} \cdots \mathbf{T}_{34}\mathbf{T}_{23}\mathbf{T}_{12} = \mathbf{I}. \tag{3}
$$

This is the matrix method in kinematics analysis for any mechanism, which considers not only the rotation, but also the translation transform among different links and joints. It should be pointed out that the matrix model [belcastro and Hull 02] is similar. However, that model can be used only for origami without the translation transformation.

For general spherical $4R$ linkage in Figure 2, the geometric parameters are

$$
a_{12} = a_{23} = a_{34} = a_{41} = 0;
$$
$$
\alpha_{12} = \alpha, \alpha_{23} = \beta, \alpha_{34} = \gamma, \alpha_{41} = \delta;
$$
$$
R_1 = R_2 = R_3 = R_4 = 0.
$$

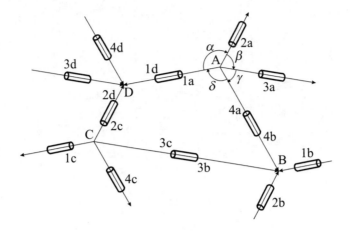

Figure 3. Assembly of four spherical $4R$ linkages.

So the translation vectors are always zero vectors and the transform matrices are in the form of

$$
\mathbf{T}_{i(i+1)} = [\mathbf{Q}_{i(i+1)}, \mathbf{0}] =
\begin{bmatrix}
C\theta_i & S\theta_i & 0 & 0 \\
-C\alpha_{i(i+1)}S\theta_i & C\alpha_{i(i+1)}C\theta_i & S\alpha_{i(i+1)} & 0 \\
S\alpha_{i(i+1)}S\theta_i & -S\alpha_{i(i+1)}C\theta_i & C\alpha_{i(i+1)} & 0 \\
0 & 0 & 0 & 1
\end{bmatrix},
\tag{4}
$$

There are four revolute variables, θ_i ($i = 1, 2, 3, 4$). Following Equatoin (3), the loop closure equation can be derived as

$$
\mathbf{T}_{41}\mathbf{T}_{34}\mathbf{T}_{23}\mathbf{T}_{12} = \mathbf{I} \quad \text{or} \quad \mathbf{Q}_{41}\mathbf{Q}_{34}\mathbf{Q}_{23}\mathbf{Q}_{12} = \mathbf{I}.
\tag{5}
$$

Since the linkage has only one DoF, the three revolute variables, θ_2, θ_3, and θ_4, can be obtained from Equation (5) when θ_1 is given as input.

Now let us consider the assembly of the four spherical $4R$ linkages shown in Figure 3, in which four spherical $4R$ linkages centered at points A, B, C, and D, respectively, are connected together by sharing common joints with the adjacent linkages. As a result, the joints 4a of linkage A and 4b of linkage B are the same joint, so are the joints 3b and 3c, 2c and 2d, and 1d and 1a. Thus,

$$
\theta_{4a} = \theta_{4b}, \quad \theta_{3b} = \theta_{3c}, \quad \theta_{2c} = \theta_{2d}, \quad \text{and} \quad \theta_{1d} = \theta_{1a}.
\tag{6}
$$

To keep the whole assembly as one DoF, the following compatibility condition must be satisfied:

$$
\mathbf{T}_{4a1a}\mathbf{T}_{AB}\mathbf{T}_{3b4b}\mathbf{T}_{BC}\mathbf{T}_{2c3c}\mathbf{T}_{CD}\mathbf{T}_{1d2d}\mathbf{T}_{DA} = \mathbf{I},
\tag{7}
$$

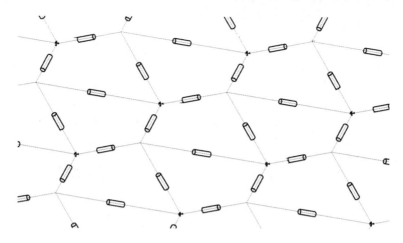

Figure 4. Assembly of a number of spherical $4R$ linkages.

in which $\mathbf{T}_{AB}$, $\mathbf{T}_{BC}$, $\mathbf{T}_{CD}$, and $\mathbf{T}_{DA}$ are the matrices of the translation between two linkages. Equation (7) can be divided into the translation and the rotation compatibility conditions. The translation condition is satisfied automatically because points A, B, C, and D form a spatial quadrilateral. So we shall focus on the rotation condition, i.e.,

$$\mathbf{Q}_{4a1a}\mathbf{Q}_{3b4b}\mathbf{Q}_{2c3c}\mathbf{Q}_{1d2d} = \mathbf{I}. \qquad (8)$$

For linkage A, Equation (5) becomes

$$\mathbf{Q}_{4a1a}\mathbf{Q}_{3a4a}\mathbf{Q}_{2a3a}\mathbf{Q}_{1a2a} = \mathbf{I}. \qquad (9)$$

Considering Equation (6) and comparing Equations (8) and (9), one of the compatibility solutions can be found by observation,

$$
\begin{aligned}
\alpha_{1a2a} &= \alpha_{1b2b} = \alpha_{1c2c} = \alpha_{1d2d} = \alpha, \\
\alpha_{2a3a} &= \alpha_{2b3b} = \alpha_{2c3c} = \alpha_{2d3d} = \beta, \\
\alpha_{3a4a} &= \alpha_{3b4b} = \alpha_{3c4c} = \alpha_{3d4d} = \gamma, \\
\alpha_{4a1a} &= \alpha_{4b1b} = \alpha_{4c1c} = \alpha_{4d1d} = \delta,
\end{aligned}
\qquad (10)
$$

which means that the four spherical $4R$ linkages are identical.

Therefore, four identical spherical $4R$ linkages can be assembled into a unit with one DoF by following the layout in Figure 3. From the successful cases of assembly of several overconstrained linkages with tiling [Chen 03, Chen and You 10], 4^4 tiling is applied here to connect a number of identical spherical $4R$ linkages into an unlimited assembly, as shown in Figure 4, which still has only one DoF.

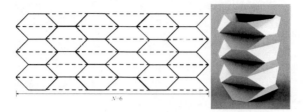

Figure 5. The rigid origami pattern (left) and closed configuration (right) with equilateral trapezoids.

3   Rigid Origami Patterns to Form Cylindrical Structures

As mentioned earlier, the rigid origami in one vertex is, in fact, a spherical linkage. So Figure 4 can be considered as a rigid origami pattern. If there is the constraint that the pattern be from a single piece of flat paper, then $\alpha + \beta + \gamma = \delta = 2\pi$ must be satisfied, which also leads to the conclusion that the identical quadrilateral in the pattern is two-dimensional. This quadrilateral pattern on a plane was first introduced by Kokotsakis [Kokotsakis 97] and later studied by Schief, Bobenko, and Hoffmann [Schief et al. 08] in the area of discrete differential geometry. Huffman also proposed it with the Gaussian curvature model [Huffman 76].

When the quadrilateral is a parallelogram, the rigid origami pattern is "Miura-ori," which generates a flat profile. Here we intend to fold one piece of flat paper into a closed patterned cylinder. Equilateral trapezoids, general trapezoids, and general quadrilaterals have been applied to the pattern in Figure 4 to get the cylindrical profile during the folding; Figures 5–7 show the respective flat paper patterns and closed configurations.

Meanwhile, to obtain the closed patterned cylinder, the geometric closure conditions must be fulfilled accordingly for all three patterns. For the

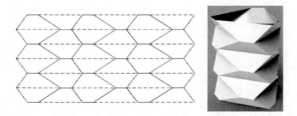

Figure 6. The rigid origami pattern (left) and closed configuration (right) with general trapezoids.

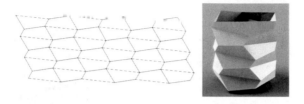

Figure 7. The rigid origami pattern (left) and closed configuration (right) with general quadrilaterals.

equilateral trapezoid and general trapezoid patterns, it is easy to derive the cylinder closure conditions, which are not discussed in detail here. We discuss only the closure condition of the general quadrilateral pattern. The pattern in Figure 8(a) shows the flat configuration. Due to the layout of identical quadrilaterals, the diagonal lines are collinear or parallel to each other. During the folding process, the points on the same diagonal line will form a helical line. After the cylinder reaches the final closed configuration, the radius of the helical line is the same as that of the cylinder on which all the vertices lie, set as R, see Figure 8(b). Set N to be the number of quadrilaterals horizontally. Then the coordinate system can be set up such that the z-axis is the axis of the cylinder and the x-axis passes point A. As a result, the cylindrical coordinates of points A, B, C, and D are

$$A\ (R, 0, 0),\ B\ (R, \theta_B, z_B),\ C\ (R, \theta_C, r\theta_C \tan \sigma),$$
$$D\ (R, \theta_B + \theta_C - \tfrac{4\pi}{N}, R\theta_C \tan \sigma + z_B),$$

where σ is the lead angle of the helix. Transforming these coordinates into Cartesian coordinate system gives

$$A\ (R, 0, 0),\ B\ (R \cos \theta_B, R \sin \theta_B, z_B),\ C\ (R \cos \theta_C, R \sin \theta_C, r\theta_C \tan \sigma),$$
$$D\ \left(R \cos \left(\theta_B + \theta_C - \tfrac{4\pi}{N}\right), R \sin \left(\theta_B + \theta_C - \tfrac{4\pi}{N}\right), R\theta_C \tan \sigma + z_B\right).$$

Therefore, the equation of plane ABC is

$$\left(R^2 \left(\sin \theta_B\right) \theta_C \tan \sigma - R z_B \sin \theta_C\right) (x - R)$$
$$+ \left(z_B R \left(\cos \theta_C - 1\right) - R^2 \theta_3 \tan \sigma \left(\cos \theta_B - 1\right)\right) y$$
$$+ R^2 \left(\sin \theta_C \left(\cos \theta_B - 1\right) - \sin \theta_B \left(\cos \theta_C - 1\right)\right) z = 0. \quad (11)$$

Point D is on the same plane because ABCD is a two-dimensional quadri-

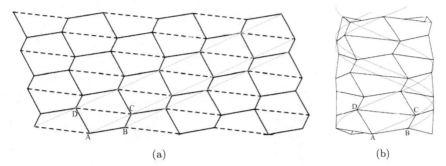

Figure 8. The rigid origami pattern with general quadrilaterals: (a) on flat paper; (b) at the closed configuration.

lateral. Thus,

$$\left(R^2 \left(\sin \theta_B \right) \theta_C \tan \sigma - R z_B \sin \theta_C \right) \left(R \cos \left(\theta_B + \theta_C - \frac{4\pi}{N} \right) - R \right)$$

$$+ \left(z_B R \left(\cos \theta_C - 1 \right) - R^2 \theta_3 \tan \sigma \left(\cos \theta_B - 1 \right) \right) \left(R \sin \left(\theta_B + \theta_C - \frac{4\pi}{N} \right) \right)$$

$$+ R^2 \left(\sin \theta_C \left(\cos \theta_B - 1 \right) - \sin \theta_B \left(\cos \theta_C - 1 \right) \right) \left(R \theta_C \tan \sigma + z_B \right) = 0,$$
$$(12)$$

which gives

$$\tan \sigma = \frac{z_B}{R \theta_C} \cdot \frac{\sin \left(\theta_B - \frac{4\pi}{N} \right) + \sin \left(\theta_C - \theta_B \right) - \sin \left(\theta_B + \theta_C - \frac{4\pi}{N} \right) + \sin \theta_B}{\sin \left(\theta_C - \frac{4\pi}{N} \right) - \sin \left(\theta_C - \theta_B \right) - \sin \left(\theta_B + \theta_C - \frac{4\pi}{N} \right) + \sin \theta_B}.$$
$$(13)$$

Equation (13) is the cylinder closure condition of the general quadrilateral pattern. From the design parameters N, R, θ_B, θ_C, and z_B, we can define not only the helix on the cylinder but also the geometry of the basic element. Figure 9 shows the motion sequence of such an example, in which

$$N = 8, \quad R/z_B = 10, \quad \theta_B = 30°, \quad \theta_C = 40°.$$

Then,

$$\sigma = 25.41°, \ \angle A = 114.21°, \ \angle B = 134.97°, \ \angle C = 62.40°, \ \angle D = 48.42°.$$

4   Conclusions and Discussion

We have pointed out that the rigid origami pattern in a single vertex is a spherical linkage. We have also shown that an identical spherical 4R linkage can be connected into an assembly with a single DoF. Correspondingly,

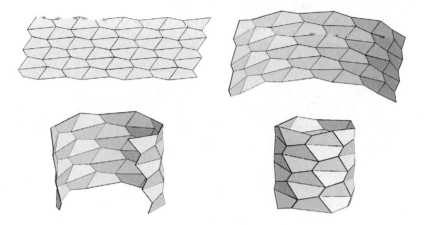

Figure 9. The motion sequence of a rigid origami pattern with general quadrilaterals.

when the quadrilaterals in the origami pattern are identical, rigidity can obtained for the pattern. The general rigid patterns with equilateral trapezoids, general trapezoids, and general quadrilaterals have been derived with special closure conditions to form the closed patterned cylinders.

The study in this paper is of interest not only in origami, but also in engineering. Cylinders with this design have been used as energy-absorbing devices for engineering application.

Finally, note that the identical quadrilateral for the mobile assembly of a spherical $4R$ linkage is only a special solution of the kinematic compatibility condition. Even though it is mathematically challenging, future work should involve a general solution, which will lead to more complicated rigid origami patterns.

Bibliography

[belcastro and Hull 02] sarah-marie belcastro and Thomas C. Hull. "Modelling the Folding of Paper into Three Dimensions Using Affine Transformations." *Linear Algebra and Its Applications* 348 (2002), 273–282.

[Buri and Weinand 08] Hani Buri and Yves Weinand. "Origami—Folded Plate Structures, Architecture." Paper presented at the 10th World Conference on Timber Engineering, Miyazaki, Japan, June 2–5, 2008.

[Chen 03] Yan Chen. "Design of Structural Mechanism." PhD dissertation, University of Oxford, Oxford, UK, 2003.

[Chen and You 10] Yan Chen and Zhong You. "Tilings for Assembly of Structural Mechanisms." Manuscript, 2010.

[Denavit and Hartenberg 55] J. Denavit and R. S. Hartenberg. "A Kinematic Notation for Lower-Pair Mechanisms Based on Matrices." *Journal of Applied Mechanics* 22:2 (1955), 215–221.

[Guest and Pellegrino 94] Simon D. Guest and Sergo Pellegrino. "The Folding of Triangulated Cylinders, Part I: Geometric Considerations." *Journal of Applied Mechanics* 61 (1994), 773–777.

[Gogu 05] Grigore Gogu. "Chebychev–Grübler–Kutzbach's Criterion for Mobility Calculation of Multi-loop Mechanisms Revisited via Theory of Linear Transformations." *European Journal of Mechanics—A/Solids* 24:3 (2005), 427–441.

[Han et al. 04] J. Han, K. Yamazaki, and S. Nishiyama. "Optimization of the Crushing Characteristics of Triangulated Aluminum Beverage Cans." *Structural and Multidisciplinary Optimization* 28 (2004), 47–54.

[Huffman 76] David A. Huffman. "Curvature and Creases: A Primer on Paper." *IEEE Transactions on Computers* C-25 (1976), 1010–1019.

[Kokotsakis 97] A. Kokotsakis. "Über bewegliche Polyder." *Journal de Mathématiques* 5:3 (1897), 113–148.

[Kresling 95] B. Kresling. "Plant 'Design': Mechanical Simulations of Growth Patterns and Bionics." *Biomimetics* 3:3 (1995), 105–122.

[Kuribayashi 04] Kaori Kuribayashi. "A Novel Foldable Stent Graft." PhD dissertation, University of Oxford, Oxford, UK, 2004.

[Lu and Yu 03] Guoxing Lu and Tongxi Yu. *Energy Absorption of Structures and Materials*. Cambridge, UK: Woodhead Publishing, 2003.

[Miura 89a] Koryo Miura. "A Note on Intrinsic Geometry of Origami." In *Research of Pattern Formation*, edited by R. Takaki, pp. 91–102. Tokyo: KTK Scientific Publishers, 1989.

[Miura 89b] Koryo Miura. "Map Fold a La Miura Style, Its Physical Characteristics and Application to the Space Science." In *Research of Pattern Formation*, edited by R. Takaki, pp. 77–90. Tokyo: KTK Scientific Publishers, 1989.

[Nojima 02] Taketoshi Nojima. "Modelling of Folding Patterns in Flat Membranes and Cylinder by Origami." *JSME International Journal, Series C* 45:1 (2002), 364–370.

[Schief et al. 08] Wolfgang K. Schief, Alexander I. Bobenko, and Tim Hoffmann. "On the Integrability of Infinitesimal and Finite Deformations of Polyhedral Surfaces." In *Discrete Differential Geometry*, Oberwolfach Seminars

38, edited by A, I. Bobenko, P. Schröder, J. M. Sullivan, and G. M. Ziegler, pp. 67–93. Basel, Switzerland: Verlag, 2008.

[Tachi 09a] Tomohiro Tachi. "Generalization of Rigid Foldable Quadrilateral Mesh Origami." *Proceedings of the International Association for Shell and Spatial Structures Symposium 2009*, edited by Albert Domingo and Carlos Lazaro, pp. 2287–2294. Madrid: IASS, 2009.

[Tachi 09b] Tomohiro Tachi. "One-DOF Cylindrical Deployable Structures with Rigid Quadrilateral Panels." *Proceedings of the International Association for Shell and Spatial Structures Symposium 2009*, edited by Albert Domingo and Carlos Lazaro, pp. 2295–2305. Madrid: IASS, 2009.

[Watanabe and Kawaguchi 09] Naohiko Watanabe and Ken'ichi Kawaguchi. "The Method for Judging Rigid Foldability." In *Origami⁴: Fourth International Meeting of Origami Science, Mathematics, and Education*, edited by Rober J. Lang, pp. 165–174. Wellesley, MA: A K Peters, 2009.

The Origami Crash Box

Jiayao Ma and Zhong You

1   Introduction

In transport vehicle design, increasing emphasis is put on safety. A common design philosophy to enhance the crashworthiness of the vehicles is to include energy absorption devices, designed to deform and absorb kinetic energy during a collision, at both the front and rear of the vehicles. Crucial to the energy absorption devices is a component called the *crash box*. It is a metallic tube with a square or rectangular section. In automobiles, a pair of boxes is mounted between the bumper and the main frame of a car to absorb energy during a low-speed collision. The primary energy absorbers in trains also adopt a similar design [Martinez et al. 04, Mayville et al. 03, Tyrell et al. 06]. Tubular devices are also found in helicopter landing gears except that they generally have a circular section [Airoldi et al. 09, Airoldi and Janszen 05]. The crash box has retained its current shape and form for a long time. More recently, various other structures have been suggested: e.g., cellular structures [Kim 02, Zhang et al. 06], metallic foam-filled tubes [Hanssen et al. 00a, Hanssen et al. 00b], and composite cylinders [Bambach et al. 09; El-Sobky and Singace 99]. Despite that, metallic square tubes are still the most extensively used type of energy absorbing devices, due to their relatively stable and predictable collapse mode, long deformation stroke, and low manufacturing costs.

The axial crushing of thin-walled tubes has for a long time been a significant research topic. Alexander [Alexander 60] was among the first to

277

study the axial crushing of relatively thick circular tubes collapsing in the concertina mode. He established a basic folding mechanism in which the folds went completely inward or outward, and derived a theoretical formula to calculate the mean crushing force, which agreed well with experimental data. Wierzbicki and colleagues [Wierzbicki et al. 92] observed from experiments that the folds did not go completely inward or outward. He introduced the eccentricity factor m to take the observation into consideration, but could determine the value of m only empirically. The theoretical value of m was later derived and experimentally validated [Singace et al. 95]. Theoretical study of the axial crushing of relatively thin circular tubes collapsing in the diamond mode was not as successful. Pugsley [Pugsley 60, Pugsley 79] found that usually only three or four lobes were finally formed around any circumference of the tube after the tube buckled in the well-known "Yoshimura pattern." He proposed two basic folding mechanisms and derived corresponding theoretical predictions of the mean crushing force. Singace [Singace 99] proposed the definition of m for the diamond mode and also obtained a theoretical expression of the mean crushing force. Notably, the number of lobes circumferentially could not be determined by either theory.

With respect to square tubes, the ratio b/t of tube width b to wall thickness t plays a key role in determining the collapse mode. Tubes with medium thickness, which are mostly used in practice, usually collapse in the symmetric mode. Wierzbicki and Abramowicz [Wierzbicki and Abramowicz 83] thoroughly studied the crushing process of square tubes, and for the first time proposed a kinematically admissible basic folding mechanism. Abramowicz and Jones [Abramowicz and Jones 84, Abramowicz and Jones 86] conducted a series of axial crush tests on square tubes and proposed the concept of effective crushing distance to reflect the fact that a tube could not be completely crushed to zero height.

Two key parameters are widely used to evaluate the performance of an energy absorber: the *specific energy absorption* (SEA), which is defined as the energy absorption per unit mass, and the *load uniformity*, which is defined as the ratio of peak force to mean crushing force. Here, the peak force is the highest force during the crushing process, and the mean crushing force is the total energy absorption divided by the final crushing distance. The current approach to reduce the peak force in order to lower the load uniformity is to introduce dents at predetermined locations on tube surface. Relevant research includes the work done by Singace and El-Sobky [Singace and El-Sobky 97] on circular tubes, and by Lee et al. [Lee et al. 99] on square tubes. When it comes to increasing the mean crushing force so as to increase the SEA, Adachi et al. [Adachi et al. 08] applied circumferential ribs as stiffeners to reduce the longitudinal wavelength of circular tubes, and they reported as much as a 30% mean crushing force increase.

A common problem among the approaches mentioned so far is that the high mean crushing force and low peak force cannot be acquired for the same design. An interesting attempt to achieve both objectives within a design is to introduce pyramid patterns on tube surface [Zhang et al. 07]. However, experimental results revealed that the tube with pyramid patterns was very imperfection sensitive, and the desired "octagonal mode" was quite difficult to trigger [Ma et al. 10]. Moreover, the pyramid patterns are nondevelopable and hence hard to make.

Here we propose a novel design with a square cross section, which, unlike conventional square tubes, contains origami patterns on its surface. The special origami pattern design has the dual functions of introducing a kind of "imperfection" on tube surface, which can lower the peak force, and of triggering a more energy-absorbing efficient failure mode similar to the diamond mode for circular tubes rather than the symmetric mode typical of square tubes. Moreover, because the origami tube is made out of developable surface, it can be manufactured out of a piece of material conveniently and accurately.

The layout of the paper is as follows. The origami pattern design is described in detail in Section 2. Section 3 concerns the finite element modeling of the crash boxes, and the numerical simulation results are presented and discussed in Section 4. Conclusions are given in Section 5.

2   Description of the Origami Pattern Design

The word *pattern*, commonly used in origami, denotes here the organization of a set of folding creases on a tube surface, which is designed to adjust the buckling and post-buckling behaviors of thin-walled tubes. Figure 1(a) shows an origami pattern design in which the solid lines represent hill folds and the dashed ones represent valley folds. If a sheet of material is folded along the folds and then the two opposite free edges are connected, the tube shown in Figure 1(b) can be obtained. The most important feature of the new tube that distinguishes it from conventional square tubes is the diamond-shaped lobes at the corners. The lobe design fulfills two functions: first, it works as a "collapse mode inducer" to direct the tube to fold following the premanufactured patterns when the tube is being crushed; second, it can also be taken as a form of "geometric imperfection" to lower the peak force of the tube at the beginning of the crash.

The origami pattern design has several desirable properties. First, the pattern is designed in a modular way, so what is presented in Figure 1 can be taken as a module, and tubes with variable lengths can be obtained by stacking the module axially. Second, the pattern is developable, and therefore the tube can be made out of a flat sheet of material with little material

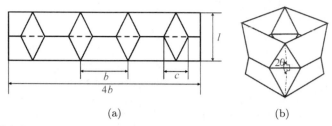

(a)   (b)

Figure 1. (a) Pattern design for the origami tube, and (b) side view of a module of the origami tube.

in-plane stretching during the manufacturing process. This property thus provides a convenient and accurate way of constructing the tube. Finally, the pattern can be easily modified to make tubes with a rectangular cross section, polygonal cross section, or tapered shape, whichever may work better in certain practical applications.

Three independent geometric parameters define the pattern: tube width b, corner width c, and module length l. The dihedral angle 2θ is determined by c and l through the following equation:

$$\cos\theta = \left(\sqrt{2} - 1\right)\frac{c}{l}. \tag{1}$$

In addition, geometric constraints of the corner width require that $c \leq b$ because otherwise the pattern would not be developable; and $c \leq (\sqrt{2}+1)l$ because $\cos\theta \leq 1$. When $c = 0$, the new tube reduces to a conventional square tube.

3   Finite Element Modeling

A series of finite element simulations were conducted to study the collapse mode and energy absorption properties of the origami tube and the effects of the geometric parameters on its performance. Twelve tubes in total, including one conventional square tube A0 and eleven origami tubes, were created and analyzed to understand the influences of b, c, and l. The tube width b and height H of the conventional square tube A0 were 60 mm and 120 mm, respectively. The origami tubes were so designed that they had tube widths and surface areas identical to those of tube A0. Parameters c and l, however, were varied systematically to study the influences of c/l and l/b on the energy absorption properties of the origami tubes. The wall thickness of all the tubes was 1.0 mm. The configurations of all the tubes are listed in Table 1. Note that M denotes the number of modules of the tubes axially. The quasistatic axial crushing process was simulated using Abaqus/Explicit [Abaqus 09]. The crushing event was modeled as two rigid

Model	c (mm)	l (mm)	M	$P_{\max}$ (kN)	$P_{\max}$ reduction	P_m (kN)	P_m increase
A0	–	–	–	40.17	–	11.98	–
A1_1	30	60	2	25.50	36.5%	18.86	57.4%
A1_2	20	60	2	26.97	32.9%	20.38	70.1%
A1_3	15	60	2	26.79	33.3%	19.03	58.8%
A2_1	20	40	3	25.00	37.8%	19.03	58.8%
A2_2	13.3	40	3	24.41	39.2%	19.35	61.5%
A2_3	10	40	3	25.35	36.9%	20.12	67.9%
A2_4	8	40	3	24.19	39.8%	20.12	67.9%
A2_5	6.7	40	3	25.37	36.8%	18.77	56.7%
A3_1	15	30	4	24.54	38.9%	19.49	62.7%
A3_2	10	30	4	25.02	37.7%	21.15	76.5%
A3_3	7.5	30	4	24.51	39.0%	20.95	74.9%

Table 1. Configurations of the tubes and numerical results.

panels crumpling the tube between them. Due to the symmetric property of the tube, only half of each tube was modeled, to reduce computational costs. Four node shell elements S4R were mainly employed to mesh the tube, supplemented by a few triangular elements to avoid excessively distorted elements. Self-contact was employed to model the contact among different parts of the tube, and surface-to-surface contact was used to model the contacts between the tube and the two rigid panels, respectively. The friction coefficient μ was taken as 0.25.

For the tube, the lower end was constrained by a pinned boundary condition, the upper end was coupled to the upper rigid panel by the three translational degrees of freedom (DOFs), and the two longitudinal edges were subject to symmetry boundary conditions. For the two rigid panels, the lower one was completely fixed, and all the DOFs of the upper one were constrained except for the translational one in the axial direction of the tube. Prescribed displacement was assigned to the free DOF of the upper rigid panel to control the loading rate and final crushing distance. The final crushing distance was chosen so that the residual height was 35 mm for all the tubes.

Mild steel, commonly used for tubular energy absorbers, was applied in the numerical simulation. The mechanical properties are listed as follows:

$$\begin{aligned}
\text{density} \quad & \rho = 7800 \text{ Kg/m}^3, \\
\text{Young's modulus} \quad & E = 210 \text{ GPa}, \\
\text{yield stress} \quad & \sigma_y = 200 \text{ MPa}, \\
\text{tensile strength} \quad & \sigma_u = 400 \text{ MPa}, \\
\text{elongation} \quad & 20.0\%, \\
\text{Poisson's ratio} \quad & \nu = 0.3, \\
\text{power law exponent} \quad & n = 0.34.
\end{aligned}$$

Convergence studies with respect to mesh size and crushing time were conducted prior to the analysis. Two principles recommended in *Abaqus Documentation* [Abaqus 09] were checked: (1) the ratio of the artificial energy to the internal energy is below 5% to make sure that an hourglassing effect would not significantly affect the results; and (2) the ratio of the kinetic energy to the internal energy is below 5% during most of the crushing process so as to ensure that the dynamic effect can be considered as insignificant. It was found that the mesh size of 1 mm and crushing time of 0.02 s satisfied these two requirements.

4   Results and Discussion

This section presents the performance of the origami tube under axial compression. Specifically, Section 4.1 shows the collapse of the conventional square tube. Section 4.2 then analyzes the axial crushing behavior of the new tube and compares it with that of the conventional one. And finally the effects of two key geometric parameters, c/l and l/b, on the failure mode and energy absorption of the new tube are investigated in Sections 4.3 and 4.4, respectively.

4.1   Axial Crushing of the Conventional Square Tube

The axial crushing of conventional square tubes that fail in the symmetric mode has been thoroughly investigated, as stated in Section 1. Here, the conventional square tube A0 is studied to set a baseline to evaluate the energy absorption enhancement of the origami tubes. Numerical simulation shows that tube A0 collapses in the symmetric mode as shown in Figure 2(a), which is typical of thin-walled square tubes with $b/t = 60$. The peak force, $P_{\max}$, and the mean crushing force, P_m, of tube A0 are listed

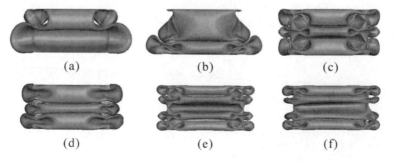

Figure 2. Crushed configurations of selective tubes. (a) A0, (b) A1_3, (c) A2_1, (d) A2_5, (e) A3_1, and (f) A3_3.

in Table 1. The value of P_m of tube A0 is calculated using the following equation:

$$P_m = \frac{\int_0^\delta P(x)dx}{\delta},$$

in which δ is the final crushing distance. It was found that P_m of tube A0 was 11.98 kN.

4.2   Collapse Mode and Energy Absorption Properties of the Origami Tube

The crushing process of tube A1_1 is shown in Figure 3, which is representative of the origami tubes. It can be seen that the tube collapses following the premanufactured origami patterns in a progressive and stable manner. At the beginning of the crushing process, both modules buckle simultaneously. Subsequently, diamond-shaped lobes begin to develop one section after another in the corner zones. Two pairs of inclined plastic hinges form in each lobe and travel opposite to each other as the tube is compressed further, sweeping across a large amount of corner area. Comparing this collapse mode with the diamond mode for circular tubes reveals that the two modes are quite close in shape. The only difference is that the origami tube is composed of flat plates instead of curved shells. The collapse mode of tube A1_1 is named the *complete diamond mode*, in which all the lobes are well formed during the crushing process.

The force-displacement curve of tube A1_1 is plotted in Figure 4. It can be seen that the peak force of tube A1_1 is considerably lower than that of tube A0. The area below the forced-displacement curve of tube A1_1, which indicates the energy absorption capability, in contrast, is much larger than that of tube A0. The numerical data in Table 1 show that, compared with the conventional square tube A0, $P_{\max}$ of the origami tube A1_1 is reduced by 36.5% and P_m is increased by 57.4%. So it can be concluded that the origami pattern design is successful in creating an energy absorber with high SEA and low load uniformity.

Figure 3. Crushing process of tube A1_1.

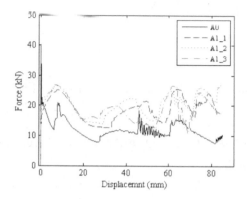

Figure 4. Force-displacement curves of tubes A0, A1_1, A1_2, and A1_3.

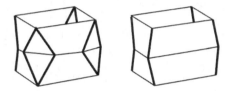

Figure 5. Schematic diagram of a partially crushed origami tube (left) and a conventional square tube (right).

The reason that the origami tube is able to absorb more energy than the conventional square tube becomes apparent when comparing the failure modes of the two tubes. Sketches of the partially crushed configurations of the two tubes are plotted in Figure 5. It can be seen that, upon crushing, there are two pairs of traveling plastic hinges, which are highlighted by bold lines in each corner of the new tube, whereas only one pair of traveling plastic hinges is formed in each corner of the conventional square tube. The additional pair of traveling plastic hinges is mainly responsible for the energy absorption increase.

4.3 Effects of the Ratio c/l

The configuration of the corner areas of the origami tube where the lobes lie is critical in inducing the desired complete diamond mode. Parameter c, or more generally c/l, determines whether the diamond mode occurs and how the inclined plastic hinges travel if it does occur. It is obvious that if this ratio is too small, the diamond mode would fail to be triggered, and the origami tube would collapse in the same way as a conventional square tube. A very large value of this ratio is also undesirable, however, as it reduces the effective crushing distance, and consequently the total energy absorption becomes lower.

For the three sets of tubes (A1_1 to A1_3, A2_1 to A2_5, and A3_1 to A3_3), each of which have identical l/b, c/l is monotonically decreased from 0.5 until the complete diamond mode ceases to appear. The crushed configurations of selective tubes are shown in Figure 2(b)–(f). Three observations are made from the numerical simulation.

- The complete diamond mode is always triggered when c/l is relatively large, but it ceases to appear when c/l is below a critical value. As can be seen in Figure 2(b), only the lobes in the upper part of tube A1_3 are fully formed, whereas the lower part deforms into the conventional symmetric mode. The collapse modes of tubes A2_5 and A3_3 are also deviated from the complete diamond mode (Figure 2(d) and (f)). We name these modes the *incomplete diamond mode*, as opposed to the complete diamond mode shown in Figure 2(c) and (e) and Figure 3. Moreover, the critical value of c/l varies with l/b. It is found that the critical value of c/l is smallest when $l/b = 0.67$, which indicates that the complete diamond mode is most easily triggered, or most stable, at this l/b value.

- For identical l/b, the force-displacement curves of the tubes are similar to each other in shape, as shown in Figure 4, tubes A1_1, A1_2, and A1_3. P_m tends to increase as c/l decreases until the critical value of c/l is reached, but the increment is quite small. This is due to the following reasons: c/l, which is related to 2θ through Equation (1), can be taken as a measure of the imperfection introduced by the pattern. The smaller c/l, the smaller the magnitude of the imperfection, and the larger the force. In contrast, because the energy absorption of a tube is determined mainly by the collapse mode it takes, as long as the collapse mode is the same, the energy absorption is similar. $P_{\max}$, contrary to P_m, shows no sign of a monotonic increase or decrease. This is also understandable because after the elimination of the very high initial buckling force, $P_{\max}$ of the origami tube does not necessarily coincide with the initial buckling force, as in the case of conventional square tubes, but could appear at any later stage during the crushing process. Again, the variation of the peak force is minor.

- The switch from the complete diamond mode to the incomplete diamond mode, such as from A1_2 to A1_3, from A2_4 to A2_5, and from A3_2 to A3_3, is always accompanied by a slight drop of P_m. This again confirms that the triggered collapse mode is mainly responsible for the energy absorption increase. If the collapse mode deviates from the complete diamond mode, the energy absorption also suffers.

4.4   Effects of the Ratio l/b

It is intuitive that the more modules included in a fixed length (i.e., the smaller the ratio l/b), the more energy the tube can absorb. This is because the total length of the horizontal folds and the number of lobes increase with the number of modules.

The force-displacement curves of tubes A1_1, A2_1, and A3_1, which have identical c/l but increasing M, the number of repeated modules, are plotted in Figure 6. It can be seen that although the shapes of the curves in terms of the number of hills and valleys differ, the areas below the curves are still very close. Comparing the P_m of the three tubes reveals that the effects of M on the energy absorption is marginal, with only a 3.3% energy absorption increase being observed from A1_1 to A3_1. Similar observations are also made by comparing tubes A1_2, A2_2, and A3_2.

A qualitative explanation of this result involves different energy dissipation mechanisms in the collapse process. There are three main sources of energy dissipation during the crushing process: the energy absorbed by the horizontal stationary plastic hinges, the inclined traveling plastic hinges, and the in-plane deformation. It can be seen from Figure 2(c) and (e) and Figure 3, that although the total length of the horizontal stationary plastic hinges and the number of lobes increase with M, the corner areas swept by the inclined traveling plastic hinges shrink because the height of each module is reduced. A large portion of the energy absorption enhancement caused by the extra horizontal stationary plastic hinges might be canceled out by the energy absorption reduction caused by the shrunk corner areas swept by the traveling plastic hinges, resulting in only minor net energy absorption gain.

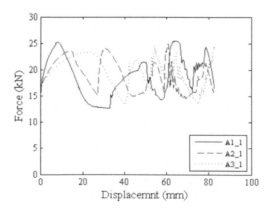

Figure 6. Force-displacement curves of tubes A1_1, A2_1, and A3_1.

5 Conclusion

A novel origami folding pattern has been proposed as a high performance tubular energy absorber. The new design has increased mean crushing force and reduced peak force compared with conventional square tubes. Numerical results show that a more efficient collapse mode in terms of energy absorption capability, which is similar to the diamond mode typical of thin-walled circular tubes, can be successfully triggered for the origami tube by the origami patterns on the tube surface. A peak force reduction of 36.5% and mean crushing force increase of 57.4% from the numerical simulation are also observed. Furthermore, the effects of the geometric parameters of the origami pattern on the collapse mode and energy absorption properties of the origami tube have been systematically investigated.

A series of static and dynamic axial crushing experiments will be carried out at the next step to validate the numerical results. A construction process has been developed to manufacture the origami tube with high quality. Moreover, exploration of new origami patterns that are able to induce more traveling plastic hinges on thin-walled tubes will also be conducted.

Acknowledgment. Jiayao Ma thanks the University of Oxford and Balliol College for financial support in the form of a Clarendon Scholarship and a Jason Hu Scholarship, which have enabled him to study at Oxford.

Bibliography

[Abaqus 09] *Abaqus Documentation*, Version 6.7. Providence, RI: Dassault Systemes Simulia Corp., 2009.

[Abramowicz and Jones 84] Wlodzimierz Abramowicz and Norman Jones. "Dynamic Axial Crushing of Square Tubes." *International Journal of Impact Engineering* 2:2 (1984), 179–208.

[Abramowicz and Jones 86] Wlodzimierz Abramowicz and Norman Jones. "Dynamic Progressive Buckling of Circular and Square Tubes." *International Journal of Impact Engineering* 4:4 (1986), 243–270.

[Adachi et al. 08] Tadaharu Adachi, Atsuo Tomiyama, Wakako Araki, and Akihiko Yamaji. "Energy Absorption of a Thin-Walled Cylinder with Ribs Subjected to Axial Impact." *International Journal of Impact Engineering* 35:2 (2008), 65–79.

[Airoldi et al. 09] Alessandro Airoldi, Janszen Gerardus, Carlo Bergamelli, and Massimiliano Sanvito. "Triggering System for the Plastic Collapse of a Metal Structural Element." US Patent 7,566,031, 2009.

[Airoldi and Janszen 05] Alessandro Airoldi and Gerardus Janszen. "A Design Solution for a Crashworthy Landing Gear with a New Triggering Mechanism for the Plastic Collapse of Metallic Tubes." *Aerospace Science and Technology* 9:5 (2005), 445–455.

[Alexander 60] J. M. Alexander. "An Approximate Analysis of the Collapse of Thin Cylindrical Shells under Axial Loading." *The Quarterly Journal of Mechanics and Applied Mathematics* 13:1 (1960), 10–15.

[Bambach et al. 09] Michael Bambach, Hussein Jama, and Mohamed Elchalakani. "Static and Dynamic Axial Crushing of Spot-Welded Thin-Walled Composite Steel-CFRP Square Tubes." *International Journal of Impact Engineering* 36:9 (2009), 1083–1094.

[El-Sobky et al. 99] Hobab El-Sobky and Abduljalil Abdulla Singace. "Profiled Polymer Pipes as Re-usable Energy Absorption Elements." *International Journal of Mechanical Sciences* 41:11 (1999), 1385–1400.

[Hanssen et al. 00a] Arve Grønsund Hanssen, Magnus Langseth, and Odd Sture Hopperstad. "Static and Dynamic Crushing of Square Aluminium Extrusions with Aluminium Foam Filler." *International Journal of Impact Engineering* 24:4 (2000), 347–383.

[Hanssen et al. 00b] Arve Grønsund Hanssen, Magnus Langseth, and Odd Sture Hopperstad. "Static and Dynamic Crushing of Circular Aluminium Extrusions with Aluminium Foam Filler." *International Journal of Impact Engineering* 24:5 (2000), 475–507.

[Kim 02] Heung S. Kim. "New Extruded Multi-cell Aluminum Profile for Maximum Crash Energy Absorption and Weight Efficiency." *Thin-Walled Structures* 40:4 (2002), 311–327.

[Lee et al. 99] Sunghak Lee, Changsu Hahn, Meungho Rhee, and Jae-Eung Oh. "Effect of Triggering on the Energy Absorption Capacity of Axially Compressed Aluminum Tubes." *Materials & Design* 20:1 (1999), 31–40.

[Ma et al. 10] Jiayao Ma, Yuan Le, and Zhong You. "Axial Crushing Tests of Thin-Walled Steel Square Tubes with Pyramid Patterns." In *51st AIAA/ASME/ASCE/AHS/ASC Structures, Structural Dynamics, and Materials Conference*, DVD-ROM. Reston, VA: AIAA, 2010.

[Martinez et al. 04] Eloy Martinez, David Tyrell, and Benjamin Perlman. "Development of Crash Energy Management Designs for Existing Passenger Rail Vehicles." In *Proceedings of the ASME International Mechanical Engineering Congress and Exposition*, CD-ROM. New York: ASME, 2004.

[Mayville et al. 03] Ronald A. Mayville, Kent N. Johnson, Richard G. Stringfellow, and David C. Tyrell. "The Development of a Rail Passenger Coach Car Crush Zone." In *Proceedings of the 2003 IEEE/ASME Joint Rail Conference*, pp. 55–61. Los Alamitos, CA: IEEE Press, 2003.

[Pugsley 60] Alfred Pugsley. "The Large-Scale Crumpling of Thin Cylindrical Columns." *The Quarterly Journal of Mechanics and Applied Mathematics* 13:1 (1960), 1–9.

[Pugsley 79] Alfred G. Pugsley. "On the Crumpling of Thin Tubular Struts." *The Quarterly Journal of Mechanics and Applied Mathematics* 32:1 (1979), 1–7.

[Singace 99] Abduljalil Abdulla Singace. "Axial Crushing Analysis of Tubes Deforming in the Multi-lobe Mode." *International Journal of Mechanical Sciences* 41:7 (1999), 865–890.

[Singace and El-Sobky 1997] Abduljalil Abdulla Singace and Hobab El-Sobky. "Behaviour of Axially cCrushed Corrugated Tubes." *International Journal of Mechanical Sciences* 39:3 (1997), 249–268.

[Singace et al. 95] Abduljalil Abdulla Singace, Hobab Elsobky, and T. Yella Reddy. "On the Eccentricity Factor in the Progressive Crushing of Tubes." *International Journal of Solids and Structures* 32:24 (1995), 3589–3602.

[Tyrell et al. 06] David Tyrell, Karina Jacobsen, Eloy Martinez, and Armand Benjamin Perlman. "A Train-to-Train Impact Test of Crash Energy Management Passenger Rail Equipment: Structural Results." In *Proceedings of the ASME International Mechanical Engineering Congress and Exposition*, CD-ROM. New York: ASME, 2006.

[Wierzbicki and Abramowicz 83] Tomasz Wierzbicki and Wlodzimierz Abramowicz. "On the Crushing Mechanics of Thin-Walled Structures." *Journal of Applied Mechanics* 50:4 (1983), 727–734.

[Wierzbicki et al. 92] Tomasz Wierzbicki, Shankar U. Bhat, Wlodzimierz Abramowicz, and D. Brodkin. "Alexander Revisited—A Two Folding Elements Model of Progressive Crushing of Tubes." *International Journal of Solids and Structures* 29:24 (1992), 3269–3288.

[Zhang et al. 07] Xiong Zhang, Gengdong Cheng, Zhong You, and Hui Zhang. "Energy Absorption of Axially Compressed Thin-Walled Square Tubes with Patterns." *Thin-Walled Structures* 45:9 (2007), 737–746.

[Zhang et al. 06] Xiong Zhang, Gengdong Cheng, and Hui Zhang. "Theoretical Prediction and Numerical Simulation of Multi-cell Square Thin-Walled Structures." *Thin-Walled Structures* 44:11 (2006), 1185–1191.

Origami Folding: A Structural Engineering Approach

Mark Schenk and Simon D. Guest

1   Introduction

For structural engineers, origami has proven to be a rich source of inspiration, and it has found its way into a wide range of structural applications. This paper aims to extend this range and introduces a novel engineering application of origami: *folded textured sheets*.

Existing applications of origami in engineering can broadly be categorized into three areas. First, many deployable structures take inspiration from, or are directly derived from, origami folding. Examples are diverse, ranging from wrapping solar sails [Guest and Pellegrino 92] to medical stents [Kuribayashi et al. 06] to emergency shelters [Temmerman 07]. Second, in contrast, folding is used to achieve an increase in stiffness at minimal expense of weight, for example in the design of lightweight sandwich panel cores for aircraft fuselages [Heimbs et al. 07]. In architecture, the principle is also applied, ranging from straightforward folded plate roofs to more complicated designs that unite an increase in strength with aesthetic appeal [Engel 68]. Third, origami patterns have been used to design shock-absorbing devices, such as car crash boxes with origami-inspired patterns that induce higher local buckling modes [Wu and You 10], and packaging materials [Basily and Elsayed 04].

In contrast to existing engineering applications, the folded textured sheets introduced in this paper use origami for a different, and slightly paradoxical, purpose· for both the flexibility and the stiffness that it provides. Origami folding patterns enable the sheets to deform easily into some deformation modes, but remaining stiff in others. This anisotropy in deformation modes is of interest, for example, for applications in morphing structures; these types of structures are capable of changing their shape to accommodate new requirements while maintaining a continuous external surface.

Section 2 introduces two example folded textured sheets, the Eggbox and Miura sheets, and highlights some of their mechanical properties of interest. Section 3 describes the mechanical model in detail, interleaved with results for the two example sheets.

2   Folded Textured Sheets

The folded textured sheets form part of ongoing research into the properties and applications of textured sheets. By introducing a "local" texture (such as corrugations, dimples, folds, etc.) to otherwise isotropic thin-walled sheets, the "global" mechanical properties of the sheets can be favorably modified. The "local" texture has no clearly defined scale, but lies somewhere between the material and the structural level and in effect forms a microstructure. The texture patterns in folded textured sheets are *inspired* by origami folding, as the resulting sheets need not necessarily be developable. The texture consists of distinct fold lines, and it is therefore better to speak of polygonal faceted surfaces. Figure 1 shows the two example sheets used in this paper: the *Eggbox* sheet and the *Miura* sheet. (The models are made of standard printing paper, and the parallelograms in both sheets have sides of 15 mm and acute angles of 60°. The Miura sheet is folded from a single flat sheet of paper; the Eggbox sheet, in contrast, is made by gluing together strips of paper, and has equal and opposite angular defects at its apexes and saddle points.)

The first obvious property of the folded sheets is their ability to undergo relatively large deformations by virtue of the folds opening and closing. Moreover, the fold patterns enable the sheets to locally expand and contract—and thereby change their global Gaussian curvature—without any stretching at material level. *Gaussian curvature* is an intrinsic measure of the curvature at a point on a surface, which remains invariant when bending but not stretching the surface [Huffman 76]. Our interest lies with the macroscopic behavior of the sheets, and we therefore consider the "global" Gaussian curvature of an equivalent mid-surface of the folded sheet. Both the Eggbox and Miura sheets are initially flat, and thus have a zero global Gaussian curvature. Now, unlike conventional sheets, both folded textured sheets can easily be twisted into a saddle-shaped configu-

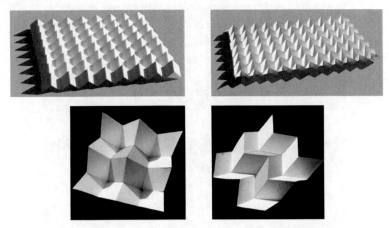

Figure 1. Photographs of the Eggbox (left) and the Miura sheet (right): (top) overview of folded textured sheets and (bottom) closeup of unit cells.

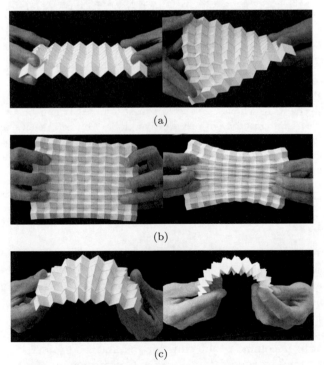

(a)

(b)

(c)

Figure 2. Mechanical behavior of the Eggbox sheet. (a) It can change its global Gaussian curvature by twisting into a saddle-shaped configuration. (b) It displays a positive Poisson's ratio under extension, but (c) it deforms either into a cylindrical or a spherical shape under bending. The spherical shape is conventionally seen in materials with a negative Poisson's ratio.

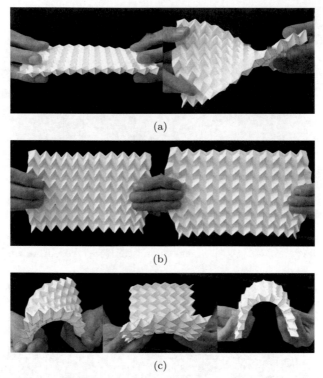

(a)

(b)

(c)

Figure 3. Mechanical behavior of the Miura sheet. (a) It can be twisted into a saddle-shaped configuration with a negative global Gaussian curvature. (b) The sheet behaves as an auxetic material (negative Poisson's ratio) in planar deformation, but (c) it assumes a saddle-shaped configuration under bending, which is typical behavior for materials with a positive Poisson's ratio.

ration which has a globally negative Gaussian curvature; see Figures 2(a) and 3(a).

The sheets' most intriguing property, however, relates to their Poisson's ratio. Both sheets have a single in-plane mechanism whereby the facets do not bend and the folds behave as hinges; by contrast, facet bending is necessary for the out-of-plane deformations. As shown in Figures 2(b) and 3(b), the Eggbox and the Miura sheet respectively have a positive and a negative Poisson's ratio in their planar deformation mode. A negative Poisson's ratio is fairly uncommon, but can, for instance, be found in foams with a reentrant micro-structure [Lakes 87]. Conventionally, materials with a positive Poisson's ratio will deform anticlastically under bending (i.e., into a saddle-shape), and materials with a negative Poisson's ratio will deform synclastically into a spherical shape. As illustrated in Figures 2(c) and 3(c), however, both folded textured sheets behave exactly opposite to what is conventionally expected, and their Poisson's ratios are of opposite sign for in-plane stretching and out-of-plane bending. This remarkable mechanical behavior has been described only theoretically for auxetic composite laminates [Lim 07] and specially machined chiral auxetics [Alderson et al. 10], but is here observed in textured sheets made of conventional materials.

2.1 Engineering Applications

Our interest in the folded textured sheets is diverse. First, they can undergo large global deformations as a result of the opening and closing of the folds. Furthermore, these folds provide flexibility in certain deformation modes, while still providing an increased bending stiffness. This combination of flexibility and rigidity is of interest in morphing structures, such as the skin of morphing aircraft wings [Thill et al. 08].

A second interesting property of the folded sheets is their ability to change their global Gaussian curvature without stretching at material level. This is of interest in architectural applications, where it may be used as cladding material for doubly-curved surfaces, or, at a larger scale, as flexible façades. Furthermore, the use of the sheets as reusable doubly-curved concrete formwork is being explored; work is still ongoing to determine the range of surface curvatures that these sheets can attain.

Applications for the remarkable behavior of the oppositely signed Poisson's ratios under bending and stretching are still being sought. Nevertheless, the folded sheets add a new category to the field of auxetic materials.

3 Mechanical Modeling Method

Available mechanical modeling methods for origami folding broadly cover rigid origami simulators [Tachi 09, Balkcom 04] or methods describing paper as thin shells using finite elements. Our purpose is not to formulate an alternative method to describe rigid origami, as we aim to obtain different information. Neither do we wish to use finite element modeling, since we are not interested in the minutiae of the stress distributions, but rather the effect of the introduced geometry on the global properties of the sheet. The salient behavior straddles kinematics and stiffness: there are dominant mechanisms, but they have a nonzero stiffness. Our method needs to cover this behavior. It should also not be limited to rigid origami, as the out-of-plane kinematics of the sheets involves bending of the facets.

Our approach is based on modeling the partially folded state of a folded pattern as a pin-jointed truss framework. Each vertex in the folded sheet is represented by a pin-joint, and every fold line by a bar element. Additionally, the facets are triangulated to avoid trivial internal mechanisms

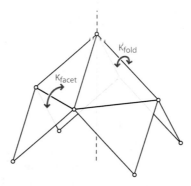

Figure 4. Unit cell of the Eggbox sheet, illustrating the pin-jointed bar framework model used to model the folded textured sheets. The facets have been triangulated to avoid trivial mechanisms and provide a first-order approximation for the bending of the facets. Bending stiffness has been added to the facets and fold lines, K_{facet} and K_{fold}, respectively.

as well as to provide a first-order approximation to bending of the facets (Figure 4).

Although the use of a pin-jointed bar framework to represent origami folding has been hinted at on several occasions, e.g., [Tachi 09, Watanabe and Kawaguchi 09], it has not been fully introduced into the origami literature. The method provides useful insights into the mechanical properties of a partially folded origami sheet, and has the benefit of an established and rich background literature.

3.1   Governing Equations

The analysis of pin-jointed frameworks is well established in structural mechanics. Its mechanical properties are described by three linearized equations: equilibrium, compatibility, and material properties. Thus,

$$\mathbf{At} = \mathbf{f}, \tag{1}$$
$$\mathbf{Cd} = \mathbf{e}, \tag{2}$$
$$\mathbf{Ge} = \mathbf{t}, \tag{3}$$

where $\mathbf{A}$ is the *equilibrium* matrix, which relates the internal bar tensions $\mathbf{t}$ to the applied nodal forces $\mathbf{f}$; the *compatibility* matrix $\mathbf{C}$ relates the nodal displacements $\mathbf{d}$ to the bar extensions $\mathbf{e}$; and the material equation introduces the axial bar stiffnesses along the diagonal of $\mathbf{G}$. It can be shown through a straightforward virtual work argument that $\mathbf{C} = \mathbf{A}^{T}$, the static-kinematic duality.

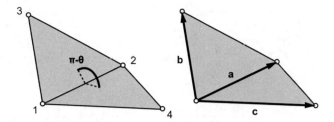

Figure 5. The dihedral fold angle θ can be expressed in terms of the nodal coordinates of the two adjoining facets. Using the vectors $\mathbf{a}$, $\mathbf{b}$, and $\mathbf{c}$, the following expression holds:

$$\sin(\theta) = \frac{1}{\sin(\gamma)\sin(\beta)} \frac{1}{|\mathbf{a}|^3 |\mathbf{b}||\mathbf{c}|} (\mathbf{a} \times (\mathbf{c} \times \mathbf{a})) \cdot (\mathbf{a} \times \mathbf{b}).$$

Here γ is the angle between $\mathbf{a}$ and $\mathbf{b}$, and β is the angle between $\mathbf{a}$ and $\mathbf{c}$.

3.2 Kinematic Analysis

The linear-elastic behavior of the truss framework can now be described, by analyzing the vector subspaces of the equilibrium and compatiblity matrices [Pellegrino and Calladine 86]. Of main interest in our case is the nullspace of the compatibility matrix, as it provides nodal displacements that—to first order—have no bar elongations: internal mechanisms,

$$\mathbf{Cd} = \mathbf{0}.$$

These mechanisms may be either finite or infinitesimal, but in general the information from the nullspace analysis alone does not suffice to establish the difference. First-order infinitesimal mechanisms can be stabilized by states of self-stress, and a full tangent stiffness matrix would have to be formulated to take into account any geometric stiffness resulting from reorientation of the members.

In the case of the folded textured sheets, the nullspace of the conventional compatibility matrix does not provide much useful information: the triangulated facets can easily "bend," which is reflected by an equivalent number of trivial internal mechanisms. The solution is to introduce additional constraints. The compatibility matrix can be reformulated as the Jacobian of the quadratic bar length constraints with respect to the nodal coordinates. This parallel can be used to introduce additional equality constraints to the bar framework. In our case, we add a constraint on the dihedral angle between two adjoining facets.

The angular constraint F is set up in terms of the dihedral fold angle θ between two facets. Using vector analysis, the angle between two facets can be described in terms of cross and inner products of the nodal coordinates $\mathbf{p}$ of the two facets (see Figure 5):

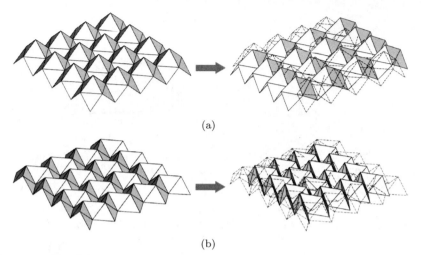

(a)

(b)

Figure 6. The (a) Eggbox and (b) Miura sheets both exhibit a single planar mechanism when the facets are not allowed to bend, as described in Section 3.2. The reference configuration is indicated by dashed lines.

$$F = \sin(\theta) = \sin(\theta(\mathbf{p})) = \dots, \tag{4}$$

and the Jacobian becomes

$$J = \frac{1}{\cos(\theta)} \sum \frac{\partial F}{\partial p_i} dp_i = d\theta. \tag{5}$$

The Jacobian of additional constraints $\mathbf{J}$ can now be concatenated with the existing compatibility matrix

$$\begin{bmatrix} \mathbf{C} \\ \mathbf{J} \end{bmatrix} \mathbf{d} = \begin{bmatrix} \mathbf{e} \\ d\theta \end{bmatrix}, \tag{6}$$

and the nullspace of this set of equations produces the nodal displacements $\mathbf{d}$ that do not extend the bars, as well as not violate the angular constraints. In effect, we have formulated a rigid origami simulator—no bending or stretching of the facets is allowed. In order to track the motion of the folded sheet, one iteratively follows the infinitesimal mechanisms while correcting for the errors using the Moore-Penrose pseudoinverse [Tachi 09]. Our interest, however, remains with the first-order infinitesimal displacements.

In the case of the two example textured sheets, the kinematic analysis provides a single degree of freedom planar mechanism; see Figure 6. In this mechanism, the facets neither stretch nor bend. This is the mechanism a rigid origami simulator would find.

3.3   Stiffness Analysis

A kinematic analysis of a framework, even with additional constraints, can clearly provide only so much information. The next step is to move from a purely kinematic to a stiffness formulation. Equations (1)–(3) can be combined into a single equation, relating external applied forces $\mathbf{f}$ to nodal displacements $\mathbf{d}$ by means of the material stiffness matrix $\mathbf{K}$:

$$\mathbf{K}\mathbf{d} = \mathbf{f}, \tag{7}$$

$$\mathbf{K} = \mathbf{A}\mathbf{G}\mathbf{C} = \mathbf{C}^T\mathbf{G}\mathbf{C}. \tag{8}$$

What is not immediately obvious is that this can easily be extended to other sets of constraints by extending the compatibility matrix:

$$\mathbf{K} = \begin{bmatrix} \mathbf{C} \\ \mathbf{J} \end{bmatrix}^T \begin{bmatrix} \mathbf{G} & 0 \\ 0 & \mathbf{G}_J \end{bmatrix} \begin{bmatrix} \mathbf{C} \\ \mathbf{J} \end{bmatrix}. \tag{9}$$

Depending on the constraint and the resulting error that its Jacobian constitutes, either a physical stiffness value can be attributed in $\mathbf{G}_J$ or a "weighted stiffness," indicating the relative importance of the constraint. In our case, the error is the change in the dihedral angle between adjacent facets. In effect, we introduce a bending stiffness along the fold line (K_{fold}) and across the facets (K_{facet}); see Figure 4. As a result, we obtain a material stiffness matrix that incorporates the stiffness of the bars, as well as the bending stiffness of the facets and along the fold lines.

Plotting the mode shapes for the lowest eigenvalues of the material stiffness matrix $\mathbf{K}$ provides insight into the deformation kinematics of the sheets. Of main interest are the deformation modes that involve no bar elongations (i.e., no stretching of the material), but only bending of the facets and along fold lines. These modes are numerically separated by choosing the axial members' stiffness of the bars to be several orders of magnitude larger than the bending stiffness for the facets and folds. In our analysis, only first-order infinitesimal modes within $\mathbf{K}$ are considered.

An important parameter in the folded textured sheets turns out to be $K_{\text{ratio}} = K_{\text{facet}}/K_{\text{fold}}$. This is a dimensionless parameter that represents the material properties of the sheet. When $K_{\text{ratio}} \to \infty$ we approach a situation where rigid panels are connected by frictionless hinges; values of $K_{\text{ratio}} \approx 1$ reflect folded sheets manufactured from sheet materials such as metal, plastic, and paper; and when $K_{\text{ratio}} < 1$, the fold lines are stiffer than the panels, which is the case for work-hardened metals or situations where separate panels are joined together, for example, by means of welding.

The results for the Eggbox and Miura sheet are shown in Figures 7 and 8, respectively. The graphs show a log-log plot of the eigenvalues versus the stiffness ratio $K_{\text{facet}}/K_{\text{fold}}$. It can be seen that the salient kinematics (the softest eigenmodes) remain dominant over a large range of the

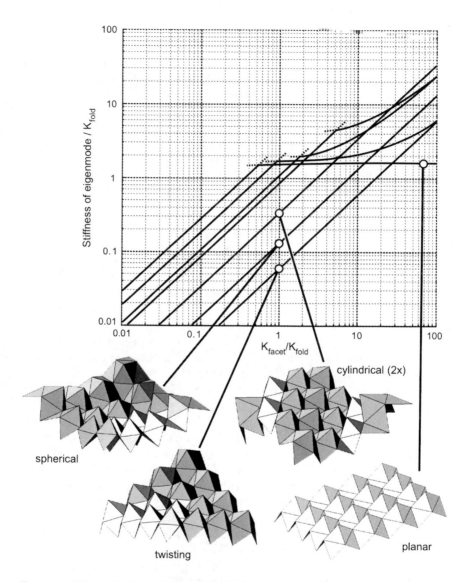

Figure 7. Plots of the relative stiffness of the nine softest eigenmodes of the Eggbox sheet. It can be seen that the twisting deformation mode remains the softest eigenmode over a large range of K_{ratio}. The spherical and cylindrical deformation modes observed in the models are also dominant. As $K_{\mathrm{ratio}} \to \infty$, the planar mechanism becomes the softest eigenmode; this corresponds with the result from the kinematic analysis.

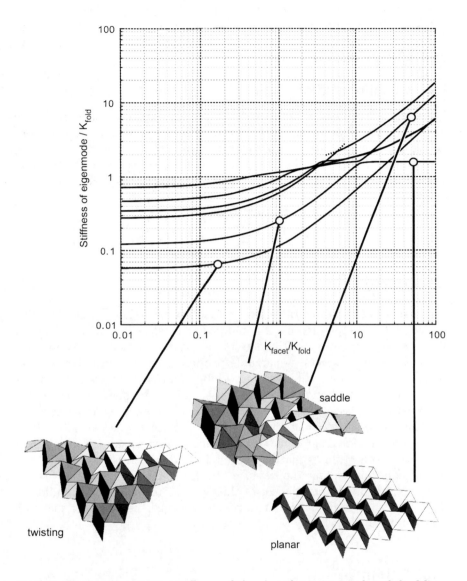

Figure 8. Plots of the relative stiffness of the six softest eigenmodes of the Miura sheet. The twisting deformation mode remains the softest eigenmode over a large range of K_{ratio}, and the saddle-shaped mode is also dominant. As $K_{ratio} \to \infty$, the planar mechanism identified in the kinematic analysis becomes the softest eigenmode.

stiffness ratio; this indicates that the dominant behavior is dependent on the geometry, rather than the exact material properties. The eigenvalues can straightforwardly be plotted in terms of a combination of different parameters, such as the fold depth and different unit cell geometries, to obtain further insight into the sheets.

3.4 Coordinate Transformation

Currently, all properties of the folded sheet are expressed in terms of the displacements of the nodal coordinates. The use of the (change in) fold angles may be more intuitive to origamists, and can improve understanding of the modes. This can be done using a coordinate transformation. The transformation matrix $\mathbf{T}$ converts nodal displacements $\mathbf{d}$ to changes in angle $d\theta$:

$$d\theta = \mathbf{Td}, \tag{10}$$

where $\mathbf{T}$ is identical to the Jacobian in Equation (5).

4   Conclusion

This paper has presented the idea of folded textured sheets, whereby thin-walled sheets are textured using a fold pattern inspired by origami folding. When considering the resulting sheets as a plate or shell, the two example sheets exhibit several remarkable properties: they can undergo large changes in shape and can alter their global Gaussian curvature by virtue of the folds opening and closing; they also exhibit unique behavior where the apparent Poisson's ratio is oppositely signed in bending and extension.

The proposed modeling method, which represents the partially folded sheet as a pin-jointed bar framework, enables a nice transition from a purely kinematic to a stiffness matrix approach, and provides insight into the salient behavior without the expense of a full Finite Element analysis. It captures the important behavior of the two example sheets, and indicates that the dominant mechanics are a result of the geometry rather than the exact material properties.

Bibliography

[Alderson et al. 10] A. Alderson, K. L. Alderson, G. Chirima, N. Ravirala, and K. M. Zied. "The In-Plane Linear Elastic Constants and Out-of-Plane Bending of 3-Coordinated Ligament and Cylinder-Ligament Honeycombs." *Composites Science and Technology* 70:7 (2010), 1034–1041.

[Balkcom 04] Devin J. Balkcom. "Robotic Origami Folding." Ph.D. thesis, Carnegie Mellon University, Pittsburg, PA, 2004.

[Basily and Elsayed 04] B. Basily and E. A. Elsayed. "Dynamic Axial Crushing of Multilayer Core Structures of Folded Chevron Patterns." *International Journal of Materials and Product Technology* 21:1–3 (2004), 169–185.

[Engel 68] Heino Engel. *Structure Systems*. Westport, CT: Praeger, 1968.

[Guest and Pellegrino 92] S. D. Guest and S. Pellegrino. "Inextensional Wrapping of Flat Membranes." In *First International Conference on Structural Morphology*, edited by R. Motro and T. Wester, pp. 203–215. Madrid: IASS, 1992.

[Heimbs et al. 07] S. Heimbs, P. Middendorf, S. Kilchert, A. F. Johnson, and M. Maier. "Experimental and Numerical Analysis of Composite Folded Sandwich Core Structures Under Compression." *Journal Applied Composite Materials* 14:5-6 (2007), 363–377.

[Huffman 76] D. A. Huffman. "Curvatures and Creases: A Primer on Paper." *IEEE Transactions on Computers* C-25:10 (1976), 1010–1019.

[Kuribayashi et al. 06] Kaori Kuribayashi, Koichi Tsuchiya, Zhong You, Dacian Tomus, Minoru Umemoto, Takahiro Ito, and Masahiro Sasaki. "Self-Deployable Origami Stent Grafts as a Biomedical Application of Ni-rich TiNi Shape Memory Alloy Foil." *Materials Science and Engineering: A* 419:1-2 (2006), 131–137.

[Lakes 87] R. Lakes. "Foam Structures with a Negative Poisson's Ratio." *Science* 235:4792 (1987), 1038–1040.

[Lim 07] Teik-Cheng Lim. "On Simultaneous Positive and Negative Poisson's Ratio Laminates." *Physica Status Solidi (b) Solid State Physics* 244:3 (2007), 910–918.

[Pellegrino and Calladine 86] S. Pellegrino and C. R. Calladine. "Matrix Analysis of Statically and Kinematically Indeterminate Frameworks." *International Journal of Solids and Structures* 22:4 (1986), 409–428.

[Tachi 09] Tomohiro Tachi. "Simulation of Rigid Origami." In *Origami⁴: Fourth International Meeting of Origami Science, Mathematics, and Education*, edited by Robert J. Lang, pp. 175–187. Wellesley, MA: A K Peters, 2009.

[Temmerman 07] Niels De Temmerman. "Design and Analysis of Deployable Bar Structures for Mobile Architectural Applications." Ph.D. thesis, Vrije Universiteit Brussel, Brussels, Belgium, 2007.

[Thill et al. 08] C. Thill, J. Etches, I. Bond, K. Potter, and P. Weaver. "Morphing Skins—A Review." *The Aeronautical Journal* 112:1129 (2008), 117–138.

[Watanabe and Kawaguchi 09] Naohiko Watanabe and Ken'ichi Kawaguchi. "The Method for Judging Rigid Foldability." In *Origami⁴: Fourth International Meeting of Origami Science, Mathematics, and Education*, edited by Robert J. Lang, pp. 165–174. Wellesley, MA: A K Peters, 2009.

[Wu and You 10] Weina Wu and Zhong You. "Energy Absorption of Thin-Walled Tubes with Origami Patterns." Paper presented at the Fifth International Meeting of Origami in Science, Mathematics, and Education, Singapore, July 14–15, 2010.

Designing Technical Tessellations

Yves Klett and Klaus Drechsler

1 Introduction: Paper as Technical Material

Folded paper as a cheap, disposable everyday material is so much part of daily life that the underlying importance and complexity of folded products in a technical context is seldomly appreciated. For example, corrugated cardboard is arguably *the* most important packing material. Less obtrusive but just as important, filters made from folded paper ensure smooth running of just about every combustion or industrial machine (folded coffee filters and tea bags do the same for their operators).

2 Lightweight Construction with Paper

Hidden from unsuspecting passengers, serious amounts of synthetic paper are already used in just about every aircraft floor and cabin paneling. These sandwich structures consist of two tough load-bearing facesheets spaced apart by a light core (Figure 1). The sandwich configuration multiplies the bending stiffness compared to a simple configuration with minimum additional weight (Figures 2 and 3), and in fact represents one central lightweight building principle.

State of the art in high-end core materials are hexagonal honeycombs consisting of glued and expanded layers of synthetic paper [Hexcel Composites 10]. Although these honeycombs offer excellent mechanical per-

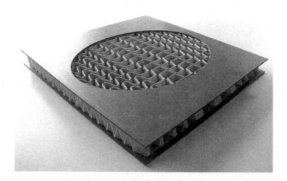

Figure 1. Aerospace sandwich structure consisting of two CFRP (carbon fiber reinforced plastic) facesheets and aramid composite MIO foldcore.

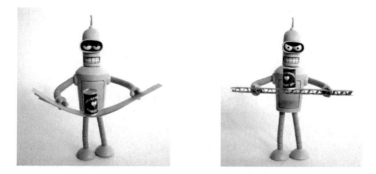

Figure 2. Three stacked layers of cardboard are floppy (left). Low-tech but high-performance, the same amount in sandwich configuration improves structural performance by an order of magnitude (right).

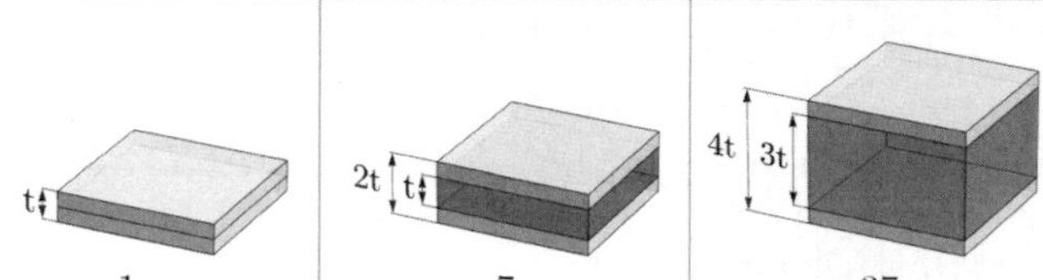

	Solid Material	Core Thickness t	Core Thickness 3t
Stiffness	1	7	37
Flexural Strength	1	3.5	9.2
Weight	1	1.03	1.06

Figure 3. Table of relative mechanical properties comparing different sandwich configurations with the solid material [Hexcel Composites 10].

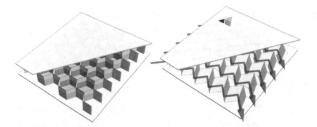

Figure 4. Honeycomb sandwich panels form closed cells that are prone to undesirable water accumulation (left). MIO foldcore panels offer comparable performance and additional integrated ventilation (right).

formance, their use is problematic, e.g., in moist environments, where the closed cells tend to fill up with water, leading to severe performance degradation and unwanted mass.

3   Modular Isometric Origami

Paper proves to be a highly effective material in the case of corrugated cardboard and honeycomb cores, so it is a fairly small step to use origami to generate alternative cellular materials with favorable properties. One material class that offers both comparable mechanical performance and additional functionality such as ventability are *modular isometric origami* (MIO) foldcores (Figure 4).

Several of these structures are well known and have already been considered for technical application and manufacture as early as 1959 [Hochfeld 59, Gewiss 60, Resch 83], but only quite recently have new large-scale sandwich concepts (Figure 5) and demand for multifunctional core materials

Figure 5. Ventable shear core (VeSCo) sandwich fuselage test assembly: 4.5 m^2 MIO core panels consisting of ca. 165,000 creases. (Image courtesy of Airbus Deutschland GmbH.)

led to breakthroughs in the material and production technology to provide foldcores in suitable quality and quantity to make them an interesting alternative to existing core types [Grzeschik and Drechsler 10].

3.1 Modularity

Compared to the production of corrugated cardboard, folding MIO structures pose a much more complex problem in terms of automation. Using a repetitive folding pattern that results in a tileable structure both in two and three dimensions reduces the complexity of folding large amounts of paper down to manageable (but not trivial) levels that can be accomplished in an automated process [Chaliulin 99, Kolax 04, Kehrle and Drechsler 04, Elsayed and Basily 09].

The smallest repetitive element in such a tessellation is the unit cell (Figure 6). Maximum simplicity of the unit cell is not only relevant for production, but also closely related to mechanical performance: usually, basic MIO unit cells provide the best performance compared to more complex tessellations. This also accounts for the fact that most MIO structures in industrial development bear a close relationship to the Miura type.

3.2 Rigid and Isometric Origami

Rigidly foldable structures with one degree of freedom (DOF) can be designed to perform significantly better mechanically than structures with multiple DOFs (e.g., corrugated cardboard) because the polygon sizes and boundary conditions defined by the additional creases can be tailored specifically to resist mechanical loads. The best known of these is the Miura map fold and related configurations, also termed Miura-ori, DDC (double developable corrugation), or herring-bone (Figure 6, top left).

Isometric (or more precisely 1-DOF isometric) origami represents a further extension of the rigidity paradigm. No stretching occurs during fold-

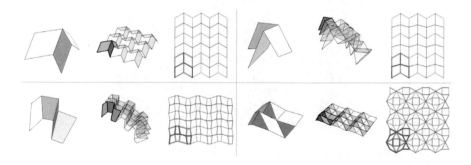

Figure 6. Different unit cells and their 3×3 3D and 2D tessellations.

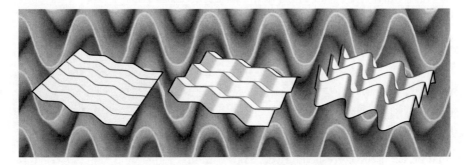

Figure 7. Sinusoidal modular isometric origami. In the background, some detail of the resulting hardware.

ing, and the geodesic distance between two arbitrary points in the paper plane remains constant at all times (Figure 7). In contrast, a typical wetfold intentionally introduces plastic deformation and is *not* isometric. Isometric origami using curved surfaces enlarges the application envelope of origami considerably, on both the artistic and design as well as the technical side [Kilian et al. 08].

The fact that during isometric folding no significant strain occurs is a central requirement for the use of high-performance materials, which usually have low breaking strains. Desirable materials such as glass, carbon, and aramid fiber composites (but also aluminum or titanium foils) can be processed only in a near-isometric fashion (Figure 8).

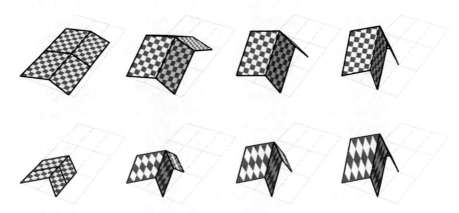

Figure 8. Comparison between a distance-preserving (isometric) folding process (top) and a deepdrawing process (bottom) to produce identical unit cells. Deepdrawing introduces plastic deformation and shear stresses.

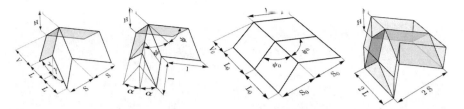

Figure 9. Parametric unit cell models used to derive geometric relations. The models show different (redundant) parameterizations of the 3D state (left and middle left), the corresponding flat state (middle right), and the unit cell core volume (right).

4 Design Strategies

For engineering purposes, designing the shape of a rigidly foldable unit cell is only a means to get a material that fulfills certain functions and requirements, which are closely related to the cell geometry. The *real* goal is to identify geometric influences on specific properties such as mechanical performance, density, feasibility, efficiency, cost and a host of application-dependent secondary functionality (e.g., ventability, optical properties, or visible surface area). To engineer cellular structures that can be folded rigidly to close specifications, two design strategies have been explored: bottom-up and top-down.

4.1 Bottom-Up

The *bottom-up* approach tries to derive optimally a simple set of parametric closed-form equations that describe the folding kinematics of a certain MIO unit cell type under the assumption of zero-thickness surfaces. Once obtained, these equations deliver a comprehensive handle on geometric and other (e.g., physical) properties of the resulting structure, hence demonstrated for the symmetric Miura unit cell.

Basic kinematics. Using the cell and dimension parameters from Figure 9, the following relations between the flat parameters (with index 0) and the 3D cell can be derived and allow easy design of the folding pattern for any set of 3D parameters (with $L > 0$, $S > 0$, $V > 0$ and $H > 0$):

$$L_0 = \sqrt{H^2 + L^2},$$

$$S_0 = \sqrt{\frac{H^2 V^2}{H^2 + L^2} + S^2},$$

$$V_0 = \frac{LV}{\sqrt{H^2 + L^2}}.$$

The kinematic equations can be formulated when one 3D parameter is treated as a control variable, determining the functions of other 3D parameters with given constant paper parameters. In the case of H as control (with $0 \leq H \leq L_0 S_0 / \sqrt{S_0^2 + V_0^2}$) from the flat state to the flat-folded or block state, this leads to

$$V(H) = \frac{L_0 V_0}{\sqrt{L_0^2 - H^2}},$$

$$S(H) = \frac{\sqrt{L_0^2 S_0^2 - H^2 \left(S_0^2 + V_0^2\right)}}{\sqrt{L_0^2 - H^2}},$$

$$L(H) = \sqrt{L_0^2 - H^2}.$$

Switching between different parametric representations (as shown in Figure 9) can be achieved by suitable substitutions, e.g.,

$$L = H \tan(\alpha) \csc(\psi), \quad S = l \sin(\psi), \quad V = l \cos(\psi), \quad L = H \tan(\gamma).$$

The different states during folding can be described by a concrete cell dimension or—usually more convenient—by a dimensionless parameter μ whose parameterization can be chosen according to user needs. In the Miura case, a useful substitution is

$$\mu_H = \frac{H}{H_{\max}}.$$

The derived equations can be used to design patterns and simulate the folding process, but also to define certain interesting physical properties such as the unit cell volume and the resulting density. It is not necessary to restrict the range for μ from 0 to 1. Depending on ease of use and the choice of reference, other ranges can be chosen (e.g., in a multivertex situation, complex restrictions may apply [Watanabe and Kawaguchi]). Especially for everyday handling, tagging the desired 3D state with μ_1 proves convenient.

Physical unit cell properties. The unit cell volume in the case of a sandwich application is defined by the space that is taken up by one unit cell between the face sheets. This volume is defined by

$$\text{Vol} = 4LSH.$$

The unit cell volume during folding results from using the kinematic equations and μ_H as a parameter:

$$\text{Vol}(\mu_H) = 4L_0^2 S_0^2 \mu_H \sqrt{\frac{1 - \mu_H^2}{S_0^2 + V_0^2}}.$$

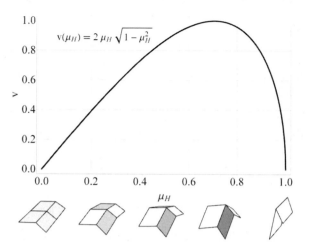

Figure 10. Plot of the specific volume v during folding.

The volume function yields a global maximum at $\mu_H = 1/\sqrt{2}$:

$$\mathrm{Vol}_{\mathrm{max}} = \frac{2L_0^2 S_0^2}{\sqrt{S_0^2 + V_0^2}}.$$

The specific volume v can be computed by normalizing with the maximum volume, leading to the surprisingly simple form of

$$v = \frac{\mathrm{Vol}}{\mathrm{Vol}_{\mathrm{max}}} = \mu_H \sqrt{1 - \mu_H^2}.$$

Accordingly, the maximum specific volume for any symmetric Miura cell will always be reached at the same μ_H, regardless of its actual shape. Obviously, for maximum volume, α will be equal to $\pi/4$. The graph of $v(\mu_H)$ is shown in Figure 10.

Another important property of a core material is the density ρ of the resulting structure. The density is closely related to the mechanical properties of the core, (i.e., the higher the density for a given raw material with grammage η, the higher the stiffness and strength of the core):

$$\rho(L, S, V, H, \eta) = \frac{L_0 S_0 \eta}{HLS} = \frac{\eta \sqrt{H^2 (S^2 + V^2) + L^2 S^2}}{HLS}. \tag{1}$$

Because initial design of a core material is often done by selecting a certain core density, it is useful to solve Equation (1) for $L(\rho, S, V, H, \eta)$ or

Figure 11. Equal volumes filled with cores with equal densities, but different ψ.

$\alpha(H, \rho, \eta)$ to easily design cores with defined densities:

$$L(\rho, S, V, H, \eta) = H\eta\sqrt{-\frac{S^2 + V^2}{S^2\left(\eta^2 - H^2\rho^2\right)}},$$

$$\alpha(H, \rho, \eta) = \tan^{-1}\left(\eta\sqrt{\frac{1}{H^2\rho^2 - \eta^2}}\right).$$

A variation for different unit cell geometries that fill the same volume and have the same density but different ψ is shown in Figure 11, ranging from $10°$–$80°$. The cores have strongly differing anisotropic mechanical in-plane shear properties (depending mainly on ψ) and slightly different out-of-plane compression properties (depending mainly on α).

Advanced kinematics. Even though the behavior of the unit cell during folding is not relevant for the mechanical properties nor the end user of a foldcore sandwich material, in-depth kinematic analysis is important for the design process. Intimate knowledge can help with optimization of the manufacturing (e.g., adapting tools to closely mimic the folding process), and also generally to design kinematic linkages with specific characteristics that can be useful (e.g., in robotic applications that need to trace specific, complex paths using just one actuator).

For this task, one particularly useful parameterization of μ can be obtained by normalizing with the width parameter S (Figure 9):

$$\mu_S = 1 - \frac{S}{S_0}.$$

The kinematic equations for μ_S then take the form of

$$V\left(\mu_S\right) = \sqrt{V_0^2 - S_0^2\left(-2 + \mu_S\right)\mu_S},$$

$$H\left(\mu_S\right) = L_0 S_0 \sqrt{\frac{\left(-2 + \mu_S\right)\mu_S}{-V_0^2 + S_0^2\left(-2 + \mu_S\right)\mu_S}},$$

$$L\left(\mu_S\right) = \frac{L_0 V_0}{\sqrt{V_0^2 - S_0^2\left(-2 + \mu_S\right)\mu_S}}$$

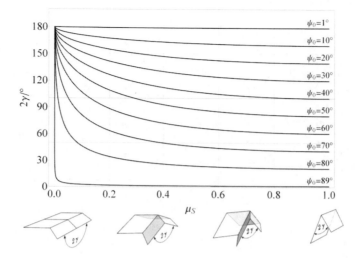

Figure 12. The $\gamma(\mu_S)$-curve set for generic symmetric four-crease vertices with different ψ_0 (top). Exemplary unit cell vertex during folding (bottom).

which leads to an interesting result for the linkage angle $2\gamma(S)$ (Figure 9), and by suitable substitution to $\gamma(\mu_S)$, for $0 < \mu_S \leq 1$ and $\psi_0 \neq \pi/2$:

$$\gamma(S) = \tan^{-1}\left(\frac{V_0}{\sqrt{S_0^2 - S^2}}\right),$$

$$\gamma(\mu_S) = \tan^{-1}\left(\frac{\cot(\psi_0)}{\sqrt{(2 - \mu_S)\,\mu_S}}\right). \tag{2}$$

Equation (2) now describes the kinematics of a generic symmetric four-crease vertex, resulting in one set of curves depending only on ψ_0 and the folding parameter μ_s. Figure 12 shows a number of these curves that characterize the vertices: vertices with ψ_0 close to 0 will move in a fairly linear fashion but cover only a small $\Delta\gamma$, whereas vertices with ψ_0 close to $\pi/2$ will move in a very nonlinear fashion but exhibit large $\Delta\gamma$ (the borderline case $\psi_0 = 0$ characterizes a couplet [Leong 11]).

Using Equation (2), the behavior of several joined vertices with defined ψ_{0_i} and L_{0_i} can now be put down in a closed form (Figure 13).

For n vertices in a planar forward kinematic chain, the vertex position is determined by addition of the components of each arm with length R_i relative to a given reference (in Figure 13 this reference is the vector $\overrightarrow{01}$). Taking care of the initial folding direction of each vertex and magnitude of

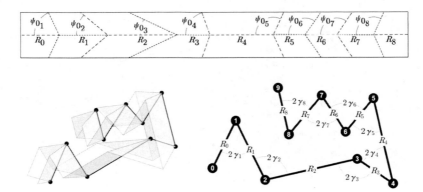

Figure 13. Multivertex chain with eight different vertices: folding pattern with characteristic parameters (left), and 3D chain and associated parameters (right).

γ_0 by adding sign factors sgn_j and $\mathrm{sgn}\left(\cos\left(\psi_{0_j}\right)\right)$ results in

$$
\begin{pmatrix} x\left(n, \mu_s\right) \\ y\left(n, \mu_s\right) \end{pmatrix} = \begin{pmatrix} \sum_{i=1}^{n} \cos\left(\sum_{j=1}^{i}\left(\pi - 2\mathrm{sgn}_j\gamma_j\right)\right) R_i \\ \sum_{i=1}^{n} \sin\left(\sum_{j=1}^{i}\left(\pi - 2\mathrm{sgn}_j\gamma_j\right)\right) R_i \end{pmatrix}
$$

$$
= \begin{pmatrix} \sum_{i=1}^{n} \cos\left(\sum_{j=1}^{i}\left(\pi - 2\tan^{-1}\left(\frac{\cot\left(\psi_{0_j}\right)\mathrm{sgn}\left(\cos\left(\psi_{0_j}\right)\right)}{\sqrt{\left(2-\mu_S\right)\mu_S}}\right)\right)\mathrm{sgn}_j\right) R_i \\ \sum_{i=1}^{n} \sin\left(\sum_{j=1}^{i}\left(\pi - 2\tan^{-1}\left(\frac{\cot\left(\psi_{0_j}\right)\mathrm{sgn}\left(\cos\left(\psi_{0_j}\right)\right)}{\sqrt{\left(2-\mu_S\right)\mu_S}}\right)\right)\mathrm{sgn}_j\right) R_i \end{pmatrix}.
$$

$$(3)$$

Formulating the complete closed-form forward kinematics for any chosen vertex in the chain now simply consists of evaluating Equation (3). One example for the resulting trace of the rather randomly chosen vertex chain from Figure 13 is shown in Figure 14, which demonstrates the complexity (and beauty) of the emergent kinematics even when using just one simple vertex type. Also, for chains of identical elements, very compact generic functions for n vertices can be derived: Figure 15 shows the result for a chain whose vertices describe circles with growing radii and x-axis offset.

This approach can also be extended into 3D nonplanar linkages by looking at larger cell arrays or the use of asymmetric vertices, usually resulting in much more complex expressions.

Obtaining inverse kinematic solutions (i.e. finding the parameters for a kinematic chain that traces a certain predefined curve with required precision) is also a very important task that can usually only be solved with complex numerical methods. Nevertheless, the derived equations may be used as a convenient starting point to model arbitrary parametric 1-DOF

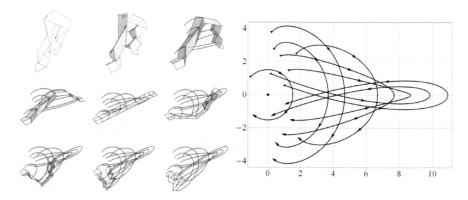

Figure 14. Folding kinematics of the multivertex chain: folding steps (left), and trace curves of the vertices in the symmetry plane (right).

Vertex	x	y
1	$-r\cos(2\cot^{-1}(\mu\tan(\psi_0)))$	$r\sin(2\cot^{-1}(\mu\tan(\psi_0)))$
2	$r - r\cos(2\cot^{-1}(\mu\tan(\psi_0)))$	$r\sin(2\cot^{-1}(\mu\tan(\psi_0)))$
3	$r - 2r\cos(2\cot^{-1}(\mu\tan(\psi_0)))$	$2r\sin(2\cot^{-1}(\mu\tan(\psi_0)))$
4	$2r - 2r\cos(2\cot^{-1}(\mu\tan(\psi_0)))$	$2r\sin(2\cot^{-1}(\mu\tan(\psi_0)))$
5	$2r - 3r\cos(2\cot^{-1}(\mu\tan(\psi_0)))$	$3r\sin(2\cot^{-1}(\mu\tan(\psi_0)))$
$\vdots$	$\vdots$	$\vdots$
n	$\frac{1}{4}r(-1+(-1)^n+2n) -$ $\frac{1}{4}r(1-(-1)^n+2n)\cos(2\cot^{-1}(\mu\tan(\psi_0)))$	$\frac{1}{4}r(1-(-1)^n+2n)$ $\sin(2\cot^{-1}(\mu\tan(\psi_0)))$

Figure 15. Generic vertex coordinate functions for a Miura cell chain with n vertices with identical ψ_0 and r (for $0 \le \mu \le 1$ and $0 \le \psi_0 \le \pi/2$).

vertex chains and tackle their inverse kinematics with standard methods (e.g., from robotics [Haddadin 09]).

4.2   Top-Down

The idea behind the top-down approach is to directly generate foldable and tileable 3D unit cells that seem suited for the desired technical application without the need for prior comprehensive analysis.

To check the correctness of the MIO-design, the 3D cell is plugged into an inverse kinematic unfolding algorithm that numerically simulates the rigid folding down to the flat state μ_0 and along the way also provides the flat-folding pattern without additional input (or work) by the user [Klett 08].

Whereas the bottom-up design has a lot in common with classical technical drawing (where the engineer needed a working model in his head to derive the two-dimensional drawing), the top-down approach is closely

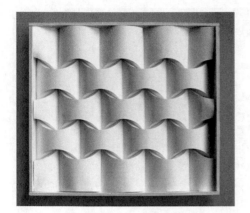
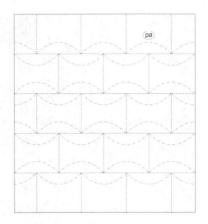

Plate I. Huffman's "Arches" design using parabolas and lines. (See page 46.)

Plate II. "Degrees of Freedom in Blue." (See page 70.)

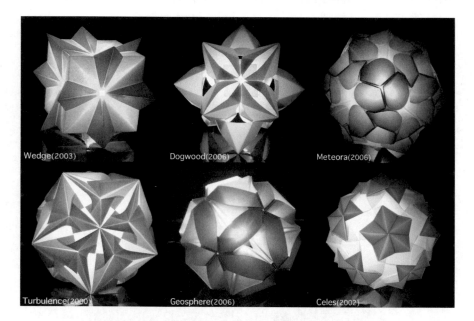

Plate III. Models by Kawamura: "Wedge," "Dogwood," "Meteora," "Turbulence," "Geosphere," and "Celes." (See page 93.)

Plate IV. Macro-modular twirl models. (See page 122.)

Plate V. Oribotics at the 2010 Ars Electronica Festival. (See page 128.)

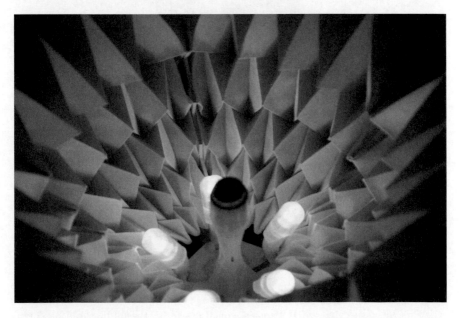

Plate VI. Detail of the blossom head from Oribotics [Futurelab]. (See page 135.)

Plate VII. A glass on the table. (See page 160.)

Plate VIII. A jumper in a pool. (See page 160.)

Plate IX. A 120-PHiZZ unit Bucky Ball by Tom Hull [Hull 06] folded by Charlene Morrow. (See page 190.)

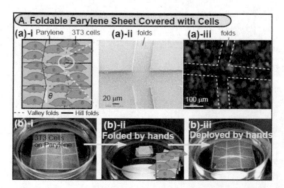

Plate X. (a)-i The schematic view of *Miura-ori* with cells ($\theta = 8$). (a)-ii Scanning electron microscope (SEM) image of the folds on the Parylene sheet. (a)-iii The 3T3 cells were cultured successfully. The green and blue colors are the cells and their nuclei, respectively. (b) The sequential images of the folding and deploying of the Parylene sheet with the cells. (See page 389.)

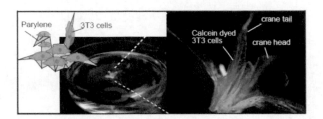

Plate XI. The Parylene origami crane (thickness 10 μm). The size of the origami was 15 mm × 15 mm. The 3T3 cells were successfully cultured on the 3D structure. (See page 389.)

Plate XII. An example design of rigid foldable origami materialized with cloth and cardboard. (See page 261.)

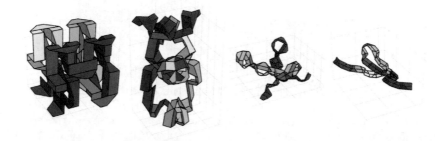

Plate XIII. Peano-Cube unfolding sequence (top) and folding pattern (bottom). Note the retained symmetry of the unfolding structure. (See page 320.)

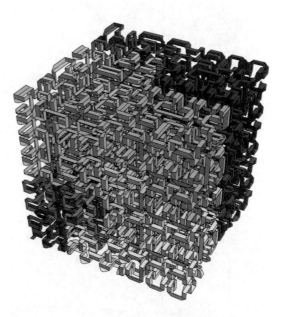

Plate XIV. The cube from Plate XIII in aluminum. (See page 320.)

Plate XV. Somewhat Borg-ish fourth iteration peanori. The second iteration is actually foldable (see Plate XIV), but realizing this one seems ambitious. (See page 320.)

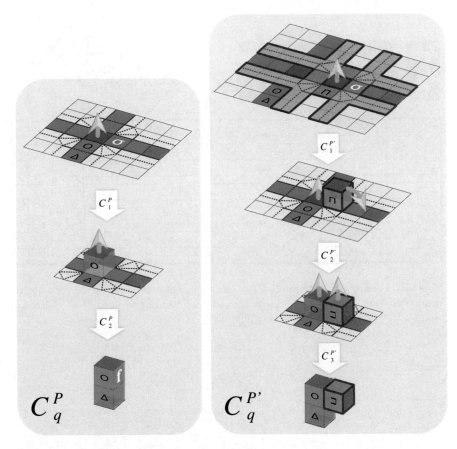

Plate XVI. Given a coalesce sequence C^P, this diagram shows the application of Lemma 1 to generate $C^{P'}$ with an additional cube. The purple regions show where additional rows and columns are inserted. (See page 412.)

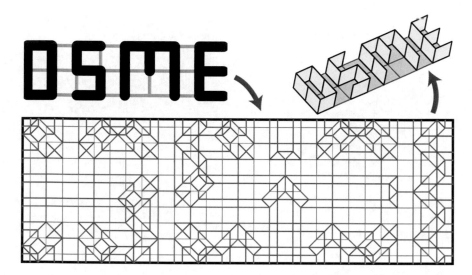

Plate XVII. Folding three-dimensional letters from a rectangle of paper. Font design from [Demaine et al. 10b]. Valleys are blue, mountains are red, 180° folds are thick, and 90° folds are thin. (See page 450.)

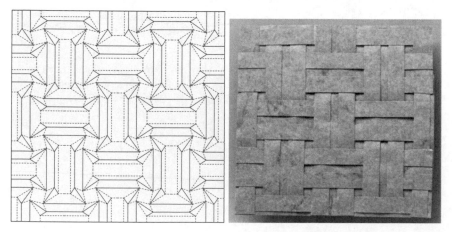

Plate XVIII. Crease pattern for a double-strip woven tessellation (left); photograph of a folded example (right). (See page 488.)

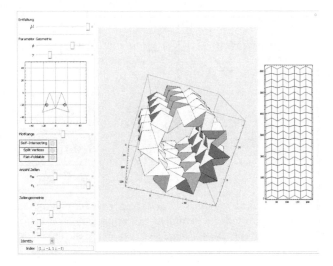

Figure 16. Graphical user interface (GUI) for top-down manipulation of MIO structures. One slider for μ controls the unfolding process. The folding pattern is displayed on the right side.

related to modern CAD systems that allow for 3D modeling and manipulation and automatically generate the drawings. The 3D unit cell geometry can be manipulated in real time without the user needing to be familiar with the involved algorithms (Figure 16).

The design of more complex unit cells becomes much easier with this approach because different working cells or cell blocks can be combined quickly without the footwork entailed by the analytic approach. The method to provide a rigidly foldable 3D unit cell is up to the user, with several methods (e.g., [Elsayed and Basily 09, Tachi 09, Lang 10, Klett 03]) already documented.

The simulation of the unfolding can, by visual or automated inspection, reveal surprising information about the rigid folding behavior or the validity of a certain unit cell.

Figure 17 demonstrates such a case: The top row shows a valid checkerboard tessellation and its unfolding sequence. The bottom row shows a derived unit cell that is the result of an affine shear transformation of the top row. While both the 3D state and the flat unfolded pattern look perfectly fine, the folding simulation reveals that the pattern cannot be folded rigidly and in fact develops large splits between some vertices in intermediate states.

Closer analysis of this case shows that each single four-crease vertex of this tessellation is rigidly foldable, but the conditions for a seamless combination of all vertices at once cannot be met during folding—only in

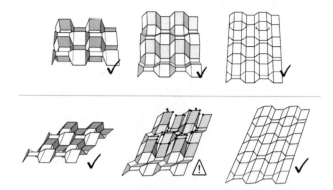

Figure 17. A checkerboard unit cell tessellation can be rigidly unfolded (top). The sheared unit cell provides a seamless 3D appearance and flat pattern, but cannot be rigidly folded in between because of splits appearing at marked vertices (bottom).

the flat and 3D design states. This fact is revealed only by simulation of the intermediate states and may otherwise be missed when looking only at the 3D and flat configurations—an important argument for the provision of easily applicable simulation tools.

Compared to a parametric analytic solution, cell properties in this process, such as density, have to be acquired on a one-by-one basis and often rely on an iterative scheme. Nevertheless, from experience, the benefit of real-time simulation of the folding accelerates the design process significantly compared to a pure bottom-up path and also allows the design of very complex structures for which the amount of analytical work would be prohibitive.

5   Not So Serious—Some Fun

Apart from doing serious work, the design methods and tools originally developed for high-performance MIOs can be applied to other configurations without constant focus on technical potential. Real-time digital modeling and simulation enables the user to explore structures that otherwise would take orders of magnitude longer to realize in hardware—or would not be foldable at all due to physical constraints.

Following parametric curves with inverse folds is one interesting way to produce beautiful structures, such as the torus braids and knots shown in Figure 18.

L-systems also lend themselves very nicely for this purpose. The top-down design approach is well suited to generate folding patterns for com-

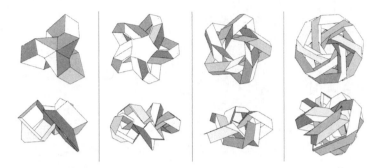

Figure 18. Torus braids and knots. Each vertex can be folded rigidly; global rigid folding leads to self-intersection.

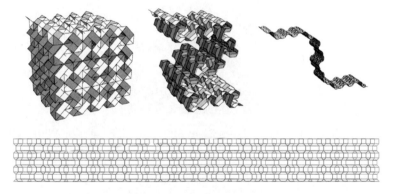

Figure 19. Cuboid rigid origami (*peanori*) derived from a second iteration 2D Peano curve. Interesting checkerboard patterns emerge when using bicolor paper.

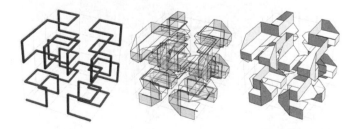

Figure 20. Hilbert-Peano space filling curve (second iteration) used to generate pseudo-fractal peanori.

Figure 21. Peano-Cube unfolding sequence (top) and folding pattern (bottom). Note the retained symmetry of the unfolding structure. (See Color Plate XIII.)

Figure 22. The cube from Figure 21 in aluminum. (See Color Plate XIV.)

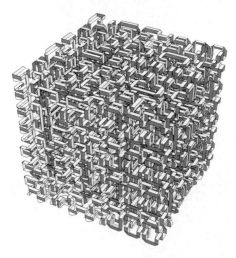

Figure 23. Somewhat Borg-ish fourth iteration peanori. The second iteration is actually foldable (see Figure 22), but realizing this one seems ambitious. (See Color Plate XV.)

plex pseudofractal origami structures up to considerable complexity: A 2D Peano curve is used as the basis for the tessellation shown in Figure 19. The resulting origami can be folded in one piece rigidly without any self-intersection. Figure 20 visualizes the design process for the Hilbert-Peano space-filling curve, Figure 21 shows the unfolding process, and Figure 22 shows a real-world implementation of the second iteration model. Finally, Figure 23 shows a higher-iteration version.

6 Conclusion

Isometric origami tessellations offer a large innovation potential for technical applications. In the case of MIO sandwich core materials, a number of different products are undergoing development and testing, but other industrial areas will also benefit from the growing know-how and the advancements in computational origami. Besides being great fun, the topic offers a huge number of interesting challenges on both the theoretical and practical side to the origami community.

Bibliography

[Chaliulin 99] I. V. Chaliulin. *Technological Schemes for Sandwich Structures Production.* ISBN 5-7579-0295-7, Kazan, Russia: Kazan State Technological University, 1999.

[Elsayed and Basily 09] E. A. Elsayed and Basily B. Basily. "Applications of Folding Flat Sheets of Materials into 3-D Intricate Engineering Designs." Working paper 07-015, Rutgers University, 2009. (Available at http://ise.rutgers.edu/resource/research_paper/paper_07-015.pdf.).

[Gewiss 60] L. V. Gewiss. "Arrangement for the Mechanical and Continuous Production of Developable Herring-Bone Structures." US Patent 2,950,656, 1960.

[Grzeschik and Drechsler 10] Marc Grzeschik and Klaus Drechsler. "Experimental parameter studies on folded cores." Paper presented at the Ninth International Conference on Sandwich Structures (ICSS-9), Pasadena, CA, June 14–16, 2010.

[Haddadin 09] Sami Haddadin. "Robotics: Inverse Kinematics." Available at http://www.robotic.de/fileadmin/robotic/haddadin/2_inverse_kinematics.pdf, 2009.

[Hexcel Composites 10] Hexcel Composites. "Honeycomb Sandwich Design Technology." Available at http://www.hexcel.com/Resources/DataSheets/Brochure-Data-Sheets/Honeycomb_Sandwich_Design_Technology.pdf, 2010.

[Hochfeld 59] H. Hochfeld. "Process and Machine for Pleating Pliable Materials." US Patent 2,901,951, 1959.

[Kehrle and Drechsler 04] R. Kehrle and K. Drechsler, "Manufacturing of Folded Core Structures for Technical Applications." In *SAMPE Europe 25th International Conference*, p. 508513. Niederglatt, Switzerland: SAMPE Europe, 2004.

[Kilian et al. 08] Martin Kilian, Simon Floery, Zhonggui Chen, Niloy J. Mitra, Alla Sheffer, and Helmut Pottmann. "Curved Folding." In *ACM SIGGRAPH 2008 Papers*, p. Article no. 75. New York: ACM Press, 2008.

[Klett 03] Yves Klett. "Auslegung der Geometrie endkonturnah gefalteter Wabenstrukturen." Diploma thesis, Institut für Flugzeugbau, Stuttgart, Germany, 2003.

[Klett 08] Yves Klett. "Cutting Edge Cores: Aerospace and Origami." Paper presented at the Ninth International Mathematica Symposium, Maastricht, The Netherlands, June 20–24, 2008.

[Kolax 04] Michael Kolax. "Concept and Technology: Advanced Composite Fuselage Structures." *JEC Composites* 10:6/7 (2004), 31–33.

[Lang 10] Robert J. Lang. "Optigami." Available at http://www.langorigami.com/science/optigami/optigami.php4, 2010.

[Leong 11] Cheng Chit Leong. "Simulation of Nonzero Gaussian Curvature in Origami by Curved-Crease Couplets." In *Origami⁵: Fifth International Meeting of Origami Science, Mathematics, and Education*, pp. 53–67. Boca Raton, FL: A K Peters/CRC Press, 2011.

[Resch 83] Ron Resch. "Construction-Element." US Patent 4,397,902, 1983.

[Tachi 09] Tomohiro Tachi. "Generalization of Rigid-Foldable Quadrilateral-Mesh Origami." *IASS Journal* 3 (2009), 173–179.

[Watanabe and Kawaguchi] Nohiko Watanabe and Kenichi Kawaguchi. "The Method for Judging Rigid Foldability." In *Origami⁴: Fourth International Meeting of Origami Science, Mathematics, and Education*, edited by Robert J. Lang. A K Peters.

A Simulator for Origami-Inspired Self-Reconfigurable Robots

Steven Gray, Nathan J. Zeichner, Mark Yim,
and Vijay Kumar

1   Introduction

The design of folded structures is closely related to the algorithmics of origami [Lang 03, Demaine and O'Rourke 08]. Traditional origami involves developable surfaces; paper can be deformed by bending but not by stretching. In contrast, foldable programmable matter (FPM) crease patterns represent rigid sheets connected by revolute joints. As a cursory study of origami would suggest, foldable sheet structures are able to approximate arbitrary shapes ranging from animals to boxes, airplanes, and boats. The limiting factor in the design is the size of the sheet and the minimum spacing between creases; it has been shown that every polyhedral surface can be folded, given a large enough sheet [Demaine et al. 00].

The aim of FPM is to create smart, self-folding sheets that can reconfigure into multiple desired forms. The foldable programmable matter editor (FPME), a tool for drawing and simulating rigid origami, has been created to facilitate the design and testing of programmable matter structures. The physical nature of folding programmable matter imposes additional requirements on the simulation. The structures consist of flat, rigid sheets of non-negligible thickness. The sheets are connected by revolute joints, which are coplanar when the structure is in its unfolded state. A variety of actuators, with their own unique properties, may be involved in the folding motion. Accordingly, geometric constraints are insufficient for this modeling. In contrast to kinematic origami simulators, the programmable matter editor utilizes the nVidia PhysX package to create a real-time dynamic simulation of the folding procedure. The editor is used to draw (or import) a crease pattern. For a given crease pattern, the user defines sequences of folds and the physical actuator properties that guide those folds. The program can then be used to visualize the folding in real time.

The work presented here begins with a discussion of FPM and its structure. The necessary data for representation are then covered in Section 3; these data are contained in XML schemas known as Programmable Matter Markup Language (PMML) and Controllable Matter Markup Language (CMML). Implementation of the editor and usage details are discussed in Section 4. Lastly, simulation and results appear in Section 5.

2 Foldable Programmable Matter

The single-vertex FPM is the basic building block of all FPM structures. By definition, all crease lines in a single-vertex FPM are incident upon one vertex. The regions surrounded by creases or boundaries are rigid. The *degree* of the single-vertex FPM refers to the number of incident creases.

Intuitively, intersecting the single-vertex FPM with a sphere yields a kinematically equivalent spherical chain. An origami vertex located *inside* the FPM results in a spherical closed chain, whereas an origami vertex located *on* the outer boundary results in a spherical open chain. Let us consider the joint angle parameterization of a single-vertex FPM. A configuration is given by *sector angles*, ω, and *dihedral angles*, θ, as shown in Figure 1. For a given sheet, the sector angles are properties of the rigid facets and thus constant. Changing configurations requires only the change in dihedral angles. The *configuration space* contains all valid angle configurations, allowing self-touching, but not self-crossing configurations.

A necessary condition for a rigidly foldable origami vertex is that it satisfies the condition for loop closure; the composition of rotation matrices representing the motion about the vertex must yield the identity matrix [belcastro and Hull 02, Demaine and O'Rourke 08]. Let matrix A represent the composition of the rotation matrices, where $R_x(\omega)$ is a rotation about the x-axis by ω and $R_z(\theta)$ is a rotation about the z-axis by θ.

$$A = R_x(\omega_n)R_z(\theta_n)\ldots R_x(\omega_2)R_z(\theta_2)R_x(\omega_1)R_z(\theta_1) = I \qquad (1)$$

The facets of the sheet remain flat and rigid; the transformation is piecewise affine. Though necessary, the closure condition is not sufficient because it allows configurations containing intersecting facets.

The valid velocities that maintain the closure configuration can be obtained by differentiation of Equation (1). The sector angles remain con-

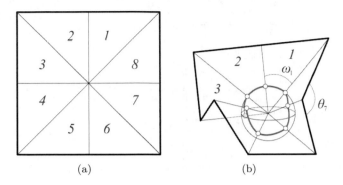

(a) (b)

Figure 1. (a) A degree-8 single-vertex FPM of perimeter 2π. Every facet remains rigid while folding about any creases. (b) Intersecting a degree-8 single-vertex FPM with a sphere yields a spherical 8-bar closed chain. Angle ω is a sector angle and θ is a dihedral angle.

stant, so the $\dot\omega_i$ terms drop out, leaving

$$\sum_{i=1}^{n} \dot\theta_i \frac{\partial A}{\partial \theta_i} A^{-1} = 0. \tag{2}$$

The partial derivative of an orthogonal matrix evaluated at identity is skew-symmetric. Thus, $\partial A/\partial\theta_i$ is a skew symmetric matrix. The nonzero components of the matrix can be represented as a vector, and each vector is an axis of rotation of the spherical linkage [McCarthy 00]. We write the rotation axis corresponding to the rotation θ_i as $\hat{\boldsymbol\theta}_i$. Valid instantaneous velocities for the linkage are the set of $\dot\theta_i$ that satisfy Equation (3) for a set of $\hat{\boldsymbol\theta}_i$:

$$\begin{bmatrix} \hat{\boldsymbol\theta}_1 & \hat{\boldsymbol\theta}_2 & \cdots & \hat{\boldsymbol\theta}_n \end{bmatrix} \begin{bmatrix} \dot\theta_1 \\ \dot\theta_2 \\ \vdots \\ \dot\theta_n \end{bmatrix} = \mathbf{0} \tag{3}$$

Multi-vertex FPM can be represented by *crease patterns*, networks that arise when origami vertices have common creases. Huffman provides an analysis of simple multi-vertex networks consisting of degree-4 vertices [Huffman 76]. The boat shown in Figure 2(a), with 4-bar spherical linkages connected in serial, is such a structure. Actuating a single DOF of any of the spherical 4-bar chains will allow the structure to fold as shown in Figure 2, assuming the linkages are not in the flat configuration and are biased in the right direction. In the flat configuration, each spherical 4-bar has 2 DOF instantaneously.

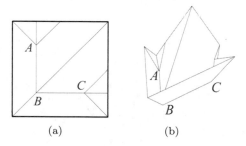

(a)   (b)

Figure 2. (a) A simple multi-vertex crease pattern, and (b) the resulting folded sheet.

Despite having results valid for all single-vertex FPM, generalization to the multi-vertex case is nontrivial except for simple examples such as the boat; the single-vertex results are only sufficient to explain local behavior of multi-vertex designs. Tachi provides a simulator for multi-vertex FPM designs in which all joints move simultaneously, based on projecting desired velocities for all joints into the constraint space [Tachi 09]. The approach is to augment the matrices of Equation (3) to include all rotation axes about all vertices and solve for the velocities. The directionality of each fold must be known beforehand. In general, whereas any two configurations of a single-vertex FPM have been shown to be connected, the same cannot be said for multi-vertex designs. The initial and goal configurations may be in disconnected components of the configuration space.

3   Representing Programmable Matter

Two XML schemas are used to represent the necessary information to create a foldable programmable matter object. The Programmable Matter Markup Language contains the physical parameters and crease pattern of the structure; the Controllable Matter Markup Language contains the folding sequence. The separation of physical structure from folding reinforces the notion of reconfigurability for a given crease pattern. Multiple CMML fold sequences may be associated with a single PMML structure. Explicit transformations between folded states are unnecessary because all configurations are connected via the flat-folded configuration.

PMML is an XML schema designed to hold the necessary information to create a specific multi-vertex FPM crease network. The PMML schema is shown in Listing 1. The associated data include the name of the sheet, the geometry of the rigid bodies, and the geometry and physical properties of the joints. The format and nomenclature have been chosen to closely

```
<PMML>
  <Title> title </Title>
  <Actor>
    <ID> actor ID </ID>
    <Triangle> x_1 y_1 x_2 y_2 x_3 y_3 </Triangle>
  </Actor>
  . . .
  <Joint>
    <ID> joint ID </ID>
    <Coords> x_1 y_1 x_2 y_2 </Coords>
    <ActorIDs> id_1 id_2 </ActorIDs>
    <SpringConst> k </SpringConst>
    <DampConst> c </DampConst>
  </Joint>
  . . .
</PMML>
```

Listing 1. The PMML schema.

represent data structures present in our PhysX-based simulator. Thus, individual rigid bodies are referred to as *actors* and are composed of triangles either created by the user or created by applying Delaunay triangulation and constraining to the rigid body outline. The triangulated bodies are used for collision detection. Joint edges are those connecting two rigid bodies. For the sake of flexure hinge manufacturability as well as crease pattern reconfigurability (one fold sequence may call for a joint to be a mountain fold and another may call for a valley fold), the joint axes are coplanar and vertically centered when the structure is in flat-folded configuration. PMML specifies the IDs of the connected bodies, coordinates that give the axis of rotation, and spring and damping constants. Thus, single-vertex FPM subunits are not specified directly; rather, they occur at the intersections of the joint axes of adjacent rigid bodies.

Folding sequences have been added using CMML, an XML schema for sequences of fold events. The CMML schema is displayed in Listing 2. For

```
<CMML>
  <Title> title </Title>
  <Fold>
    <EventNumber> event number </EventNumber>
    <ID> joint ID </ID>
    <FoldAngle> desired final angle </FoldAngle>
    <Velocity> desired velocity </Velocity>
  </Fold>
  . . .
</CMML>
```

Listing 2. The CMML schema.

each fold, the joints involved and their desired final angles are specified. There is no limit to the number of joints that can be moved concurrently, nor to the number of fold steps. A desired speed is stored to be used when motor actuation, rather than spring actuation, is used.

Fold sequences can be replayed in the simulation environment. Any motion planning algorithm for FPM can be simulated if the output complies with the CMML format, and the PMML document precisely describes the geometry of the given FPM.

4 Editor Implementation and Usage

The interface of the FPME was created to facilitate the design of multi-vertex FPM. Fundamentally, it is a visual tool to create, save, and modify PMML and CMML. Graphics and interface design are handled using OpenGL and FLTK, respectively; the program is coded in C++. The program has two modes: the editor and the simulator.

While in editor mode, shown in Figure 3(a), users interact with a two-dimensional grid representing the initial flat-folded sheet. The editor encompasses various tools for inputting information for PMML: drawing, translation, rotation, scaling, mirroring, angle and length measurement, merging vertices, merging edges, splitting edges, copying and pasting triangles, creating a grid or circular lattice, and grouping triangles into rigid bodies.

The primary toolbar is vertical and located on the left of the display. The top-most text entry box allows users to enter a name to be used for reference when compiling PMML and CMML together. The largest text box below displays information about the currently selected tool or most recently selected drawing element. Information is updated as objects are selected, including the angle between two creases, positions of endpoints of creases, direction of fold of a crease, etc. Continuing downward, the radio buttons provide easy access to the most commonly used tools in the program. Below the radio buttons lies a quick tool for moving objects. Entering the translation into the x- and y-coordinates will translate the currently selected objects. Settings for folding are located on the bottom of the toolbar. Here users can select the spring and damping constants, which physically change how the model will fold during the simulation process. In addition, the user can specify desired angles for each crease at the fold step.

The menu bar offers the same functionality as the toolbar while adding some operations that designers will find useful. The File menu handles the current document and the PMML and CMML files associated with the crease pattern. The Tools menu allows the user to change between tools.

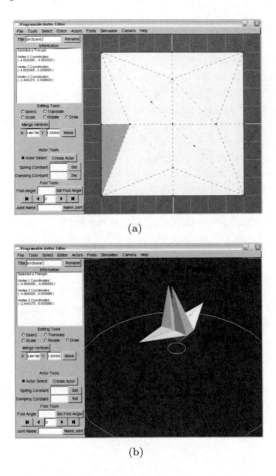

(a)

(b)

Figure 3. Example of the bird base in (a) the editor and (b) the simulator.

The Select menu allows for quick selection of different parts of the crease pattern. Users may select all of the creases, vertices, or rigid portions of a design. The Editor menu contains all of the major operations used for editing the crease pattern. Here users will find merging vertices, merging edges, creating lattice patterns for tessellations, copy and paste, splitting edges, and more. The Actors menu handles operations for determining which parts of the crease pattern remain rigid for folding and which do not. The Simulator menu allows the user to enter the simulator mode and hide the editor. The Camera menu allows users to center the camera on their current crease pattern and pan the camera around the two-dimensional grid. The Help menu provides documentation on each operation.

4.1   Drawing and Creating Crease Patterns

Users create their desired crease patterns with the aid of the tools described above. By using the draw tool, users can click three points to draw a single rigid-body triangle. Rigid bodies, or actors, are currently created by drawing individual triangles and grouping them together. By drawing a triangle that shares a vertex or an edge, the program will automatically merge the common elements. Users then specify the bodies they want to be rigid during folding by selecting the triangles that are all adjacent and making them a single rigid body. Once two rigid bodies sharing a common edge have been created, the edge is automatically made into a revolute joint. This method allows for the creation of arbitrary concave or convex crease patterns, and the triangulation allows for algorithms involving convex shapes to be used during collision detection. Larger, regular crease patterns may be generated with ease using the circular or gridded lattice tools.

Once the layout of the flat-folded FPM has been entered, the user is able to specify actuation. Selecting one or more creases, the user enters a fold angle, spring constant, and damping constant. Different line styles are used to display mountain (positive angle) and valley (negative angle) folds. The user can step through a fold sequence, entering different fold angles for each joint at each step. If no new fold angle is entered, the angle from the previous step is carried through. For larger FPM, the preferred method of CMML generation is to have separate motion planning software generate CMML as output.

4.2   Runtime Analysis

The current implementation of the editor uses a brute force data structure, scaling linearly with the number of triangles present. Each triangle points to three edges, and each edge points to the two vertices it contains. Although edges and vertices may be shared between triangles, they are maintained separately for each triangle. Several operations (marquee selection, lattice creation, copy, paste) run in $O(n^2)$, where n is the number of triangles in an actor. Adding, removing, and merging vertices and edges are $O(n)$ operations. The editor performs well with crease patterns up to hundreds of triangles before suffering noticeable performance degradation. Upcoming revisions to the editor will feature a half-edge data structure, reducing the runtimes for most functions and making larger designs practical.

5   PhysX Simulator and Integration

An important feature of the FPME is the ability to simulate and display fold sequences stored in CMML, as shown in Figure 3(b). Additionally, the simulator may directly load PMML and CMML templates for evaluation without using the editor interface. The simulator provides designers with the ability to view and evaluate fold sequences in real time in three dimensions. The user is then able to use feedback about collision points and velocities to adjust PMML properties such as geometry and spring constants, or CMML properties such as angular velocities, desired fold angle, and fold order. The fold simulator incorporates nVidia's PhysX technology for dynamic physics simulation. GPU-based acceleration is available when using nVidia hardware. As numerical error is a concern, simulation properties have been adjusted so that length scales are on the order of one. The PhysX environment automatically handles collision checking and resolution, enforces constraints such as the revolute joints between rigid bodies, solves for closed-chain spring systems, and propagates the system dynamics.

The initial stage of simulation is to convert the planar data from the editor or PMML file into rigid bodies in three dimensions. The data structure used in the editor is traversed, and planar shapes are converted into PhysX rigid bodies by extrusion. PhysX rigid bodies are set up similar to those in the editor in that the rigid bodies are made of triangular prisms. As stated in Section 3, CMML defines the desired angle for each crease in a fold sequence. The user may select whether the actuators are motors or torsion springs. The motors are velocity controlled, exerting up to a maximum torque on a revolute joint to reach a desired angular velocity, and rotating until they reach a desired position. Spring-based actuation

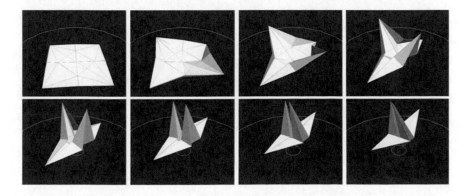

Figure 4. Snapshots from the bird-base folding process.

is meant to mimic the action of tensile actuators such as Nitinol wire. To physically simulate the folding, the rest angle of each angular spring is changed to the desired angle for the current fold. The rigid bodies will naturally fold along the creases. Implementation for the bird base can be seen in Figure 4.

6   Conclusion

We have created a robust editor and simulator to facilitate the design of programmable matter structures. We have created PMML and CMML schemas to standardize representations of crease pattern data. With re-configurability in mind, fold sequences have been separated from the crease pattern; multiple fold sequences may be applied to the same crease pattern. The editor interface allows for intuitive creation and modification of PMML and CMML files. The PhysX simulator allows users to view their crease patterns folding in real time to help determine and resolve potential issues.

Additional work remains to bridge the gap between simulation and reality. For programmable matter, small-scale actuation and the presence of multiple overlapping layers of material is a challenge. Material thickness issues are also exacerbated by overlapping layers; in practice, nesting multiple folds leads to alignment issues and failure to reach desired fold angles. Future revisions of the FPME will address these issues.

Bibliography

[belcastro and Hull 02] s.-m. belcastro and T. Hull. "A Mathematical Model for Non-flat Origami." In *Origami³: Proceedings of the Third International Meeting of Origami Science, Mathematics, and Education*, edited by Thomas Hull, pp. 39–51. Natick, MA: A K Peters, 2002.

[Demaine and O'Rourke 08] Erik D. Demaine and Joseph O'Rourke. *Geometric Folding Algorithms: Linkages, Origami, Polyhedra*. New York: Cambridge University Press, 2008.

[Demaine et al. 00] Erik D. Demaine, Martin L. Demaine, and Joseph S. B. Mitchell. "Folding Flat Silhouettes and Wrapping Polyhedral Packages: New Results in Computational Origami." *Computational Geometry Theory and Applications* 16:1 (2000), 3–21.

[Huffman 76] D. A. Huffman. "Curvature and Creases: A Primer on Paper." *IEEE Transactions on Computers* C-25:10 (1976), 1010–1019.

[Lang 03] Robert J. Lang. *Origami Design Secrets: Mathematical Methods for an Ancient Art*. Natick, MA: A K Peters, Ltd., 2003.

[McCarthy 00] J. M. McCarthy. *Geometric Design of Linkages*. New York: Springer, 2000.

[Tachi 09] Tomohiro Tachi. "Simulation of Rigid Origami." In *Origami⁴: Fourth International Meeting of Origami Science, Mathematics, and Education*, edited by Robert J. Lang, pp. 175–187. Wellesley, MA: A K Peters, 2009.

A CAD System for Diagramming Origami with Prediction of Folding Processes

Naoya Tsuruta, Jun Mitani, Yoshihiro Kanamori, and Yukio Fukui

1   Introduction

In recent years, many methods for designing origami pieces based on mathematical theories have been developed. One of the most powerful approaches has been implemented in the software system TreeMaker [Lang 06]. Although origami pieces designed using this software are restricted to a particular structure referred to as a uniaxial origami base, many complex and realistic works of art have been created. Describing the sequence of folds used to create a complex work is difficult. However, traditional origami diagrams, namely, a sequence of step-by-step illustrations, are still used, even today.

One problem in using such diagrams is that drawing diagrams manually is time consuming; as an origami piece becomes more complex, the cost for drawing the diagrams increases. Traditional origami pieces, such as a crane, are folded by approximately a dozen steps. But, many recent complex pieces need more than 100 steps, thus requiring more than 100 diagrams each to explain how to fold them. To avoid this cost, a crease pattern can sometimes be used instead of a diagram. A *crease pattern* consists of line segments that show the crease lines that appear on the paper when an origami piece is unfolded. Although drawing crease patterns is much easier than drawing diagrams, folding from crease patterns is more difficult for nonexperts because crease patterns do not contain any proce-

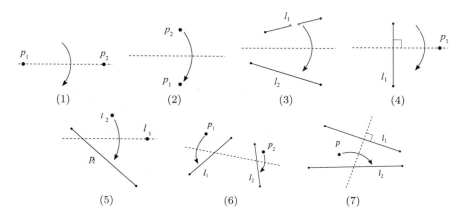

Figure 1. Huzita-Hatori (Huzita-Justin) axioms.

dural information. Therefore, for nonexperts, origami diagrams are still generally required.

In this paper, we propose a computer-aided design (CAD) system that alleviates the problem of drawing diagrams of flat-foldable origami pieces. Our system predicts possible candidate folds that can occur in the current state of an origami piece and lists these candidates. This prediction is made based on the axioms of the Huzita-Hatori (Huzita-Justin) axioms [Hatori 10], a series of seven basic methods for folding a sheet of paper by referring to points and lines (Figure 1). All possible folding lines are generated by applying these axioms in our system, and then folded shapes are calculated and listed as candidates for the next step. When the user selects one of these candidates, the shape is automatically appended to the origami diagram as the latest step. To moderate the explosion of the number of candidates and to simplify the implementation of the system, we limit the axioms to the first four and limit the types of folding to valley folds only. Since the axioms we use for the prediction fold one line at a time, the list of candidates does not include some major folds that require multiple noncollinear lines to be folded simultaneously, such as rabbit-ear-folds and petal-folds. Although this might be a limitation of our system, we provide a user interface for specifying folding manually when the intended folds are not among the candidates. Details of the prediction feature are discussed in Section 3.

We also show the result of enumerating origami pieces folded by applying Huzita-Hatori (Huzita-Justin) axioms multiple times using the prediction feature recursively. The result of this enumeration suggests the possibility of discovering new interesting origami pieces designed by a computer.

2   Related Work

Miyazaki developed an origami simulator with which a user could fold a simple origami piece on a PC by using standard mouse and keyboard devices [Miyazaki et al. 96]. Although this simulator was groundbreaking, it did not facilitate the production of origami diagrams. The Foldinator [Szinger 02] and the eGami [Fastag 09] systems are dedicated origami simulators that were developed to reduce the cost of drawing by hand. A series of figures are generated by simply specifying lines where a piece of origami is folded step by step. Moreover, these programs automatically add symbols, such as arrows, that are needed to explain how to form the fold. We believe the prediction function proposed in this paper is also applicable to these systems.

As mentioned in the previous section, because drawing a diagram for a complex origami piece is difficult, sometimes just a crease pattern is used. Mitani focused on drawing crease patterns and developed an origami pattern editor (ORIPA) [Mitani 08] that allows users to design crease patterns quickly. This editor can also calculate the shape of a folded origami piece from the crease pattern.

Our system is not the first CAD system to implement predictions. Although not related to origami, Igarashi and Hughes developed a 3D modeling system [Igarashi and Hughes 01] that shows a list of 3D shapes generated automatically as candidates for the next operation. The user can simply select one of these shapes if the user's intention is included in the list. If not, the user continues manually. Our approach of using prediction for drawing diagrams was motivated by this CAD system.

3   Our Proposed System

Figure 2 shows the prototype of our system interface. The upper (main) window contains three panels. The left panel has a toggle button to switch between types of folding (mountain, valley, inside-reverse, and outside-reverse) and rotation/flip buttons. The center panel displays the state of the current origami piece. The user can input a folding line by clicking two positions. The right panel shows the history of steps carried out from the initial to the current state. The lower window is a suggestion window, which shows a list of candidates for the next step. The user proceeds to the next step by clicking one of the candidates or manually inputting a folding line. The flow of the prediction process we propose is as follows:

1. Enumerate all possible folding lines by applying axioms to the current state.

2. Enumerate all possible ways to fold along each line.

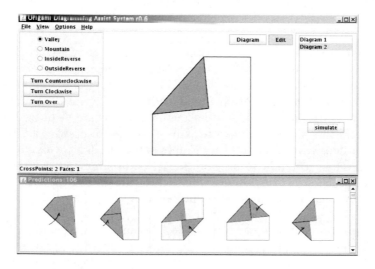

Figure 2. Our proposed system interface.

3. List shapes of origami pieces generated by applying all possible foldings as candidates.

4. Remove duplicate candidates.

5. Assign a score to each candidate.

6. Display a list of candidates, sorted by score.

The prediction function requires considerable computational time; it
contains the following processes: arranging folding lines, assigning moun-
tain/valley status to a folding line, and calculating the folded configuration. Therefore, we use only the first four axioms of Huzita-Hatori (Huzita-
Justin). This simplification drastically reduces computational time by limiting the number of possibilities.

We believe the effect of this simplification on the accuracy of the prediction is small because the last three axioms are rarely used in practical
origami construction. Furthermore, we limit the types of folding to just
valley folding and making a crease line to simplify the system (Figure 3).
The latter operation, making a crease line, is the fold-and-unfold operation used to make a mark for subsequent folds. By flipping the origami
piece, candidates that are equivalent to mountain folding can be obtained.
Details of the prediction process are discussed in the following sections.

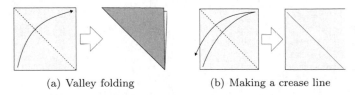

(a) Valley folding   (b) Making a crease line

Figure 3. Folding operations.

3.1 Listing Candidates by Applying Possible Foldings

To list all possible foldings, the system first generates folding lines by applying the four axioms, using all possible points and lines in the current state. We use a simple brute-force approach that tries all possible combinations of points and lines. If multiple folding lines are created at the same position by different axioms (as can occur), then we retain one and discard the others to make each folding line unique.

Next, the system enumerates the possible ways of folding along each folding line. When a folding line passes over multiple layers, there are multiple possibilities. This is because one must choose how many layers are folded at the same time. To simplify the problem, we stipulate that only the topmost n layers may be folded at the same time along the folding line, where n can take on the value from 1 to the total number of layers. Since we apply two types of folding, valley folding and making a crease line, for the folding line, there are at most $2n$ candidates. For example, there are four possibilities when a folding line is applied to the triangle-shaped origami piece illustrated in Figure 4.

The algorithm used to enumerate candidates by the approach described is shown in Listing 1. Impossible configurations may be generated when an ith layer is connected to a lower jth layer by edge e, and e intersects the folding line. As long as only this simple algorithm is used, other impossible

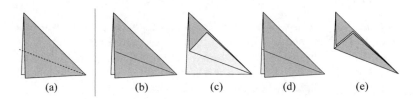

(a)   (b)   (c)   (d)   (e)

Figure 4. The four folding possibilities for an origami piece with one folding line: (a) the origami piece and a folding line (a dotted line) is applied to the piece; (b) "making a crease line" is applied only to the topmost layer; (c) "valley folding" is applied only to the topmost layer; (d) "making a crease line" is applied to all (two) layers; (e) "valley folding" is applied to all (two) layers.

```
foreach (i = 1 to n) {
  fold topmost i layers at the same time along the folding line
  if(an impossible configuration results){
    discard the result;
    continue;
  } else {
    add the result of applying valley folding
      to topmost i layers to candidates
    add the result of applying making a crease line
      to topmost i layers to candidates
  }
};
```

Listing 1. The algorithm for enumerating candidates.

cases, such as those that arise because one part of the origami piece would penetrate another part, are eliminated. Furthermore, the cases in which flaps are tucked inside other layers are eliminated.

3.2 Removing Duplicate Candidates

The list of candidates obtained by applying the process described in the previous section may contain duplicate elements once rotating and mirroring are taken into account. These duplicates are removed before the list of candidates is displayed. When considering two pieces for which the following three conditions are all satisfied, we recognize those pieces as having the same shape and configuration:

1. The numbers of polygonal parts forming the pieces are the same.

2. The sums of distances from the barycenters to each vertex are the same.

3. The lists of pairs of physically connected polygonal parts in the shapes are the same.

For the third check, we assign IDs to polygons according to the stacking order of the folded shape. Then we can check whether a connected pair listed for one candidate is also included in the list for the other candidate.

3.3 Ranking of Candidates

To make it easy to detect the intended shape from a list of candidates, appropriate sorting of the list is important. The number of candidates increases dramatically as the number of foldings increases, so it is desirable that candidates that have a high probability of being selected by the user be near the top of the list.

To assign scores to each candidate, we consider the angles between a horizontal line (x-coordinate of the screen) and the folding lines applied to the current state. The angles range from $0°$ to $180°$. Angles of $90°$, $45°$, and $22.5°$ are often observed in common origami pieces, so we assign higher scores to candidates that have folding lines at these angles. Furthermore, as most origami pieces are symmetric and repetition is often observed, we assign higher scores to candidates that are folded along a folding line based on an angle related to those of the preceding folding step. Two angles are related if one is obtained from the other by reversing an angle through horizontal, vertical, or diagonal lines. Specifically, we assign scores according to the angles of the folding line to the following, in descending order:

- an angle related to those of the preceding folding;

- $0°$, $90°$, or $180°$;

- $45°$ or $135°$;

- $22.5°$, $67.5°$, $112.5°$ or $157.5°$;

- anything else.

4   Results and Discussion

In this section, we consider the results, focusing upon the efficiency of the prediction function and the enumeration of candidates. We implemented our system in Java and ran it on a PC with a Core 2 Duo 2.66 GHz processor.

4.1   Efficiency of the Prediction Function

We evaluated our system by drawing a diagram for a *kabuto*, a Japanese traditional origami helmet. Figure 5 shows the diagram created by our system. The numerical values are shown in Table 1. The second column shows the ranking of selected candidates. A smaller number indicates a better prediction.

Of the eight steps needed to fold the kabuto, the desired figures were found in the list predicted by the system in five of the steps. We found that our prediction was effective, especially at the early stages of folding. One problem is that it can become difficult to find the desired figure in the generated list, especially in later steps, because the number of candidates increases in these steps. Although we proposed a method to assign scores to each origami piece, it is still difficult to assign appropriate scores to make

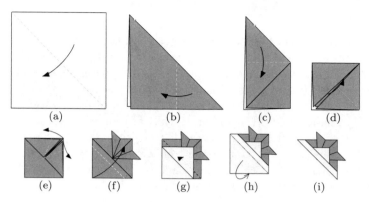

Figure 5. Diagrams for kabuto (helmet).

precise predictions. Furthermore, as the number of candidates increases, more candidates tend to be assigned the same score. To solve this problem, the strategy for assigning scores to candidates will need to be improved. For example, it would be appropriate to assign a higher score to a fold along a line located close to the previous folding line because continuous folds are usually made near each other, such as in steps 6 and 7 and in steps 2 to 5 (formation of the flaps). Adding weights to scores according to the particular axiom used would be another solution. The first and second axioms are those most often used in common origami pieces. For example, all candidates selected for drawing the diagram of the kabuto use only the first two axioms.

The computational time increases in later steps because the number of candidates can increase dramatically, depending upon the complexity of the shape. If it takes several seconds to display candidates, the system lacks interactivity. Therefore, the system would become more usable by both improving the score and reducing the number of candidates.

Number of step	Position of selected figure	Number of candidates	Processing) time (sec)
1	2	4	0.016
2	2	8	0.015
3	1	44	0.031
4	6	23	0.023
5	manual input	44	0.016
6	manual input	595	0.546
7	1	868	0.735
8	manual input	1048	0.859

Table 1. Result for kabuto (helmet).

Number of foldings	Number of candidates	Processing time
1	8	0.017 (sec)
2	1,149	0.273 (sec)
3	1,476,913	5.8 (min)
4	more than 6 billion	terminated

Table 2. The number of foldings and the number of candidates.

4.2   Enumeration of Simple Origami Shapes

By using the prediction function, it is possible to enumerate all variations of the origami pieces that can be made using the axioms. Although the system described above predicts the next folded state from the current state, it can predict multiple steps at one time by applying the prediction procedure recursively. We examined the number of pieces that recursive prediction can generate. The relation between the number of foldings and the number of variations is shown in Table 2. For this analysis, we used all seven axioms to increase the number of variations. For four folds, though, due to the large number of candidates, we could not calculate the exact value, so we estimated the number of candidates.

As expected, the number of variations increases exponentially. There are eight pieces that can be generated from a single fold, as shown in Figure 6. The number of origami pieces generated by folding twice is 1,149. Some are shown in Figure 7. When three or four foldings are made, many more variations are obtained. We were able to find some pieces that resemble recognizable shapes, such as animals, objects, and symbols, among the list of candidates. We show some examples, which we have named, in Figures 8 and 9.

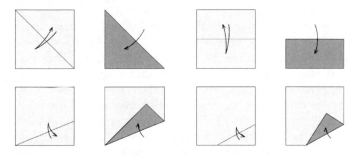

Figure 6. Eight shapes constructed using a single fold. The lower four apply axioms 5 and 6, allowing a point to be on a line.

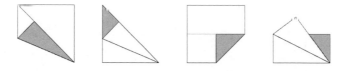

Figure 7. Example of a two-fold shape (number 4 out of 1,149).

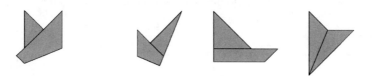

Figure 8. Examples of three-fold shapes. From left to right, fox, tick (check mark), yacht, and arrowhead (number 4 out of 13,957,372).

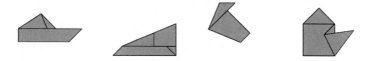

Figure 9. Examples of four-fold shapes. From left to right, boat, iron (appliance), dog, and teapot.

5 Conclusion and Future Work

We have proposed a new system for drawing diagrams using predictions for folding. The system generates folded shapes from a particular state based on the axioms and displays them as candidates. We found that the prediction is especially effective for simple origami figures that are folded in a few steps. It became clear, however, that it is difficult to apply our system to complex origami.

As noted in the Section 1, there are limitations with the prediction algorithm. We use only single-fold axioms, so the system drops many possible foldings that are used in real-world manipulations. However, reducing the number of candidates is important both for computational cost and to simplify user selection. Thus, there is a trade-off between the explosion of the number of candidates and the loss of correct ones. One solution would be to limit the area on which the fold lines are placed. Alternatively, if there was an origami database, we could use the experimental data in this database to improve the accuracy of prediction.

Another area of future work could be the automatic generation of origami pieces. Origami pieces can become complex, as we noted in Section 1, but many pieces emphasize simplicity, such as the "2-fold Santa" [Versnick 10]. This piece expresses Santa Claus by folding a square sheet of paper only twice. Our system has the potential to generate many such simple origami pieces.

Bibliography

[Fastag 09] Jack Fastag. "eGami: Virtual Paperfolding and Diagramming." In *Origami⁴: Fourth International Meeting of Origami Science, Mathematics, and Education*, edited by Robert J. Lang, pp. 273–284. Wellesley, MA: A K Peters, Ltd., 2009.

[Hatori 10] Koshiro Hatori. "K's Origami: Origami Construction." Available at http://origami.ousaan.com/library/conste.html, accessed May 17, 2010.

[Igarashi and Hughes 01] Takeo Igarashi and John F. Hughes. "A Suggestive Interface for 3D Drawing." In *Proceedings of the 14th Annual ACM Symposium on User Interface Software and Technology*, pp. 173–181. New York: ACM Press, 2001.

[Lang 06] Robert J. Lang. "TreeMaker." Available at http://www.langorigami. com/science/treemaker/treemaker5.php4, 2006.

[Mitani 08] Jun Mitani. "ORIPA: Origami Pattern Editor." Available at http: //mitani.cs.tsukuba.ac.jp/pukiwiki-oripa/index.php, 2008.

[Miyazaki et al. 96] Shinya Miyazaki, Takami Yasuda, Shigeki Yokoi, and Jun ichiro Toriwaki. "An Origami Playing Simulator in the Virtual Space." *Journal of Visualization and Computer Animation* 7:1 (1996), 25–42.

[Szinger 02] John Szinger. "The Foldinator Origami Modeler and Document Generator." In *Origami³: Proceedings of the Third International Meeting of Origami Science, Mathematics, and Education*, edited by Thomas Hull, pp. 129–136. Natick, MA: A K Peters, Ltd., 2002.

[Versnick 10] Paula Versnick. "Orihouse." Available at http://home.tiscali.nl/ gerard.paula/origami/orihouse.html, 2010.

Development of an Intuitive Algorithm for Diagramming and 3D Animated Tutorial for Folding Crease Patterns

Hugo Akitaya, Matheus Ribeiro,
Carla Koike, and Jose Ralha

1   Introduction

Origami designers employ both diagrams and crease patterns (CPs) as tools for documenting and sharing their creations. It would be desirable to build a diagram given a CP, but this process is not straightforward. Each experienced folder has a preferred sequence of folding a specific CP, and frequently they just collapse all the folds together at once. But, by observing diagrams, it is possible to recognize recurrent sequences used to fold specific patterns of folds within a CP.

This paper presents an algorithm based on these observations designed to assist the creation of step-by-step folding sequences given an origami CP. It aims to help both origami designers, by making the diagramming process less time consuming, and folders, by giving an alternative option to fold more complex models without needing to collapse all the folds together.

This paper is organized as follows. Section 2 introduces the development of modern origami and its relation to the birth of computational origami. Section 3 deals with some concepts regarding origami design and flat foldability used by the algorithm. Section 4 describes the algorithm itself. Section 5 discusses the scope of applicability, and Section 6 presents an example of the application on a CP manually. Section 7 presents some results of a developing software implementation.

2 Computational Origami

Origami is the centuries-old traditional Japanese art of paper folding. Although it is an ancient art, only recently has the number of new models significantly increased, both in quantity and in complexity. Origami has reached other frontiers, such as science, technology and mathematics. The popularization of origami started when Akira Yoshizawa published his first book, *Atarashi Origami Geijutsu* (*New Origami Art*), in 1954, introducing a new notation to express how to fold an origami model through diagrams with step-by-step instructions [Yoshizawa 54]. This notation uses arrows and lines to represent the positions of folds and the movements performed when folding the paper (see Figure 1 for some symbols). It became so popular that origami designers from all over the world still use it to document their works [Lang 00].

Humiaki Huzita and Jacques Justin provided, in 1989, the first formal description of mathematical principles of paper folding, the Huzita-Justin Axioms [Huzita 92, Justin 91]. Soon, mathematical modeling made it possible for simple shapes to become multilegged insects, multihorned dragons, and multiwinged creatures. The work of Toshiyuki Meguro, Jun Maekawa, Fumiaki Kawahata, and Robert Lang, among others, has made an enormous contribution to the development of technical origami modeling [Lang 98].

Although the CP is not a new tool, it gained more importance in this context. Most of modern design techniques results in a CP of the desired origami base. Besides being easy to draw, CPs contain very useful information under the structure and design techniques of a base through only one picture. Also, it is difficult to find a step-by-step sequence for the desired

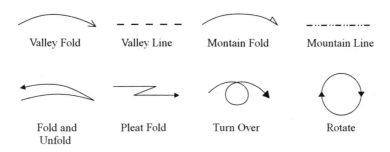

Figure 1. Common symbols used in origami diagrams.

base. CPs can be helpful in situations in which the sequence does not exist, and the folders collapse all the folds simultaneously.

The development of mathematical origami also gave birth to another field of study: computational origami, which deals with development of algorithms and their implementation, based on the mathematical modeling of origami and its design techniques. The software TreeMaker, for example, employs tree theory, which is based on the circle/river packing technique [Lang 96]. This software computes the best scale and arrangements for the circles and builds the CP for the desired base automatically.

Another example of origami software is ORIPA [Mitani 2007]. ORIPA is a CP drawing tool that shows a preview of the folded CP and exports it to a computer-aided design (CAD) file.

3 Basic Definitions on Technical Origami

A *crease pattern* is the set of creases on a square that defines the structure of an origami. Crease patterns, unlike diagrams, don't show any arrows, and it is common to see CPs that don't even differentiate mountain folds from valley folds. That is because the origami base resulting from many CPs are flat, and thus there is no more than one way to fold each crease in most cases. If a CP results in a planar base, it is said to be flat-foldable. Flat-foldability of a CP is related to Maekawa's theorem and Kawasaki's theorem [Hull 94], and the CP must satisfy noncrossing conditions [Justin 97].

Another aspect of CPs is that most of the time they represent only the base of the origami. A base is a folded state of the paper that has the same structure as the subject, but without the details that characterize the final model. The detail folds are added later in a process called *shaping*.

Lang describes mathematical algorithms and techniques to design uniaxial bases with an arbitrary arrangement of flaps known as *circle packing* and/or *tree theory* [Lang 98].

In the circle-packing technique, the base is designed in the shape of a specified tree graph with the desired edge lengths and configuration, and then the problem is reduced to packing the circles inside the square of the paper. This technique results in a uniaxial base with a determined configuration and length of flaps. A uniaxial base is a base that has all its flaps lying on a single axis and all its hinges perpendicular to this axis [Lang 96].

This determined configuration and length of flaps can be expressed by a tree graph in which the edges of the leaf nodes represent the flaps in position and length. On the square of paper, the leaf nodes are represented by circles and the edges that don't connect any leaf node are represented by rivers. By packing these two elements in the square, one can determine the portions of paper that each part of the base will consume. Active paths are lines that connect the center of the touching circles or of the circles that have no space between them apart from the space occupied by rivers. The active paths result in folds that lie along the axis (axial folds) and determine the perimeter of polygons within the square. These polygons could be filled in with *molecules*, which are partial CPs that will bring all the active paths together and will produce the flaps with the desired length. The designed CP is then obtained by filling all the polygons with molecules [Lang 03a].

Packing the circles and rivers within a square often produces unused portions of paper at the corners of the square. The effective CP is the group of all the polygons, excluding these unused areas.

4   Turning a CP into a Folding Sequence

The construction of this algorithm was made by observing some patterns of creases and how we would fold them, as well as following some examples of diagrams and our own experience. We tried then to make some rules modeling our way of folding. As input, it has a circle-packed CP, and as output, it generates partial CPs that lead to intermediary flat states between the square of paper and the folded base, resembling a series of progressive CPs.

This algorithm focuses only on circle/river-packed CPs of uniaxial bases constructed with the universal molecule. This would allow the software implementation to use as input the output of TreeMaker [Lang 98], which already has the major part of information regarding the tree theory [Lang 03a] on which this algorithm relies.

Because the algorithm tries to follow a single strategy, the sequences generated might not be the most intuitive way for folding some specific CP.

As part of the modeling, we created the concept of "maneuvers." A *maneuver* is a set of common steps used to fold specific parts of a CP. The next sections describe the maneuvers employed in the proposed algorithm and their structures; Table 1 summarizes them. Through the paper, it is shown that these parts of the CP are defined regarding the leaf nodes location and the molecules that connect these nodes to other leaf nodes. In the flowcharts, the word *node* refers to these leaf nodes. An *internal node* refers to a circle the center of which does not lie on the border of the effective CP, and a *border node* refers to a circle the center of which lies on the border of the effective CP.

Due to space restrictions, the whole algorithm is not presented here; all of the flowcharts can be found at www.origamiracle.cic.unb.br.

Maneuver	Usage	Description
1	Used in "Choose and fold next internal node" block.	Collapses two symmetrical neighbor triangular molecules. The molecules must have collinear hinges.
	Ex.:	
2	Used in "Choose and fold next internal node" block.	Folds small stubs. It is subdivided into three smaller maneuvers according to the stub characteristics.
	Ex.:	
3	Used in "Choose and fold next molecule" block.	Reverse fold or crimp fold.
	Ex.:	
4	Used in "Choose and fold next molecule" block.	Petal fold.
	Ex.:	
5	Used in "Choose and fold next molecule" block.	Maneuver to fold gusset molecules. It must be a corner molecule and have a symmetry line that intercepts the effective CP twice.
	Ex.:	

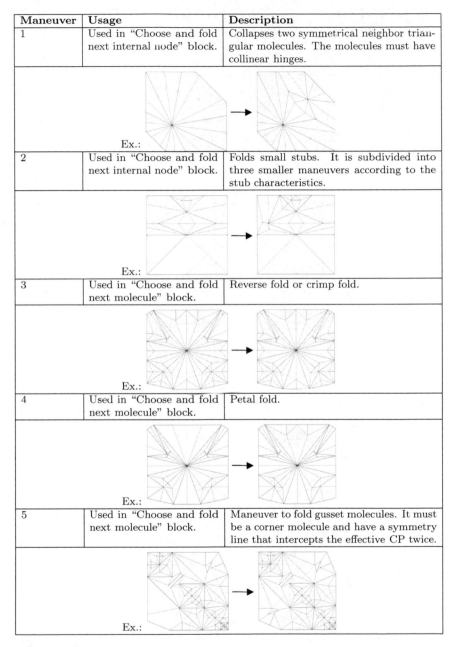

Table 1. Summary of maneuvers. CP examples are Maneuvers 1, 7, and 8 [Lang 04a]; Maneuvers 2 and 9 [Lang 03b] ("Periodical Cicada," Opus 377); Maneuvers 3, 4, 6, and 10 [Lang 03b] ("Scorpion Varileg," Opus 379); and Maneuver 5 [Lang 04b].

Maneuver	Usage	Description
6	Used in "Choose and fold next internal node" block.	Similar to maneuver 1, but with asymmetrical molecules (with collinear hinges). Requires more prefolding than maneuver 1.
	Ex.:	
7	Used in "Choose and fold next internal node" block.	Used to fold asymmetrical triangular molecules that do not have collinear hinges.
	Ex.:	
8	Used in "Choose and fold next internal node" block.	Normally, as a result of other maneuvers, a node could be sunk. This maneuver is used to unsink these nodes by folding molecules with the same hinges after propagation.
	Ex.:	
9	Used in "Choose and fold next molecule" block.	Rabbit-ear fold.
	Ex.:	
10	Used to fold the first node and after some maneuvers.	Squash folds.
	Ex.:	

Table 1. (Continued.)

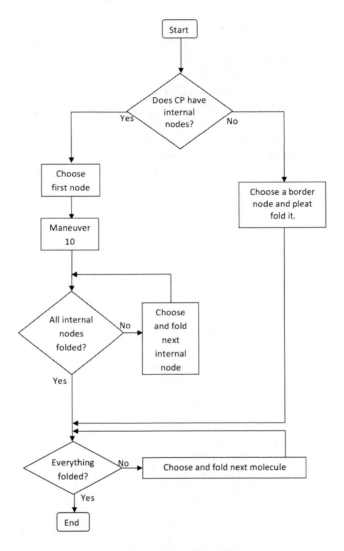

Figure 2. Macro view of the algorithm.

4.1 Main Algorithm

The process of folding consists of transferring some folds of the input CP to a temporary CP. Then, the model is flattened by a subalgorithm called "Propagate Folds" and is output. This way, the sequences of folds can be seen through the series of CPs created.

The proposed algorithm is shown in the flowchart in Figure 2. Initially, a choice is made that depends on whether there is any internal node in the

CP. If there are internal nodes in the CP, a subalgorithm chooses which internal node will be folded next, and folds it. After folding all internal nodes, the border molecules should be folded, completing the base. If there is not an internal node, the folds that come from the center of a border node should be pleat-folded, and then the remaining molecules should be folded. The process of folding is based on ten maneuvers, each one with a different complexity and generating a different number of steps.

A node is said to be folded if all the creases that come from the node's location are already transferred to the temporary CP. A molecule is said to be folded if all its creases are already transferred to the temporary CP.

4.2 Choose First Node

The "Choose First Node" subalgorithm (Figure 3) gives scores to the internal nodes according to their size and position in order to decide which node will be folded first.

The "size" of a leaf node is referred here as the length of the edge connected with it. Thus, the largest node is the one whose edge has the biggest length. When deciding which of the nodes is going to be folded first, the largest node is the natural choice because it covers a large area

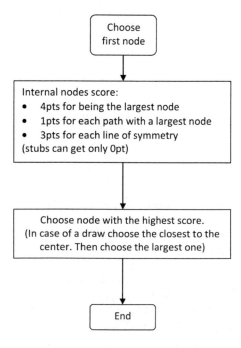

Figure 3. Choose First Node flowchart.

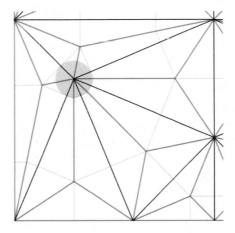

Figure 4. CP extracted from "Periodical Cicada" [Lang 03b, pp. 17, 118]. Highlighted node is considered a stub.

in the CP and usually has a lot of creases emanating from it. Postponing the fold of the largest node may lead to particularly difficult unsink folds, as a fold of another node can cause this node to be sunk.

However, symmetry is often a reason that it sometimes is easier to begin with other nodes. A node on the symmetry line is preferable if there are active paths connecting it to the largest nodes symmetrically because the fold of the largest nodes will proceed shortly after the fold of this node and will be replicated symmetrically.

We emphasize that the proposed system models the patterns of a particular folder, so it won't necessarily be the best way. The point assignment for each situation was made by observing a series of CPs: the first node that was chosen by the folder so that the algorithm's choice would be similar to that of the folder for the analyzed situations.

A *stub* is a new leaf node added (together with a new edge) to the conceptual tree of the base used to break down higher-order polygons into several lower-order polygons [Lang 03a]. This usually produces a small circle connected with a bigger one through an active path that does not cross any river. We also observed that the molecules that have this active path as a ring path usually have an opening angle greater than 90 degrees.

The strategy here is to find the equivalent arrowhead molecule. So we consider as stubs only the leaf nodes that have three consecutive paths with an angle greater than 90 degrees from each other. Figure 4 shows an example of what is considered a stub by the algorithm.

4.3   Maneuver 10

This maneuver is used to complete the fold of a node after the execution of other maneuvers or to fold a node when there is still nothing folded on the partial CP. It uses a series of squash folds to transfer all the folds that come from the node location to the partial CP.

This subalgorithm receives as input a set of neighbor molecules that have a leaf node in common. Then, molecules and groups of molecules are grouped in pairs iteratively until there is just one group. This big molecule corresponds to the initial state of the paper before the execution of the squash folds. Next, the squash folds are made, dividing grouped molecules until all the original molecules are partially folded (have one ridge and two paths transferred to the partial CP).

This grouping process tries to minimize the addition of unnecessary marks to the paper by merging neighbor molecules that have the same opening angle and folding along the symmetry line, aiming at using creases of the original CP as the new ridge of the groups of molecules. Nevertheless, there could be still some undesirable marks, and the diagrammer will be able to choose to take the first and the last output CP and collapse all the folds together.

4.4   Choose and Fold Next Internal Node

To continue folding the internal nodes, the "Choose and Fold Next Internal Node" subalgorithm searches within the CP for folding patterns related to the known maneuvers. The maneuvers to be executed are then chosen, prioritizing the occurrence of simpler maneuvers before the more complex ones.

This subalgorithm tries to fold the remaining nodes from the inside first, and then to the borders; imagine that there is a circle inflating from the first node folded. When the perimeter of the circle touches an unfolded node, it checks whether it can be folded by Maneuver 8, 1, 6, 7, or 2 (in this order) and folds it. If the entire CP (still with uncompleted internal nodes) gets in the circle without applying any maneuver, this CP is not supported by the proposed set of maneuvers. More details can be found in Section 5.

Maneuver 8 is an unsink fold. The reason its execution is preferred over the other maneuvers (which are easier to execute) is that the more folded molecules a sunken node has, the more difficult it is to perform Maneuver 8. We choose to increase the number of unsinks, thereby decreasing their individual difficulty. On the other hand, we have Maneuver 7 that has also an unsink fold, but is more unusual than Maneuver 8. By being executed after Maneuver 1 and 6, a node previously foldable through Maneuver 7 might be able to be folded with Maneuver 8, which is simpler.

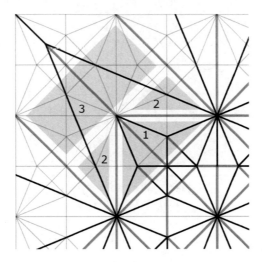

Figure 5. (1) Folded molecules; (2) partially folded molecules; (3) not-yet-folded molecules.

4.5   Maneuvers 1, 2, 6, 7, and 8

Maneuvers 1, 2, 6, 7, and 8 are used to fold internal nodes. Each maneuver folds one node at a time. Its goal is then to transfer the creases that come from the location of this node to the partial CP. Generally, two paths divide the folded molecules and the not-yet-folded molecules connected to a certain leaf node. It is also noticeable that the two not-yet-folded molecules that have these paths as ring paths are partially folded (they already have a ridge fold transferred to the partial CP). (See Figure 5.)

Maneuvers 1, 2, 6, 7, and 8 transfer two ridge folds and a set of hinge folds of the partially folded molecules (in most cases completing their fold) and partially fold the other molecules.

In the cases of Maneuvers 1, 6, and 7, those two paths are coincident, and there are no folded molecules attached to this node. Maneuver 1 collapses two symmetrical neighbor triangular molecules. Maneuver 6 is similar to Maneuver 1, but with asymmetrical molecules (with collinear hinges). The main difference is that Maneuver 6 requires more prefolding than Maneuver 1. Maneuver 7 folds asymmetrical triangular molecules that don't have collinear hinges.

Maneuver 8 folds sunken nodes for which the partially folded molecules have a coincident path of hinges (Figure 6). Maneuver 2 folds an arrowhead molecule equivalent to the molecules attached to the stub, and then completes it with a spread sink.

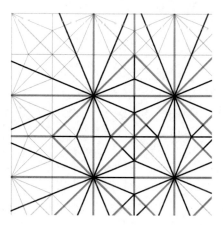

Figure 6. Application of Maneuver 8.

4.6   Choose and Fold Next Molecule

Once all internal nodes are folded, there remain only border molecules or molecules with a path for which the two vertices lie on the border of the effective CP. The algorithm tries to fold them by using petal folds (Maneuver 4), reverse folds and crimp folds (Maneuver 3), or rabbit-ear folds (Maneuver 9). Also there is Maneuver 5 to fold some cases of gusset molecules (border).

Since the execution of Maneuver 3, 4, or 5 does not affect the others, they can be applied arbitrarily. So, to minimize movements of flaps, the choice is made by the proximity to the unfolded axial path (more detail in Section 4.8). Maneuver 9, in contrast, must be preceded by Maneuver 3 or 4.

4.7   Maneuvers 3, 4, 5, and 9

Maneuvers 3, 4, 5, and 9 are used to fold two kinds of molecules: border molecules and internal molecules that have a path for which the two vertices lie on the border. The second kind of molecules is always connected to a border molecule through the previously referred path. This border molecule should be folded with Maneuver 9 after the application of Maneuvers 3 or 4 to fold the internal molecule.

Maneuver 5 was added later to offer support to border gusset molecules that are symmetrical to the line that connects two opposite border nodes in the polygon. Figure 7 shows the molecules foldable through this maneuver and its axis of symmetry.

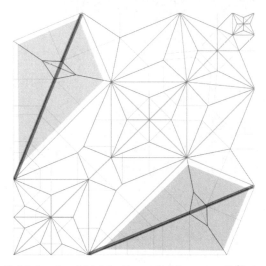

Figure 7. Molecules foldable through Maneuver 5 [Lang 04b].

4.8 Organize Flaps

The partial CP will always have at least one path of unfolded axial creases that comes from and goes to the edge of the paper. An example is shown in Figure 8. The execution of the maneuvers depends on the relative position of the unfolded axial path to the molecules that will be folded. There could also be some hinge fold or its prolongation that might cross one of the molecules to be folded.

The goal of the "Organize Flaps" subalgorithm is to dislocate the unfolded axial creases and the hinge paths of folds to enable the execution of the maneuvers.

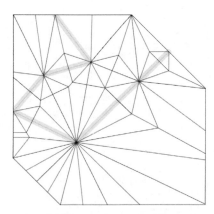

Figure 8. Path of unfolded axial creases highlighted [Lang 04a].

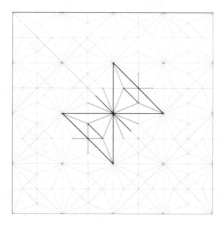

Figure 9. CP after execution of Maneuver 1.

4.9 Propagate Folds

Figure 9 shows the partial CP after execution of Maneuver 1, with the folded nodes highlighted.

This partial CP is not flat-foldable because there are vertices that don't satisfy Maekawa's and Kawasaki's theorems. The "Propagate Folds" algorithm generates a set of provisory folds to make this partial CP flat-foldable. Figure 10 shows the results of the application of this algorithm on the CP shown in Figure 9.

The already folded portion of the CP is uniaxial and has flat molecules. Every maneuver and "organize flaps" application modifies the layer order-

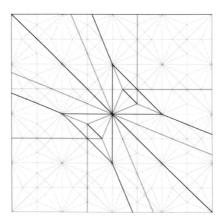

Figure 10. CP with propagated creases.

ing in some specific way. We haven't modeled these ways yet, so there is still no guarantee of noncrossing conditions. In the future, we will try to fill this gap to implement three-dimensional simulation.

5 Applicability Scope

The scope of this algorithm is related to the set of documented maneuvers. A maneuver is applicable only to specific situations, so some CPs may not be supported. The major part of the documented maneuvers was designed for folding triangular molecules and later expanded to some quadrangular molecules. So far, a CP is foldable through this algorithm if it is circle packed; has only rabbit-ear, waterbomb, sawhorse, and/or gusset molecules; and obeys the following rules:

- If it has gusset molecules, they must be border molecules and have the symmetry explained in Section 4.7.

- While there are still internal nodes to be folded, the CP must have at least one unfolded internal node foldable through the set of maneuvers. This doesn't happen when all the pairs of partially folded molecules lack coincident hinge paths toward each other and the other two hinge paths that come from the same ridge intercept closest to the node already folded go through more than one molecule to the edge of the paper.

In Figure 11, the hinge folds of the pair of partially folded molecules m1 and m2 don't meet. But the other two hinge paths of m1 (which has

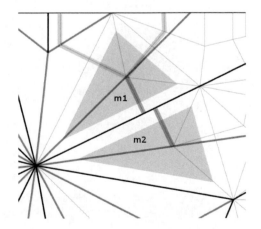

Figure 11. Molecules foldable through Maneuver 7 [Lang 04a].

the closest ridge interception to the node already folded among the two molecules) go to the edge of the paper through only one molecule. So this model is foldable through Maneuver 7.

It is important to notice that the increase in complexity of the CP results in a need for more specific maneuvers. Adding only a few supported molecules would require a lot more maneuvers than we have already documented.

Also notice that the algorithm can handle CPs with an arbitrary number of internal nodes as long as the CP respects the two rules above. Actually, the number and disposition of rivers is more decisive to the applicability of the algorithm. The algorithm probably will not work for CPs with a greater number of rivers.

The project website has an example of the folding of an infinite grid of bird bases. There we can see that, as the CP doesn't have any rivers at all, there are only a few situations that occur repeatedly and can be folded without problems.

6   Example

In this section, we describe an example of the manual application of the algorithm into a CP. We are still working on providing an example for every case supported by each flowchart on the website. Consider the CP shown in Figure 12.

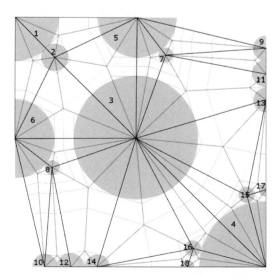

Figure 12. Anteater's CP [Akitaya 10].

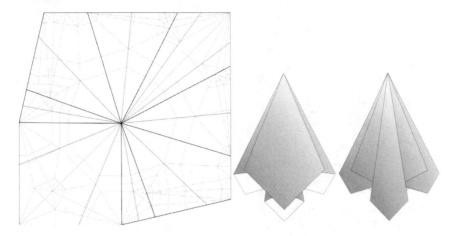

Figure 13. Maneuver 10 applied to node 3.

The Choose First Node subalgorithm chooses node 3, which scores 7 points for being the largest and on a symmetry line. After the execution of Maneuver 10, the resulting partial CP is as shown in Figure 13.

Next, node 2 is foldable through Maneuver 1. The ridge folds of the two partially folded molecules attached to node 3 are transferred to the partial CP as mountain folds. Then their coincident hinge path is transferred as a valley fold. Then, the axial folds of such molecules are transferred to the partial CP (one as a valley and the other unfolded). An added provisory mountain fold bisects those paths to flatten the model. Results are shown in Figure 14.

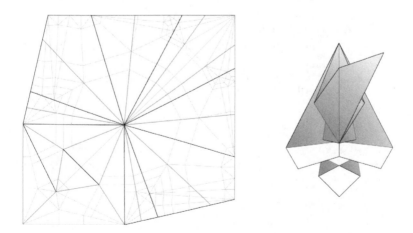

Figure 14. Maneuver 1 applied to node 1.

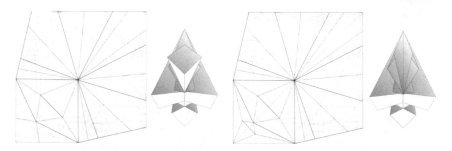

Figure 15. Maneuvers 10 (left) and 4 (right).

Inside the algorithm of Maneuver 1, Maneuver 10 is executed to partially fold the other molecules attached to node 3 (Figure 15, left). And then Maneuver 4 folds the two symmetrical partially folded molecules (Figure 15, right).

Now, the "Choose and Fold Next Internal Node" subalgorithm chooses node 7 or 8 arbitrarily. These nodes are connected to border nodes (5 and 6) through triangular molecules. A stub is said to be connected to a node if there is no river between that node and the stub. Maneuver 2.2 folds molecules attached to the stub to make an arrowhead molecule, as shown in Figure 16.

Then, Maneuver 2 completes with a squash fold, by transferring the remaining ridge folds of the molecules (the one that comes from the stub is a valley fold) and the hinge path they have in common as a mountain fold. Results are shown in Figure 17.

The "Choose and Fold Next Internal Node" subalgorithm chooses node 15 or 16. They are folded with Maneuver 2 also. Results are shown in Figure 18.

Having all internal nodes folded, the algorithm does Maneuver 3 (reverse fold) to complete the remaining molecules. Results are shown in Figure 19.

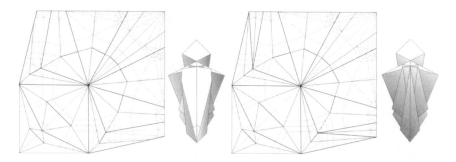

Figure 16. Maneuver 2.2 (nodes 7 and 8).

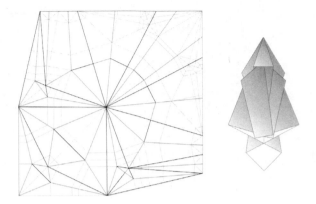

Figure 17. Maneuver 2.2 spread sink.

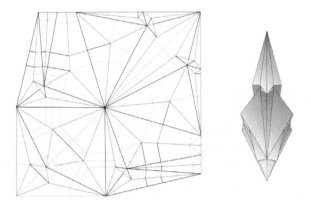

Figure 18. Maneuver 2.2 (nodes 15 and 16).

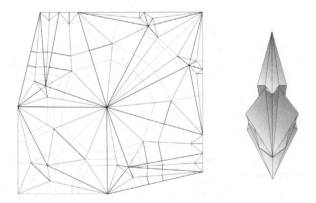

Figure 19. Completed CP.

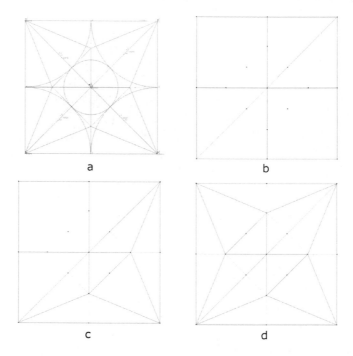

Figure 20. Bird base CP created with TreeMaker and temporary CPs: (a) CP created with TreeMaker; (b) first temporary CP, created after Maneuver 10; (c) second temporary CP, created after Maneuver 4; (d) last temporary CP, created after Maneuver 4.

7   Software Implementation

The software implementation is meant to receive a CP file created with TreeMaker as input, create several temporary CPs, and save them as ORIPA files. This way, we can assume that the TreeMaker files are circle/river packed and that the temporary CPs can be viewed and folded in ORIPA. TreeMaker can also triangulate the CP so that all molecules would be triangular or quadrangular, one of the conditions required by the algorithm, and ORIPA can also create .dxf CAD files.

So far, we have started Maneuvers 10 and 4. With these maneuvers, we're able to fold the first steps of some origamis. One of them is the bird base. The original CP created with TreeMaker and the temporary CPs created by the software are shown in Figure 20.

The bird base has only one internal node. It is folded using Maneuver 10 (Figure 20(b)). Then there are four symmetrical molecules that can be folded using Maneuver 4 (Figure 20(c) and (d)). ORIPA can fold these

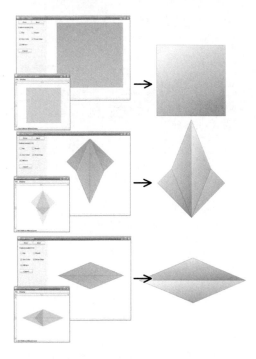

Figure 21. Temporary CPs folded by ORIPA and SVG images created based on ORIPA folding.

temporary CPs and, by using the points of the folded CPs, we can create vectorized images in scalable vector graphic (SVG) format. The folded CPs and the vectorized images created are shown in Figure 21.

The result is a vectorized image, so it can be edited easily on any vector graphics editor. The diagrammer is allowed to change the ordering of facets to change line strokes (to create dashed lines, for example), and to change the drawing size. An example of a diagram created this way is shown in Figure 22. Steps 2 through 5 were done with the SVG images. Step 1 is just a square, and each step from 6 to 9 copies the previous step with some small edits.

8 Conclusion

This work describes an algorithm to fold a certain type of CP constructed by a circle/river packing technique. Although it is very strict about the type of CP, the results obtained show that, when implemented, this algorithm would be able to reduce substantially the work spent on diagramming.

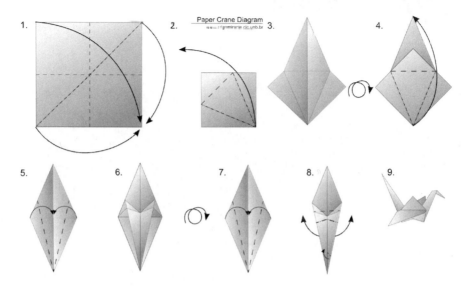

Figure 22. Paper crane diagram.

Combined with the TreeMaker software, it would offer a rapid solution for an origami design, helping the designer to test fold and accelerating the task of diagramming.

As future work, we will go on with the implementation of the proposed algorithm in software. Also, we will add new maneuvers and their rules of application to expand the group of covered molecules. Because the algorithm is basically an expert system, it reflects the experience of only one folder. To suppress this limitation, a system could be built to help an origami designer to create a diagram for a CP, not only proposing maneuvers at each step of the folding process, but also allowing the user to add his or her own maneuvers when the proposed ones are not acceptable. We can expect that the addition of new maneuvers, as well as the rules for their application by many designers and folders, will expand the algorithm so that it would be able to fold any flat origami base.

Bibliography

[Akitaya 10] Hugo Akitaya. "Giant Anteater." In *5OSME Diagram Book*, pp. 147–158. Singapore: 5OSME, 2010.

[Hull 94] Thomas Hull. "On the Mathematics of Flat Origamis." *Congressus Numerantium* 100 (1994), 215–224.

[Huzita 92] Humiaki Huzita. "Understanding Geometry through Origami Axioms." In *Proceedings of the First International Conference on Origami in Education and Therapy (COET91)*, edited by J. Smith, pp. 37–70. London: British Origami Society, 1992.

[Justin 91] Jacques Justin. "Resolution par le pliage de l'equation du troisieme degre et applications geometriques." In *Proceedings of the First International Meeting of Origami Science and Technology*, edited by H. Huzita, pp. 251–261. Padova, Italy: Dipartimento di Fisica dell'Università di Padova, 1991.

[Justin 97] Jacques Justin. "Towards a Mathematical Theory of Origami," In *Origami Science and Art: Proceedings of the Second International Meeting of Origami Science and Scientific Origami*, edited by K. Miura, pp. 15–29. Shiga, Japan: Seian University of Art and Design, 1997.

[Lang 96] Robert J. Lang. "A Computational Algorithm for Origami Design." In *Proceedings of the 12th Annual ACM Symposium on Computational Geometry*, pp. 98–105. New York: ACM Press, 1996.

[Lang 98] Robert J. Lang. "TreeMaker 4.0: A Program for Origami Design." Available at http://www.langorigami.com/science/treemaker/TreeMkr40.pdf, 1998.

[Lang 00] Robert J. Lang. "Origami Diagramming Conventions: A Historical Perspective." Available at http://www.langorigami.com/info/diagramming_series.pdf, 2000.

[Lang 03a] Robert J. Lang. *Origami Design Secrets: Mathematical Methods for an Ancient Art*. Natick, MA: A K Peters, Ltd., 2003.

[Lang 03b] Robert J. Lang. *Origami Insects*. Tokyo: Gallery Origami House, 2003.

[Lang 04a] Robert Lang. "Maine Lobster." In *British Origami Society Spring Convention*, p. 114. London: British Origami Society, 2004.

[Lang 04b] Robert Lang. "Silverfish." In *Origami Tanteidan 10th Convention*, pp. 248, 285. Tokyo: Origami House, 2004.

[Mitani 07] Jun Mitani. "Development of Origami Pattern Editor (ORIPA) and a Method for Estimating a Folded Configuration of Origami from the Crease Pattern" (in Japanese). *Transactions of Information Processing Society of Japan* 48:9 (2007), 3309–3317.

[Yoshizawa 54] Akira Yoskizawa. *Atarashi Origami Geijutsu (New Origami Art)*. Tokyo: Origami Geijutsu-Sha, 1954.

Hands-Free Microscale Origami

Noy Bassik, George M. Stern, Alla Brafman,
Nana Y. Atuobi, and David H. Gracias

1 Introduction

The art of folding provides a unique means to project two-dimensional (2D) structures into three dimensions (3D). This ability is most commonly demonstrated in the world of origami, where complex 3D structures are created from a single material, most commonly paper. Resulting designs differ solely based on the number and type of corrugations and folds. The principles for folding origami apply at a wide range of scales and to a variety of materials.

We have combined microscale patterning techniques and origami folding principles at a small size scale, where folding by hand is challenging or impossible. Our structures are first precisely patterned using photolithography to create 2D patterns. The structures are released from the underlying substrate and fold up spontaneously when exposed to triggers such as heat or specific chemicals. We refer to these structures as *self-folding* and to the methodology as *hands-free origami*. We have demonstrated this approach by creating microscale structures with up to thousands of folds. Our approach combines *rigid origami*, a subtype of origami using flat paper and linear fold lines, with *kirigami*, where folding is combined with paper cutting. Structures can be constructed for a wide range of practical applications; it is possible to use any nonsquare starting pattern and materials such as metals and polymers.

Currently, methods to construct micro-origami involve expensive serial processes, including those folded by hand with a microscope and tweezers

[Hui et al. 00]. More advanced methods involve stercolithography, 3D multiphoton polymerization, or top-down micromachining with miniaturized tools. Our hands-free method of folding is based upon the utilization of forces present in differentially stressed thin films. These thin-film materials are selectively patterned with varying thickness, stiffness, and strain energy to create different regions, which will become mountain folds, valley folds, and rigid panels. We can selectively control many parameters to adjust the degree of localized curvature. Other techniques have allowed for curving and bending based on thin-film stress, but we have extended the technique to use origami designs from the world of paper folding.

The major advantage of our approach is that all of the patterning takes place in 2D and uses lithography, which is a precise and mature manufacturing paradigm. On the appropriate trigger, the flat pattern folds up to a 3D structure without human intervention. Additionally, the process can be used to simultaneously pattern and fold large structures with thousands of components, and to generate structures with folding in two directions. We have also shown versatility in material composition by incorporating metals, polymers, and dielectrics.

Using these methods, we have designed and folded various patterns into cylinders and coils using unidirectional folding. We explored bidirectional folding and constructed cubic cores derived from paper origami designs. We have further designed a diameter-changing stent and miniaturized paper airplanes smaller than 1 mm in length. All fold up spontaneously and are truly hands-free.

2   Stress-Based Microscale Folding

Like paper folding, hands-free origami requires an energy source to bend its folds. In the absence of a human or machine to fold the material, this energy needs to be carried onboard to fold itself. This energy comes from materials undergoing relaxation and dimensional changes from a stretched state. Conceptually, the folding is similar to the bending of a bimetallic-strip thermostat in response to a change in temperature. Instead of the bending being caused by a difference in the thermal expansion coefficients, however, our thin films bend due to differential stresses that have developed during their deposition. Certain thin films develop either tensile or compressive stresses during and after deposition. The stress can be dormant, or cause the material to delaminate, bend, or fracture [Doerner and Nix 88]. Delamination is commonly seen in old paint that dries and becomes stressed, then peels from the wall. We use this property *intentionally* to create a controlled bending that precisely folds our hinges to a desired angle.

A single layer of material patterned with stress would merely relax on release from the substrate. We pattern a multilayer with a differential stress in order to generate out-of-plane bending and allow a transition from 2D to 3D [Arora et al. 06]. Multiple layers can be used to generate areas with specific radii (Figure 1(a)). A simple bilayer would not produce any meaningful structures because the continuous plane film would tear itself apart and roll into cylinders. To overcome this limitation, we devised a method of stress patterning and extended it to an array of rigid panels and flexible hinges. Rigid panels are akin to paper that remains flat, and flexible hinges are the folds. The flexible hinges have a specific length to ensure the correct bending angle.

In Figure 1(b) and (c), we show a flexible hinge made of chromium (Cr), copper (Cu), and a polymer as well as a rigid panel made of Cr, Cu, and nickel (Ni). Here, the Cr is the stressed metal layer that drives folding. The Cu is the unstressed metal layer that comprises the other component of the bilayer. The Cu film serves as a continuous lower film that supports the fabrication of both flexible hinges and rigid panels. A relatively soft polymer film patterned on top of the bilayer does not restrict folding, whereas a thick, stiff metal such as Ni arrests bending. When both

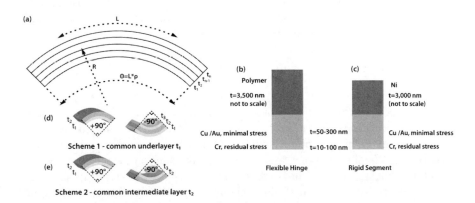

Figure 1. (a) Schematic of multilayer films bending to an equilibrium angle. (b) A flexible hinge constructed from stressed Cr, unstressed Cu or Au, and a soft polymer. (c) A rigid panel constructed from stressed Cr, unstressed Cu or Au, and stiff Ni, where the overlying metal keeps the panel from bending. (d) One method for achieving folding in a bidirectional manner: use two stressed layers with thicknesses t_1 and t_3, surrounding an unstressed layer of thickness t_2. The resulting system of equations can be solved for the desired fold angles. (e) A conceptually simpler method: use two layers for each fold. (Panels B and C from [Bassik et al. 08]. Copyright Wiley-VCH Verlag GmbH & Co. KGaA. Reproduced with permission. Panel D reprinted with permission from [Bassik et al. 09]. Copyright 2009 American Institute of Physics.)

mountain and valley folds are desired (bidirectional folding), the order of
the stressed and unstressed metals can be reversed, or a multilayer stack
can be designed for explicit bend angles (Figure 1(d) and (e)).

2.1   Modeling of Multilayer Thin-Film Curvatures

Thin-film stress was first characterized by G. G. Stoney in 1909, a century
ago, by observing the peeling behavior of sub-millimeter films during elec-
trodeposition [Stoney 09]. The resulting equation has found widespread
use as a relation between substrate bending and thin-film stress, and forms
the basis for the modeling we use to design our structures. By relaxing
some of Stoney's assumptions, and extending the equations to multilay-
ers, a complete description of the bending behavior of elastic thin films
has been developed. One of these is the modified Stoney relation by G.
P. Nikishkov [Nikishkov 03]. The equations combine the intrinsic strain
values of each film, the mechanical properties of the films (elastic moduli,
Poisson's ratios), and their thicknesses, to provide the radius of curvature
of the combined films. We use this model by starting with parameters that
we control such as film thickness, modulus, and stress, to solve for a desired
bending radius. It is important to note that the underlying initial strain
can come from a variety of sources [Nix 89, Floro et al. 02]. As long as a
correct strain can be calculated and mechanical properties are known, the
model will be able to predict the bending angle.

2.2   Methods and Materials of Construction

We have developed a photolithographic process to construct miniaturized
origami via a 2D bilayer approach. Since ultraviolet (UV) light is used
to expose an entire wafer in parallel, it is straightforward to fabricate
thousands of devices simultaneously. Our processing begins with a flat
substrate. We then deposit a sacrificial layer, such as a polymer or easily etched metal, which will be dissolved to release the structures. Other
materials are then patterned via thermal evaporation, spin-coating, sput-
tering, or electro-deposition, and removed via etching. The materials used
in our miniaturized micro-origami largely depend on intended application,
but there are several design requirements. A flexible hinge must include at
least one layer that is highly stressed. Because the overall thickness and
stiffness of the hinge multi-layer must be also minimized, all other materials
used must be either thin or soft. In contrast, materials that are used for
rigid panels need to be relatively stiff (high modulus) and thick. Note that
for these microstructures a stiff "thick" material has a typical thickness of
about 5 microns.

Figure 2(a)–(h) demonstrates a typical process flow and resulting structures, on the order of 1 mm in size. After removing the sacrificial layer,

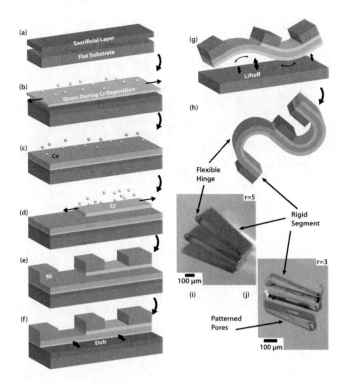

Figure 2. Schematic for generating origami with mountain and valley folds via lithographic patterning: (a) A sacrificial layer is spin-coated onto a rigid flat substrate. (b) A Cr thin film, with a tensile residual stress, is deposited onto the sacrificial layer by thermal evaporation. (c) A Cu film is then deposited, with minimal residual stress, onto the Cr film by thermal evaporation. (d) Additional Cr is deposited in selected regions via lift-off metallization. These areas will become valley folds. (e) Thick, rigid Ni is patterned using photolithography and electro-deposition. (f) Exposed Cu and Cr are removed to create hollow regions in the sheet. (g) The sacrificial layer is dissolved, causing the sheet to assemble while lifting off the substrate. (h–j) Assembled structures with positive and negative fold angles of $\pm 180°$. (Reprinted with permission from [Bassik et al. 09]. Copyright 2009, American Institute of Physics.)

the structure is released and folds spontaneously. In Figure 2(a)–(j), both mountain and valley folds patterned on a linear strip fold into an accordion shape with rigid panels held in a succession of folds. For more complex structures, rigid panels can be connected by arrays of hinges to create patterns that initially appear as 2D arrays (Figure 3(a)). When the sacrificial layer dissolves rapidly, the structure assembles in seconds.

A wide variety of materials such as metals and polymers have been used by our group to fabricate hands-free origami. Relevant data regarding such

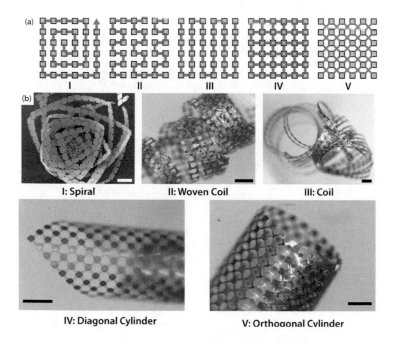

Figure 3. Unidirectional folding allows for many different types of structures. (a) Hinge designs interconnecting identical arrays of rigid panels as modeled on a 6 × 6 array. (b) Folded structures with hinge patterns presented in (a), corresponding to roman numerals. Note that cylinders IV and V have differing axes of rolling as set only by their hinges. Scale bars are in 250 μm. (From [Bassik et al. 08]. Copyright Wiley-VCH Verlag GmbH & Co. KGaA. Reproduced with permission.)

materials and others are available from literature [Sasaki et al. 04, Bassik et al. 08, Chua et al. 03, Klokholm and Berry 68].

2.3   Triggers of Folding and Response to Environmental Stimuli

There is wide interest in structures that can respond to changes in local conditions for use as sensors, micromachines, or medical devices. Our origami platform may be useful in creating structures that reconfigure in such conditions. This reconfiguration would be accomplished by changing the bend angle in response to a stimulus or signal. There are two methods for achieving this goal with hands-free origami. One method is to change the stress within a component of a hinge, causing either bending or straightening [Arora et al. 09]. The other is to prevent bending by patterning a specific rigid material above the hinge, and then softening it or removing it to allow actuation. We have used the latter technique in a variety of small folded structures such as containers [Leong et al. 08] and microgrippers [Leong et al. 09].

Recently, we have developed photopatternable biopolymers that feature the ability to crosslink in UV light but retain biochemical components that are degraded by enzymes. We have shown that different biopolymers will respond only to select triggers, allowing specified hinges to actuate in sequence [Bassik et al. 10]. Two orthogonal triggers then enable three states: (1) a flat state, lifted off; (2) an intermediate state with some hinges folded; and (3) a final state with all hinges folded. We discuss a design using these ideas next.

3   Miniaturized Microscale Origami Structures: Unidirectional Folding

We began by exploring how arrays of rigid panels could be interconnected by different patterns of hinges. One of the major design requirements with the simultaneous, hands-free method is the necessity of folds occurring at the same time. This feature differs from most paper origami, where folds are made in a specific order. Our current method lends itself to tessellated origami [Nojima and Saito 06], pop-up books, and macroscale sheet metal designs. If desired, voids in the crease patterns allow space for the structure to fold on top or adjacent to itself, similar to the *Miura fold*.

We began using thin film stresses to fold structures with only one direction of curvature. The first structures we fabricated were those of spirals, cylinders, and coils [Bassik et al. 08]. These structures were made using an array of 30×30 square rigid panels with flexible interconnecting hinges. Figure 3(a) displays the type of hinge patterns we used based on a miniaturized 6×6 array with lines indicating the flexible hinge panels. We noted that differently connected hinge patterns formed markedly diverse structures, or reliably set the axis of rolling for a cylinder.

4   Bidirectional Microscale Folding Using Thin Metal Films

We examined how thin film stresses would be used to fold hands-free structures with curvature of both types—mountain and valley folds. The simple approach would be to place stressed layers on both sides of a thin unstressed membrane. The challenge lies in devising a fabrication process that allows for patterning of multiple stressed materials. Due to material limitations and an achievable metal stress of approximately 1 GPa, hinges that fold in both directions to 90° are approximately 50 microns in length.

4.1   Self-Folding of Micropatterned Cubic Cores

We adapted and miniaturized a tessellated paper origami structure to the microscale using our design principles [Bassik et al. 09]. Based on Nojima's origami designs that use the Miura fold pattern, we began with a simple cubic core [Nojima and Saito 06]. This design features square plates that rotate out of the plane to generate a 2D array of cubic hollows and enclosures. A critical step was converting a paper-fold diagram to a lithographically compatible pattern. In our design, each hinge needed a specific width. In the macroscopic world of paper folding, the folder makes a tight radius of curvature and miniscule fold lengths. We experimented with several ratios of hinge length to rigid panel length, and found that increasing this ratio created a cubic core that more closely resembled the paper structure (Figure 4). Each cubic element was approximately 150 μm per side for 3:1 rigid:hinge ratio; depending on the initial array size, the total device size was anywhere from approximately 1 mm to nearly 1 cm. The lithographic process allowed us to orient features on each rigid panel in a preferred plane—either parallel to the original 2D structure or on a vertical axis. Without the combination of lithography and bidirectional folding, it is extremely difficult to orient thousands of metallic structures on two axes in an organized array.

4.2   Self-Folding of Micropatterned Cylindrical Stents with Changing Radii

Using bidirectional hands-free origami, we sought to engineer new designs for biomedical applications. Stents are a valuable medical device, used to keep arteries open when they have been blocked. Existing stents reach their final configurations via use of an expandable balloon, which presses the stent open after it is guided into position with a catheter. The balloon is then deflated and the catheter threaded out, while the stent remains pressed against the interior walls of the vessel. We sought to create such a structure that would change shape on its own, with no need for a balloon.

An existing origami stent [Kuribayashi et al. 06] uses planar origami folding patterns in order to more efficiently compress the stent into its compact form. The shape memory alloy stent could be deployed slowly by immersing the stent in liquid nitrogen, then preheating it with warm air at approximately 46°C, and finally placing it at body temperature to fully deploy. We sought to find a technique that did not require large temperature changes.

Using bidirectional origami, we were able to design a stent that operated under physiologic temperatures (Figure 5). The stent would fold in three stages: Stage 1 would be a flat 2D sheet, Stage 2 would be a tight radius to allow for mobility, and Stage 3 would be a larger radius to lock in

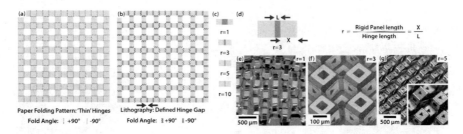

Figure 4. Schematic origami fold diagram for a cubic core that can be realized in paper and by hands-free origami. Hatched lines represent $+90°$ mountain folds, solid lines represent $-90°$ valley folds, solid gray squares represent rigid panels, and white squares represent hollow areas where all material has been removed. (a) Paper fold diagram. (b) Lithographic mask used to fabricate a cubic core with r = 3, as defined by the schematic and equation (c) and (d). (e) Optical image of a core with r = 1 (f) Electron micrograph of a core with r = 3 featuring pores on different faces. (g) Optical image of a core with r = 5 featuring pores on different faces. (Reprinted with permission from [Bassik et al. 09]. Copyright 2009, American Institute of Physics.)

place. The transition from Stage 2 to Stage 3 occurs by activating a set of hidden hinges located inside the cylinder. These add surface area and increase the outer radius. Select triggers are needed to fold selectively at each stage. The stent would be lifted off flat with two polymers specifically patterned to inhibit the stent from folding. The key is orthogonal actuation of the polymers: each must be triggered individually. We chose to use a commercial polymer as the first trigger and a designed custom biopolymer as the second trigger. In other work, we have shown that this dual-trigger approach is extendable to orthogonal biopolymers, such as polypeptides and polysaccharides, where only enzymatic action would be required for triggering. This initial design used a simple 1D corrugation to achieve folding, but with a designed bending radius at each stage. Future designs may utilize 2D pleated folding for a more compact intermediate phase, such as the cubic core described above.

The critical aspect of designing an origami stent via this approach is precise control of the individual hinge angles. To model actuation of the stent, the bending angle of all hinges was designed using a multilayer mechanics computer simulation. This simulation modeled a linear series of hinges H and rigid panels R, and displayed a 2D cross section of each chain. The stent was designed using regular repeating units, each of which would start flat and then transform into two states of differing dimensions. By connecting a specific number of these stent subunits in series, the desired radii of curvature could be achieved. Each subunit featured 7 hinges and 8 rigid panels, meaning a 14-unit array had 210 components in series.

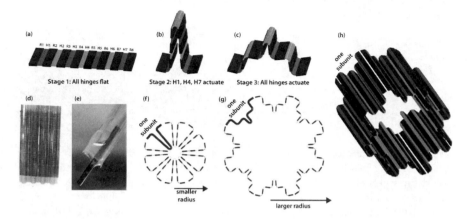

Figure 5. Stent design showing one subunit. Initial flat state (a) is a series of rigid panels R and hinges H. Some hinges will become mountain folds and others valley folds. In the intermediate stage (b) three hinges actuate, creating a bend with specific radius. The final stage (c) has all hinges actuated, increasing the surface area and length of the unit. (d–e) Fabricated stent lifted off in flat initial configuration and stent manually folded and manipulated onto the end of a plastic pipette. (f–g) A schematic of an array of subunits that forms a cylinder with a specified radius, in both the intermediate and final stages of folding. (h) A 3D schematic of the final stent as a cylinder. Each hinge is 65 μm in width with each subunit a 1.7 mm square.

The simulation used equilibrium thin film folding equations for each hinge. Because each hinge type could be constructed separately, they were modeled as metallic layers with overlying polymers. The model showed a 2D cross section of the stent and allowed for changes in hinge length and thin film stress to vary the bending angle.

Fabrication of the stent proceeded as described above, but challenges emerged in self-folding. The combined weight of the panels as the stent folded was too high, and one hinge could not rotate 15 rigid panels in series. This issue did not emerge in previous designs because no single hinge was lifting more than one rigid panel of similar size out of the plane. The stent showed a change in curvature with each trigger, but would fold only if manipulated so gravity did not prevent folding. The solution to this is either to increase the torque available from the hinge or decrease the moment mass.

4.3   Self-Folding of Micropatterned Paper Airplanes

Motile 3D structures are of great interest to the origami enthusiast and the scientific community. A well-known toy is the paper airplane. On the

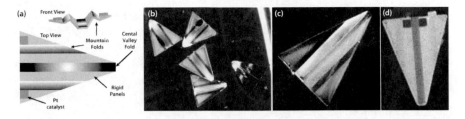

Figure 6. (a) Design of a miniaturized self-folding origami paper airplane. This design uses five rigid panels, two mountain folds, and one valley fold. Two squares of Pt are patterned to provide propulsion by gas decomposition of peroxide. (b–d) Lithographic patterning allows for fabrication of many structures with different designs simultaneously. The paper airplanes use Cr and Au layers and are 1 mm in length.

microscale, scientists are always searching for designs that can move autonomously [Paxton et al. 06], and the paper airplane presents an attractive design for miniaturization.

We sought to construct a paper airplane, but miniaturized beyond where a human being could easily fold it (Figure 6). Motility was enabled by patterning thin platinum in specific regions of the paper airplane. By placing the device in a dilute solution of peroxide, autonomous motion could be achieved as the platinum catalytically decomposed the peroxide and liberated oxygen gas. The resulting flow and momentum transfer propelled the structure. The structure also auto-oriented due to higher drag forces at the larger end of the triangle shape, similar to a weather vane or a dart.

All models were created using precisely designed mountain and valley folds. We targeted maximum dimensions of 1 mm square for the 2D precursor to the paper airplane. The maximum height and width were 1000 μm, with hinge widths of 60 μm, although the hinge placement varied. The angles of the wings with respect to the trailing end were also variable. Since photomasks for 2D lithography are generated using computer aided design, it is trivial to add more rudders or fins and also vary their dimensions.

The miniaturized paper airplanes were fabricated using techniques similar to those of previous devices. Copper was used as a sacrificial layer, with hinges of Cr and Au, and rigid panels were 5 μm of Au. The thickness of the sputtered Pt was kept minimal, as only the exposed surface was catalytic.

We verified the activity of the platinum catalyst in a 0.5% solution of hydrogen peroxide. When a paper airplane was floated on the surface and held up due to surface tension, it translated at several cm/sec, or approximately 30 lengths/sec. For comparison, a paper airplane from an 8.5″ × 11″ piece of paper traveling at 30 lengths/sec would travel at 8.3 m/sec, or more than 18 mph. Unfortunately, once surface tension was broken, the density of the miniaturized paper airplanes caused them to sink in solution, and movement was hampered due to friction. This behavior can be avoided by patterning the Pt catalyst on the underside, which we expect to generate gas under the paper plane surface, minimizing friction. We also expect that the rigid panel density reduction and buoyancy increase with polymers will assist with mobility.

5 Conclusion and Future Possibilities

In conclusion, we have described techniques for producing origami patterns on the millimeter- and microscale that fold up without human intervention. The ability to integrate lithographic processing with miniaturized self-folding origami allows for relative ease of fabrication. We have shown that thousands of identical subunits can be patterned and interconnected with both mountain and valley folds on the microscale. This combination allows for placement of many subunits in precise 3D relationships, and is critical for making complex structures. We have applied theoretical models of multilayer stress to this folding and used them to design several schemes for fabrication with metals and polymers.

Note that at the present time it is extremely challenging to create precisely patterned microstructures. This challenge severely limits engineering capabilities at these small-size scales. Miniaturized origami principles help overcome this pressing challenge in the creation of precisely patterned 3D structures at small-size scales. We believe that hands-free miniaturized origami will find applications in diverse fields, as mechanical components for microactuators and electronic elements in metamaterials.

Bibliography

[Arora et al. 06] William J. Arora, Anastasios J. Nichol, Henry I. Smith, and George Barbastathis. "Membrane Folding to Achieve Three-Dimensional Nanostructures: Nanopatterned Silicon Nitride Folded with Stressed Chromium Hinges." *Applied Physics Letters* 88:5 (2006), 053108.

[Arora et al. 09] William J. Arora, Wyatt E. Tenhaeff, Karen K. Gleason, and George Barbastathis. "Integration of Reactive Polymeric Nanofilms into a Low-Power Electromechanical Switch for Selective Chemical Sensing." *Journal of Microelectromechanical Systems* 18:1 (2009), 97–102.

[Bassik et al. 08] Noy Bassik, George M. Stern, M. Jamal, and D. H. Gracias. "Patterning Thin Film Mechanical Properties to Drive Assembly of Complex 3D Structures." *Advanced Materials* 20 (2008), 4760–4764.

[Bassik et al. 09] Noy Bassik, George Stern, David H. Gracias "Microassem-
    bly Based on Hands Free Origami with Bidirectional Curvature" *Applied
    Physics Letters* 95:9 (2009), 091901/1–3.

[Bassik et al. 10] Noy Bassik, Alla Brafman, Aasieyeh M. Zarafshar, Mustapha
    Jamal, Delgermaa Luysanjay, Florin Selaru, and David Gracias. "Enzymat-
    ically Triggered Actuation of Miniaturized Tools." *Journal of the American
    Chemical Society* 132:46 (2010), 16314-16317.

[Chua et al. 03] Christopher L. Chua, David K. Fork, Koenraad Van Schuylen-
    bergh, and Jeng-Ping Lu. "Out-of-Plane High-Q Inductors on Low-
    Resistance Silicon." *Journal of Micromechanical Systems* 12:5 (2003), 989–
    995.

[Doerner and Nix 88] Mary F. Doerner and William D. Nix. "Stresses and De-
    formation Processes in Thin Films on Substrates." *CRC Critical Reviews in
    Solid State and Material Sciences* 14:3 (1988), 225–268.

[Floro et al. 02] Jerrold A. Floro, Eric Chason, Robert C. Cammarata, and
    David J. Srolovitz. "Physical Origins of Intrinsic Stresses in Volmer-Weber
    Thin Films." *MRS Bulletin* 27:19 (2002), 19–25.

[Hui et al. 00] Elliot E. Hui, Roger T. Howe, and M. Steven Rodgers, "Single-
    Step Assembly of Complex 3-D Microstructures." In *Proceedings of the Thir-
    teenth Annual International Conference on Micro Electro Mechanical Sys-
    tems (MEMS 2000)*, pp. 602–607. Los Alamitos, CA: IEEE Press, 2000.

[Klokholm and Berry 68] E. Klokholm and B. S. Berry. "Intrisic Stress in Evap-
    orated Metal Films." *Journal of the Electrochemical Society* 115 (1968),
    823–826.

[Kuribayashi et al. 06] Kaori Kuribayashi, Owen K. Tsuchiya, Zhong You, Da-
    cian Tomus, Ernest M. Umemoto, T. Ito, and M. Sasaki. "Self-Deployable
    Origami Stent Grafts as a Biomedical Application of Ni-rich TiNi Shape
    Memory Alloy Foil." *Materials Science and Engineering A—Structural Ma-
    terials Properties Microstructure and Processing* 419:1–2 (2006), 131–137.

[Leong et al. 08] Timothy G. Leong, Brian R. Benson, E. K. Call, and David H.
    Gracias. "Thin Film Stress Driven Self-Folding of Microstructured Contain-
    ers." *Small* 4:10 (2008), 1605–1609.

[Leong et al. 09] Timothy G. Leong, Christina L. Randall, Bryan R. Benson,
    Noy Bassik, George M. Stern, and David H. Gracias. "Tetherless Thermo-
    biochemically Actuated Microgrippers." *Proceedings of the National of the
    National Academy of Sciences* 106:3 (2009), 703–708.

[Nikishkov 03] Gennadiy P. Nikishkov. "Curvature Estimation for Multilayer
    Hinged Structures with Initial Strains." *Journal of Applied Physics* 94
    (2003), 5333–5336.

[Nix 89] William. D. Nix. "Mechanical-Properties of Thin-Films." *Metallurgical Transactions A—Physical Metallurgy and Materials Science* 20:11 (1989), 2217–2245.

[Nojima and Saito 06] Taketoshi Nojima and Kazuya Saito. "Development of Newly Designed Ultra-Light Core Structures." *JSME International Journal Series A—Solid Mechanics and Material Engineering* 49:1 (2006), 38–42.

[Paxton et al. 06] Walter F. Paxton, Shakuntala Sundararajan, Thomas E. Mallouk, and Ayusman Sen. "Chemical Locomotion." *Angewandte Chemie* 45:33 (2006), 5420–5429.

[Sasaki et al. 04] Minoru Sasaki, Danick Briand, Wilfred Noell, Nicholaas F. de Rooij, and Kazuhiro Hane. "Three-Dimensional SOI-MEMS Constructed by Buckled Bridges and Vertical Comb Drive Actuator." *IEEE Journal of Selected Topics in Quantum Electronics* 10:3 (2004), 455–461.

[Stoney 09] G. G. Stoney. "The Tension of Metallic Films Deposited by Electrolysis." *Proceedings of the Royal Society of London Series A—Containing Papers of a Mathematical and Physical Character* 82:553 (1909), 172–175.

Foldable Parylene Origami Sheets Covered with Cells: Toward Applications in Bio-Implantable Devices

Kaori Kuribayashi-Shigetomi and Shoji Takeuchi

1 Introduction

This paper describes a method for producing foldable micro-sized origami structures by using poly(p-xylyene) (Parylene), a biocompatible and flexible polymer, combined with cells (Figure 1). The foldable structures with cells would be useful for highly biocompatible implantable devices. These foldable structures can be folded into a compact size along mountain and valley folds like origami for delivery and be deployed when necessary; well-known examples in daily life are tents and umbrellas. They are also used in the aeronautical fields, where devices such as solar arrays and antennas must be packed into small bundles for transport, and then expand to become usable in space [Miura 93, You and Pellegrino 97].

Recently, foldable origami structures have been applied in the medical field, for example, the stent and stent graft [Kuribayashi et al. 06, You and Kuribayashi 09]. The stent and stent graft are types of flexible tubular medical devices that are capable of being folded into small dimensions for minimum invasive surgery and then expanded to open up a blocked organ and also protect a weakened wall in the human body. They are used for treatment of such diseases as stenosis, aneurysm, or esophageal cancer [Serruys et al. 94, Chung and Qadir 98]. Compared with traditional methods (e.g., open heart surgery, radiotherapy, or chemotherapy), minimum

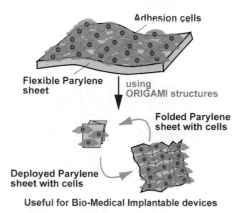

Figure 1. Conceptual illustration of Parylene origami sheets covered with cells.
They can be folded or deployed.

invasive surgery using the foldable structures of the stents and stent grafts
causes less pain and scarring and reduces recovery time for the patients,
as well as providing lower health-care costs. The stents and stent grafts
are generally made of metals, such as a stainless steel and shape memory
alloy, that are foreign to the human body, resulting in the development of
re-stenosis. To prevent the problem of re-stenosis, drug eluting stents have
been produced, but they are still in development.

Here, using both micro-electro-mechanical systems (MEMS) and origami
folding techniques, we produce micro-sized foldable and deployable structures with cells for biocompatible medical devices. To build the structures,
we use the material Parylene, because it is biocompatible and easy to produce the folds in micro-size and to peel off from the substrate. Parylene
is widely used to coat implantable devices and has emerged as a promising material to generate miniaturized devices due to its biocompatibility
and inertness [Fontaine et al. 96, Wolgemuth 00, Stieglitz et al. 00]. The
US Food and Drug Administration (FDA) has approved the Parylene film
for use in human implantable devices. The Parylene sheet with micro- and
nano-sized holes is also widely used to pattern biomaterials such as cells and
proteins as a stencil mask for drug discoveries, biosensor developments, and
tissue-engineering research [Wright et al. 07, Jinno et al. 08, Kuribayashi et
al. 10, Tan et al. 10]. In this research, we produce micro-sized origami folds
on the Parylene sheet and culture the adhesion cells to produce foldable
three-dimensional (3D) structures with the cells. Key advantages of our
foldable structures with the cells include the ability to move and handle
adherent cells in a sheet form, and fold it in a compact size for later expansion, which would find applications in foldable bio-implantable medical
devices.

2 Materials and Methods

This section describes the preparation and manipulation of the Parylene sheet and the methods for culturing the cells.

2.1 Preparation of a Parylene Sheet with Micro-sized Origami Folds

Figure 2 shows the process flow for preparing a Parylene sheet with micro-sized folds. First, we deposited a sheet of 10-μm thickness Parylene film onto the gelatin-coated layer. Parylene forms a thin, flexible, and transparent film when deposited by chemical vapor deposition (CVD) with a Parylene deposition machine (LABCOTER PDS2010, Specialty Coating Systems, USA) (Figure 2(i)). The deposition process consists of three steps: a dimer (di-p-xylene) is vaporized at approximately 150°C, a monomeric form is created at 690°C, and then polymerization occurs at room temperature, forming a uniform Parylene film on the surface of the substrate. The thickness of the Parylene film depends on the initial amount of the dimer used.

Next, we patterned photoresist (PR) and aluminum (Al) on the Parylene film using a standard photolithography technique (Figure 2(ii)) to define etching regions for origami folds (hinges). The Al layer was evaporated, and the positive PR (S1818 Shipley, USA) was then spin-coated onto the substrate. After being exposed to ultraviolet light through a glass mask, the PR was developed using an alkaline solution (NMD developer,

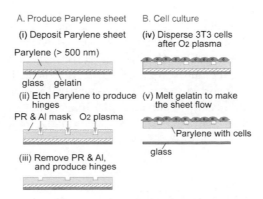

Figure 2. Process flow schematic of a Parylene sheet with folds and then covered by cells: (A) Parylene sheet with micro-sized origami folds was produced with MEMS technique. (i) Parylene sheet was deposited onto a glass substrate spin-coated by a gelatin solution and then (ii–iii) etched by O₂ plasma with a PR and Al mask. (B) Cells were cultured onto the sheet. It is detached from the glass substrate before folding.

Tokyo Ohka, Japan), and the underlying Al layer was etched away using an Al etchant (Wako, Japan). After that, the half-thickness of the Parylene film was then etched with O_2 plasma (10 ml/min, 25 W; RIE FA-1, SAMCO, Japan) to produce the folds (Figure 2(ii)).

Finally, the Parylene sheet with the folds for culturing the cells was obtained after removing the PR and Al layers (Figure 2(iii)).

2.2   Preparation of Cells onto the Parylene Sheet, Folding, and Deploying the Sheet

Adhesion cells were seeded on the Parylene sheet with the folds (Figure 2(iv)). In general, the Parylene is hydrophobic, so it is difficult to culture and adhere the cells [Chang et al. 07]. To enhance the adhesion between the cells and the sheet, we applied O_2 plasma onto the sheet to make it hydrophilic before culturing the cells. We used mouse fibroblast cells (3T3) between the 14th and 25th passage. The cells were maintained in Dulbecco's modified Eagle's medium (DMEM) supplemented with 10% fetal bovine serum (FBS) (JBS-5441, Japan Bioserum, Japan), 100 U/mL penicillin, and 100 μg/mL streptomycin at 37°C under a humidified atmosphere of 5% CO_2. Subconfluent 3T3 were harvested by incubation for 5 min at 37°C with 0.25% trypsin and 0.02% EDTA in Ca^{2+}- and Mg^{2+}-free PBS. The cells were resuspended in DMEM and plated onto the prepared Parylene sheet at a concentration of 2×10^6 cells/mL. After 24–48 hours, the sheet with the cells was detached easily from the glass substrate. Parylene can be easily peeled off from the substrate, and melting the gelatin underlying the sheet made the process of peeling easier. (Figure 2(v)). Then, the sheet was folded and deployed by hand with tweezers.

We used the fluorescent imaging kit LIVE/DEAD Viability/Cytotoxicity to determine cell viability (Invitrogen, USA), which was performed in accordance with the manufacturer's instructions. The cell morphology was observed using an inverted optical microscope with phase contrast with fluorescence (IX70, Olympus, Japan). The images were captured using a CCD camera (AxioCam HRc, Carl Zeiss, Germany) and image software (AxioVision ver.4.7.1, Carl Zeiss, Germany).

3   Results and Discussion

Figure 3 shows the results of the Parylene sheet covered with the 3T3 cells. This folding pattern is called *Miura-ori*, one of the most famous foldable structure designs [Miura et al. 80, Miura 93]. The half thickness (5 μm) of the Parylene was etched for producing the folds. The width of the etched folds was 50 μm (Figure 3(a)-ii).

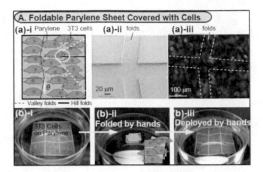

Figure 3. (a)-i The schematic view of *Miura-ori* with cells ($\theta = 8$). (a)-ii Scanning electron microscope (SEM) image of the folds on the Parylene sheet. (a)-iii The 3T3 cells was cultured successfully. The green and blue colors are the cells and their nuclei, respectively. (b) The sequential images of the folding and deploying of the Parylene sheet with the cells. (See Color Plate X.)

We succeeded in seeding and covering the Parylene sheet with the cells (Figure 3(a)-iii). The cells adhered and became a sheet. We labeled the 3T3 cells with a fluorescent dye using the LIVE/DEAD assay kit, which stains living cells green and dead cells red after the folding and deploying processes. The cells were successfully labeled, and we confirmed that all of the cells on the Parylene sheet were alive (Figure 3(a)-iii).

Figure 3(b)i–iii shows photographs of 3T3 cells after four days of culture on the Parylene sheet. The sheet was peeled off from the glass substrate after cell culturing. With the guidance of the folds, we were able to easily fold and deploy the sheet with the *Miura-ori* pattern. We also labeled the cells with the LIVE/DEAD assay. We found that these folding and deploying processes were achieved without significant damage. Further research on culturing the cells after these processes for a long period is necessary to understand cellular behavior under various conditions and apply this technique to real medical products.

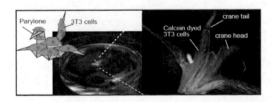

Figure 4. The Parylene origami crane (thickness 10 μm). The size of the origami was 15 mm $\times$ 15 mm. The 3T3 cells were successfully cultured on the 3D structure. (See Color Plate XI.)

Using the Parylene sheet, we can make complicated 3D structures. We demonstrated that the 3T3 cells could be cultured successfully on a 3D crane made by using the Parylene sheet (Figure 4). The cells can be cultured either before or after folding the structures.

4   Conclusions

We presented the method of producing origami foldable and deployable structures using the biocompatible polymer of the Parylene sheet with cells. The adhesion cells were successfully cultured onto the Parylene sheet with origami folds. Our method does not require any harmful chemical to detach the cells on the Parylene sheet from the substrate, and we can fold and expand the produced Parylene structures with the cells.

Currently, in the field of regenerative medicine, a functional cell sheet has been developed using a method of combining biomaterials with living cells to treat esophageal cancer, cardiac failure, tracheal resection, and corneal dysfunction [Yang et al. 07]. The cultured cell sheet is harvested from a substrate and transformed, to cover a an area damaged by disease. We believe our method of flexible and compact foldable origami can be applied to produce a foldable cell sheet to be used in highly biocompatible implantable medical devices for minimum invasive surgeries and treatments.

Acknowledgments. This research was partly supported by ERATO and Science and Engineering Entrepreneurship Development Program for Vigorous Research from Japan Science and Technology Agency (JST), Japan.

Bibliography

[Chang et al. 07] Tracy Y. Chang, Vikramaditya G. Yadav, Sarah De Leo, Agustin Mohedas, Bimal Rajalingam, Chia-Ling Chen, Selvaprab Selvarasah, Mehmet R. Dokmeci, and Al Khademhosseini. "Cell and Protein Compatibility of Parylene-C Surfaces." *Langmuir* 23:23 (2007), 11718–11725.

[Chung and Qadir 98] S. Chung and A. Qadir. "Endoscopic Management of Advanced Oesophageal Cancer." *Eur J Gastroenterol Hepatol* 10:9 (1998), 737–739.

[Fontaine et al. 96] A. B. Fontaine, K. Koelling, S. D. Passons, J. Cearlock, J. Hoffman, and D. G. Spigos. "Polymeric Surface Modifications of Tantalum Stents." *J Endovasc Surg* 3:3 (1996), 276–283.

[Jinno et al. 08] Satoshi Jinno, Hannes-Christian Moeller, Chia-Ling Chen, Bimal Rajalingam, Bong Geun Chung, Mehmet R. Dokmeci, and Ali Khademhossein. "Microfabricated Multilayer Parylene-C Stencils for the Generation of Patterned Dynamic Co-cultures." *J Biomed Mater Resh A* 86:1 (2008), 278–288.

[Kuribayashi et al. 06] Kaori Kuribayashi, Koichi Tsuchiya, Zhong You, Dacian Tomus, Minoru Umemoto, Takahiro Ito, and Masahiro Sasaki. "Self-Deployable Origami Stent Grafts as a Biomedical Application of Ni-rich TiNi Shape Memory Alloy Foil." *Mater Sci Eng. A* 419 (2006), 131–137.

[Kuribayashi et al. 10] Kaori Kuribayashi, Yukiko Tsuda, Hajime Nakamura, and Shoji Takeuchi. "Micro-patterning of Phosphorylcholine-Based Polymers in a Microfluidic Channel." *Sensor Act B* 149 (2010), 177–183.

[Miura et al. 80] K. Miura, M. Sakamaki, and K. Suzuki. "A Novel Design of Folded Map." Paper presented at the 10th International Conference of the International Cartographical Association, Tokyo, Japan, August 25–September 5, 1980.

[Miura 93] Koryo Miura. "Concepts of Deployable Space Structures." *Int J Space Struct* 8 (1993), 3–16.

[Serruys et al. 94] Patrick W. Serruys et al. "A Comparison of Balloon-Expandable-Stent Implantation with Balloon Angioplasty in Patients with Coronary Artery Disease." *N Engl J Med* 331:8 (1994), 489–495.

[Stieglitz et al. 00] T. Stieglitz, S. Kammer, K. P. Koch, S. Wien, and A. Robitzki. "Encapsulation of Flexible Biomedical Microimplants with Parylene C." Paper presented at the 7th Annual Conference of the IFESS, Ljubljana, Slovenia, June 25–29, 2000.

[Tan et al. 10] Christine P. Tan, Benjamin R. Cipriany, David M. Lin, and Harold G. Craighead. "Nanoscale Resolution, Multicomponent Biomolecular Arrays Generated by Aligned Printing with Parylene Peel-Off." *Nano Lett* 10:2 (2010), 719–725.

[Yang et al 07] Joseph Yang, Masayuki Yamato, Tatsuya Shimizu, Hidekazu Sekine, Kazuo Ohashi, Masato Kanzaki, Takeshi Ohki, Kohji Nishida, and Teruo Okano. "Reconstruction of Functional Tissues with Cell Sheet Engineering." *Biomaterials* 28 (2007), 5033–5043.

[You and Pellegrino 97] Zhong You and Sergio Pellegrino. "Foldable Bar Structures." *Int J Solids Structures* 34:15 (1997), 1825–1847.

[You and Kuribayashi 09] Zhong You and Kaori Kuribayashi. "Expandable Tube with Negative Poisson's Ratio and Their Application in Medicine." In *Origami⁴: Fourth International Meeting of Origami Science, Mathematics, and Education*, edited by Robert J. Lang, pp. 117–127. Wellesley, MA: A K Peters, 2009.

[Wolgemuth 00] Lonny Wolgemuth. "Assessing the Performance and Suitability of Parylene Coating." *Med Dev Diagn Ind* (August 2000), 42.

[Wright et al. 07] L Dylan Wright, Bimalraj Rajalingam, Selvapraba Selvarasah, Mehmet R. Dokmeci, and Ali Khademhosseini. "Generation of Static and Dynamic Patterned Co-cultures Using Microfabricated Parylene-C Stencils." *Lab Chip* 7 (2007), 1272–1279.

Part IV

Mathematics of Origami

Introduction to the Study of Tape Knots

Jun Maekawa

1 Introduction

If you tie paper tape methodically, the result can be a regular pentagonal knot. This fact is well known. For example, some Japanese family crests are based on this shape. These designs are called *Fumi* (letters) *Mon* (crest), and they derive from traditional forms of letter folding (Figure 1).

The origin and the history of this letter-folding custom are not clear. Nor is it known who was the first person to consider it as a geometrical problem. It appears to be a very old problem. In the modern era, it was described by Coxeter as an mathematical exercise [Coxeter 61].

Figure 1. Pentagonal knots in Japanese family crests.

2 Why a Knot in a Tape Forms a Regular Pentagon

The reason why a knot in a tape shapes a regular pentagon is really a simple mathematical exercise, as shown by the following proof, illustrated in Figure 2:

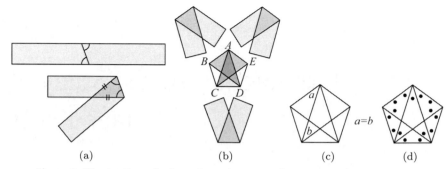

Figure 2. Illustration of why a knot in a tape forms a regular pentagon.

1. The overlapping area of a once-folded tape is an isosceles triangle because the alternate angles of parallel lines are equal to each other (Figure 2(a)).

2. A single knot is formed with three such folds.

3. In Figure 2(b), triangle ABD is equal to triangle AEC because they are symmetric about the center line.

4. In Figure 2(b), triangle ACD is equal to triangle ABD and AEC because it shares its two sides with triangle ABD and AEC.

5. In Figure 2(c), angle a is equal to angle b because they are alternate angles of the parallel lines.

6. In Figure 2(d), angles shown as dots are all equal to each other.

7. The knot is a regular pentagon.

3 Regular Odd-Sided Polygonal Knots

Although pentagonal knots are well known, less known is the fact that a knot can form a regular heptagon or nonagon in a similar fashion, as shown in Figures 3(a) and 3(b)) I found these "odd-gonal" knots in an enlightening book of mathematics by Takano [Takano 71].

A heptagonal knot is based on a double knot, and a nonagon one on a triple knot. We can generalize them to other odd-sided polygons (odd-gons), as shown in Figure 3(c). The principle is simple. A side and the opposite vertex of a regular odd-gon form an isosceles triangle, which is the overlapping area of the folded tape (Figure 3(d)).

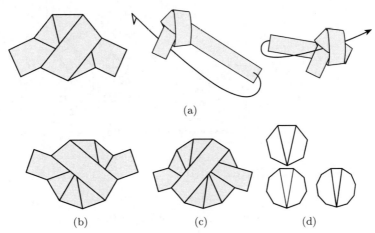

Figure 3. Regular odd-gonal knots.

4 Regular Even-Sided Polygonal Knots

Naturally, the next question is whether we can make even-gonal knots. It is possible when we consider another type of isosceles triangle that appears in polygons. It is an isosceles triangle given by a side of a regular polygon and the center point (Figure 4(a)).

Figure 4(b) shows an octagon [Maekawa 96], and Figure 4(c) shows a dodecagon. Both a square and a hexagon are also possible, although these forms are not knots in the topological sense (Figure 4(d)).

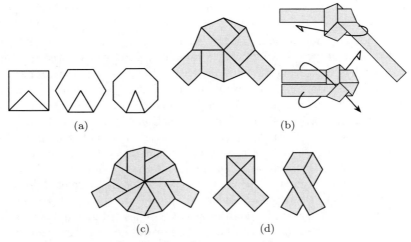

Figure 4. Regular even-gonal knots.

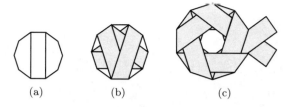

Figure 5. Regular decagonal knots.

5 Regular Decagonal Knots

I have excluded the decagon from the previous discussion. When we regard a tape knot as a side-to-side connection of the polygon, the reason for the exclusion will be clear.

A pentagonal knot is obtained by connecting every second side of a pentagon, and a heptagonal knot by connecting every third side. In general, when we connect the sides of a regular n-gon, we may connect every mth side, where $m < (n - 2)/2$.

In the case of a decagon, we may connect every first, second, third, or fourth side. We must also impose the necessary condition that all connections must be identical. Otherwise, the width of the tape will conflict with the folding pattern.

The every-fifth-side connection does not create a complete knot because it connects only a side with the opposite side; it does not visit all sides (Figure 5(a)). The every-fourth-side connection also does not create a regular decagon because the tape will return to the starting side with only five folds (Figure 5(b)). Even though the every-first-side, every-second-side, and every-third-side form of connection (Figure 5(c)) make decagons, they have holes.

The side-to-side-connection is the basis of a comprehensive understanding of regular polygonal knots. For a regular n-gon, we can create a complete knot by connecting all the sides with an every-mth-side connection, where m is less than n and coprime to n. That is equivalent to drawing with one stroke a stellar polygon that is based on a regular n-gon.

Brunton has previously studied the relationship between tape knots and stellar polygons [Brunton 61]. Using Euler's totient function $\Phi(n)$, which returns the number of the natural numbers between 1 and n that are coprime to n, Brunton obtained the result that the number of possible knots of a regular n-gon is $(1/2)\Phi(n) - 1$. He divided by 2 because clockwise and counterclockwise knots are identical, and he subtracted 1 because we cannot make a knot when we connect adjacent sides. In his analysis, he did not distinguish mountain and valley folds, nor did he consider the order of the layers.

Brunton also considered the condition that permits knots without holes: either the edges of the folded tape intersect at the center of the polygon or the layers of the folded tape overlap on the center of the polygon. It is always possible to make a knot without a hole for any regular odd-gon. For regular even-gons, we can make knots without holes by connecting the second farthest sides when $m = (1/2)n - 1$ is coprime to n. In the case of regular decagons, m is 4 and has a common factor of 2 with 10, which means we cannot make a knot without a hole with a regular decagon. In general, we cannot make a knot with an n-gon where n is of the form $4k+2$ for integer k because $m = 2k$.

6   Stability of Knots

So far, I have not referred to details of knot theory or topology while using the term *knot*. Now I include them in the consideration of the stability of the folded knots. (As a general reference for knot theory, see [Kauffman 91].)

There can be many different patterns that are made with a specified side connection pattern. For example, there are nine different regular pentagon patterns (when we consider mirror images to be identical) that vary in the directions of the folds and the orders of the layers. (See Figure 6 for several examples. Note that not all possible patterns are shown in the figure. For example, *V-V-V* is the mirror image of *M-M-M*.) Among all of the patterns in Figure 6, only one is a topological knot. The knot

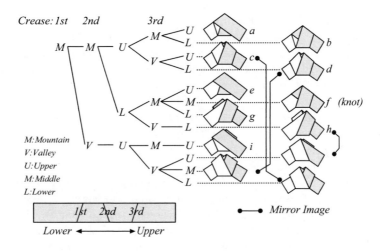

Figure 6. Pentagonal patterns made with tapes.

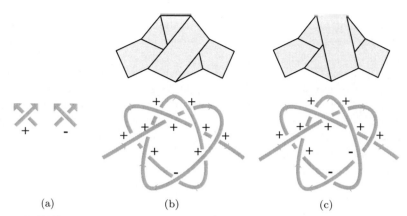

Figure 7. Different writhe numbers in two heptagonal knots: (a) key, (b) writhe number $= 9 - 1 = 8$, (c) writhe number $= 8 - 2 = 6$.

"naturally" becomes a regular pentagon. In other words, the knot is the most stable pentagon pattern. Moreover, those patterns that are not knots vary in their stability. The most stable pattern is f, which is a topological knot, and the least stable one is i, which is a mountain-valley-mountain pattern.

To evaluate the stability of a tape pattern, one can use the *writhe number* from knot theory. Consider the pattern to be made with an oriented line rather than a tape. Figure 7 shows two of the heptagonal knots represented in this way. Even though both are knots, their stability is different. The writhe number of a knot is defined as the sum of the crossing patterns, each of which is either $+1$ or -1, of the line taking into account the orientation of both lines [Kauffman 91]. We use the absolute value of the writhe number because we consider the mirror images of two tape patterns (which would have opposite writhe numbers) to be identical. By way of example, the writhe number in Figure 7(b) is 8, and the writhe number in Figure 7(c) is 6. In fact, the knot in Figure 7(b), which has the larger writhe number, is more stable. That observation suggests that knots with larger writhe numbers are more stable, although some issues are still left, such as how to measure or define "stability."

The writhe number is relevant to the stability of knots from tapes as well as those from lines. In Figure 8, the two edges of a tape are represented as distinct lines when the tape is folded in a pentagon pattern. Figures 8(a), 8(b), and 8(c) correspond to patterns f, I, and c in Figure 6, respectively. Note that we consider a tape to be a Möbius strip so that the lines from each of the two edges will be joined to each other in such a ways as to form a single closed loop. The pattern in Figure 8(a) is a knot, while those in Figures 8(b) and 8(c) are not. These two patterns have the same writhe

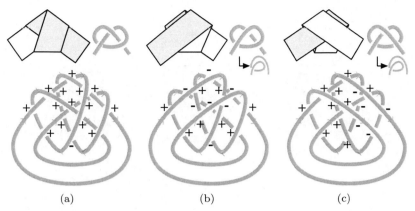

Figure 8. Examples of writhe numbers of pentagonal knots: (a) writhe number = 14 − 1 = 13; (b) writhe number = 8 − 7 = 1; (c) writhe number = 10 − 5 = 5.

number 1 when we consider them as tapes from a single line. That is, the two patterns are both lines with one loop each. The pattern in Figure 8(c), however, is more stable than that in Figure 8(b), and that fact is indicated by the writhe numbers of line-pairs.

Figure 9 shows another clear example: links made with two folds. Among the three possible patterns, the one in Figure 9(a) has the largest writhe number and is the most stable.

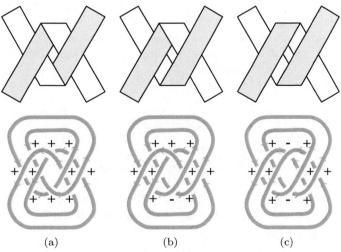

Figure 9. Pairs of folded tapes and their writhe numbers: (a) writhe number = 10; (b) writhe number = 8; (c) writhe number = 6.

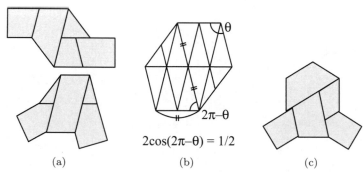

$$2\cos(2\pi-\theta) = 1/2$$

(a)   (b)   (c)

Figure 10. Nonregular polygonal knots.

7   Nonregular Polygonal Knots

Although we have seen only regular polygons so far, we can generalize regular polygonal knots to non-regular polygonal ones. Figures 10(a) and 10(c) show examples [Maekawa 96]. One of the angles in these hexagonal knots is "special" with the angle θ chosen such that $\cos\theta = -1/4$ (Figure 10(b)), which is the Marardi angle of four dimensions; that is, the angle between each regular tetrahedron in a four-dimensional regular pentachoron. Even though I do not think there is a deep relation between paper knots and four-dimensional geometry, I was delighted with this coincidence.

To make these nonregular hexagonal knots, one connects the sides in an irregular order. If we number the sides in sequence, the connecting order of the knots shown can be described as 1-3-5-2-6-4 and 1-4-2-6-3-5. There are an infinite number of such polygons and a correspondingly infinite number of nonregular polygonal knots. In addition, although the connections in these hexagons are cyclic, as in regular polygonal knots, in some cases the angles does not allow connecting the first and the last sides when we connect all the sides (Figure 10(c)).

8   Conclusion and Further Research

This paper has described a preliminary study of tape knots. I believe this subject is useful as educational material for mathematics classes. Moreover, I expect there could be several further lines of inquiry. They include a more thorough study of the "stability" of knots, counting of all less-than-n-gon patterns made with tapes, and further development of conditions for nonregular polygonal knots.

Acknowledgement. I would like to thank Hatori Koshiro for helping me translate this paper into English.

Bibliography

[Brunton 61] James K. Brunton. "Polygonal Knots." *The Mathematical Gazette* 45:354 (1961), 299–302.

[Coxeter 61] Harold S. M. Coxeter. *Introduction to Geometry.* New York: John Wiley & Sons, 1961.

[Kauffman 91] Louis H. Kauffman. *Knots and Physics.* River Edge, NJ: World Scientific Publishing, 1991.

[Maekawa 96] Jun Maekawa. "Oru ni Mijikashi" (in Japanese). *Oru Magazine* 16 (1996), 38–39.

[Takano 71] Kazuo Takano. *Shakaijin no Sugaku* (in Japanese). Tokyo: Morikita. 1971.

Universal Hinge Patterns for Folding Orthogonal Shapes

Nadia M. Benbernou, Erik D. Demaine,
Martin L. Demaine, and Aviv Ovadya

1   Introduction

An early result in computational origami is that every polyhedral surface can be folded from a large enough square of paper [Demaine et al. 00]. A recent algorithm for this problem even attains practical foldings [Demaine and Tachi 11]. But each polyhedral surface induces a completely different crease pattern. Is there a single hinge pattern for which different subsets fold into many different shapes?

Our motivation is to develop programmable matter out of a foldable sheet [Hawkes et al. 10]. The idea is to statically manufacture a sheet with specific hinges that can be creased in either direction, and then dynamically program how much to fold each crease in the sheet. Thus, a single manufactured sheet can be programmed to fold into anything that the single hinge pattern can fold.

We prove a universality result: an $N \times N$ square tiling of a simple hinge pattern can fold into all face-to-face gluings of $O(N)$ unit cubes (polycubes). Thus, by setting the resolution N sufficiently large, we can fold any 3D solid up to a desired accuracy.

The proof is algorithmic: we present the *Cube Extrusion Algorithm* which converts a given polycube into a crease pattern (a subset of the universal hinge pattern) and a 3D folded state in the shape of that polycube, with seamless faces. Figure 1 shows a simple example.

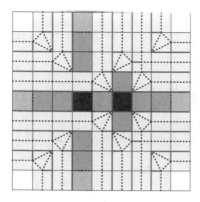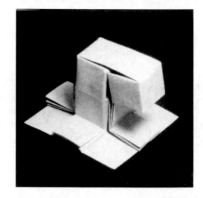

Figure 1. Folding a bend-shaped polycube with a square base via the Cube Extrusion Algorithm. For simplicity, the mountain-valley pattern in this and other figures does not exactly show the reflected creases when multiple layers are folded.

At the core of our algorithm is the notion of a *cube gadget*, which folds a cube in the middle of a sheet of paper. Such foldings of a single cube have been independently developed by origamists over the years; the first documented design we are aware of was created by David A. Huffman in 1978.[1] The novelty is in the way we combine multiple cube gadgets to form a desired polycube. We present three different cube gadgets, one of which is the gadget independently created by Huffman, each with its own advantages and disadvantages when combined to fold a polycube.

We also describe an implementation of the algorithm that can be used to automate experimentation and design of geometric origami using a cutting plotter or laser cutter to score the paper.

2 Definitions

We start here with a few definitions about origami, specified somewhat informally for brevity. For more formal definitions, see [Demaine and O'Rourke 07, Chapter 11].

For our purposes, a *piece of paper* is a connected collection of flat polygons in 3D joined along shared edges (a polyhedral complex; note that we can have multiple polygons in the same place, but with different connections and a specified stacking order). A notable special case is a single $m \times n$ rectangle of paper for integers m and n (but in general, a piece of paper does not have to be flat; for example, a polyhedron is a piece of paper). We index the unit squares of such a rectangle in the style of matrices: $s_{i,j}$

---

[1] Personal communication with the Huffman family. The fourth author independently developed this gadget in middle school circa 2000.

refers to the unit square in the ith row and jth column, and $s_{1,1}$ is in the upper-left corner.

A *hinge* is a line segment drawn on a piece of paper that is capable of being creased in either direction. A *hinge pattern* is a collection of hinges drawn on a piece of paper. The hinge patterns we consider in this paper are all based on subdivisions of the unit-square grid, adding a finite number of hinges within each unit square. The unit squares of the hinge pattern correspond to the unit squares of a rectangle of paper.

An example of a hinge pattern is the *box-pleated pattern* (known in geometry as the *tetrakis tiling*), which is formed from the unit-square grid by subdividing each square in half vertically, horizontally, and by the two diagonals, forming eight right isosceles triangles. The upper-left corner of Figure 3 shows an example for four unit-squares.

An *angle pattern* is a hinge pattern together with an assignment of a real number in $[-180°, +180°]$ to each hinge, specifying a fold angle (negative for valley, positive for mountain). We allow a hinge to be assigned an angle of 0, in which case we call the hinge *trivial*, even though we do not draw trivial hinges in most figures. A hinge with a nonzero angle is called a *crease*. The *crease pattern* is the subgraph of the hinge pattern consisting of only the creases.

An angle pattern determines a 3D geometry called *folded geometry*, which maps each face of the crease pattern to a 3D polygon via Euclidean isometry (by the composition of rotations at creases). More explicitly, a folded geometry is a map from all points of the piece of paper to $\mathbb{R}^3$ that satisfies constraints as specified in [Demaine and O'Rourke 07, Chapter 11]—and there is an obvious mapping from angle patterns to folded geometries.

A *folded state* consists of such a folded geometry together with an ordering λ, which is a partial function over the touching points in the folded geometry, in our case describing the stacking relationship among polygons of the crease pattern that touch in the folded geometry. We define the *starting sheet* of a folded state to be the original piece of paper, that is, the domain of the folded geometry.[2] A *folding sequence* is a sequence of folded states $F_1, F_2, \ldots, F_k$ from the same starting sheet. The last folded state in a folding sequence is called the *final folded state*.

We define the *number of layers at a point q* to be the number of non-crease points in the piece of paper that get mapped to q by the folded geometry. The *number of layers of a folded state* is the maximum number of layers over all points.[3]

---

[2] Note that "sheet" is not to suggest that the piece of paper needs to be flat; it can be any polyhedral complex.

[3] This measure is a simple way to bound the effect of paper thickness, but in practical origami there are other quantities that could be measured.

Next, we define the notion of "coalescing," which lets us ignore certain details of a folded state. A *coalesce folded state* is a folded state augmented with a *coalesce set*, which is a subset of the starting sheet. If we take the starting sheet and identify (glue together) all pairs of points in the coalesce set that are collocated by the folded geometry, then we obtain a metric space called the *coalesce result*. This coalesce result is also a piece of paper under our definition and therefore can be the domain of a new coalesce folded state. A *coalesce sequence* is a sequence $C_1, C_2, \ldots, C_k$ of coalesce folded states, where each C_k is a folding of the coalesce result of C_{k-1}.

One can generate a folding sequence from a coalesce sequence by letting $F_1 = C_1$ and $F_k = F_{k-1} \circ C_k$, and then composing the geometry and ordering functions in the obvious way. Note that the starting sheet of each F_k is the starting sheet of C_1, whereas the starting sheets of the other C_ks can be any shape folded from that starting sheet. The *final folded state of a coalesce sequence* is the final folded state of the generated folding sequence. We say that a folding sequence or coalesce sequence *folds a piece of paper π into a shape σ* if the starting sheet of the first folded state is the piece of paper π and the image of the last folded geometry is the shape σ.

In this paper, we allow all but the last folded state in a folding sequence or coalesce sequence to have crossings. All folded states are reachable from the starting sheet by the continuous folding motion [Demaine et al. 04], so the final folded state is still reachable. We use folding sequences as tools to construct the final folded state, not as instructions for folding.

Now that we can describe how to fold a shape, we define our target shapes. A *polycube P* is a union of unit cubes on the unit-cube lattice with a connected *dual graph*; the dual graph has a vertex for each unit cube and an edge between two vertices whose corresponding cubes share a face. The *faces* of the polycube are the (square) faces of the individual cubes that are not shared by any other cubes.

A *folding of a polycube* is a folded state that covers all faces of the polycube, and nothing outside the polycube. In fact, some of our foldings of polycubes also include the internal squares, the faces shared by multiple cubes, and some of our foldings do not put anything else interior to cubes— but in general we do not require either property. A face of a folded polycube is *seamless* if the outermost layer of paper covering it is an uncreased unit square of paper. Our foldings will generally be seamless.

3   Cube Gadgets

We now introduce the notion of a cube gadget (see Figure 2). For positive integers r and c, an *$[r, c]$-cube gadget* is a method of extruding a cube from a rectangular piece of paper at a specified location. The input to the cube

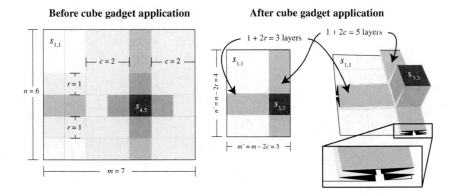

Figure 2. Abstract effect of applying a $[1, 2]$-cube gadget at square $s_{4,5}$ of a 6×7 rectangle of paper. The two leftmost diagrams are top views before and after folding. The right diagram is a stylized perspective view of the folded state.

gadget is an $m \times n$ rectangle of paper, for integers $m > 2r$ and $n > 2c$, as well as a unit square $s_{i,j}$ on the paper, where $r < i < m - r$ and $c < j < n - c$. The output of the cube gadget is a folding of the $m \times n$ rectangle into the shape of a cube sitting on a smaller $(m - 2r) \times (n - 2c)$ rectangle of paper. The cube sits on the square $s_{i-r,j-c}$ in the smaller sheet of paper. All six faces of the cube are seamless except for the bottom face. The top face of the cube is covered by square $s_{i,j}$ from the original piece of paper. The boundary of the original $m \times n$ rectangle paper is mapped onto the boundary of the smaller $(m - 2r) \times (n - 2c)$ rectangle.

The cube gadgets in this paper achieve the folding by making horizontal pleats in the r rows above and below row i, and making vertical pleats in the c columns left and right of column j. The pleats are called *half-square pleats* because they are composed of unit squares folded in half.

Each pleat adds two layers to the row or column it is under, so the number of layers of the folded state is at least $1 + 2\max\{r, c\}$ (one for the row or column plus two for each pleat). Another property of our foldings is that all folding is within the $2r + 1$ rows and $2c + 1$ columns surrounding square $s_{i,j}$. Thus, the quadrant of paper consisting of rows $< i - r$ and columns $< j - c$ is not folded and is incident to the top-left corner of the cube, and similarly for the other four quadrants.

In this paper, we give three different cube gadgets based on three different hinge patterns, as shown in Figure 3. The three cube gadgets are based, respectively, on the box-pleated pattern, the slit pattern, and the arctan $\frac{1}{2}$ pattern.

The arctan $\frac{1}{2}$ gadget and slit gadget are $[1, 1]$-cube gadgets, and the box-pleated gadget is a $[1, 2]$-cube gadget. The advantage of the box-pleated

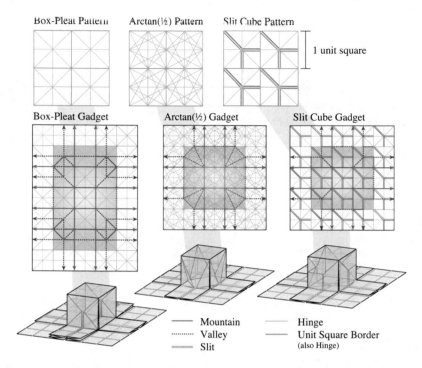

Figure 3. The hinge patterns (top), mountain-valley patterns (middle), and semi-transparent folded states (bottom) for the three cube gadgets. The highlighted region of each mountain-valley pattern has dimensions $(2r + 1) \times (2c + 1)$ and is the region used to actually fold the cube (half-square pleats extend out from those regions).

and slit gadgets is that the hinge pattern is simpler: box pleating has all creases with angles at integer multiples of $45°$. The slit gadget attains higher efficiency than seems possible with regular box pleating by adding a regular pattern of slits in the paper. The arctan $\frac{1}{2}$ gadget attains higher efficiency by using more hinges, some of which are at angles of arctan $\frac{1}{2}$. We use the arctan $\frac{1}{2}$ gadget in all figures for consistency.

Next, we show how a cube gadget can be used to modify an existing folding, which will be the key construction in our folding of general polycubes. Figure 4 provides some intuition for how existing cubes move as new cubes are folded and Figure 5 provides a formal example of the lemma below. Note that we use the arctan $\frac{1}{2}$ cube gadget in all figures from here onwards for consistency.

Lemma 1 (Gadget Application). *Let C^P be a coalesce sequence for a polycube P from an $m \times n$ rectangle of paper. Let f be a face of the polycube P that is seamless in the final folded state of C^P. Then there is a coalesce*

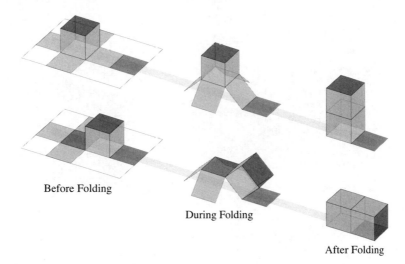

Before Folding

During Folding

After Folding

Figure 4. Given a piece of paper with a cube already folded on it, this diagram shows in an abstract manner how the paper moves when a new cube is folded, depending on whether the old cube is on the top face (above) or a side face (below) of the new cube.

sequence $C^{P'}$ for the polycube P' consisting of P plus a cube extruded from face f, from an $(m + 2r) \times (n + 2c)$ rectangle of paper. The construction is parameterized by a cube gadget.

Proof: Let $C^P = C_1^P, C_2^P, \ldots, C_q^P$, and let F_q^P be the final folded state (for P). Let $s_{i,j}$ be the square in the starting sheet of F_q^P that is mapped to f via F_q^P. We use σ to refer to this square in an abstract sense—when we add rows or columns to the start sheet σ will move with the insertions—so the square coordinates referred to by σ may change.

We construct a new coalesce folded state $C_1^{P'}$ that we will prepend to C^P. We define $C_1^{P'}$ to have the same starting sheet as that of F_q^P, except that we insert r rows above σ, r rows below σ, c columns left of σ, and c columns right of σ. So the starting sheet is a rectangle of paper of size $(m + 2r) \times (n + 2c)$, and σ refers to the square $s_{i+r,j+c}$ in this enlarged sheet of paper. We define this entire enlarged rectangle to be the coalesce set of $C_1^{P'}$. Now we define the folded state of $C_1^{P'}$ to be the given cube gadget applied at σ. The result looks like a cube sitting on the square $s_{i,j}$ of an $m \times n$ rectangle of paper. (The paper does have additional layers in some places from the pleats, but it folds flat except at the cube, and these layers are all coalesced because the entire sheet is in the coalesce set.)

Now, for each coalesce folded state C_k^P, we create a modified coalesce folded state $C_{k+1}^{P'}$ with σ replaced by a cube of paper. Here we use the

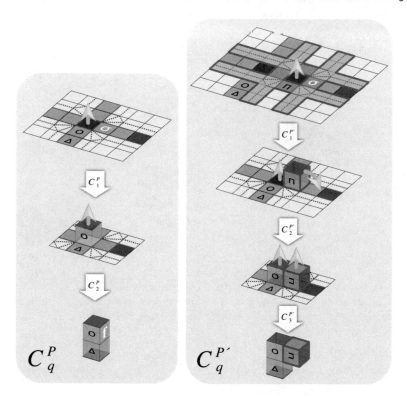

Figure 5. Given a coalesce sequence C^P, this diagram shows the application of Lemma 1 to generate $C^{P'}$ with an additional cube. The purple regions show where additional rows and columns are inserted. (See Color Plate XVI.)

fact that σ never gets folded throughout the sequence (since it is always the face of a cube), and thus corresponds to a seamless square of paper in the starting sheet of each coalesce folded state C_k^P. Note that we add the five new faces of the cube to the starting sheet, but we do not add these faces to the coalesce set of $C_{k+1}^{P'}$; the latter will remain a rectangle. We also add any polygons of paper internal to the cube that appear in $C_1^{P'}$, in the same orientation. Because the coalesce result of $C_k^{P'}$ is the starting sheet of $C_{k+1}^{P'}$, we have thus generated a new coalesce sequence $C^{P'} = C_1^{P'}, C_2^{P'}, \ldots C_q^{P'}, C_{q+1}^{P'}$.

Note that these added cubes may create intersections in the coalesce folded states $C_k^{P'}$ (as mentioned in Section 2). However, the final folded state $F_{q+1}^{P'}$ of $C^{P'}$ (as well as $C_{q+1}^{P'}$ itself) is guaranteed not to have intersections. This follows because F_q^P had no self=intersections, the application $C_1^{P'}$ of the cube gadget has no self-intersections, and adding the cube of paper to make P into P' cannot create intersections. (See Figure 6.)

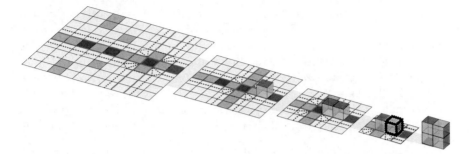

Figure 6. This sequence of folded states for a particular coalesce sequence shows an example of self-intersection in part of a coalesce folding sequence. The self-intersection occurs at the highlighted cube and is resolved in the final step. The self-intersection can be avoided in this case by making additional simple folds.

4   Folding Polycubes

Our Cube Extrusion Algorithm for folding any polycube is parameterized by an arbitrary cube gadget and consists of repeated application of the gadget according to Lemma 1. We describe the recursive algorithm by way of an inductive proof.

Theorem 1 (Cube Extrusion Algorithm). *Any polycube of N cubes can be folded with all faces seamless from a $(2rN+1) \times (2cN+2)$ rectangle of paper by a sequence of N applications of an $[r, c]$-cube gadget plus one additional fold.*

Proof: We prove by induction that any polycube P' of N cubes can be folded seamlessly by a coalesce sequence from a $(2rN + 1) \times (2cN + 2)$ rectangle of paper. We arbitrarily choose a "bottom face" f_b of P', and let b be the unique (bottom) cube having f_b as a face.

The base case is $N = 1$, when P' consists of the single cube b. We can use the cube gadget directly at square $s_{r+1,c+1}$ to obtain a folding of the single cube from a $(2r + 1) \times (2c + 2)$ rectangle of paper. The folded state is a cube next to a (pleated) unit square of paper. Note that the bottom face of the cube corresponds to f_b and is adjacent to the square of paper. By definition of cube gadgets, all faces of the cube except the bottom face are seamless. We fold the extra square of paper over to seamlessly cover the bottom face, thus making a seamless one-cube polycube. The resulting folded state forms the first and only step in a coalesce sequence.

It remains to prove the inductive step. Let T be a spanning tree of the dual graph of P'. Because every tree has at least two leaves, T has a leaf corresponding to a cube $l \neq b$. Let u be the unique cube sharing a face with l, and let f_{ul} be the face shared by u and l.

Now consider the polycube $P = P' \setminus \{l\}$, with $N - 1$ cubes. Because $l \neq b$, f_b remains a face of P. By induction, there is a coalesce sequence C^P that folds a $(2r(N-1)+1) \times (2c(N-1)+2)$ rectangle of paper into P. By Lemma 1, we extrude from f_{ul} to obtain a new coalesce sequence $C^{P'}$ for P' from a rectangle of size $((2r(N-1)+1)+2r) \times ((2c(N-1)+2)+2c) = (2rN+1) \times (2cN+2)$.

The final folded state of the inductively obtained coalesce sequence is the desired folded state from the rectangle of paper into the polycube.   □

Without the concern for a seamless bottom face, we can reduce the +2 in the rectangle bound down to +1.

The algorithm runs in polynomial time. The bottleneck is in converting the coalesce sequence into its final folded state. Each of the N cube gadgets causes the creation of at most $O(N)$ creases, because the piece of paper at that point has size $O(N) \times O(N)$ with $O(1)$ existing creases per square.

The size bound of an $O(N) \times O(N)$ rectangle of paper is tight up to constant factors for square paper. Specifically, folding a $1 \times 1 \times N$ tower of cubes requires starting from a square of side length $\Omega(N)$ in order to have diameter N, as folding can only decrease diameter.

4.1   Hinge Pattern Completeness

Next we show that the Cube Extrusion Algorithm does not create creases that stray from the given hinge pattern.

For a rectangular piece of paper, the *tile* $t_{i,j}$ of a crease pattern is the set of creases within the unit square $s_{i,j}$. The *tile set* of a cube gadget is the set of all distinct tiles that can be generated by the cube gadget. The hinge pattern *generated* by a tile is the result of replicating the tile in a unit-square grid.

Proposition 1. *Given a cube gadget with a finite tile set and half-square pleats as the only folded structure outside of the cube, if we add to an empty tile a hinge for every crease of each tile in the tile set for every 2D orthogonal orientation (rotations and reflections), then the resulting tile generates the hinge pattern required to fold a polycube with the Cube Extrusion Algorithm.*

This proposition is nontrivial because it is possible that some combination of cube gadgets would create new tiles that are not present in any single gadget, which are thus not in the tile set.

Proof: We prove that no other hinges are needed beyond those found in the constructed generator tile.

There are two types of folded tiles used by cube gadgets to make polycubes as described in the proposition: *inner tiles* which make up the non-visible parts inside the cubes and *pleat tiles* that make up the non-visible part of the pleats (outside of the cube). Inner tiles are never folded once they are made part of a cube as our foldings never modify the inner structure of an existing cube, so they do not require any additional hinges beyond those in the tile set. Pleat tiles may be folded again — but they are all half-square pleats, which means that they simply reflect a crease along the midline of the tile — but each of the reflected halves would already have existed in reflections of that tile of the cube gadget, so this also does not create additional hinges. □

4.2 Paper Dimensions

We now show the specific bounds on the dimensions of the required rectangle of paper for each of the three cube gadgets considered in this paper.

Corollary 1 (arctan $\frac{1}{2}$ Universality Lemma). *Any polycube of N cubes can be folded with all faces seamless from an* arctan $\frac{1}{2}$ *hinge pattern on a* $(2N + 1) \times (2N + 2)$ *rectangle of paper using the Cube Extrusion Algorithm.*

Proof: The arctan $\frac{1}{2}$ gadget has $r = 1$ and $c = 1$. Plugging these constants into Theorem 1 yields a process for folding the polycube from a rectangle of paper of size $(2N + 1) \times (2N + 2)$. □

Corollary 2 (Slit Universality Lemma). *Any polycube of N cubes can be folded with all faces seamless from a slit hinge pattern on a* $(2N + 1) \times (2N + 2)$ *rectangle of paper using the Cube Extrusion Algorithm.*

Proof: Same as for Corollary 1. □

The previously discussed gadgets create foldings from a rectangle of paper that is within an additive constant of being square. As we show now, directly applying the Cube Extrusion Algorithm with the tetrakis cube gadget generates a folding with a ratio within a constant of 1×2, but a slight modification allows us to use an approximately square sheet of paper.

Corollary 3 (Box-Pleated Universality Lemma). *Any polycube P of N cubes can be folded with all faces seamless from a box-pleated (tetrakis) hinge pattern on a* $(2N + 1) \times (4N + 2)$ *rectangle of paper using the Cube Extrusion Algorithm. A slight modification of the Cube Extrusion Algorithm uses a* $(3n + 1) \times (3n + 2)$ *rectangle of paper for even N, and a* $(3N) \times (3N + 3)$ *rectangle of paper for odd N.*

Proof: The box-pleated gadget has $r = 1$ and $c = 2$. Plugging these constants into Theorem 1 yields a process for folding P from a rectangle of paper of size $(2N + 1) \times (4N + 2)$.

Now we describe the modified approach. Define the *transpose of a cube gadget* to be the cube gadget with r and c interchanged, so that now we insert r columns to the left and right of the column and c rows below and above the specified row. We alternate the box-pleated gadget and its transpose for a polycube of N cubes such that the box-pleated gadget is applied $\lceil N/2 \rceil$ times and its transpose is applied $\lfloor N/2 \rfloor$ times. This yields a final folded state F_n with a starting sheet of size $(2r \cdot \lceil \frac{N}{2} \rceil + 2c \cdot \lfloor \frac{N}{2} \rfloor + 1)$ $\times (2c \cdot \lceil \frac{N}{2} \rceil + 2r \cdot \lfloor \frac{N}{2} \rfloor + 2)$, which simplifies slightly to $(2 \cdot \lceil \frac{N}{2} \rceil + 4 \cdot \lfloor \frac{N}{2} \rfloor + 1)$ $\times (4 \cdot \lceil \frac{N}{2} \rceil + 2 \cdot \lfloor \frac{N}{2} \rfloor + 2)$. For even N, this bound is $(3N + 1) \times (3N + 2)$, and for odd N, it is $(3N) \times (3N + 3)$.   □

4.3   Number of Layers

Proposition 2. *For any polycube of N cubes, the Cube Extrusion Algorithm produces a folding that uses $O(N^2)$ layers.*

Proof: The folded state produced by Theorem 1 has a starting sheet of size $(2rN+1) \times (2cN+2)$, which is clearly $O(N^2)$ for constants r and c. Because each square of a hinge pattern contains $O(1)$ hinges (by definition), there can be at most $O(N^2)$ layers in the folding (even if we folded the paper up into the smallest unit of area allowable by the hinge pattern).   □

Unfortunately, this quadratic bound on the number of layers is tight in the worst case:

Proposition 3. *For any N and any cube gadget G, there exists a polycube of N cubes for which the Cube Extrusion Algorithm yields a folding requiring $\Omega(N^2)$ layers.*

Proof: Without loss of generality, assume that N is odd. The example we use is a horizontal L-shaped polycube, as shown in Figure 7. To construct it, we take a single cube b and two faces that share an edge of the cube. We extrude $(N - 1)/2$ cubes from each of the faces. (If N were even, we would extrude $(N/2) - 1$ cubes from one of the faces and $N/2$ cubes from the other face.)

We now show that our folding algorithm would construct a folding having $\Omega(N^2)$ layers.

Let the bottom cube of the folding algorithm be b, and take the bottom face f_b to be one of the faces parallel the plane spanned by the legs of the L.

Now consider the face in the resulting folded state opposite to f_b: it is seamless, but hidden beneath it are pleats from prisms of both legs of the L. There are $\Omega(N)$ pleats from each leg, and the pleats are orthogonal to each other. This results in $\Omega(N^2)$ layers.   □

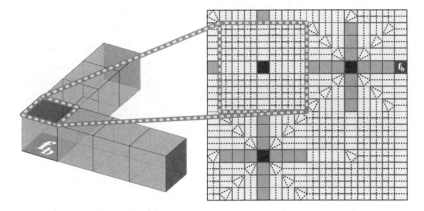

Figure 7. This horizontal L-shaped polycube uses $\Omega(N^2)$ layers when folded by the Cube Extrusion Algorithm.

5 Implementation

A simplification of this algorithm was implemented in Ruby (Figure 8). It allows the user to define a polycube through extrusions and displays the corresponding folding sequence to the screen as a series of angle patterns, but it does not generate a final folded state. It can also show a *simplified composition of the folding sequence*, which can be saved to a PostScript file. The simplified composition just shows the original angle pattern for each square tile on the full sheet of paper (i.e., the angle pattern from the step when the square was inserted as part of a row or column), and does not take into account any nonorthogonal creases made as a result of later steps.

The PostScript file can be sent to a cutting plotter, which can etch the simplified composition onto a sheet of paper. Alternatively, one can use a laser cutter to score the paper (see Figure 8). Only a single cube gadget is foldable at each step when folding along visible etchings of the simplified

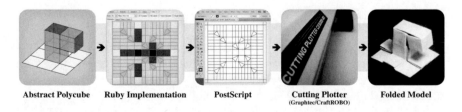

Abstract Polycube **Ruby Implementation** **PostScript** **Cutting Plotter** **Folded Model**
(Graphtec/CraftROBO)

Figure 8. The process by which paper origami can be constructed using the described algorithms.

composition, so one can simply fold cube gadgets until there are no visible etchings.

If the foldings generated by the Cube Extrusion Algorithm are applied to a larger sheet of paper, it appears that the polycubes are "lying on top of" or "extruded" from the sheet. Artistically this effect is generally preferable to having a polycube closed off by a small flap. In particular, it prevents the pleats from expanding as much, and in some cases enables multiple polycubes to be extruded from a single sheet.

6   Rigid Foldability and Self-Folding Sheets

The Cube Extrusion Algorithm can be used to make artistic paper origami, but to get from one folded state to the next may require curving the paper or introducing temporary nonhinge creases. This becomes an issue for controlling a self-folding programmable matter sheet, where the polygons of a hinge pattern are (nearly) rigid.

An open question concerns which polycubes are rigidly foldable from a particular cube gadget, although it seems through simple empirical testing that our cube gadgets fold rigidly in isolation (when making one-cube polycubes). Polycubes with more than one cube may nonetheless not be rigidly foldable, and polycubes with intermediary folded states that self-intersect are almost certainly not rigidly foldable. Our intuition suggests that the slit pattern has the broadest potential for programmable matter.

Acknowledgments. A. Ovadya would like to thank a host of friends who have not complained about his incessant folding of cubes, and in particular Simone Agha, Tucker Chan, Lyla Fischer, Andrea Hawksley, Robert Johnson, and Maria Monks for being sounding boards, folding, and/or commenting. E. Demaine thanks the Huffman family—Elise, Linda, and Marilyn—for access to David A. Huffman's notes, which include the arctan $\frac{1}{2}$ cube gadget. We also thank Jason Ku and Scott Macri for assisting with Figure 3 and Figure 4, respectively.

Bibliography

[Demaine and O'Rourke 07] Erik D. Demaine and Joseph O'Rourke. *Geometric Folding Algorithms: Linkages, Origami, Polyhedra.* Cambridge, UK: Cambridge University Press, 2007.

[Demaine and Tachi 11] Erik D. Demaine and Tomohiro Tachi. "Origamizer: A Practical Algorithm for Folding Any Polyhedron." Manuscript, 2011.

[Demaine et al. 00] Erik D. Demaine, Martin L. Demaine, and Joseph S. B. Mitchell. "Folding Flat Silhouettes and Wrapping Polyhedral Packages: New

Results in Computational Origami." *Computational Geometry: Theory and Applications* 16:1 (2000), 3–21.

[Demaine et al. 04] Erik D. Demaine, Satyan L. Devadoss, Joseph S. B. Mitchell, and Joseph O'Rourke. "Continuous Foldability of Polygonal Paper." In *Proceedings of the 16th Canadian Conference on Computational Geometry*, pp. 64–67. Montréal, Canada: CCCG, 2004.

[Hawkes et al. 10] E. Hawkes, B. An, N. M. Benbernou, H. Tanaka, S. Kim, E. D. Demaine, D. Rus, and R. J. Wood. "Programmable Matter by Folding." *Proceedings of the National Academy of Sciences of the United States of America.* Published online before print, June 28, 2010. Available at http://www.pnas.org/cgi/doi/10.1073/pnas.0914069107.

A General Method of Drawing Biplanar Crease Patterns

Herng Yi Cheng

1   Introduction

Much work has been done on the subject of extrusion origami, where polyhedra are extruded from a flat sheet via grafting. Notable works include the extrusion of orthogonal mazes [Demaine et al. 11] and the algorithm to extrude any *polycube* (a polyhedron composed of identical cubes) [Benbernou et al. 11]. The polycube algorithm is based on drawing the net of a cuboid on the paper, after which any paper between adjacent (in the cuboid) faces is folded and hidden inside the cuboid. Similar in approach is the Origamizer program [Tachi 09], which folds a polyhedral surface by breaking it up into surface polygons that are distributed on a flat sheet and then brought back together by hiding the paper between them behind the surface using tucks.

Common to all these algorithms is the partitioning of paper into polygons and empty space between them that is hidden away using specialized and recurring crease patterns, or *gadgets*. The cuboid gadget from the polycube algorithm has been independently discovered on many occasions, and has been generalized and used to fold prismatoids, which are polyhedra with their vertices among two parallel planes [Natan 10].

This paper describes a generalization of the prismatoid gadget to fold *biplanars*, which are polyhedra with their vertices among two planes. An algorithm has been derived to draw the crease patterns of certain convex biplanars. The algorithm proceeds in three steps:

1. The target biplanar is defined.

2. Its net is drawn using vector and angle calculations.

3. Its crease pattern is drawn by inserting biplanar gadgets into the net.

A computer program that implements this algorithm automatically has been written; it provides an interface that allows users to define the target biplanar, after which it performs the necessary calculations and gives the crease pattern as output.

Section 2 sets up the relevant biplanar-related terminology to be used throughout this paper. The algorithm and related mathematical derivations for drawing crease patterns are presented in Sections 3 through 5. The exact limitations on biplanars that the presented algorithm can fold will be enumerated in Section 6, as their necessity will be clarified by the algorithm itself.

2   Mathematical Conventions

The term *biplanar* refers here to *convex* biplanars. All biplanars exist in $\mathbb{R}^3$, with O as the origin. The two planes on which a biplanar's vertices lie are referred to as the *roof plane* and *base plane*, denoted by π_R and π_B, respectively (see Figure 1). By convention, π_B is taken to be the xy plane. The faces of the biplanar in the roof and base planes are the *roof* and *base*, denoted as R and B, respectively. Naturally, $R \subset \pi_R$ and $B \subset \pi_B$. The other faces of the biplanar are *walls*.

The *projection* of a point X is the shadow it casts on the base plane. The projection of a set of points is its "aerial view" (i.e., the set of shadows cast by the points on the base plane). The projection of the base would thus be the base itself:

$$(x, y, z) \xrightarrow{\text{project}} (x, y, 0). \tag{1}$$

The *boundary line* of a biplanar is denoted by $\ell(R, B)$ and is defined as

$$\ell(R, B) = \begin{cases} \text{undefined}, & \pi_R \parallel \pi_B, \\ \pi_R \cap \pi_B, & \pi_R \nparallel \pi_B. \end{cases} \tag{2}$$

Finally, the *net* of a biplanar is obtained by removing the base, cutting along the edges not lying on the roof plane, and flattening the result.

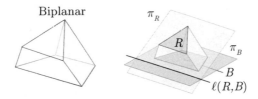

Figure 1. Biplanar notation.

3 Defining the Biplanar

Since users would find it hard to draw a 3D figure, the dimensions are "split up"; that is, they first draw the roof projection and the base in the program's 2D interface, then specify the heights (distance from π_B) of three of the roof vertices. The roof projection would give the x- and y-coordinates of the roof vertices, and the heights would give their z-coordinates, thus fully defining the biplanar. Only the heights of three vertices, X_1, X_2, and X_3, are needed because they define the other heights as well. (See Figure 2.)

Given another roof vertex $X_4(x, y, z)$, the values of x and y are known from the roof projection. Hence,

$$X_1 \in \pi_R \tag{3}$$

$$\implies \pi_R : \boldsymbol{r} \cdot \boldsymbol{n} = \overrightarrow{OX_1} \cdot \boldsymbol{n} \text{ where } \boldsymbol{n} = \overrightarrow{X_1X_2} \times \overrightarrow{X_1X_3} \tag{4}$$

$$\implies \overrightarrow{OX_4} \cdot \boldsymbol{n} = \overrightarrow{OX_1} \cdot \boldsymbol{n} \tag{5}$$

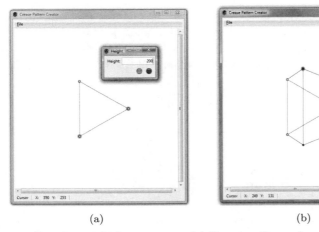

(a)   (b)

Figure 2. Drawing with the program: (a) Drawing the roof projection and specifying the heights. (b) Drawing the base together with the roof projection.

The only unknown in Equation (5) is the z-coordinate, which can be solved for. (For simplicity, n is considered to be pointing out of the biplanar; i.e., it has a positive z-coordinate. If otherwise, n can simply be negated.)

4   Drawing the Net

Once the coordinates of the vertices of the roof have been found, the net can be drawn easily by unfolding the biplanar's walls. That is, each quadrilateral wall connected to the roof along the top edge is rotated along its top edge as the axis until it is coplanar with the roof; see Figure 3.

The angle α by which the wall is rotated is simply the angle between the normal vectors of the roof and the wall being rotated:

$$\alpha = \cos^{-1}(\hat{n} \cdot \hat{n}_W), \text{ where } n_W = \overrightarrow{R_2 B_2} \times \overrightarrow{R_1 B_1}. \tag{6}$$

(For simplicity, n_W is considered to be pointing out of the biplanar. This can be achieved by monitoring the clockwise/anticlockwise orders of the biplanar vertices.)

The new position of the wall's bottom edge can thus be obtained via multiplication by the appropriate rotation matrices:

$$B_1' = R_u(\alpha) B_1 \text{ and } B_2' = R_u(\alpha) B_2 \text{ where } u = \frac{\overrightarrow{R_1 R_2}}{\|\overrightarrow{R_1 R_2}\|}. \tag{7}$$

Using Equations (6) and (7), one can unfold every quadrilateral wall. However, walls can also be triangular, either pointed at the top or at the bottom, both of which can be found in *antiprisms*. As for triangular walls that are pointed at the top, $R_1 = R_2$, leaving the axis of rotation, u, undefined. At this stage, such walls are treated as *degenerate* quadrilaterals, with an infinitesimal top edge that lies on the intersection of the roof and the wall. Hence, the direction of the axis is still fixed:

$$u = \frac{n_W \times n}{\|n_W \times n\|}. \tag{8}$$

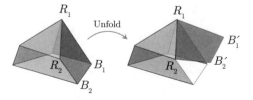

Figure 3. Unfolding a quadrilateral wall.

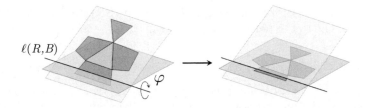

$\ell(R,B)$

Figure 4. Rotating the net.

Hence, all the walls of the biplanar can be unfolded such that they are coplanar with the roof; this "net" lies flat on π_R. The final step is to rotate the whole net such that it lies flat on π_B instead so that its 2D coordinates may be obtained for display in the graphical user interface (GUI). If $\pi_R \parallel \pi_B$, then rotation is unnecessary because the net is already horizontally flat. Otherwise, the axis of rotation is $\pi_R \cap \pi_B = \ell(R, B)$. (See Figure 4.)

The angle of rotation, φ, is the dihedral angle between π_R and π_B, or the angle between their normal vectors:

$$\boldsymbol{n} = \begin{bmatrix} a \\ b \\ c \end{bmatrix} \implies \varphi = \cos^{-1}\left(\hat{\boldsymbol{n}} \cdot \begin{bmatrix} 0 \\ 0 \\ 1 \end{bmatrix}\right) = \cos^{-1}\left(\frac{c}{\sqrt{a^2 + b^2 + c^2}}\right). \quad (9)$$

The direction vector $\boldsymbol{u}$ of $\ell(R, B)$ can be found as follows:

$$\boldsymbol{u} = \boldsymbol{n} \times \begin{bmatrix} 0 \\ 0 \\ 1 \end{bmatrix} = \begin{bmatrix} b \\ -a \\ 0 \end{bmatrix}. \quad (10)$$

Hence, the horizontally flat net can be obtained by multiplying each vertex by the rotation matrix $R_{\hat{\boldsymbol{u}}}(\varphi)$.

5   Drawing the Crease Pattern

From this section onwards in this paper, all geometry is considered in the plane of the net/crease pattern, i.e., $\mathbb{R}^2$. The gadgets drawn on the paper between adjacent walls allow the walls to close up upon folding, forming the biplanar. The algorithm for drawing them is presented in Section 5.1. The purpose of the gadget is to hide paper inside the biplanar. The paper from different gadgets, however, may clash inside the biplanar. An algorithm for solving that problem is presented in Section 5.2. This problem also poses a restriction on foldable biplanars, which is explained at the end of that section. The conditions allowing the paper surrounding a folded biplanar to be flat arc scrutinized in Section 5.3. Finally, Section 5.4 elaborates on an adjustment to the net-drawing algorithm.

5.1 The Gadget

For each pair of neighboring walls, the *inner angle*, denoted by θ (Figure 5), is defined as the angle between the walls. Given any biplanar, θ can be measured from its net.

The *outer angle*, denoted by ϕ (Figure 6), is the angle between the paper extending from the walls after folding. If, nearest to ϕ, the lower internal angles of the walls are β_1 and β_2 and the internal angle of the base is γ, then ϕ can be calculated using the following formula:

$$\phi + (\pi - \beta_1) + (\pi - \beta_2) + \gamma = 2\pi \implies \phi = \beta_1 + \beta_2 - \gamma. \tag{11}$$

Let the bottom vertices (beside θ) of the walls be B_1 and B_2, and let the extension of the wall edges beside θ be ℓ_1 and ℓ_2. Let the positions of the lines after folding be ℓ_1' and ℓ_2', respectively. Clearly, ϕ is the angle between ℓ_1' and ℓ_2'. Thus, focusing on one pair of adjacent walls, the crease pattern should look something like Figure 7.

Fold X_1Y, the fold that maps ℓ_1 to ℓ_1', has to be the angle bisector of the angle between them. Fold X_1Y also maps B_1 to some point B_1' on ℓ_1', so X_1Y is the perpendicular bisector of B_1B_1'. Similarly, X_2Y is the perpendicular bisector of B_2B_2'. However, to close the gap between the walls, B_1 and B_2 must be mapped to the same point. Hence, the condition $B_1' = B_2' = \ell_1' \cap \ell_2'$ must be fulfilled. Let $Z = \ell_1' \cap \ell_2'$.

Theorem 1. *In the net of a biplanar, given two adjacent walls with bottom vertices B_1 and B_2 beside the inner angle, let B_1 and B_2 be mapped to Z after folding. The locus of valid positions of Z is an arc A outside the net such that*

$$A = \left\{ Z_0 \in \mathbb{R}^2 : \angle B_1 Z_0 B_2 = \pi - \frac{\theta + \phi}{2} \right\}. \tag{12}$$

Furthermore, the final crease pattern is unique to the choice of Z.

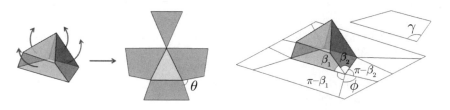

Figure 5. Completed net.   Figure 6. Folded biplanar.

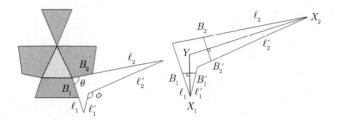

Figure 7. Crease pattern framework.

Proof: Let Z_0 be some valid position of Z, and let $\delta_1 = \angle B_1 X_1 Z_0$ and $\delta_2 = \angle B_2 X_2 Z_0$. Thus,

$$\theta + \delta_1 + \delta_2 = \phi. \tag{13}$$

Since $X_1 Y$ and $X_2 Y$ bisect the angles δ_1 and δ_2,

$$\angle X_1 Y X_2 + \frac{\delta_1}{2} + \frac{\delta_2}{2} = \phi, \tag{14}$$

and Equation (13) implies

$$\angle X_1 Y X_2 = \frac{\theta + \phi}{2}. \tag{15}$$

Let $M_1 = B_1 Z_0 \cap X_1 Y$ and $M_2 = B_2 Z_0 \cap X_2 Y$. Consider the sum of internal angles in quadrilateral $M_1 Y M_2 Z_0$:

$$\begin{aligned}
\angle Y M_1 Z_0 + \angle Y M_2 Z_0 &= \pi = 2\pi - \angle M_1 Y M_2 - \angle M_1 Z_0 M_2 \\
\implies \angle B_1 Z_0 B_2 &= \angle M_1 Z_0 M_2 \\
&= \pi - \angle M_1 Y M_2 \\
&= \pi - \angle X_1 Y X_2 = \pi - \frac{\theta + \phi}{2}.
\end{aligned}$$

Thus, the locus of Z_0 must be the arc specified in the theorem statement.

The crease pattern is determined from the positions of ℓ_1' and ℓ_2'. For any feasible Z_0, $\ell_1' = Z_0 X_1$, where X_1 is the intersection of ℓ_1 and the perpendicular bisector of $B_1 Z_0$. Similarly, ℓ_2' can be determined from Z_0. Hence, the crease pattern is unique to the choice of Z_0. $\qquad\square$

For simplicity, Z is chosen symmetrically such that it lies on the angle bisector of θ. The rest of the algorithm is relatively straightforward (Figure 8). (Note that for crease patterns in this paper, solid lines represent mountain folds and dashed lines represent valley folds.) The final product is shown in Figure 9.

1. The walls' bottom edges are reflected along ℓ_1 and ℓ_2. Their intersections with the angle bisectors are noted.

2. The intersections are connected to the roof vertex by valleys. All lines are designated as mountains or valleys as shown if relevant or erased if not.

3. Repeat all these steps between every pair of adjacent walls.

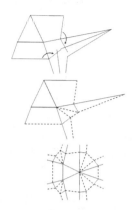

Figure 8. The final steps.

Figure 9. Final product.

5.2 Overlapping Inside the Biplanar

Figure 10(a) shows the flipside of the folded biplanar; the black lines mark the position of the base. The paper between adjacent walls is hidden inside the biplanar after folding, and it takes up space from the base edges. If too much space is taken up, the biplanar cannot be folded.

Figure 10(b) circles the line segments in the crease pattern that take up space. Each base edge has two line segments taking up space, one from

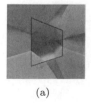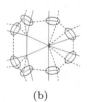

(a)   (b)

Figure 10. Overlapping along the base: (a) Flipside view of final product. (b) Line segments that take up space.

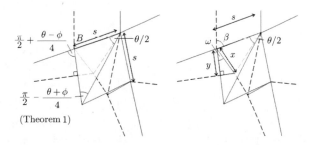

Figure 11. Calculating the Amount of Space Taken Up

each end. Given one end of a base edge, B, such that the internal angle of the corresponding wall at B is β (Figure 11), Equation (16) determines the length x of the line segment taking up space from that end, where ϕ is the outer angle:

$$\omega = \pi - \left(\frac{\pi}{2} + \frac{\theta - \phi}{4}\right) - \beta = \frac{\pi}{2} - \frac{\theta - \phi}{4} - \beta,$$

$$\frac{2y}{\sin \frac{\theta}{2}} = \frac{s}{\sin\left(\frac{\pi}{2} - \frac{\theta + \phi}{4}\right)} = \frac{s}{\cos \frac{\theta + \phi}{4}} \implies y = \frac{s \cdot \sin \frac{\theta}{2}}{2 \cos \frac{\theta + \phi}{4}}$$

$$\frac{y}{x} = \cos \omega = \sin\left(\beta + \frac{\theta - \phi}{4}\right)$$

$$\therefore x = \frac{y}{\sin(\beta + \frac{\theta - \phi}{4})}$$

$$= \frac{s \cdot \sin \frac{\theta}{2}}{2 \cos \frac{\theta + \phi}{4} \sin(\beta + \frac{\theta - \phi}{4})}$$

$$= \frac{s \cdot \sin \frac{\theta}{2}}{\sin(\beta + \frac{\theta}{2}) + \sin(\beta - \frac{\phi}{2})}. \tag{16}$$

Comparing the two x values of a base edge (one for each end) to its length predicts any overlapping. The *Shrinking Algorithm* (Figure 12) has been derived to solve most cases of overlapping along *quadrilateral walls*. This algorithm is denoted by A_n, where n is the *shrinkage factor* that decides the degree of reduction in the occupied space. A_n reduces the occupied space to a value $x(n) < x$.

Since A_1 produces no change, we have, from Equation (16),

$$x(1) = \frac{s \cdot \sin \frac{\theta}{2}}{\sin(\beta + \frac{\theta}{2}) + \sin(\beta - \frac{\phi}{2})}. \tag{17}$$

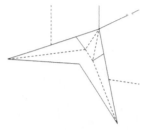
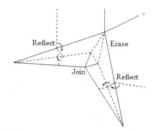

1. Dilate the crease pattern between the walls by factor $\frac{1}{n}$, and the center at the roof vertex. Draw the lines (the new ℓ_1' and ℓ_2') beside the new outer angle as mountains.

2. Edit the crease pattern as shown. This will fold into a biplanar with $\frac{1}{n}$ of the original height; the remaining height will be filled by copies of a smaller biplanar.

3. Draw in the crease pattern of a biplanar whose inner and outer angles are both ϕ. If $n > 2$, dilate the added crease pattern by a factor of $\frac{1}{n-1}$ and stack on another $(n-2)$ copies of the dilated crease pattern.

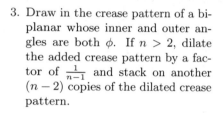
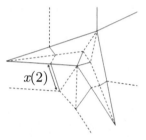

Figure 12. The Shrinking Algorithm (A_2).

In addition, for all $n \in \mathbb{N}$ such that $n > 1$,

$$x(n) = \frac{1}{2n} \times \frac{s \cdot \sin \frac{\phi}{2}}{\sin(\beta + \frac{\phi}{2}) + \sin(\beta - \frac{\phi}{2})} = \frac{s \cdot \tan \frac{\phi}{2}}{4n \sin \beta} \implies x(n) \propto \frac{1}{n}. \quad (18)$$

In the case of triangular walls, the algorithm can also be applied by treating the wall as a degenerate quadrilateral, as demonstrated in Section 4.

Figure 13 shows the results of using higher shrinkage factors.

If overlapping occurs along triangular walls pointed at the top, no amount of shrinking can solve it because overlapping will always occur near the top vertex. As for triangular walls that are pointed at the bottom, overlapping is also unsolvable because $x(n) \neq 0$. This means that biplanars with triangular walls that are pointed at the bottom cannot be folded. For this reason, the term *triangular walls* in this paper will refer only to those pointed at the top.

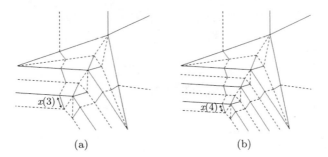

(a) (b)

Figure 13. Further shrinking: (a) $n = 3$, (b) $n = 4$.

5.3 Flat Surrounding Paper

In Figure 7, X_1Y and ℓ_1 form a pair of creases that radiate away from the net in the final crease pattern. X_2Y and ℓ_2 form another pair. Because Z (from Theorem 1) was chosen symmetrically, the pairs are congruent. Each pair will intersect at some point outside the folded biplanar if and only if $\phi > \theta$; the pair will be parallel if and only if $\phi = \theta$. Once they intersect, the paper will form a "cone" at the intersection (sum of angles $< 2\pi$), disrupting the flatness of the paper. Furthermore, the greater the difference $\phi - \theta$, the closer the intersection will be to the biplanar. The following lemma investigates the conditions allowing a biplanar to have perfectly flat surrounding paper (an advantageous situation in most cases).

Lemma 1. *A biplanar has perfectly flat surrounding paper* $\iff \theta = \phi$ *for every gadget.*

Proof: Every wall of the biplanar can be assumed to be a quadrilateral since one could imagine the roof to sink slightly lower (while the other faces remain stationary), turning triangular walls that are pointed at the top into quadrilateral walls (triangular walls pointed at the bottom are not considered). Let the gadgets be numbered from 1 to n (i.e., there are n walls). Clearly, both the roof and base are n-gons. Let their internal angles be $\rho_1, \rho_2, \cdots, \rho_n$ and $\gamma_1, \gamma_2, \cdots, \gamma_n$, respectively, such that

$$\sum \rho_i = \sum \gamma_j. \tag{19}$$

Let the upper internal angles of the walls be $\alpha_1, \alpha_2, \cdots, \alpha_{2n}$ and the lower internal angles be $\beta_1, \beta_2, \cdots, \beta_{2n}$. Considering the sum of all the

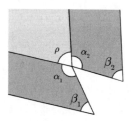

Figure 14. A prismatoid net.

internal angles of the walls,

$$\sum \alpha_i + \sum \beta_j = 2n\pi \tag{20}$$

$$\implies \sum \theta_i = 2n\pi - \sum \rho_j - \sum \alpha_k$$

$$= 2n\pi - \sum \gamma_j - \left(2n\pi - \sum \beta_k\right)$$

$$= \sum \beta_k - \sum \gamma_j = \sum \phi_i.$$

For no pairs of radiating creases to intersect, $\phi \le \theta$ for every gadget. Therefore, $\phi = \theta$ for every gadget, or else $\sum \phi < \sum \theta$. ☐

Theorem 2. *A prismatoid has perfectly flat surrounding paper.*

Proof: For any given prismatoid, it can be assumed that it has only quadrilateral walls (See Lemma 1). This means that the corresponding (linked by an edge) internal angles of the roof and base are equal, and that the top and bottom edges of the walls are parallel (Figure 14).

The base angle is, as explained above, equal to the roof angle ρ:

$$\theta = 2\pi - \rho - \alpha_1 - \alpha_2$$

$$= 2\pi - \rho - (\pi - \beta_1) - (\pi - \beta_2)$$

$$= \beta_1 + \beta_2 - \rho = \phi.$$

Hence, both pairs of radiating creases are parallel. This holds for every gadget of the prismatoid, thus its surrounding paper is perfectly flat. ☐

Whether being a prismatoid is a necessary condition for perfectly flat surrounding paper is still a conjecture, but this paper presents the conjecture's equivalence to another geometrical problem.

Conjecture 1. A biplanar with perfectly flat surrounding paper must be a prismatoid.

Given any biplanar with roof and base planes π_R and π_B, respectively, let π'_R be the plane passing through the lowest roof vertex and parallel to π_B. Again, the walls can be assumed to be quadrilaterals. A prismatoid is bounded by the walls of the biplanar, π'_R and π_B, and it has the same set of outer angles as the biplanar. By Theorem 2, the prismatoid's surrounding paper is totally flat, so if the same can be said of the biplanar, then the biplanar must also have the same set of inner angles as the prismatoid. This means that at every corresponding roof vertex of the prismatoid and biplanar, the sum of the roof angle and wall angles adjacent to it is the same. This seems unlikely but is as yet neither proved nor disproved. Disproving it would in turn prove Conjecture 1.

5.4   Positioning Triangular Walls

In the net shown in Figure 5, a triangular wall is connected to the roof by a point. Triangular walls that are pointed at the top are actually free to rotate their top vertex (which is a vertex of the roof) as long as they do not touch the other walls; all of the resultant nets are valid and fold to the same biplanar (Figure 15). Touching that vertex of the roof would be two quadrilateral walls and, between them, any number of triangular walls. Between every pair of adjacent walls would be a gadget; the outer angle of each gadget would be fixed because it depends only on the target biplanar, but the inner angle would vary as the triangular walls pivot along their top vertices.

However, the sum of all the inner angles would be fixed. If the inner and outer angles of gadget i are θ_i and ϕ_i, respectively, then

$$\sum \theta_i = 2\pi - \rho - \sum \alpha_i. \tag{21}$$

Thus, the positions of the triangular walls determine how the "quota" of inner angles is distributed among the gadgets. Assuming it would be disadvantageous to have some intersections between each pair of radiating creases being far from the biplanar yet some being close, the "angle deficit"

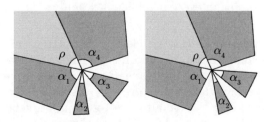

Figure 15. Two equivalent nets.

$(\sum \phi_j - \sum \theta_j)$ should be distributed evenly among the gadgets. This is done by positioning the triangular walls such that θ_i is proportional to ϕ_i:

$$\theta_i = \phi_i \left(\frac{2\pi - \rho - \sum \alpha_j}{\sum \phi_k} \right). \tag{22}$$

This concludes the adjustment to the net-drawing algorithm, as well as the whole algorithm for folding biplanars.

6 Conclusion

To sum up, an algorithm has been derived using vector and angle calculations to fold biplanars which fulfill the following conditions:

1. The biplanar is convex.

2. The biplanar has no triangular walls that are pointed at the bottom.

3. During the application of the algorithm, no overlapping occurs along triangular walls.

The biplanar is first unfolded into a net, after which gadgets are drawn between adjacent walls. Overlapping along the base that results in non-foldable biplanars can be predicted by Equation (16), and solved in most cases by the Shrinking Algorithm. However, the Shrinking Algorithm also adds layers to the paper when applied, which could lead to paper drift. A computer program has been written to execute the algorithm automatically.

It has been shown that prismatoids have completely flat surrounding paper, and it is conjectured that biplanars must be prismatoids to have completely flat surrounding paper as well. Possible future research includes proving or disproving Conjecture 1 and relaxing any of the conditions limiting foldable biplanars—that is, deriving an algorithm for concave biplanars or solving overlapping in every possible situation.

Acknowledgments. Because this article is the outcome of a mathematics project and its continuation, I would like to thank my two teacher-supervisors, Samuel Lee and Cheong Kang Hao, for their mentorship. Their patient guidance throughout both projects was indispensable to the completion of the projects. I would also like to thank the fellow presenters at the 5OSME, who have provided me with exposure to the field of extrusion origami.

Bibliography

[Benbernou et al. 11] Nadia M. Benbernou, Erik D. Demaine, Martin L. Demaine, and Aviv Ovadya. "Universal Hinge Patterns to Fold Orthogonal Shapes." In *Origami⁵: Fifth International Meeting of Origami Science, Mathematics, and Education*, edited by Patsy Wang-Iverson, Robert J. Lang, and Mark Yim, pp. 405–419. Boca Raton, FL: A K Peters/CRC Press, 2011.

[Demaine et al. 11] Erik D. Demaine, Martin L. Demaine, and Jason Ku. "Folding Any Orthogonal Maze." In *Origami⁵: Fifth International Meeting of Origami Science, Mathematics, and Education*, edited by Patsy Wang-Iverson, Robert J. Lang, and Mark Yim, pp. 449–454. Boca Raton, FL: A K Peters/CRC Press, 2011.

[Natan 10] Carlos Natan. "Origami 3D Tesselations." Available at http://www.flickr.com/photos/origamiz/sets/72157606559615966/, 2010.

[Tachi 09] Tomohiro Tachi. "3D Origami Design Based on Tucking." In *Origami⁴: Fourth International Meeting of Origami Science, Mathematics, and Education*, edited by Robert J. Lang, pp. 259–272. Wellesley, MA: A K Peters, 2009.

A Design Method for Axisymmetric Curved Origami with Triangular Prism Protrusions

Jun Mitani

1 Introduction

When we fold a sheet of paper, we usually fold it flat along straight lines, so most traditional origami pieces are flat when completed. In recent decades, many researchers have studied the geometry of such flat-foldable origami [Lang 03]. As a result of these studies, several origami design methods have been established. Today, we can design quite complicated origami pieces with the help of computers [Lang 06]. Since these approaches are based on the theory of flat-foldability, the designed shapes remain essentially flat. However, several origami artists have tried to make non-flat origami pieces, which we call "3D origami." Although most 3D origami pieces are designed by trial and error, some pieces are designed based on mathematical theory. This paper presents methods for designing two families of 3D origami, a conical type and a cylindrical type. The details of these origami types are described in Section 2.

Three-dimensional origami represents the surface of a 3D shape with a sheet of paper, so the first step for designing the crease pattern (an unfolded pattern for origami with fold lines) is to unfold the 3D shape onto a 2D plane by figuratively cutting it along some subset of its edges and unfolding the result to lie flat. When we unfold a 3D shape, we usually cannot avoid the existence of empty spaces between the unfolded pieces. If we cut away the paper from the 2D pattern in those empty spaces, the result is no longer

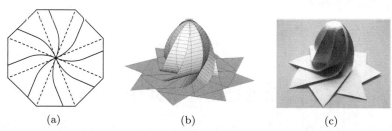

Figure 1. A curved 3D origami designed by the method proposed by [Mitani 09]: (a) the crease pattern, (b) 3D CG model, (c) photo of the origami.

origami; instead, we call it papercraft. To make the unfolded pattern work for origami, we have to lay out the unfolded pieces appropriately and add fold lines in the empty spaces to fill the gaps between the facets of the 3D shape.

Tachi proposed a method for generating a crease pattern in which so-called "tucks" are placed in the empty spaces so that the target 3D shape can be made from a single sheet of paper without any cuts [Tachi 09]. The novel point of this approach is that the tucks are folded toward the inside of the 3D shape, so they are invisible from the outside. The shape of the folded 3D origami has exactly the same appearance as the target 3D shape. However, the crease pattern tends to be complicated because the tucks are placed around all the polygons that make up the surface. When we use many polygons to approximately represent a smooth surface, the crease pattern becomes too complicated to be folded by human hands.

Another approach is to place tucks outside the folded surface. Such origami have been designed by several origami artists (e.g., [Palmer 01, Bell 07]). I showed that it is possible to make curved 3D origami with a simple crease pattern if the target shape is axisymmetric and the tucks are placed outside the shape [Mitani 09]. At about same time, Lang published a *Mathematica* package on the Internet for similar purposes [Lang 09]. Now we can make egg-shaped origami, as shown in Figure 1(c) with a simple crease pattern (Figure 1(a)) by using this method.

This paper proposes a new method that is similar to, but different from, the method proposed earlier [Mitani 09]. The shape of the origami designed with the new method is shown in Figure 2. As with the other method [Mitani 09], the shape is such that a sheet of paper wraps around an axisymmetric 3D object. The new feature of the present method is that the designed shape has 3D tucks with a triangular cross section. Even though the tuck has a 3D shape, it is still possible to design smooth curved origami in an approximate way by using many thin polygons.

In prior work, Mosely designed an interesting origami shape that she called a "Bud," which uses the type of 3D tuck that is the subject of this

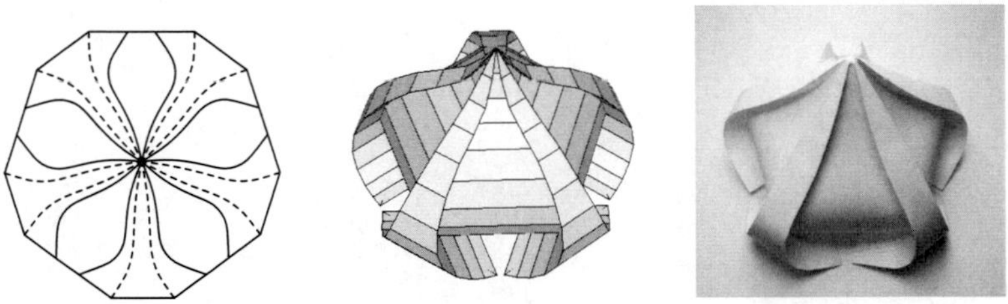

Figure 2. Curved 3D origami designed by the method proposed in this paper: (a) crease pattern, (b) 3D CG model, (c) photo of the origami.

paper. Although she stated that "I calculated the equations governing its shape in space" [Mosely 09], the details of how she introduced the crease pattern were not described or generalized. In this paper, I show how to design the crease pattern from a discrete representation of any axisymmetric 3D model. This approach is simple and applicable for polyhedral and polyhedrally approximated curved surfaces.

2   Shape of the Target 3D Origami

The shape of the target 3D origami resembles cloth wrapping around a 3D object. Although there are some variations in the wrapping, we propose two types, which we call the "conical type" and "cylindrical type" (Figure 3).

Figure 4 shows an example of the conical type of our 3D origami. The wrapped shape is axisymmetric, as shown in Figure 4(b), and is generated by rotating the input 2D polyline shown in Figure 4(a) by $360/N$ degrees around the z-axis ($N = 6$ in this case). The 3D origami shape, shown in Figure 4(c), has external triangular prism protrusions. The 3D origami is made from the crease pattern shown in Figure 4(d). The crease pattern has no gaps and is inscribed in a regular N-gon.

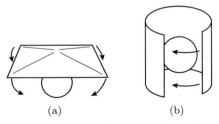

Figure 3. Two types of wrapping: (a) conical type, (b) cylindrical type.

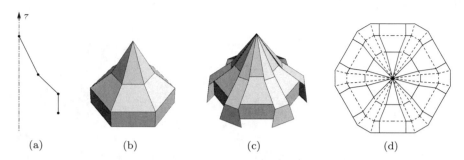

Figure 4. Conical type 3D origami: (a) input 2D polyline, (b) axial symmetrical shape generated with (a), (c) shape of the 3D origami, (d) the crease pattern.

Figure 5 shows an example of the other type of 3D origami, the cylindrical type. As in the conical type, the wrapped shape is axisymmetric, as shown in Figure 5(b), and is generated by rotating the input 2D polyline shown in Figure 5(a) by $360/N$ degrees around the z-axis ($N = 6$ in this case). The origami shape shown in Figure 5(c) has external triangular prism protrusions. Figure 5(d) shows the crease pattern of the origami. The different points on the conical type show that the 2D input line does not have to touch the z-axis, and the crease pattern is rectangular.

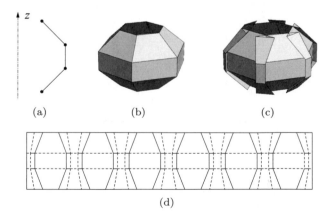

Figure 5. Cylindrical type 3D origami: (a) input 2D polyline, (b) axial symmetry shape generated with (a), (c) shape of the 3D origami, (d) crease pattern.

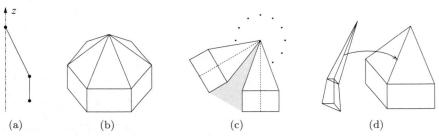

(a)   (b)   (c)   (d)

Figure 6. (a) Example of input 2D polyline. (b) Axisymmetric shape generated with (a). (c) A portion of the unfolded pattern of the 3D shape. (d) The triangular prism protrusion placed outside the 3D shape.

3 Designing the Crease Pattern

This section describes in detail the methods for designing the crease patterns of the conical type and cylindrical type.

3.1 Conical Type

Let us consider the input polyline shown in Figure 6(a). The 3D shape generated by rotating the input around the z-axis in N discrete steps is shown in Figure 6(b). The input polyline here has only two segments, and we set $N = 6$, although these values are arbitrary.

Figure 6(c) illustrates a portion of the unfolded pattern of the 3D shape shown in Figure 6(b), in which polygons are arranged with rotational symmetry about the central point. The triangular prism protrusion made from the empty space (colored light gray in this figure) will be placed outside the 3D shape, as illustrated in Figure 6(d).

Now let us consider the generation of the crease pattern, that is, how the fold lines are arranged in the light gray area shown in Figure 6(c). We show two panels of the 3D shape, corresponding to the second line segment of the input polyline, along with the section of the triangular prism between them in Figure 7(a). P and P' are polygons that are neighboring panels of the 3D shape. Points C_0, C_0', C_1, and C_1' are the center points of their respective edges. Points C_0, B_0, and C_0' lie on a common horizontal plane, as do C_1, B_1, and C_1' (on a different plane). The unfolded pattern of this portion is illustrated in Figure 7(b).

The symbols appearing in Figure 7(a) also appear in Figure 7(b) to indicate the correspondences between the 3D origami and the unfolded pattern. The symbols appearing only in Figure 7(b) are as follows:

- B_0' and B_1' are located at the same position as B_0 and B_1, respectively, in Figure 7(a).

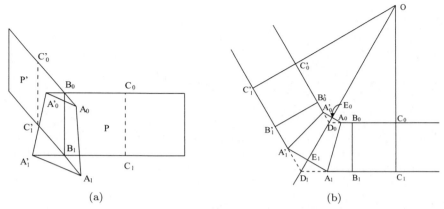

Figure 7. (a) A portion of 3D origami with triangular prism protrusion. (b) Unfolded pattern of (a).

- O is the intersection of the extended line of C_1C_0 and the extended line of $C_1'C_0'$.

- D_0 is the intersection of the extended lines of C_0B_0 and $C_0'B_0'$.

- D_1 is the intersection of the extended lines of C_1B_1 and $C_1'B_1'$.

- E_0 is the intersection of A_0A_0' and OD_1.

- E_1 is the intersection of A_1A_1' and OD_1.

Since the unfolded pattern in Figure 7(b) is N-fold rotationally symmetrical, the angle $\angle C_1OD_1$ is $180/N$ degrees ($N = 6$). We call this angle α hereinafter.

We see that triangles $B_0A_0A_0'$ and $B_1A_1A_1'$ on Figure 7(a) (cross sections of the triangular prism) lie on a common horizontal plane with $C_0B_0C_0'$ and $C_1B_1C_1'$, respectively. This is because we can construct an axisymmetric 3D origami by stacking layers, each layer being a cross section of the complete structure. Now the problem is to define the positions of A_0, A_0', A_1, and A_1' because B_0, B_1, C_0, C_1, B_0', B_1', C_0', and C_1' are automatically defined from the input polyline and the length of OC_0.

When line segment B_0B_1 is folded, line A_0A_1 rotates around B_0B_1. To satisfy the condition that A_0 lies on the same plane as $B_0C_0C_0'$, A_0 must lie on the extended line of $C_0'B_0$. This is because the circular locus of A_0 intersects the plane at only two positions, and one is obviously where A_0 lies on the extended line of C_0B_0.

Figure 8 illustrates the cross-sectional plane passing through C_0, B_0, and C_0' of the 3D solid. O' is the intersection of the rotational axis and

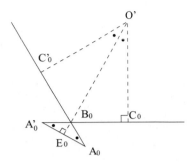

Figure 8. Cross section of 3D origami that passes through C_0, B_0, and C_0'.

the cross-sectional plane. E_0 is the intersection of the extended lines of $O'B_0$ and A_0A_0'. In this cross section, $C_0O'B_0$ and $E_0A_0B_0$ are similar triangles because $\angle O'B_0C_0 = \angle A_0'B_0E_0 = \angle A_0B_0E_0$. Since $\angle B_0O'C_0 = \alpha$, $\angle B_0A_0E_0 = \alpha$. Therefore $|A_0E_0| = |B_0A_0|\cos\alpha$.

Now we need to place point A_0 in Figure 8(b) so that the length of $|A_0E_0|$ becomes $|B_0A_0|\cos\alpha$. Since $\angle D_0A_0E_0 = \alpha$ (because $A_0E_0D_0$ and OC_0D_0 are homologous triangles), the condition $|A_0E_0| = |B_0A_0|\cos\alpha$ is satisfied by simply placing A_0 at the center of B_0D_0.

The discussion above addresses the cross-sectional plane that passes through A_0, B_0, and A_0'. Another plane passing through A_1, B_1, and A_1' can be analyzed in the same manner. Thus, point A_1 must be placed at the center of B_1D_1 in the crease pattern.

Replicating this pattern with the desired N-fold rotational symmetry results in a ring of panels and triangular prisms in the 3D folded form and a ring of corresponding facets in the crease pattern. Each ring corresponds to one segment of the input polyline.

The length of $|OC_0|$ is a free variable, so it can take any value within the condition $|OC_0| > |B_0C_0|/\tan\alpha$. As $|OC_0|$ becomes larger, the size of the resulting triangular prism becomes larger. Because length $|OC_0|$ is adjustable, we can make a gapless unfolded pattern of the 3D origami by appropriately connecting the rings of each polyline segment to the layers of the next ring in both the folded form and crease pattern. Both lengths $|B_0D_0|$ and $|OC_0|$ go to zero when the layer is the topmost layer (i.e., the center of the pattern).

3.2 Cylindrical Type

In this section, we describe the methods for designing the crease patterns of the cylindrical type. The approach is almost the same as that for the conical type. We consider the input polyline shown in Figure 9(a). The 3D shape generated by rotating the input around the z-axis in N discrete

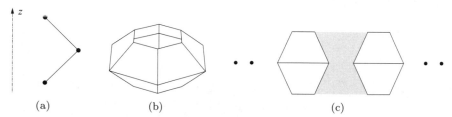

Figure 9. (a) Example of input 2D polyline. (b) Axisymmetric shape generated with (a). (c) A portion of the unfolded pattern of (b).

steps is shown in Figure 9(b). Although we use an input polyline that has only two segments and we set $N = 6$, these values are for the example only and can be chosen arbitrarily.

Figure 9(c) shows a portion of the unfolded pattern from the 3D shape in Figure 9(b), in which the polygons are arranged at even intervals along the rectangle. The triangular prism protrusion, which is made from the empty space in the crease pattern (the light gray region) will be placed outside the 3D shape.

Now let us consider the construction of the crease pattern—that is, how the fold lines are to be placed in the light gray area. A portion of the 3D origami, consisting of two adjacent panels and the triangular prism between them, is illustrated in Figure 10(a). The corresponding unfolded pattern of this portion is illustrated in Figure 10(b).

Here, we see that the triangles $B_0 A_0 A_0'$ and $B_1 A_1 A_1'$ in the cross section of the complete 3D shape (Figure 10(a)) lie in a common horizontal plane with $C_0 B_0 C_0'$ and $C_1 B_1 C_1'$, respectively, just as they did for the conical type. The problem now becomes how to define the positions of A_0, A_0', A_1, and A_1'. We can freely adjust the distance between points B_0 and B_0' in Figure 10(b) subject to the condition that the polygons do not intersect each other. When the distance is longer, the size of the triangular prism

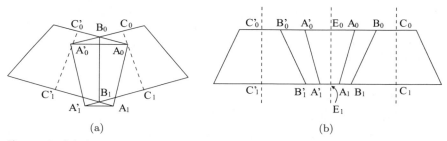

(a)   (b)

Figure 10. (a) A portion of the 3D origami with the triangular prism protrusion. (b) Unfolded pattern of (a).

becomes larger. The cross-sectional plane that includes C_0, B_0, and C_0' takes the same form as that of the conical type, shown in Figure 8, with the same similarity of triangles. Therefore, we can say that $|A_0 E_0| = |B_0 A_0| \cos \alpha$, as before. Thus, we can choose to place A_0 in Figure 10(b) so that the length $|A_0 E_0|$ becomes $|B_0 A_0| \cos \alpha$.

In the conical type, each input segment of the polyline gave rise to a ring of panels and prisms in the 3D form and a ring of facets in the crease pattern. Now, in the cylindrical type, each input segment of the polyline gives rise to a ring of panels and prisms in the 3D form, but to a rectangular strip in the crease pattern. By connecting the pattern of each layer from each segment of the input line to the layer resulting from the next segment of the polygon, we can obtain a gapless unfolded pattern for the 3D origami.

4   Examples and Discussions

We implemented the method described in Section 3 on a personal computer. The system automatically generates the crease pattern and the 3D model from the 2D input polyline. The user can switch between the conical type and the cylindrical type. In addition, the user can change the rotational symmetry (the value of N) and can specify the size of the triangular prism for the cylindrical type. This system enables users to preview the 3D origami shape on a screen before trying to fold it with real materials. Figure 11 shows examples of the 3D origami designed with our method: (a) and (b) are conical types with $N = 4$, $N = 6$, respectively; and (c) and (d) are cylindrical types with $N = 4$, $N = 6$, respectively. Patterns (a) and (c) are generated from the same 2D input polyline.

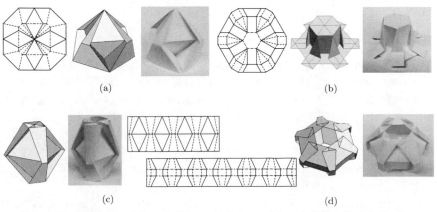

(a)   (b)

(c)   (d)

Figure 11. Examples:(a) and (b) conical type, (c) and (d) cylindrical type.

We observe that the width of the protrusions of the conical type grow wider according to their distance from the topmost point. We can also observe that the width of the protrusions of the cylindrical type becomes wider when the protrusions approach the rotational axis. Theoretically, there are no intersections between the protrusions when $N \leq 4$. However, we may have intersections between the protrusions when $N > 5$. Because most of the generated shapes are not stable when they are made with a sheet of paper, we sometimes need to fix them from the inside.

Although our description involves discrete polygons, we can use the same method to create apparently curved shapes by approximating the curved line with a set of short line segments. From such an approximated curved line, we can obtain an approximately curved 3D origami, as shown in Figure 2. In this case, we can omit the horizontal fold lines because they are nearly flat. As a result, we can make a curved 3D origami from a simple crease pattern.

5 Conclusion

We described two methods for designing 3D origami, in which the crease line wraps around an object to form an axisymmetric shape, and triangular prism protrusions are located on the outside surface. We showed examples of the conical type and cylindrical type, which were designed on a personal computer. It was confirmed that we can preview the 3D origami before we attempt to actually make it. This preview can significantly reduce the trial-and-error time in origami design. To create the shape of the 3D origami, the proposed method wraps the paper around an object inside. Hence, the design method may be applicable for designing packages, lamp shades, gift boxes, and so on.

Bibliography

[Bell 07] Phillip Chapman Bell. "Origami—A Set on Flickr." Available at http://www.flickr.com/photos/oschene/sets/1457200/with/432538232/, 2007.

[Lang 03] Robert J. Lang. *Origami Design Secrets: Mathematical Methods for an Ancient Art.* Natick, MA: A K Peters, Ltd., 2003.

[Lang 06] Robert J. Lang. "TreeMaker." Available at http://www.langorigami.com/science/treemaker/treemaker5.php4, 2006.

[Lang 09] Robert J. Lang. "Origami Flanged Pots." Available at http://demonstrations.wolfram.com/OrigamiFlangedPots/, 2009.

[Mitani 09] Jun Mitani. "A Design Method for 3D Origami Based on Rotational Sweep." *Computer-Aided Design and Applications* 6:1 (2009), 69–79.

[Mosely 09] Jeannine Mosely. "Sufrace Transitions in Curved Origami." In *Origami⁴: Fourth International Meeting of Origami Science, Mathematics, and Education*, edited by Robert J. Lang, pp. 143–150. Wellesley, MA: A K Peters, Ltd., 2009.

[Palmer 01] Chris Palmer. "PolyPouches." Available at http://www.shadowfolds.com/polypouches/, 2001.

[Tachi 09] Tomohiro Tachi. "3D Origami Design Based on Tucking Molecule." In *Origami⁴: Fourth International Meeting of Origami Science, Mathematics, and Education*, edited by Robert J. Lang, pp. 259–272. Wellesley, MA: A K Peters, Ltd., 2009.

Folding Any Orthogonal Maze

Erik D. Demaine, Martin L. Demaine, and Jason S. Ku

1 Introduction

In most real-world origami, the final folded model is only a small factor smaller than the original piece of paper. This property is obviously useful for practical folding, but we are far from understanding what makes it possible mathematically. The universality result for origami by [Demaine et al. 00] uses an extremely large scale factor, and the computational origami design techniques of TreeMaker [Lang 03] and Origamizer [Demaine and Tachi 10] solve their nonlinear optimization problems via heuristics and so cannot definitively say which models require large scale factor (although in practice they work well for many models of interest).

In this paper, we prove that a wide family of *origami maze* designs can be folded with a small scale factor. We develop an algorithm to fold a square of paper into any orthogonal maze, consisting of vertical walls protruding equal heights out of a square floor. More precisely, given an orthogonal graph drawn on an $n \times n$ square grid, we fold a $(2h+1)n \times (2h+1)n$ square of paper into the square with the orthogonal graph extruded orthogonally to a specified (uniform) height h. The zero-thickness ridges could form a path like the Hilbert curve, a maze or labyrinth, troughs for liquid distribution, or letters of the alphabet (as in Figure 1).

How good is the scale factor of $2h+1$? The answer depends on the target orthogonal graph. The scale factor always must be at least 1, because the intrinsic diameter of the target shape (even for an empty graph) is at least

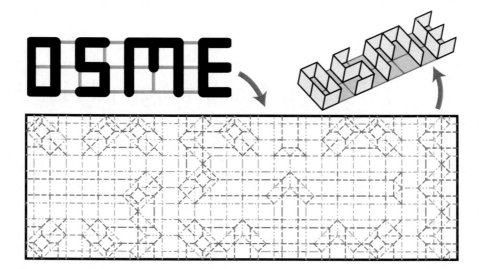

Figure 1. Folding three-dimensional letters from a rectangle of paper. Font design from [Demaine et al. 10b]. Valleys are dashed, mountains are dot-dashed, 180° folds are thick, and 90° folds are thin. (See Color Plate XVII.)

$\sqrt{2}\,n$, and because intrinsic diameter can only decrease by folding. If the graph is connected and spans all $(n+1)^2$ points, then it has $(n+1)^2 - 1$ edges, so at least one of the $2n$ rows or columns must contain at least $[(n+1)^2 - 1]/(2n) = n/2 + 1$ edges of the graph, which induces an intrinsic straight line of length $n + (n/2 + 1)(2h) = n(h+1) + 2h$ (going up and down each ridge of height h). The diameter of the paper must be at least this long, proving that the scale factor must be at least $(h+1)/\sqrt{2}$, so our algorithm is guaranteed to be a factor of slightly less than $2\sqrt{2}$ away from optimal. If a row or column has all $n+1$ edges, then our algorithm is definitely within a factor of $\sqrt{2}$ from optimal. For the complete $n \times n$ grid graph, we suspect that our folding is optimal (at least among watertight foldings). We also suspect that our folding is very close to optimal for "most" (e.g., random) subgraphs of the $n \times n$ grid.

A particularly practical situation, used in our examples and implementation, is $h = 1$. In this case, we start with a square just three times larger in dimensions than the final shape. Furthermore, The number of layers of paper that come together at any point is bounded by a constant.

Our foldings are *watertight* [Demaine and Tachi 10]: the boundary of the paper maps to the boundary of the model. In contrast, the original algorithm for folding any polyhedral surface [Demaine et al. 00] is inefficient and not watertight. Origamizer [Demaine and Tachi 10] provides a family of foldings that may include similarly efficient (and watertight) foldings, but

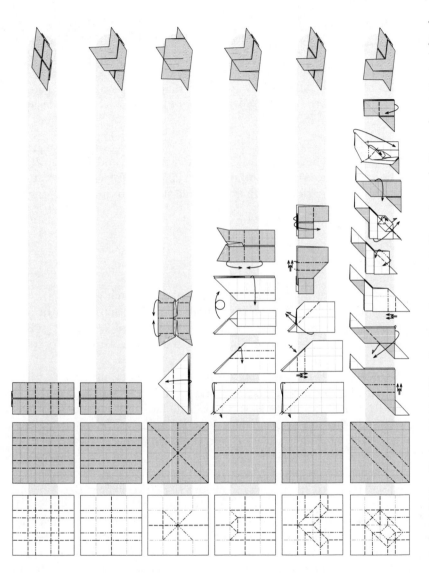

Figure 2. Diagrams establishing the local foldability of each vertex gadget. From top to bottom: degree 0, straight, degree 4, degree 3, degree 1, and corner. Valleys are dashed, mountains are dot-dashed, 90° folds are gray, and 180° folds are black.

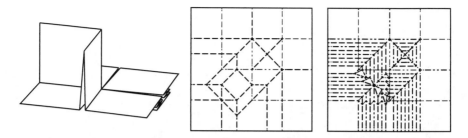

Figure 3. Corner gadget: 3D folded state (left), simple crease pattern (center), and thinned crease pattern (right). Valleys are dashed, mountains are dot-dashed, 180° folds are thick, and 90° folds are thin.

does not provide an efficient algorithm to find such a good folding. The box-pleating techniques of [Benbernou et al. 09] are most closely related, but applied in a straightforward matter, would use a square of side length $\Theta(n^2)$.

2 Algorithm

Our algorithm uses an appropriate folding gadget at each grid point, depending on which of the four incident edges should be extruded. In all, there are six gadgets up to rotation, as shown in Figure 2. The step-by-step diagrams are not part of the algorithm, but rather serve to describe the layering of the final folded state.

The algorithm tiles the crease pattern and corresponding folded state, for each vertex, in a grid pattern according to the orthogonal graph. Figure 1 shows a simple example. Here we use the fact that all gadgets have a consistent interface on each of their four sides (depending on whether the side is a ridge or floor), allowing gadgets be combined in an arbitrary combination.

Depending on the height h of extrusion, the gadgets in Figure 2 may get placed very close to each other. Consider the 90° corner gadget shown in the middle of Figure 3. The crease pattern extends outside the central 2×2 square reserved for the gadget, in the lower left. If $h > 1$, then the crease pattern may overlap an adjacent gadget, which is invalid. To fix this problem, we thin the excess structure near the floor (nonridge) edges of every gadget by sufficiently many sink folds, as shown on the right of Figure 3. This thinning is necessary only for large extrusion heights h.

The algorithm runs in linear time and is easy to implement. We have implemented the algorithm as a freely available web application [Demaine et al. 10a]. You can design an orthogonal graph or generate a random maze

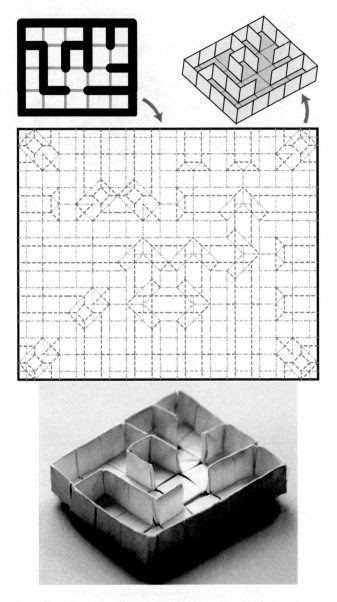

Figure 4. A 4 × 5 origami maze. (Folding by Christopher Chin.)

(see Figure 4), and the application produces a crease pattern, which you can print and fold into your design. The application assumes that $h = 1$. Also implemented is a simple orthogonal-graph font [Demaine et al. 10b] for writing messages such as Figure 1.

Figure 5. Folding of a Hilbert curve from a square of paper by Jason Ku, using an early version of the corner and straight gadgets. This photo appeared in *Technology Review* and *Popular Science* (October 8, 2009).

Acknowledgments. We thank Erez Lieberman for posing an artistic challenge, folding a Hilbert curve, that ended up inspiring this work; see Figure 5. We thank Aviv Ovadya and Tomohiro Tachi for helpful comments.

Bibliography

[Benbernou et al. 09] Nadia Benbernou, Erik D. Demaine, Martin L. Demaine, and Aviv Ovadya. "A Universal Crease Pattern for Folding Orthogonal Shapes." Preprint, arXiv:0909.5388, 2009. (Available at http://arxiv.org/abs/0909.5388.)

[Demaine and Tachi 10] Erik D. Demaine and Tomohiro Tachi. "Origamizer: A Practical Algorithm for Folding Any Polyhedron." Manuscript, 2010.

[Demaine et al. 00] Erik D. Demaine, Martin L. Demaine, and Joseph S. B. Mitchell. "Folding Flat Silhouettes and Wrapping Polyhedral Packages: New Results in Computational Origami." *Computational Geometry: Theory and Applications* 16:1 (2000), 3–21.

[Demaine et al. 10a] Erik D. Demaine, Martin L. Demaine, and Jason Ku. "Maze Folder." Available at http://erikdemaine.org/fonts/maze/, 2010.

[Demaine et al. 10b] Erik D. Demaine, Martin L. Demaine, and Jason Ku. "Origami Maze Puzzle Font." Paper presented at the 9th Gathering for Gardner, Atlanta, Georgia, March 24–28, 2010.

[Lang 03] Robert J. Lang. *Origami Design Secrets: Mathematical Methods for an Ancient Art.* Natick, MA: A K Peters, 2003.

Every Spider Web Has a Simple Flat Twist Tessellation

Robert J. Lang and Alex Bateman

1 Introduction

Origami tessellations are a broad field of folded structures with deep roots in origami, mathematics, and computational geometry. The term is usually understood to refer to a flat (or nearly so) folded shape in which some region of the plane is partitioned into a collection of sets by the pattern of the folded edges. Unlike conventional representational origami, in which the silhouette of the fold is paramount, what matters in origami tessellations is the pattern of the folded edges within the boundary of the folded shape. More often than not, the folded edges partition the plane into a mathematical tiling—and sometimes, those tiles include, or entirely comprise, regular polygons. In an early attempt to provide a mathematical description, Kawasaki and Yoshida [Kawasaki and Yoshida 88] defined "crystallographic flat origami" as a flat fold with certain crystallographic symmetries. Over time, the boundaries of the field have been extended to include figures with surface relief, 3D forms, and even curved creases, as well as patterns that are highly geometric but do not exhibit strict mathematical symmetry.

The field of origami tessellations has two distinct sets of roots. One lies within the Japanese art of origami and its mid-twentieth-century period of expansion. Although most Japanese origami was representational, several of the great masters explored regular patterns of folds, including Yoshihide Momotani [Momotani 84] and Shuzo Fujimoto [Fujimoto 82]. Their work, and in particular that of Fujimoto, inspired a generation of origami artists, notably Chris K. Palmer around the turn of the twenty-first century [Palmer 97], whose work largely inspired the present investigation. Some of the richness of what followed can be found in Gjerde [Gjerde 09]. But significant work on periodic folded forms existed in other fields, ranging from pleated fabrics dating back to the early twentieth century (and before) in Europe and investigations into 2D periodic folds by computer scientists and artists such as Ron Resch [Resch 68], David Huffman [Huffman 76], and Paulo Taborda Barreto [Barreto 97].

A key element of many origami tessellations is the *simple flat twist*, composed of a polygon (usually, but not always, regular) with pleats radiating away from the polygon. Following the appearance of Palmer's work at origami conventions in the mid-1990s, there was a burst of interest in the origami world in the field of tessellations of all types, but in particular in the area of simple flat twist tessellations. At that time, we began looking into techniques for creating origami tessellations from mathematical tilings and found a relatively simple algorithm for their creation [Bateman 02]. One began with a conventional tiling of polygons that, for aesthetic reasons, were typically regular polygons. Each polygon was shrunk by some factor ρ about its centroid and rotated through some angle β. Each original edge of the tiling was thereby split into two edges, each associated with the polygon on either side of the original edge. By connecting corresponding vertices with new lines, one arrived at a new tiling in which the original tiles were shrunken and rotated, each edge had been transformed into a parallelogram, and there was a new set of polygons that corresponded to the original vertices of the tiling. Remarkably, this simple algorithm very often results in a flat-foldable origami crease pattern whose overall appearance and symmetry follows that of the original tiling.

This simple algorithm—shrink and rotate each polygon in the tiling— can be used to create a wide range of tessellations: those that arise from k-uniform tilings of regular polygons as well as tilings that are not periodic. An example of the latter is a tessellation based on the Goldberg spiral [Goldberg 55], shown in Figure 1, designed and folded by Alex Bateman [Bateman 10a].

With the realization that this and many other tessellations could be constructed programmatically by this shrink-and-rotate algorithm, Bateman implemented the algorithm in 2000 in a program called Tess [Bateman 10b] and described the algorithm in 2001 at the Third International Meeting of Origami in Science, Mathematics, and Education (3OSME) [Bateman 02].

We illustrate the shrink-and-rotate algorithm in Figure 2. We start with a tiling, shrink each polygon, rotate each polygon by the same angle, then connect shrunken-and-rotated vertex pairs that came from the same initial vertex of the polygon with lines. The result is a flat-foldable crease pattern

Figure 1. Alex Bateman's Goldberg spiral origami tessellation.

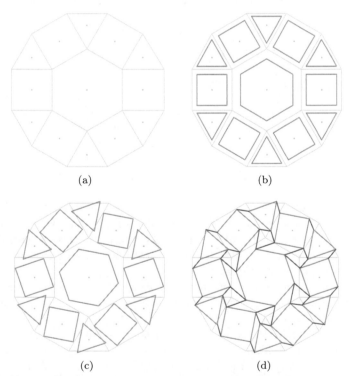

(a)   (b)

(c)   (d)

Figure 2. Progressive constructions of a shrink-rotate tessellation crease pattern: (a) The original tiling. (b) Shrink the polygons by a factor $\rho = 0.756$. (c) Rotate them all through the same angle $\beta = 19.10°$. (d) Connect vertices with new creases.

(or at least it can be, depending on the specifics of the shrinkage, rotation, and crease assignment). Finally, we assign the creases (i.e., choose whether each crease is mountain or valley).

The allure of the shrink-and-rotate algorithm is clear: it allows the creation of tessellations from tilings that include a wide variety of shapes, not just simple, regular tilings. In fact, it could potentially allow an origami artist to transform literally any line drawing into a tessellation, opening up a vast new world of artistic expression. However, it does raise the question: does the shrink-and-rotate algorithm work for every possible tiling? If not, which ones does it work on, and why?

In this paper, we show that, in fact, the shrink-and-rotate algorithm does *not* work for every possible tiling; we show how to tell if it does work and how to choose the centers of shrinkage and rotation. Along the way, we explore a number of topics in geometry, computational geometry, and mathematics, and remarkably, find that the key principle for the whole field traces its lineage back over 100 years, to a work by physicist James Clerk Maxwell. And we explain the somewhat enigmatic title of this paper.

2   Shrink and Rotate

2.1   Twist and Aspect Ratio

The first question that arises if we say that we're going to "shrink and rotate" each polygon is, "rotate about what point?" Triangles, for example, have at least four points that can be considered the "center" for purposes of rotation: the centroid, incenter, circumcenter, and orthocenter. (For definitions of these terms, see any good text on geometry, or, e.g., [Weisstein 10].)

For the moment, let us consider only tilings composed of regular polygons, in which case symmetry removes much of the ambiguity; the obvious choice to use as the center of rotation, as well as the center of shrinkage, is the *centroid*, or center of mass, of the polygon (which, for a regular polygon, is also its center of rotational symmetry).

We can characterize a shrink-and-rotate action by two parameters, which we now introduce. We define the ratio between the shrunken polygon and the original as the scaling, and denote it by ρ. And we define the rotation angle to be the angle by which the shrunken polygon is rotated relative to the original, and denote it by β.

Rotation angle β is not the same as the twist angle α, which is the angle between the edges of a twist polygon and their adjacent pleats. Remarkably, for any tiling with regular polygons, if we apply the same scaling factor and rotation angle to every polygon in the tiling, the result is a crease pattern in which every vertex satisfies the Kawasaki-Justin condition (KJC) for

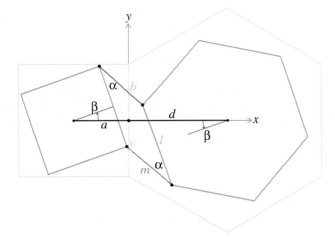

Figure 3. Configuration of a single edge, before (light) and after (dark) the shrink and rotate construction.

flat-foldability (which is itself somewhat unexpected); furthermore, every polygon becomes a twist with identical twist angles.

Let us zoom in on one edge of the original tiling and the two centers of rotation, one for each tile, as shown in Figure 3. With regular polygons, the line between the rotation centers is perpendicular to the tile edge, so without loss of generality we can place the intersection of those two lines at the origin of our coordinate system with both rotation centers on the x-axis. Then the tile edge lines up on the y-axis, and we can characterize the two vertices of the tile edge and the two centers of rotation by their distances from the origin. We denote these distances in the figure by the letters a–d.

The relationship between the scaling/rotation (ρ, β) and the resulting twist angle α turns out to be

$$\alpha = \cos^{-1}\left(\frac{\sin\beta}{\sqrt{\rho^2 - 2\rho\cos\beta + 1}}\right). \tag{1}$$

This is remarkable because none of the distances a–d appear anywhere within the expression. The twist angle depends *only* on the shrinkage and rotation.

If we solve for the lengths of the sides of the parallelogram, we find that

$$l = (b + c)\rho, \tag{2}$$

$$m = (a + d)\sqrt{\rho^2 - 2\rho\cos\beta + 1}, \tag{3}$$

so that the ratio between the two adjacent sides of the parallelogram is

$$\frac{l}{m} = \frac{b+c}{a+d} \cdot \frac{\rho}{\sqrt{\rho^2 - 2\rho\cos\beta + 1}}. \tag{4}$$

In this expression, the first multiplicative factor depends solely on the geometry of the original tiling, and the second depends solely on the scaling and rotation. We call this second factor the *aspect ratio parameter* γ because it controls the aspect ratio of every parallelogram in a shrink-rotate tessellation:

$$\gamma \equiv \frac{\rho}{\sqrt{\rho^2 - 2\rho\cos\beta + 1}}. \tag{5}$$

We now have two new parameters by which we can characterize the tessellation: γ, the parallelogram aspect ratio parameter, and α, the twist angle. These provide an alternative way of characterizing the twist, since given (ρ, β), we can calculate (γ, α). We can go in the other direction, too: given a pair (γ, α) that characterizes the aspect ratio of the parallelogram and the twist angle, we can compute the scaling and rotation that gives rise to this pair, which are

$$\rho = \frac{\gamma}{\sqrt{\gamma^2 + 2\gamma\sin\alpha + 1}}, \tag{6}$$

$$\beta = \pm\cos^{-1}\left(\frac{\gamma + \sin\alpha}{\sqrt{\gamma^2 + 2\gamma\sin\alpha + 1}}\right), \tag{7}$$

where the sign of β is positive for a counterclockwise twist and negative for a clockwise twist. It turns out that while the parameter pair (ρ, β) is necessary to *construct* a shrink-and-twist tessellation, the pair (γ, α) is a bit more useful to *characterize* such a tessellation, for a couple of reasons.

First, the possible valid crease assignments for a simple flat twist depend on the twist angle. When it comes to assigning creases to a shrink-rotate tessellation crease pattern, we'll need to know the twist angle α for each polygon in order to know whether a crease assignment is possible.

But second, there is yet another surprise to be had in this algorithm, as we see next.

2.2 Crease Pattern/Folded Form Duality

In a single simple flat twist, the twist angle in the crease pattern falls within the range $\alpha \in (0°, 180°)$, with $\alpha \in (0°, 90°)$ giving a counterclockwise twist and $\alpha \in (90°, 180°)$ giving a clockwise twist. If we construct the pattern with a negative value for α, then the pattern lines overlap the central polygon, and, in fact, we obtain the geometrical pattern of the lines of the folded form of the twist.

The same situation applies to a shrink-rotate tessellation. If we create a pattern with a constant negative twist angle α, then the pattern of lines, at least, will be the line pattern of the folded form of some shrink-rotate tessellation. The question is, which one?

Consider the following: if we construct two shrink-rotate tessellations from a given tiling using a twist angle α for one of them and $-\alpha$ for the other, then all of the polygons in the former will be similar (in the strict geometric sense) to the corresponding polygon in the other, with a single scaling constant between every pair of polygons. What's more, in the corresponding parallelograms, the angles that become twist angles are equal and opposite—which means the same relation must be true for the opposite pair of angles in each parallelogram.

If, furthermore, the ratio between adjacent sides of each parallelogram in one tessellation is the same as it is in the other, then all parallelograms, and indeed, all polygons in the first tessellation are similar to the corresponding polygons in the other tessellation—except that the parallelograms in the $-\alpha$ tessellation are reflections of the parallelograms in the $+\alpha$ tessellation.

This is all to say that if this pair of conditions is satisfied—the twist angles are equal and opposite, and corresponding parallelograms have the same ratio between adjacent sides—then the $-\alpha$ tessellation pattern is similar—the same up to a scaling constant—to the folded form of the crease pattern given by the $+\alpha$ tessellation pattern.

But here's the clincher: the aspect ratio for each parallelogram is the product of a constant term $\frac{b+c}{a+d}$ that depends only on the tiling and the coefficient γ, which depends only upon the shrinking/rotation. The two aspect ratios will be the same for every pair of corresponding parallelograms if and only if both patterns are characterized by the same value γ. This brings us to the following theorem.

Theorem 1 (Crease Pattern/Folded Form Duality). *If a shrink-rotate tessellation crease pattern is constructed from a tiling using aspect ratio parameter and twist angle (γ, α), then the folded form of that tessellation is given to within a scaling constant by the same construction using $(\gamma, -\alpha)$.*

This symmetric property was identified by Bateman, who used it within his *Tess* program [Bateman 10b]; if one describes a tessellation crease pattern by the parameters (γ, α) rather than some other pair of parameters, then one can easily construct the line pattern of the folded form of any of the shrink-rotate tessellations by inverting the sign of α. This inversion gives the proper locations of the folded edges; it does not, of course, tell which edges are visible (which depends on the crease assignment—which, in turn, may or may not exist, depending on the twist angles and polygons involved).

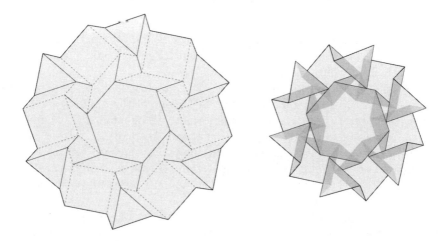

Figure 4. The crease pattern (left), and the folded form (right).

We show both crease pattern and folded form for the tiling and crease pattern of Figure 2 in Figure 4.

3   Nonregular Polygons

3.1   A Broken Tessellation

The shrink-and-rotate algorithm works for a wide range of tilings—all tilings composed of regular polygons, among others—but it doesn't work for all tilings. In fact, it breaks down on a relatively simple tiling composed of rhombuses, as shown in Figure 5.

The pattern in Figure 5(b), unfortunately, is not flat-foldable for any crease assignment because it does not satisfy the KJC. This is easily seen by considering a single rhombic twist from the pattern, as shown in Figure 5(c). Since the vertices are degree 4, in order to fold flat, opposite sector angles must add up to 180°, and this is clearly not the case.

The reason for this can be traced to the positions of the centers of rotation. These centers, plus the lines between them, constitute a second graph, called a *dual graph*, of the original tiling. (To be precise, it is an *interior dual*, since the dual contains only lines that cross the interior lines of the original.) The dual graph is shown in black in Figure 5(a). Observe in Figure 5(a) that the edges of the dual graph cross the edges of the original angle at angles other than 90°.

In our previous analysis of how the parallelogram ratio γ and twist angle α depended upon the shrink factor ρ, rotation angle β, and the dimensions

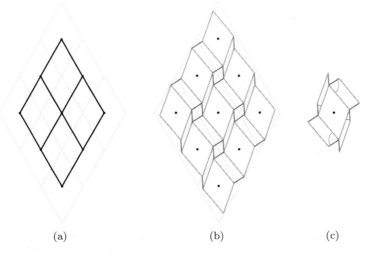

(a)   (b)   (c)

Figure 5. A rhombus-tiling-based crease pattern using shrink-and-rotate. (a) The initial rhombus tiling. (b) The crease pattern constructed by shrink-and-rotate about the rhombus centers. (c) A single twist from the crease pattern.

of the tiling, we assumed that both rotation centers lay on the x-axis with the tiling edge on the y-axis; in other words, we assumed perpendicularity. Perpendicularity is not guaranteed; if we re-do the analysis of each parallelogram for an arbitrary angle of intersection between the lines of the dual graph and the lines of the original angle, we find that the twist angle depends on this angle as well. If we want a common twist angle at every polygon, then each line of the original graph and its corresponding line in the dual graph should make the same angle with one another.

With this definition in hand, we can now pose the problem of creating a shrink-and-rotate tessellation in a more abstract way:

> For a shrink-and-rotate tessellation, the centers of rotation of the tiling (primal) graph should be the vertices of a straight-line embedding of the interior dual graph for which the angles are constant between the primal edges and their corresponding dual edges.

The only requirement is that the angle between all corresponding primal-dual pairs of edges be constant (e.g., perpendicular to one another). (In fact, for any dual graph whose edges make constant angles with their corresponding primal edges, it can be rotated so that the constant angle is 90°.) So all we need to find is an embedding where all edge pairs are perpendicular to one another, which we call an *orthogonal* embedding of the dual. But, how can we find such an embedding?

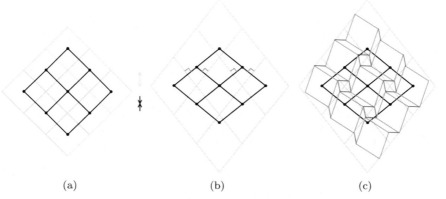

(a) (b) (c)

Figure 6. (a) A square tiling (light gray) and its dual graph (black). (b) The tiling is stretched vertically by a factor 1.25; the dual graph is compressed vertically by the same factor. (c) The crease pattern using the vertices of the dual graph as the centers of rotation.

3.2 A Valid Rhombus Tessellation

Let us return to the rhombus tiling. We can think about this as a tiling of rhombuses, of course, but we can also think of it as a tiling of squares that has been stretched vertically by some factor m. If we start with both the square tiling and its interior dual and we stretch them both by the same amount, then we end up with the situation we have already seen, where the vertices of the interior dual wind up in the centers of the tiles but the edges of the dual no longer cross the edges of the original tiling at constant angles.

But what if we stretched the dual by a different amount? This strategy, in fact, works; as shown in Figure 6, if we *compress* the interior dual graph vertically by the same amount as we stretched the original tiling vertically, the edges of the dual graph are all perpendicular to their corresponding edges in the primal graph. If we carry out the full shrink-and-rotate algorithm, then, we find that this procedure gives a tessellation that does, indeed, fold flat (at least locally, at each vertex, and by a suitable choice of twist angle, it can be made to fold flat globally as well).

This technique works for the rhombus tiling, but, in fact, it can be readily shown that it works for any tiling; if we have a tiling and a straight-line embedding of the interior dual graph such that the edges of the primal and dual graphs are perpendicular, then if we stretch the primal graph in one direction and compress the dual graph in the same direction, the edges of the new tiling will remain perpendicular, and so a shrink-and-rotate tessellation composed from the new tiling will be locally flat-foldable.

Now, in Figure 6(b), the centers of rotation are no longer in the same place in each rhombus. In fact, the periodicity of the dual graph is different from the periodicity of the original tiling, so we can see that if we were to extend the tiling farther in the vertical direction, the dual graph vertices would move closer and closer to the corners of their corresponding rhombi, and each would, in fact, eventually fall outside of its corresponding rhombus. This is not a problem; in fact, as can be readily verified, the dual graph and original graph need not overlap at all for the shrink-and-rotate algorithm to work, as long as the correspondence between the lines of the original and dual graphs is maintained.

So, once we have an embedding of the interior dual graph, we can move it around as we like and use its vertices as the centers of rotation for a shrink-rotated tessellation; the result will, in fact, be the exact same tessellation, but with a different position relative to the original tiling. Given that dual graph (or any translated/uniformly scaled version), we can construct a shrink-and-twist tessellation that satisfies the Kawasaki-Justin condition at every interior vertex. Thus,

> A shrink-rotate tessellation can be constructed that satisfies KJC at all vertices if there exists an orthogonal embedding of its interior dual graph.

4   Maxwell's Reciprocal Figures

As we have seen from the examples thus far, we can construct a shrink-rotate tessellation that satisfies KJC at all vertices from a plane graph by finding an embedding of the interior dual graph of the original tiling whose edges are all perpendicular to the edges of the original graph. For certain special classes of plane graphs, we can construct this orthogonal dual graph in a straightforward way. But what about the general case? Is it possible to construct an orthogonal dual graph for a general plane graph? And if not, when is it possible?

It turns out that this question has already been asked, and answered, in a field far removed from origami tessellations—removed by both topic and time. The topic of the orthogonal dual was considered in the field of mechanical engineering more than a hundred years ago. And it was addressed by none other than one of the greatest physicists and mathematicians of the nineteenth century, James Clerk Maxwell [Kappraff 02].

The paper that provides the key to making shrink-rotate tessellations was titled "On Reciprocal Figures and Diagrams of Forces," and was published by Maxwell in the *London, Edinburgh, and Dublin Philosophical Magazine and Journal of Science* in 1864. In this paper, Maxwell introduced the concept of a *reciprocal figure*: two figures are reciprocal if "the properties of the first relative to the second are the same as those of the second relative to the first" [Maxwell 64]. That, in a nutshell, describes the relationship between the primal graph and the dual graph that gives rise to a KJC-satisfying shrink-rotate tessellation. The vertices of the primal graph correspond to polygons of the dual graph and vice versa. And the edges of the one are perpendicular to the edges of the other or can have any constant angle we choose, as long as every edge makes the same angle with its opposite number. The dual graph is reciprocal (in the sense of Maxwell) to the primal graph; it is called, then, the *reciprocal diagram* of the primal graph.

In the nineteenth century, there was a very practical application for reciprocal diagrams. If a planar figure represented a set of beams connected at their ends, such as a bridge truss, it was critical to know the distribution of tensile and compressive forces in the various members of this truss in order to be sure that no single element was stressed beyond its breaking point. What Maxwell showed was that the distribution of forces could be described by a reciprocal diagram: draw the truss structure; then construct a reciprocal diagram, a graph whose edges are perpendicular to the the edges of the original figure. Then the relative edge lengths of the reciprocal diagram would be proportional to the forces in the individual truss members.

4.1   Indeterminateness and Impossibility

In a general reciprocal pair of figures, there is a one-to-one correspondence between each edge of the primal graph and the dual graph, so the two graphs have the same number of lines. Our situation is slightly different: our dual graphs, which we call the interior dual, have fewer lines than the primal graph because the dual does not contain edges that cross the boundary lines of the primal graph. We can, however, add a few lines to the interior dual to convert it to a proper reciprocal diagram of the original graph: for every polygon incident to the boundary of the primal graph, we can add an edge from the dual vertex corresponding to the boundary polygon that is perpendicular to the boundary edge of the primal graph.

It should be clear that it is always possible to add edges to the interior dual to convert it to a proper reciprocal figure. Conversely, if we have a proper reciprocal figure that satisfies orthogonality for every edge of the primal graph, we can "snip off" the boundary-crossing edges of the dual graph to realize the desired interior dual. So an orthogonal interior dual exists if and only if a reciprocal figure exists, as defined by Maxwell.

But does the reciprocal figure exist? Not always, it turns out. Suppose, for example, that we have a primal graph consisting of V vertices, E edges, and F faces, with E_b edges on the boundary (and $E_i = E - E_b$ edges in the interior). These numbers are not entirely independent; they must satisfy the Euler relation,

$$V + F - E = 1. \tag{8}$$

The number of boundary vertices V_b must be equal to the number of boundary edges, $V_b = E_b$, so the number of interior vertices is given by $V_i = V - E_b$.

Now, the interior dual graph has a vertex for every face of the primal graph, an edge for every nonboundary edge of the primal, and a face for every interior vertex of the primal. If we define the dual quantities by a primed variable, we find that

$$
\begin{aligned}
V' &= F, \\
E' &= E - E_b, \\
F' &= V - E_b.
\end{aligned}
\tag{9}
$$

Of course, we can see directly that the interior dual graph satisfies its own Euler relation $V' + F' - E' = 1$ as well.

Now, let us count how many degrees of freedom there are in this system and how many equations must be satisfied. The degrees of freedom in the interior dual graph are the coordinates of its vertices, so with two coordinates per vertex, there are $2V'$ degrees of freedom available.

As we have seen, a uniform translation in any direction or a uniform scaling will preserve the orthogonality relation between the primal and dual graphs, so we can choose those three degrees of freedom arbitrarily; for example, the x, y-coordinates of any one vertex plus the length of one of the edges.

Then there is a condition on each of the edges of the dual graph: each edge must be perpendicular to its corresponding edge in the primal graph. (In Section 4.2, we see what this condition is explicitly.) So there are E' conditions that must be satisfied for orthogonality, making $E'+3$ conditions in total on the $2V'$ degrees of freedom that are available to us.

If the number of conditions is exactly equal to the number of degrees of freedom, then there is precisely one solution for the reciprocal dual (apart from translation and scaling). If, however, the number of conditions is less than the number of degrees of freedom, then the system is indeterminate, and there are multiple solutions for the dual graph. If the number of conditions exceeds the number of degrees of freedom, then the system is impossible—it is not possible to construct the orthogonal dual graph.

Let us define the "excess" degrees of freedom as the difference between the number of degrees of freedom and the number of conditions. Then we have

$$
\begin{aligned}
DOF &\equiv 2V' - (E' + 3) \\
&= 2F - E + E_b - 3.
\end{aligned}
\tag{10}
$$

The condition on solvability is that the excess degrees of freedom is nonnegative, i.e., $DOF \geq 0$. This places a condition on the dual graph

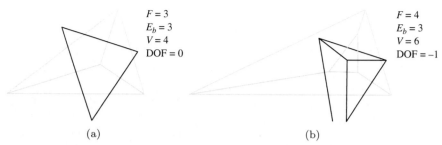

Figure 7. Two primal graphs and their attempted reciprocal figures. (a) This figure satisfies the Maxwell condition with no leftover degrees of freedom. (b) This figure does not satisfy the Maxwell condition, and there is no way to close the outer triangle that preserves orthogonality.

that

$$2V' - E' \geq 3. \tag{11}$$

This is a well-known requirement for the existence of a reciprocal figure, and it is commonly known as the *Maxwell condition* for that reason.

We can transform the Maxwell condition into a condition on our primal graph by using the relationships in Equation (10); the Maxwell condition becomes, for the primal graph,

$$2F + E_b - E \geq 3. \tag{12}$$

This, in turn, can be expressed in several equivalent ways by substituting the Euler relationship back into Equation (12), giving, for example,

$$F + E_b - V \geq 2. \tag{13}$$

Equation (13) displays most clearly the requirements for the existence of a solution: more faces than vertices and a many-edged border give solvability. Figure 7 shows two examples: one that is perfectly determined and one that is insolvable.

Note, however, that Equation (13) does not mean that a solution is *never* possible if this condition isn't satisfied; this is the condition that must be satisfied for there to be a solution for *every* embedding of the primal graph. There can be (and often are) *particular* embeddings that are sufficiently symmetric that orthogonality can be satisfied for multiple edges in ways that use up fewer degrees of freedom. An example of such a primal-dual combination is shown in Figure 8.

One apparent implication of Equation (13) is that with a sufficiently large number of edges on the border (i.e., with E_b sufficiently large), the Maxwell condition will always be satisfied. This seems counterintuitive: if I have a small graph for which there is no reciprocal figure, surely it would

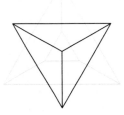

Figure 8. An embedding of the primal graph of Figure 7 that has an orthogonal interior dual graph.

not be possible to create a reciprocal figure by breaking up the border of the graph into more pieces?

Indeed, there is no contradiction: the Maxwell condition must apply to *every possible subgraph* of the primal graph. Every subgraph of the primal graph has an interior dual that is a subgraph of the overall interior dual; for the overall interior dual to exist, every possible subgraph of the primal graph must have an interior dual. Consequently, any subgraph of the primal graph that violates the Maxwell condition, no matter how small, insures that there is no solution for the overall graph.

4.2 Positive and Negative Edge Lengths

There is another issue, however, that can prevent the construction of a shrink-rotate tessellation, and it is illustrated in Figure 9.

The primal graph (in light gray) has $F = 6$, $E_b = 4$, and $V = 8$, so that $DOF = 0$; the dual graph is uniquely defined (to within translation and scaling). For this particular embedding of the primal graph, however,

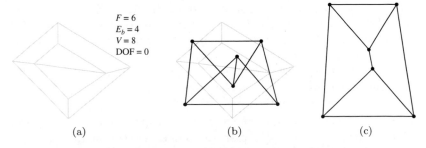

Figure 9. A primal graph with a crossing dual graph: (a) This figure satisfies the Maxwell condition with no leftover degrees of freedom. (b) The dual graph contains crossings. (c) A topologically proper (but nonorthogonal) embedding of the dual graph.

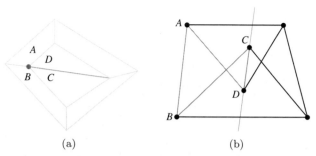

Figure 10. (a) The primal graph with four faces labeled *A–D* in cyclic order. (b) The dual graph with corresponding vertices marked.

the dual graph is not a plane graph: it has two crossings of its edges. That means that two of the polygons in the dual graph are improperly drawn. Figure 9(c) shows the dual graph drawn with noncrossing edges; it is topologically correct, but is no longer orthogonal to the primal graph.

This makes it impossible to turn this into a shrink-rotate tessellation crease pattern; recall that the resized/rotated versions polygons of the dual graph show up in the tessellation pattern, so that the crease pattern would contain these crossed polygons; the crossings would be additional vertices in the crease pattern that would not satisfy the Maekawa-Justin condition (they would have two mountains and two valleys or all four creases of one type). So the existence of crossed edges in the dual graph is sufficient to prevent turning the primal graph into a shrink-rotate tessellation.

In fact, remember that the Maxwell condition applies to any subgraph of the primal graph. Any primal graph that contains the graph shown in Figure 9 as a subgraph would have to contain the crossed dual graph as a subgraph of its dual. Thus, the presence of such a structure in any primal graph is enough to "poison" the entire tessellation.

There is a physical interpretation of this situation. Suppose we pick a vertex of the primal graph and count the faces around it in counterclockwise order, as shown in Figure 10. Each face of the primal graph corresponds to a vertex of the dual and vice versa; so cycling through faces around a vertex in the primal corresponds to cycling through vertices around a face of the dual. Thus, the marked vertex in the primal is the vertex that corresponds to the crossing polygon in the dual.

Because the primal and dual edges must be orthogonal, line CD in the dual must be orthogonal to the shaded edge between faces C and D of the primal graph. If we fix the position of vertex C, vertex D must lie somewhere on the line through CD. For the polygon to be noncrossing, vertex D would have to lie *above* vertex C on this line. Any noncrossing rendering of the polygon would have vertex D somewhere above C.

We normally talk about the length of an edge in a plane graph as the magnitude of the distance between the vertices—a number that is always positive. But we can associate a *signed* length with an edge, by calling the signed length positive if the direction of circulation of the dual polygon is the same as the direction of circulation around the vertex (e.g., vertex D lies above vertex C on the orthogonal line) and calling the signed length negative if the direction of circulation goes the opposite direction.

This definition carries over to the force model of Maxwell. If the primal graph is a planar beam construction under some set of stresses, the lengths of the edges of the dual graph give the stress (compressive or tensile) in each corresponding member of the mechanical structure. In fact, this correspondence refers to the *signed* length as we have defined it here. If the primal graph is placed under tension, then the signed length of each edge of the dual graph gives the tension in each member of the primal graph. If the signed length is negative, then the tension is negative; or in other words, the member is under compression, not tension.

Thus, for a primal graph to be turned into a tessellation, it must have an orthogonal interior dual graph—a reciprocal diagram—in which there are no edge crossings, i.e., all edges must have positive signed length. This means that there must be some application of tensile forces to the primal graph that makes the tension in each edge of the graph strictly positive.

Such a graph has a name in the field of static analysis: it is called a *spiderweb* [Kappraff 02]. The analogy is fairly direct. If we can create the graph from thread (or spider silk) and apply tensile stresses so that there are no slack threads, then the dual graph can be properly noncrossing with edges orthogonal to the edges of the primal graph. Using the vertices of the dual graph, one can then shrink and rotate each of the polygons of the primal tessellation to realize a crease pattern that satisfies KJC. By choosing the shrinkage and rotation angles (ρ, β) to keep the twist angle α sufficiently small, it is possible to assign creases to every parallelgram, and thus, every twist, to make them individually flat-foldable; and so, the entire crease pattern can be made flat-foldable.

Or, put succinctly, every spiderweb can be turned into a flat-foldable shrink-rotate tessellation, and the key is to use the vertices of the orthogonal interior dual as the centers of rotation.

The construction of such an orthogonal interior dual graph for a given embedding of the primal graph is a straightforward computational problem in statics analysis (it can be addressed using linear programming) and can be solved with well-known algorithms. As a demonstration of this theory, we have taken an actual spiderweb and turned it into a tessellation. Figure 11 shows a graph traced from an actual spiderweb (with some modifications around the edges to ensure that the positive-tension condition is satisfied; unlike theoretical spiderwebs, real spiderwebs can include "slack"

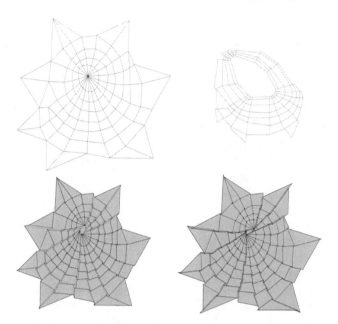

Figure 11. A spiderweb tessellation: the initial spiderweb graph (top left), the orthogonal interior dual (top right), the crease pattern (with $\gamma = 5, \alpha = 20°$) (bottom left), and a translucent rendering of the folded form (bottom right).

lines). We have calculated a reciprocal diagram with positive edge lengths, and from that, the orthogonal interior dual. Using the vertices of the interior dual as the centers of rotation, we construct the crease pattern and, as well, the folded form. We leave as an exercise for the reader the folding of the crease pattern from physical paper.

Bibliography

[Barreto 97] Paulo Taborda Barreto. "Lines Meeting on a Surface: The 'Mars' Paperfolding." In *Origami Science and Art: Proceedings of the Second International Meeting of Origami Science and Scientific Origami*, edited by K. Miura, pp. 343–359. Shiga, Japan: Seian University of Art and Design, 1997.

[Bateman 02] Alex Bateman. "Computer Tools and Algorithms for Origami Tessellation Design." In *Origami³: Proceedings of the Third International Meeting of Origami Science, Mathematics, and Education*, edited by Thomas Hull, pp. 121–127. Natick, MA: A K Peters, Ltd., 2002.

[Bateman 10a] Alex Bateman. "Gallery (of Origami Tessellations)." Available at http://www.papermosaics.co.uk/gallery.html, accessed June 6, 2010.

[Bateman 10b] Alex Bateman. "Tess: Origami Tessellation Software." Available at http://www.papermosaics.co.uk/software.html, accessed June 6, 2010.

[Fujimoto 82] Shuzo Fujimoto. *Seizo Soru Origami Asobi no Shotai (Creative Invitation to Paper Play)*. Tokyo: Asahi Culture Center, 1982.

[Gjerde 09] Eric Gjerde. *Origami Tessellations: Awe-Inspiring Geometric Designs*. Natick, MA: A K Peters, Ltd., 2009.

[Goldberg 55] Michael Goldberg. "Central Tessellations." *Scripta Mathematica* 21 (1955), 253–260.

[Huffman 76] David A. Huffman. "Curvature and Creases: A Primer on Paper." *IEEE Transactions on Computers* C-25:10 (1976), 1010–1019.

[Kappraff 02] Jay Kappraff. *Connections: The Geometric Bridge between Art and Science*. River Edge, NJ: World Scientific Publishing Co. Ltd., 2002.

[Kawasaki and Yoshida 88] Toshikazu Kawasaki and Masaaki Yoshida. "Crystallographic Flat Origamis." *Memoirs of the Faculty of Science, Kyushu University, Series A, Mathematics* XLII:2 (1988), 153–157.

[Maxwell 64] James Clerk Maxwell. "On Reciprocal Figures and Diagrams of Forces." *London, Edinburgh, and Dublin Philosophical Magazine and Journal of Science* 27 (Jan-Jun):24 (1864), 514–525.

[Momotani 84] Yoshihide Momotani. "Wall." In *BOS Convention 1984 Autumn*. London: British Origami Society, 1984.

[Palmer 97] Chris K. Palmer. "Extruding and tessellating polygons from a plane." In *Origami Science and Art: Proceedings of the Second International Meeting of Origami Science and Scientific Origami*, edited by K. Miura, pp. 323–331. Shiga, Japan: Seian University of Art and Design, 1997.

[Resch 68] Ronald D. Resch. "Self-Supporting Structural Unit Having a Series of Repetitions Geometric Modules." US Patent 3,407,558, 1968.

[Weisstein 10] Eric W. Weisstein. "Triangle Center." *MathWorld–A Wolfram Web Resource*. Available at http://mathworld.wolfram.com/TriangleCenter.html, 2010.

Flat-Unfoldability and Woven Origami Tessellations

Robert J. Lang

1 Introduction

The field of *origami tessellations* has seen explosive growth over the last 20 years. Interpreted broadly, an "origami tessellation" is a figure folded from a single sheet of paper in which the surface is divided up (tessellated) into a highly geometric pattern that is created by the folded edges and/or the transmission image of the varying layers (if folded from translucent paper and backlit), so that the pattern of the folded edges, rather than the outline of the figure, provides the dominant aesthetic. Though many origami tessellations are derived by regular tilings of the plane, the field of such 2D and 3D patterns is large and diverse. Similarly diverse are the algorithms by which they may be constructed.

For flat origami tessellations, one of the key constraints in their design is the *Kawasaki-Justin condition* (KJC), a condition found and described independently by Toshikazu Kawasaki [Takahama and Kasahara 85] and Jacques Justin [Justin 86]. There are several equivalent formulations, but the most common is the following: A crease pattern can be folded flat only if, at every interior vertex,

$$\alpha_1 - \alpha_2 + \alpha_3 - \alpha_4 \ldots = 0, \tag{1}$$

where the $\{\alpha_i\}$ are the sector angles around the vertex, numbered cyclically. The Kawasaki-Justin condition is not *sufficient* for flat-foldability;

additional conditions apply to the crease assignment and layer ordering (also formulated by Justin, see [Justin 97]). However, in many situations, the primary challenge in designing an origami crease pattern is assuring that it satisfies KJC.

The design of an origami figure, be it tessellation or representational, is typically framed as an *inverse problem*: the desired folded form or some aspect(s) of it are specified, and then the crease pattern that will produce that folded form is constructed.

For a flat origami figure, if we know the folded form in its entirety, then the design problem is essentially done: we can simply unfold the folded form, either algorithmically or physically, using real paper. But in the design process, we rarely know the complete folded form at inception. Rather, we specify some elements of the folded form based on design criteria and then choose other elements of the folded form to assure *validity*. Validity, in this case, means that the folded form comes from a flat sheet of paper, as opposed to some unusual, nonflat shape.

KJC applies to a *crease pattern*; it ensures that a crease pattern results in a flat form when it is folded. To solve an inverse problem, however, we want to find a condition on a *folded form*: what are the conditions on its mathematical description that ensure that it *unfolds* to a flat sheet of paper. Such a condition is called a *flat-unfoldability* condition. This paper introduces and describes a broadly useful flat-unfoldability condition.

The utility of a flat-unfoldability condition is this: if we can design a folded form that achieves a particular design goal and that satisfies flat-unfoldability conditions, then we can simply mathematically unfold the folded form to realize the crease pattern that we need to start with. We demonstrate this approach with the solution to a previously vexing origami design problem: the design of a general woven origami tessellation.

2   Woven Tessellations

One of the simplest origami tessellations is the *alternating simple flat twist tessellation* composed of twisted squares in which adjacent squares rotate in opposite directions. In such a tessellation, it is possible to assign creases so that all of the squares are surrounded by mountain folds (i.e., they are all in cyclic form). An example of such a tessellation is shown in Figure 1.

If you invert the crease assignment in such a tessellation (or turn it over), a surprise is in store: the pattern of edges looks uncannily like a set of woven strips, as shown in Figure 2. This is a pleasing illusion because origami tessellations are, by definition, folded from a single sheet, but this structure looks like it was constructed from many separate strips of paper (plus a background field).

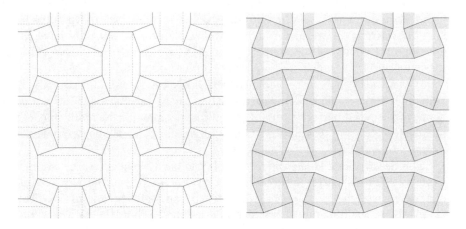

Figure 1. Crease pattern for an alternating simple flat twist tessellation consisting of square twists with cyclic mountain crease assignment (left). The folded form of this tessellation with translucent paper (right).

This particular pattern suggests that there might be a family of such patterns as we start to consider generalizations (several people have done so; see [Bateman 10a, Bateman 10b] for some examples). In the square woven tessellation, the strips run up and down and side to side and are evenly spaced. But one could envision patterns in which the strips run at other angles, or other spacings—or even at no particular angle, with

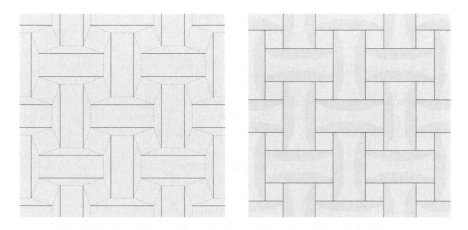

Figure 2. Crease pattern for an alternating simple flat twist tessellation consisting of square twists with cyclic valley crease assignment (left). The folded form of this tessellation with nearly opaque paper (right). This crease assignment displays the appearance of continuous woven strips.

every strip running in a different direction. In addition, if we think of such patterns as "patterns formed by woven strips," we not only have the spacing and angles of the strips to choose from: we can also choose, at every intersection, which strip appears to be on top.

We can also choose the width of the apparent strips, and we can, in principle, choose them all independently. These variables outline a vast space of potential patterns, any of which may or may not be realizable as a single-sheet origami tessellation. Going forward, I call such patterns *woven origami tessellations.*

The descriptions of the patterns possible with woven origami tessellations parallel the patterns possible in textile weaving: in both cases, one can vary the strips (analogous to the warp and weft threads in textiles), their widths, angles, and the pattern of crossings, or how the various strips and threads cross over one another. The simplest woven pattern is called a *plain weave*, or a *simple over-and-under* weave, in which any given strip, followed along its length, alternately goes over and under the strips that it crosses. If we further stipulate that

- no more than two strips cross at a given point,

- every strip travels in a straight line,

then we can narrow the field of possible woven tessellations considerably. I call any tessellation that uses a simple over-and-under weave and that satisfies these two conditions a *simple woven tessellation.*

The question then arises: what are the possible simple woven tessellations, and how are they folded using origami?

3   Simple Woven Patterns

To further simplify matters (and to make an aesthetic choice), let us assume that all strips are the same width. Then we can construct a simple woven pattern by the following prescription, illustrated in Figure 3.

1. Choose some pattern of straight lines such that no more than two intersect at any vertex.

2. Fatten each line to the desired width of the strips.

3. Add a boundary to the pattern to define the background field.

4. At each crossing, erase two of the four lines at the crossing to create a simple over-and-under weave.

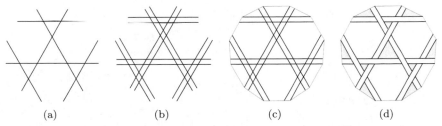

Figure 3. Construction sequence for a simple weave: (a) begin with a pattern of lines for which no more than two intersect at any given point; (b) thicken each line; (c) add a border to the field; (d) selectively erase crossings based on a two-coloring of the polygons between the woven edges.

How to do this last step is, perhaps, not entirely obvious, but a simple procedure suggests itself. Note that all interior vertices of the line pattern have degree 4, which means that the pattern can be two-colored, as shown in Figure 3. Each polygon is surrounded on all interior sides by partial strips. If we give each polygon a counterclockwise (CCW) circulation, each side of the polygon has a *head*, which is the end of the strip segment in the CCW direction, and a *tail*, which is the end of the strip segment at the other end. We can use these definitions, plus the two-coloring, to create the over-and-under woven pattern as follows:

- In colored polygons, the head of one strip segment goes *under* the tail of the next strip segment.

- In white polygons, the head of one strip segment goes *over* the tail of the next.

How do we turn the woven pattern into an origami figure? We can get an idea of this by looking at the square pattern again, but this time let's give the paper some translucency so that we can see the hidden layers of paper—which is where all the action is—and we'll zoom in a bit. Figure 4 shows the crease pattern and folded form.

Pay particular attention to the highlighted region in Figure 4. We see that this shaded trapezoid reappears throughout the pattern; in fact, every crossing of two strips has one of these trapezoids. Every trapezoid has two obtuse-angle vertices, which are the (hidden) ends of a going-under strip, and two acute-angle vertices, which are the endpoints of a covering-up strip. Then, of course, each of the acute-angle vertices of a trapezoid is incident to an obtuse-angle vertex of an adjacent trapezoid. When the pattern is unfolded, each trapezoid appears explicitly in the crease pattern.

In principle, we could imagine that any woven tessellation pattern might be realized by placing some version of this same trapezoidal structure at

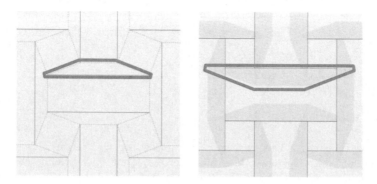

Figure 4. Crease pattern (left) and folded form (right) with the key trapezoidal structure highlighted.

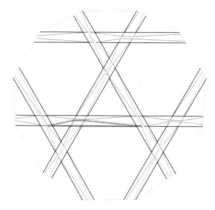

Figure 5. A hypothetical folded form of the woven-strip pattern from Figure 3.

every strip crossing in a woven tessellation. Using our test pattern of strips from Figure 3, a hypothetical example is shown in Figure 5.

The next question is: can this folded form pattern be folded from a flat sheet of paper?

4   Flat-Unfoldability

The question just posed is a specific example of a more general problem: if we are given a description of a flat-folded form (complete or partial), what are the conditions that ensure that it can be folded from a simple flat sheet of paper? This question is an inverse problem, and its answer would be the opposite of the more well-known flat-foldability conditions that determine whether a crease pattern can be folded into a flat-folded

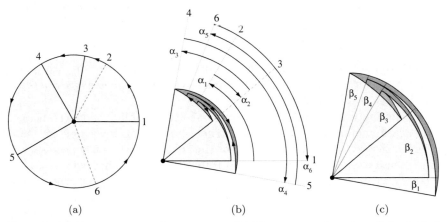

Figure 6. (a) Crease pattern for a vertex. (b) The measured angles of the folded form from one vertex to the next. CCW angles are positive; CW angles are negative. (c) Angle β_i is the angular extent from the ith fold to the next.

form. Those conditions are KJC (that the alternating sum of the angles at any interior vertex equals zero) and Justin's layer-ordering conditions (that disallow self-intersection). KJC is a *metric* condition; it ensures that the folded form lies flat when constructed from nonstretchy (isometric) material. Justin's conditions are combinatorial, and because they describe the layer ordering of the folded form, they are, implicitly, a condition on the folded form.

If a folded form satisfies Justin's conditions—we have chosen a layer ordering that avoids self-intersection—then the only condition still to be met is some isometric condition, and the particular isometric condition would be that the sector angles around each interior vertex, when the paper is unfolded, must sum to 360° so that the vertex lies flat.

Given a folded vertex and the information about each layer, we can define the sector angles of the vertex, in order, as the rotational angle from each folded edge to the next within each separate layer. If we adopt the usual convention that CCW rotation is a positive angle, then we will find that successive sector angles in the folded form alternate in sign: first positive (CCW), then negative (clockwise, or CW), then positive, and so forth, until we reach the folded edge at which we started. If we label these *folded* sector angles $\alpha_1, \alpha_2, \ldots$, as shown in Figure 6(b), then the condition that we end up where we started is simple:

$$\alpha_1 + \alpha_2 + \alpha_3 + \alpha_4 \ldots = 0. \tag{2}$$

But if this folded vertex arose from a flat sheet of paper, when we unfold the vertex, we should get 360 degrees of angle around the unfolded vertex, and so the condition that the folded form sector angles must satisfy will be

$$\alpha_1 - \alpha_2 + \alpha_3 - \alpha_4 \ldots = \pm 360°, \tag{3}$$

where the sign of the result depends on whether we started with a positive or negative angle. This equation is the *flat-unfoldability condition*, analogous to KJC for flat-*foldability*.

To apply Equation (3), one must be able to identify the layers incident to each folded edge in order to construct the cyclic ordering of the sector angles. However, sorting the order of the folded edges is not necessary. If we examine the folds that emanate from the vertex, they will always lie strictly within a 180° arc. Starting at one end, we take β_i to be the angular extent of the ith arc, and we denote by n_i the number of layers of paper *that are incident to the vertex* (i.e., we don't count layers that are not part of the vertex figure). Then, since every layer incident to the vertex must form part of the flat, unfolded vertex, it must be the case that

$$\sum_i n_i \beta_i = 360°. \tag{4}$$

For the commonly encountered degree-4 vertex, the pattern of fold lines falls into one of three possible patterns, and it is possible to construct special cases of Equation (4) that apply to the line pattern of the folded form. Every flat-foldable degree-4 vertex falls into one of the following cases (Figure 7):

1. all four sector angles are distinct,

2. the sector angles come in two pairs of equal angles,

3. all four sector angles are equal to 90°.

If all four sector angles are distinct, then there are four distinct lines in the line pattern of the folded form, and the largest angle in the line pattern $< 180°$ will be one of the four sector angles, as shown in Figure 7(a). If we number the three visible angles within this angle by $\{\beta_1, \beta_2, \beta_3\}$, then the four sector angles in the crease pattern will be, respectively,

$$(\beta_2), (\beta_1 + \beta_2), (\beta_2 + \beta_3), \text{ and } (\beta_1 + \beta_2 + \beta_3), \tag{5}$$

and so the flat-unfoldability condition applicable to the visible angles in the folded form is

$$2\beta_1 + 4\beta_2 + 2\beta_3 = 360°, \tag{6}$$

or, equivalently,

$$\beta_1 + 2\beta_2 + \beta_3 = 180°. \tag{7}$$

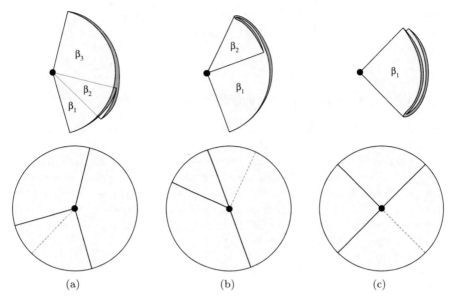

Figure 7. Folded form (top) and crease pattern (bottom) for the three distinct configurations of a degree-4 vertex: (a) all four sector angles are distinct, (b) sector angles form two pairs of equal angles, (c) all four sector angles are equal to 90°.

If the sector angles come in two pairs of equal angles, then there will be three distinct lines in the line pattern of the folded form, and the largest angle in the line pattern less than 180° will be equal to two of the four sector angles, as shown in Figure 7(b). Numbering the two visible angles within this angle by $\{\beta_1, \beta_2\}$, the four sector angles in the crease pattern will be, respectively,

$$(\beta_1 + \beta_2), (\beta_1 + \beta_2), (\beta_2), \text{ and } (\beta_2), \tag{8}$$

and so the flat-unfoldability condition applicable to the visible angles in the folded form is

$$2\beta_1 + 4\beta_2 = 360°. \tag{9}$$

And finally, if all four sector angles are equal, there are two distinct lines in the line pattern of the folded form; the angle in this pattern less than 180° will be equal to each of the sector angles of the crease pattern, as shown in Figure 7(c), and so the relationship between the sector angles of the crease pattern and the corresponding condition on the visible angle $\{\beta_1\}$ of the folded form will be

$$\alpha_1 = \alpha_2 = \alpha_3 = \alpha_4 = \beta_1 = 90°. \tag{10}$$

Thus, given the line pattern of a folded form composed of degree-4 vertices and a valid layer ordering, the folded form can be unfolded to a flat sheet of paper if and only if for every interior vertex of the line pattern, one of Equations (6)–(10) (as appropriate) is satisfied (and there is no self-intersection).[1]

5   Parameterizing the Woven Tessellation

To satisfy the flat-unfoldability conditions, there must, of course, be some variables whose values can be adjusted to satisfy the equations. (One hopes that there are enough adjustable variables to satisfy *all* of the equations that must be satisfied.) In the woven tessellation, at each strip crossing, there are two trapezoids, each of which has four vertices. Each vertex, however, appears in two different trapezoids (as an obtuse vertex of one trapezoid and an acute vertex of the other). Thus, if N_C is the number of strip crossings, then there will be

$$\frac{1}{2} \times 4 \times N_C = 2N_C$$

flat-unfoldability conditions to be satisfied. That suggests that we should have at least that many variables in a parameterization of the crease pattern.

As it turns out, there is not total freedom in parameterizing the crease pattern because the base line (and thus the two acute vertices) of each trapezoid is required to lie on the "over" edge of a woven strip at a crossing and each of the obtuse vertices must lie on an "under" edge at a crossing. For each vertex, there is only one free parameter, which we can take to be, for example, the perpendicular distance of that vertex from the edge covering it. We will call this distance the *inset distance* d_i for the vertex, as illustrated in Figure 8. If we do the same counting of variables, we find that the number of variables associated with each full trapezoid is also $2N_C$. So there are exactly as many variables arising from interior vertices as we have flat-unfoldability conditions.

In fact, though, there are somewhat more variables. Along the boundary of the folding pattern, we have partial trapezoids that are defined with more inset distances, but there is no need to satisfy a flat-unfoldability equality for vertices on the boundary of the folded pattern. (There is a requirement that the total angle be less than *or* equal to 360°, but that will not be an issue here.) So, in addition to the $2N_C$ variables associated with each of the crossings, we have $2N_B$ extra variables to work with, where N_B is the number of strips that hit the boundary in the folded form.

The problem, then, is underconstrained; there are more variables than there are equations, and it is necessary to choose some additional criteria

---

[1] There is no guarantee, of course, that any such pattern can be *rigidly* unfolded.

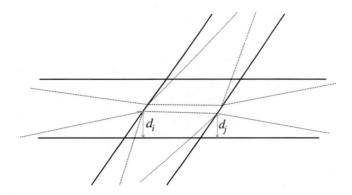

Figure 8. Schematic of a crossing vertex. d_i is the inset distance at each vertex.

that allow one to solve for a particular solution. One could theoretically identify exactly the number of equalities needed to solve for all of the variables, but identifying the proper set can be a challenge. A more robust approach is to add one or more equality and/or inequality conditions that address any additional (perhaps aesthetic) criteria, then perform a multi-dimensional optimization with a suitable figure of merit to "soak up" any remaining degrees of freedom.

One additional criterion that would be useful is to set a minimum value on the inset distance; this prevents the creation of trapezoids that are too skinny to be easily folded. Such a requirement takes the form of a set of inequality constraints:

$$d_i \geq d_{\min} \text{ for all } i. \tag{11}$$

Once a minimum inset distance is set, then we would also prefer that each of the trapezoids not be much *wider* than the minimum size, in order to minimize the chances that two trapezoids overlap in such a way as to violate a self-intersection condition. This can be accomplished by introducing a slack variable, $d_{\max}$, setting inequality constraints

$$d_i \leq d_{\max} \text{ for all } i, \tag{12}$$

and then taking $d_{\max}$ as the figure of merit to be minimized.

Another approach that has a similar effect of keeping the pleat size down and has the benefit of applying force to all vertices is to take the figure of merit to be the root-mean-square (RMS) sum of all inset distances and minimize that. In practice, I have found that this merit function works well and results in aesthetically pleasing patterns.

6   Conclusion

I have implemented the algorithm described above using Mathematica 7.0.1. The algorithm takes as input a pattern of lines that defines a woven pattern and a desired strip width (which must be small enough that there are no points where three or more woven strips overlap; auxiliary functions in the Mathematica notebook let the user solve for the maximum strip width for a given pattern of lines). The program finds all intersections between pairs of lines to turn the line pattern (plus a specified boundary curve) into a plane graph; it then two-colors the plane graph to determine the over-and-under pattern of the strips. At each strip crossing, it constructs the two trapezoids, suitably parameterized on the inset distances $\{d_i\}$. The user specifies a minimum inset distance (commonly half of the strip width), and then the inset distances are solved for, using the RMS-minimization optimization algorithm. The result is a folded form satisfying the flat-unfoldability conditions, which therefore can be unfolded to a flat pattern. Another Mathematica function takes the embedded graph of the folded form and algorithmically unfolds it to realize the crease pattern that gives rise to the desired folded form.

Figure 9 shows an example of this series, including the computed crease pattern, the computed folded form, and a folded example. Both the folded form and crease pattern in Figure 9 are computed, so there is no guarantee (beyond the mathematical arguments given above) that these two patterns really go together. However, this is, in fact the case; Figure 9 shows a photograph of a model folded from the crease pattern.

As a side note, these patterns tend to be very difficult to fold, and even though the starting state and ending states are both flat, the partially folded intermediate state is typically highly convoluted with the facets bent and/or even somewhat crumpled (one attempts to minimize the crumpling, of course). It is unlikely that one could find and successfully fold even a moderately complex woven pattern by trial and error, but with this algorithm, any simple woven tessellation is now possible.

Since developing this algorithm, I have designed and folded a variety of woven tessellations of this family; a representative sampling may be found at [Lang 10]. One can also envision various generalizations and variations; for example, instead of using an unconstrained boundary, one could implement periodic boundary conditions to realize a woven tessellation pattern that tiles the plane. But even beyond woven tessellations, I feel that the technique used here of applying flat-unfoldability conditions to a partially defined folded form pattern is a broadly useful tool for the design of geometric origami figures such as origami tessellations.

I close with a somewhat more complicated woven pattern, a "double-weave" pattern that appears to be composed of pairs of woven strips, shown

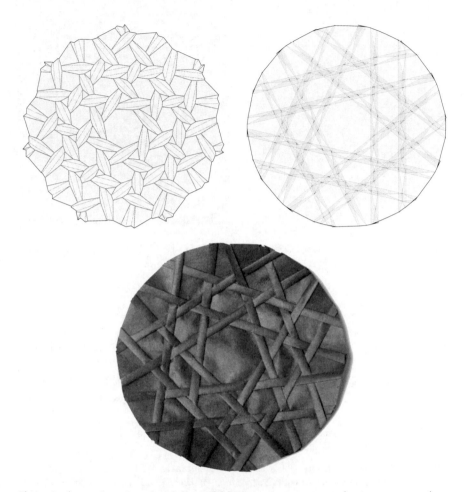

Figure 9. A woven pattern with sevenfold symmetry: computed crease pattern (top left); computed folded form, with translucent rendering to show the trapezoids (top right); a folded example (bottom).

in Figure 10. This pattern is more complicated than the simple woven pattern: there are both degree-4 and degree-6 vertices in the line pattern of the folded form (and therefore, of course, in the crease pattern). Nevertheless, it, and others like it, can be constructed in the same way, by applying flat-unfoldability conditions to a partially defined desired crease pattern. I expect that many more origami designs may be realized using this technique.

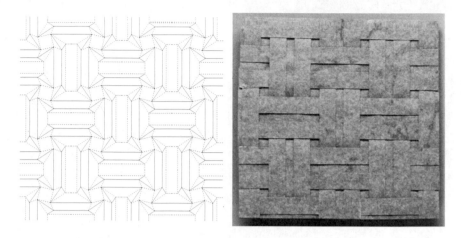

Figure 10. Crease pattern for a double-strip woven tessellation (left); photograph of a folded example (right). (See Color Plate XVIII.)

Bibliography

[Bateman 10a] Alex Bateman. "Not a Rattan Weave Tessellation." *Flickr.com.* Available at http://www.flickr.com/photos/42673512@N00/1513700768/, 2010.

[Bateman 10b] Alex Bateman. "Thin Origami Rattan Weave." *Flickr.com.* Available at http://www.flickr.com/photos/42673512@N00/1611849065/, 2010.

[Justin 86] Jacques Justin. "Mathematics of Origami, Part 9." *British Origami* 118 (1986), 28–30.

[Justin 97] Jacques Justin. "Towards a Mathematical Theory of Origami." In *Origami Science and Art: Proceedings of the Second International Meeting of Origami Science and Scientific Origami,* edited by K. Miura, pp. 15–30. Shiga, Japan: Seian University of Art and Design, 1997.

[Lang 10] Robert J. Lang. "Gallery (of Origami Tessellations)." *Flickr.com.* Available at http://www.flickr.com/photos/langorigami/, 2010.

[Takahama and Kasahara 85] Toshie Takahama and Kunihiko Kasahara. *Top Origami.* Tokyo: Sanrio Publications, 1985.

Degenerative Coordinates in 22.5° Grid System

Tomohiro Tachi and Erik D. Demaine

1 Introduction

The crease patterns for many origami models are designed within an angular grid system of $90°/n$, for a nonnegative integer n. Precisely, in this system, every (pre)crease passes through an existing reference point in the direction of $m(90/n)°$ for some integer m, and every reference point is either $(0,0)$, $(1,0)$, or an intersection of already constructed (pre)creases. For example, 45° $(n = 2)$, 30° $(n = 3)$, 22.5° $(n = 4)$, 18° $(n = 5)$, and 15° $(n = 6)$ grid systems are known to be useful for the design of origami.

In particular, the 22.5° grid system has been used for centuries—one of the oldest examples is the classic origami crane—and the system keeps producing complex but organized origami expressions such as "Devil" (1980) by Jun Maekawa [Maekawa and Kasahara 83, Maekawa 07, pp. 146–154] and "Wolf" (2006) by Hideo Komatsu [Komatsu 06]. Toshikazu Kawasaki calls this system *Maekawa-gami*. Figure 1 shows a square filled with several precreases in the 22.5° grid system.

Why are these angular grid systems so useful? A striking feature of Figure 1 is that there are many ways to construct the same point and, as a consequence, many alignments among points and lines. Intuitively, this *degeneracy* of the construction system helps tame the complexity of crease patterns designed within the system.

In this paper, we formalize this notion of degeneracy and organized complexity by characterizing the coordinates of reference points in the 22.5°

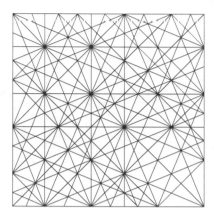

Figure 1. Maekawa-gami: 22.5° grid.

grid system of the unit square as those points (x, y) with $x, y \in \mathcal{D}_{\sqrt{2}}$, where

$$\mathcal{D}_{\sqrt{2}} = \left\{ \frac{m + n\sqrt{2}}{2^\ell} \;\middle|\; \text{integers } m, n, \text{ and } \ell \geq 0 \right\}.$$

In particular, we establish degeneracy by proving that all constructible points fall into $\mathcal{D}_{\sqrt{2}}^2 = \mathcal{D}_{\sqrt{2}} \times \mathcal{D}_{\sqrt{2}}$, and establish universality by proving that all points in $\mathcal{D}_{\sqrt{2}}^2$ can be constructed. In the latter result, the number of required operations is linear in the bit complexities of x and y, where the *bit complexity* of a number $(m + n\sqrt{2})/2^\ell \in \mathcal{D}_{\sqrt{2}}$ is $\lg(2+|m|)+\lg(2+|n|)+\ell$.

2   Model

More precisely, we consider the following models of 22.5° grid construction.

The initial set of points can be either two marks on the x-axis, $\{(0,0), (1,0)\}$, or all four corners of the square, $\{(0,0), (1,0), (0,1), (1,1)\}$. The choice between these two options does not affect the results; for the strongest results, we use the former set for our construction, and the latter set for proving degeneracy.

The *grid-line construction* is to draw a line through an existing point, at an angle of $k\,22.5°$ with respect to the x-axis, for an integer $k \in \{0, 1, \ldots, 7\}$. This line also defines newly constructed points by its intersections with all other drawn lines.

A grid-line construction can be simulated by $O(1)$ applications of Huzita–Justin axioms. Recall that the Huzita–Justin axioms [Huzita and Scimemi 91, Justin 91, Demaine and O'Rourke 07, Chapter 19] include the ability to fold the line through two given points, fold the perpendicular bisector of two points, fold the angular bisector of two given lines, fold the perpendicular to a given line passing through a given point, and two operations constructing tangents to parabolas. In fact, we need only two of these axioms: folding the angular bisector of two lines and folding through a point and perpendicular to a line. Then we can perform a grid-line construction by constructing two lines through the point perpendicular to the two axes, then bisecting one of the 90° angles once or twice to obtain the desired integer multiple of 22.5°.

To avoid this constant-factor overhead, our construction will simultaneously adhere to the constraints of both the 22.5° grid system and these two Huzita–Justin axioms. Thus, every operation we perform in the construction will bisect an angle at an existing point, or be perpendicular to an existing line and through an existing point, and furthermore the angle the constructed line makes with respect to the x-axis will be an integer multiple of 22.5°. Note that these two operation types are indeed special cases of grid-line constructions in addition to corresponding to real origami constructions. We call these operations *grid-line axioms.*

3   Construction

In this section, we give a universal construction algorithm for points in $\mathcal{D}^2_{\sqrt{2}}$:

Theorem 1. *We can construct any point in $\mathcal{D}^2_{\sqrt{2}}$ by a sequence of grid-line axioms whose length is linear in the bit complexities of the two coordinates.*

The algorithm constructs each coordinate of the target point separately. Thus we focus mostly on the construction of a single number in $\mathcal{D}_{\sqrt{2}}$, as measured by the distance from the origin of a point along the x-axis. To perform such a construction, we combine several gadgets (Figure 2) for constructing individual numbers and performing arithmetic on numbers:

1. *Root gadget:* Construct the number $\frac{\sqrt{2}}{2}$. The gadget essentially reflects the diagonal down to the axis (Figure 2(a)).

2. *Half gadget:* Given a positive number a, construct $a/2$. The gadget uses 45° lines to construct the midpoint (Figure 2(b)). Note that, given all of the Huzita–Justin axioms, we could instead simply use a perpendicular bisector to construct the desired point.

3. *Double gadget:* Given a positive number a, construct $2\,a$. The gadget essentially reflects a copy of a (Figure 2(c)).

4. *Add gadget:* Given two positive numbers a and b, construct $a + b$. The gadget essentially reflects a copy of the smaller integer to the right of the larger integer (Figure 2(d)).

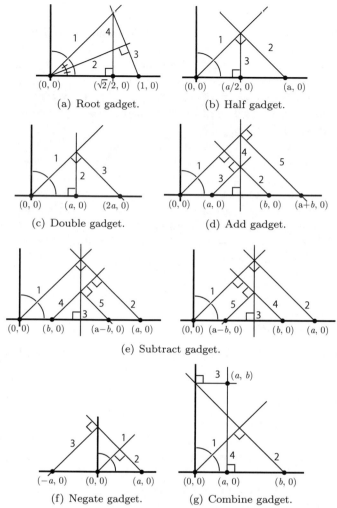

(a) Root gadget.

(b) Half gadget.

(c) Double gadget.

(d) Add gadget.

(e) Subtract gadget.

(f) Negate gadget.

(g) Combine gadget.

Figure 2. Gadgets: before applying any gadgets, we construct the x-axis from the given points $(0,0)$ and $(1,0)$, and construct the y-axis as perpendicular to the x-axis through $(0,0)$.

5. *Subtract gadget:* Given two positive numbers a and b with $a > b$, construct $a-b$. The gadget computes $a/2$ and then essentially reflects b around $a/2$, which gives $2(a/2) - b = a - b$ (Figure 2(e)). In fact, the construction has two cases, depending on whether $b \leq a/2$ or $b \geq a/2$.

6. *Negate gadget:* Given a positive number a, construct $-a$. The gadget essentially reflects a to the left of the y-axis (Figure 2(f)).

Lemma 1. *We can construct any number in $\mathcal{D}_{\sqrt{2}}$ by a sequence of grid-line axioms whose length is linear in the bit complexity of the number.*

Proof: Consider a number $x = (m + n\sqrt{2})/2^\ell \in \mathcal{D}_{\sqrt{2}}$.

We construct $|m|$ using a standard repeated doubling trick (analogous to repeated squaring [Cormen et al. 09, Section 31.6]). If $|m|$ is even, recursively construct $|m|/2$ and then use the double gadget. If $|m|$ is odd, recursively construct $(|m| - 1)/2$, then use the double gadget to obtain $|m| - 1$, and then use the add gadget to add 1. In the base cases, we already have the constants 0 and 1. The number of operations is $O(\lg(2 + |m|))$.

Similarly, we can construct $|n|\sqrt{2}$ by using $\sqrt{2}$ instead of 1 as a base case. We can construct $\sqrt{2}$ via the root gadget followed by the double gadget. The number of operations is $O(\lg(2 + |n|))$.

Finally, we combine the two values $|m|$ and $|n|\sqrt{2}$ using an add or subtract gadget to obtain $|m + n\sqrt{2}|$, then use the half gadget ℓ times to construct $|x|$, and then use the negate gadget if x is negative. (We can negate only at the end because the other gadgets are designed for positive numbers.) □

Proof (of Theorem 1): Consider a point $p = (x, y) \in \mathcal{D}^2_{\sqrt{2}}$, where $x = (m_x + n_x\sqrt{2})/2^{\ell_x})$ and $y = (m_y + n_y\sqrt{2})/2^{\ell_y})$. We use Lemma 1 to construct the numbers x and y along the x-axis. Then we copy the value y onto the y-axis using a 45° line, and then find the intersection of two perpendiculars to find the point p, as shown in Figure 2(g). □

For points in the unit square, we can restrict work to within the square of paper:

Lemma 2. *We can construct any number in $[0, 1] \cap \mathcal{D}_{\sqrt{2}}$ by a sequence of grid-line axioms, with all intermediate points in $[0, 1]$, whose length is linear in the bit complexity of the number.*

Proof: Consider a number $x = (m + n\sqrt{2})/2^\ell \in \mathcal{D}_{\sqrt{2}}$ with $0 \le x \le 1$. Let $s = 1 + \max\{\lg|m|, \lg(|n|\sqrt{2})\}$ be the smallest integer such that $|m|/2^s \le \frac{1}{2}$ and $(|n|\sqrt{2})/2^s \le \frac{1}{2}$. First we construct $S = 1/2^s$ by s half gadgets. Then we apply the construction in Lemma 1 but use S in place of 1. Because all numbers in this construction are at most $|m| + |n|\sqrt{2}$, but everything is scaled by $1/S$, all intermediate values are in $[0, 1]$. Thus we obtain $(m + n\sqrt{2})/2^{\ell+s}$. Finally, we scale back up using s double gadgets. (For practicality, we could have saved $\min\{\ell, s\}$ half/double gadget pairs, but this affects only the constant factor.) The total number of operations is $O(\lg(2 + |m|) + \lg(2 + |n|) + \ell)$ because $s = O(\lg(2 + |m|) + \lg(2 + |n|))$. □

Theorem 2. *We can construct any point in $[0, 1]^2 \cap \mathcal{D}^2_{\sqrt{2}}$ by a sequence of grid-line axioms, with all intermediate points in $[0, 1]^2$, whose length is linear in the bit complexities of the two coordinates.*

Proof: Simply follow the proof of Theorem 1 but use Lemma 2 in place of Lemma 1.   □

4   Degeneracy

In this section, we show the degeneracy in the grid system: all constructible points are restricted within $\mathcal{D}^2_{\sqrt{2}}$.

Theorem 3. *Every point constructible by a sequence of grid-line construc-tions, starting from the corners of a unit square, is in $\mathcal{D}^2_{\sqrt{2}}$.*

The proof is by induction. In the base case, we start from the four points at the corners of the square, $\{(0,0),(1,0),(0,1),(1,1)\}$, which are in $\mathcal{D}^2_{\sqrt{2}}$. For the induction step, we prove that any newly constructed points by grid-line constructions from points in $\mathcal{D}^2_{\sqrt{2}}$ are also in $\mathcal{D}^2_{\sqrt{2}}$. In order to cull duplicate combinations, we first extend the system to allow 45° rotation.

Lemma 3. *Points in $\mathcal{D}^2_{\sqrt{2}}$ are closed under $k\,45°$ rotation about the origin, for any $k \in \{0, 1, \ldots, 7\}$.*

Proof: For a point $(x, y) \in \mathcal{D}^2_{\sqrt{2}}$, its 45° rotation (x', y') is given by

$$\begin{bmatrix} x' \\ y' \end{bmatrix} = \begin{bmatrix} \frac{\sqrt{2}}{2} & -\frac{\sqrt{2}}{2} \\ \frac{\sqrt{2}}{2} & \frac{\sqrt{2}}{2} \end{bmatrix} \begin{bmatrix} x \\ y \end{bmatrix}.$$

Because $\mathcal{D}_{\sqrt{2}}$ is closed under addition, subtraction, and multiplication, x' and y' are in $\mathcal{D}_{\sqrt{2}}$. By induction, a point produced by $k\,45°$ rotation is in $\mathcal{D}^2_{\sqrt{2}}$.   □

Lemma 4. *Two lines made by grid-line constructions from points in $\mathcal{D}^2_{\sqrt{2}}$ have their intersection point in $\mathcal{D}^2_{\sqrt{2}}$.*

Proof: Consider two grid-line constructions from points in $\mathcal{D}^2_{\sqrt{2}}$, say, line L_0 from point (x_0, y_0) and line L_1 from point (x_1, y_1). For $i \in \{0, 1\}$, we can define line L_i as points (x, y) satisfying $s_i(x - x_i) + t_i(y - y_i) = 0$, where (s_i, t_i) is a vector perpendicular to the line and thus t_i/s_i is the slope. For the lines to have an intersection, they must not be parallel, i.e., $s_0 t_1 - s_1 t_0 \neq 0$. The intersection point (x, y) is given by

$$\begin{cases} s_0(x - x_0) + t_0(y - y_0) = 0, \\ s_1(x - x_1) + t_1(y - y_1) = 0, \end{cases} \tag{1}$$

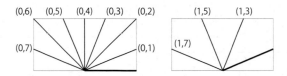

Figure 3. Possible combinations of directions after suitable $k\,45°$ rotation. Labels refer to indices into the list of possible (s,t) vectors (see Equation (3)), starting at index 0.

which can be represented in matrix form as

$$\begin{bmatrix} s_0 & t_0 \\ s_1 & t_1 \end{bmatrix} \begin{bmatrix} x \\ y \end{bmatrix} = \begin{bmatrix} s_0 x_0 + t_0 y_0 \\ s_1 x_1 + t_1 y_1 \end{bmatrix}.$$

By Cramer's Rule, the solution is given by

$$\begin{bmatrix} x \\ y \end{bmatrix} = \frac{1}{s_0 t_1 - t_0 s_1} \begin{bmatrix} t_1 & -t_0 \\ -s_1 & s_0 \end{bmatrix} \begin{bmatrix} s_0 x_0 + t_0 y_0 \\ s_1 x_1 + t_1 y_1 \end{bmatrix}. \tag{2}$$

Because $\mathcal{D}_{\sqrt{2}}$ is closed under multiplication and addition, it suffices to show that $1/(s_0 t_1 - t_0 s_1)$ is in $\mathcal{D}_{\sqrt{2}}$ for any combination of vectors (s_0, t_0) and (s_1, t_1).

There are eight possible orientations for each vector (s_i, t_i) in the 22.5° system, given by representative vectors

$$\begin{aligned} (s,t) \;\in\; \{&(1,0),(1,-1+\sqrt{2}),(1,1),(-1+\sqrt{2},1),\\ &(0,1),(1-\sqrt{2},1),(-1,1),(-1,-1+\sqrt{2})\}. \end{aligned} \tag{3}$$

Note that we do not need to list the negations of these vectors, as it suffices to capture all line slopes, not signed directions, for the line equations in Equation (1).

Instead of checking every pair of slopes ($\binom{8}{2} = 28$ patterns), we can reduce the possible combinations down to 10 cases by using Lemma 3 to perform $k\,45°$ rotation in advance. Namely, if one of the lines has angle $k\,45°$ for an integer k, then we rotate by $-k\,45°$ to give that line orientation $(1,0)$, and obtain seven possible cases for the other line (Figure 3, left); and if both lines have angles that are not integer multiples of 45°, then we rotate so that one of them is 22.5° and obtain three cases for the other line (Figure 3, right).

Table 1 computes $1/(s_0 t_1 - t_0 s_1)$ for each of these ten cases. In all cases, the value is in $\mathcal{D}_{\sqrt{2}}$, so $(x,y) \in \mathcal{D}_{\sqrt{2}}^2$. □

Proof (of Theorem 3): Consider a sequence of grid-line constructions $\ell_1, \ell_2, \ldots, \ell_n$. By definition, each ℓ_k is a line through an existing point,

case	(s_0, t_0)	(s_1, l_1)	$1/(s_0 l_1 - l_0 s_1)$
$(0,1)$	$(1,0)$	$(1, -1 + \sqrt{2})$	$1 + \sqrt{2}$
$(0,2)$	$(1,0)$	$(1,1)$	1
$(0,3)$	$(1,0)$	$(-1 + \sqrt{2}, 1)$	1
$(0,4)$	$(1,0)$	$(0,1)$	1
$(0,5)$	$(1,0)$	$(1 - \sqrt{2}, 1)$	1
$(0,6)$	$(1,0)$	$(-1,1)$	1
$(0,7)$	$(1,0)$	$(-1, -1 + \sqrt{2})$	$1 + \sqrt{2}$
$(1,3)$	$(1, -1 + \sqrt{2})$	$(-1 + \sqrt{2}, 1)$	$(1 + \sqrt{2})/2$
$(1,5)$	$(1, -1 + \sqrt{2})$	$(1 - \sqrt{2}, 1)$	$(2 + \sqrt{2})/2^2$
$(1,7)$	$(1, -1 + \sqrt{2})$	$(-1, -1 + \sqrt{2})$	$(1 + \sqrt{2})/2$

Table 1. Case analysis of slopes: the leftmost column refers to labels in Figure 3.

either an original corner of the square $\{(0,0), (1,0), (0,1), (1,1)\}$, or defined by an intersection between two lines ℓ_i and ℓ_j for $i, j < k$.

We claim by induction that each line ℓ_k is a grid-line construction from a point in $\mathcal{D}^2_{\sqrt{2}}$. If ℓ_k is constructed from a corner of the unit square, this claim follows because $(0,0), (1,0), (0,1), (1,1) \in \mathcal{D}^2_{\sqrt{2}}$. Otherwise, ℓ_k is constructed from the intersection of two lines ℓ_i and ℓ_j with $i, j < k$. By induction on k, both ℓ_i and ℓ_j are grid-line constructions from points in $\mathcal{D}^2_{\sqrt{2}}$. Thus Lemma 4 applies, and the intersection point defining ℓ_k is in $\mathcal{D}^2_{\sqrt{2}}$.

Finally, the points formed by the sequence of grid-line constructions $\ell_1, \ell_2, \ldots, \ell_n$ are the intersections between two lines ℓ_i and ℓ_j. By the claim above, Lemma 4 applies to show that every such point is in $\mathcal{D}^2_{\sqrt{2}}$. $\square$

5 Conclusion

We have characterized the degeneracy of points constructible in the 22.5° grid system. The restricted form we establish for constructible coordinates indicates that there are many possible ways to construct a point, which should tend to lead to fortuitous alignments of creases. For example, these alignments make it easier to choose creases so that they meet at vertices of degree at least 4, as necessary for flat foldability, whereas without a grid system, generically chosen lines would meet only in pairs and would not satisfy Kawasaki's condition for local flat foldability. Furthermore, our algorithms show that any desired point in the grid system can be constructed efficiently using just two types of origami operations, in a sequence of length linear in the bit complexity of the coordinates. These results provide mathematical support for why practical origami design uses grid systems.

A simple extension of our theory is to the situation of 22.5° grid-line constructions starting from a length ratio of $(m + n\sqrt{2})/(m' + n'\sqrt{2})$, for some integers m, n, m', n'. For example, Maekawa's "Wani" (alligator/crocodile) [Maekawa and Kasahara 83] starts by dividing the paper side in thirds, and then works on the 22.5° grid. This situation commonly results from the origami design process (e.g., grafting or adjusting flap lengths), because it effectively adjusts the size of the square of paper. Our theory can capture these situations simply by viewing the square as having side length $m' + n'\sqrt{2}$ instead of 1, that is, by scaling all coordinates by a factor of $m' + n'\sqrt{2}$, thereby placing the constructed points back on the 22.5° grid.

An obvious direction for future research is characterizing $90°/n$ grid systems for $n \neq 4$. In particular, modern origami design has explored the 15° grid system lately. For results in this direction, see the follow-up paper [Butler et al. 11].

Bibliography

[Butler et al. 11] Steve Butler, Erik D. Demaine, Ron Graham, and Tomohiro Tachi. "Constructing Points through Folding and Intersection." Manuscript, 2011.

[Cormen et al. 09] Thomas H. Cormen, Charles E. Leiserson, Ronald L. Rivest, and Clifford Stein. *Introduction to Algorithms*, Third edition. Cambridge, MA: MIT Press, 2009.

[Demaine and O'Rourke 07] Erik D. Demaine and Joseph O'Rourke. *Geometric Folding Algorithms: Linkages, Origami, Polyhedra.* Cambridge, UK: Cambridge University Press, 2007.

[Huzita and Scimemi 91] Humiaki Huzita and Benedetto Scimemi. "The Algebra of Paper Folding (Origami)." In *Proceedings of the First International Meeting of Origami Science and Technology*, edited by H. Huzita, pp. 215–222. Padova, Italy: Dipartimento di Fisica dell'Università di Padova, 1991.

[Justin 91] Jacques Justin. "Resolution par le Plaige de L'equation du Troisieme Degre et Applications Geometriques." In *Proceedings of the First International Meeting of Origami Science and Technology*, edited by H. Huzita, pp. 251–261. Padova, Italy: Dipartimento di Fisica dell'Università di Padova, 1991.

[Komatsu 06] Hideo Komatsu. "Wolf." *Origami Tanteidan Magazine* 17:3 (2006), 22–32.

[Maekawa and Kasahara 83] Jun Maekawa and K. Kasahara. *Viva Origami!* (in Japanese). Tokyo: Sanrio Publications, 1983.

[Maekawa 07] Jun Maekawa. *Genuine Origami.* Tokyo: Nichibo Shuppansha, 2007. English version published in 2008.

Two Folding Constructions

Robert Orndorff

1 Introduction

This paper describes two origami constructions: (a) given an arbitrarily large piece of paper and two adjacent collinear segments of unit length and length n, construct a segment of length $\sqrt{n}$; and (b) given a unit square piece of paper, construct the square roots of the reciprocals of the integers. Although extremely straightforward, perhaps these constructions will be of interest to readers. They can be seen as variations, or "distant cousins," of extant folding construction methods. For example, one makes use of a folded flap for a benchmark, but it does so in a new manner. The first construction is based on Descartes' method for constructing square roots. Of the two, mine is perhaps the more useful because it is iterative, concise, and self-contained and begins with a unit square piece of paper. The square roots of all of the reciprocals of the integers can be found with this construction alone.

A list of pertinent folding construction methods and references would include the following: binary folding algorithm [Lang 10, p. 6]; binary approximation [Brunton 73, p. 26; Lang 10, p. 6]; error-halving approximations [Brunton 73, p. 26; Fujimoto and Nishiwaki 82, p. 170; Hilton and Pederson 83; Hull 06, p. 15; Veenstra 09]; crossing-diagonal methods [Fujimoto and Nishiwaki 82, p. 170; Lang 10, p. 13]; Fujimoto-Mosely method [Fujimoto and Nishiwaki 82, p. 170; Lang 10, p. 17]; Noma method [Lang 10, p. 19]; continued fractions [Lang 10, p. 24]; equilateral triangle variation [Fujimoto and Nishiwaki 82, p. 170]; the Haga theorems, note-worthy here for their use of a folded flap [Haga 08, pp. 1–32; Hull 06, p. 75; Lang 10, p. 21]; and traditional folds [Lang 10, p. 28; Row 01]. Several of these methods resemble in some aspects the constructions described in this paper. Only one construction is directly related: Row's right triangle construction [Row 01, p. 23].

2   Method A: Descartes' Construction

Descartes described a square root construction in the first book of *Geometria* [Descartes 83, p. 8] (Figure 1).

As Descartes said: "In order to calculate the square root of segment GH, first add a segment FG of length *one* to GH. Bisect segment FH to locate point K. Draw a circle of radius KH at center point K. Raise a perpendicular from point G to the circle. Segment GI is the square root."

Descartes' method is a special case of Euclid's *Elements*, Book VI, Proposition 13, of which he was no doubt aware: "To find a mean proportional (i.e., a geometric mean) between two given straight lines" [Playfair 45, p. 131]. That is, if FG and GH (Figure 1) are two given line segments (neither need be of unit length here), then GI is their geometric mean. And, as Descartes noted, the geometric mean of one and any number is that number's square root.

Descartes' construction can be readily folded (Figure 2). The method is obvious but perhaps worthy of mention. Given an arbitrarily large piece of paper and two adjacent collinear line segments FG of length 1 and GH of length n, construct a segment of length $\sqrt{n}$. Crease line FH onto itself such that the crease line goes through point G, constructing a perpendicular through point G. Crease line FH onto itself such that the crease line passes through point F, constructing a perpendicular through point F. Crease point F to corner point H, making a crease mark at point K, bisecting

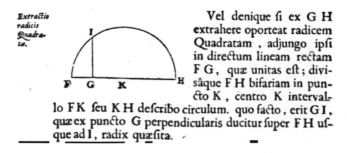

Figure 1. The square root construction of René Descartes.

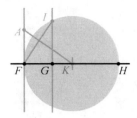

Figure 2. Folding the square root construction of René Descartes.

segment FH. Crease point F to perpendicular G such that the new crease line passes through point K, constructing the segment AK and point A. Crease segment AK onto itself such that the new crease line passes through point F, constructing new segment FI. Segment GI will be of length $\sqrt{n}$.

3 Method B: My Construction

With my construction, one begins with a rectangle of proportions $1 : 1/\sqrt{n}$ and constructs a rectangle of proportions $1 : 1/\sqrt{(n+1)}$. A $1 : 1/\sqrt{1}$ rectangle (i.e., a square) is thus used to construct a $1 : 1/\sqrt{2}$ rectangle, the $1 : 1/\sqrt{2}$ rectangle is used to construct a $1 : 1/\sqrt{3}$ rectangle, and so on (Figures 3 and 4.)

That this is valid can be shown in several ways. For example, in Figure 5, we assign segment lengths $BC = 1$ and $AB = 1/\sqrt{n}$. The bottom edge BC of rectangle $ABCD$ has been folded to diagonal crease AC, making crease line CE and flap CEF. What is the length of segment CG?

Because segments AD, BC, and FG are parallel, $\triangle ABC \sim \triangle CGF$. If edge $BC = 1$, then the projection of the same edge onto the paper underneath, segment FC, has the same length. Using the Pythagorean

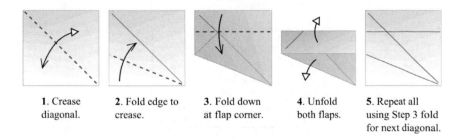

| 1. Crease diagonal. | 2. Fold edge to crease. | 3. Fold down at flap corner. | 4. Unfold both flaps. | 5. Repeat all using Step 3 fold for next diagonal. |

Figure 3. Instructions for folding the first iteration of my construction.

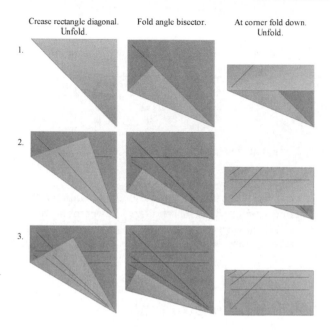

Crease rectangle diagonal.   Fold angle bisector.   At corner fold down.
Unfold.   Unfold.

Figure 4. Three iterations of the construction.

theorem,

$$AC = \sqrt{\frac{n+1}{n}}. \tag{1}$$

Also, the ratios of corresponding sides to hypotenuses must be equal:

$$\frac{CG}{1} = \frac{1/\sqrt{n}}{\sqrt{(n+1)/n}} = \frac{1}{\sqrt{n+1}}. \tag{2}$$

Perhaps, finally, two related items are worth mentioning. Euclid's *Elements*, Book I, Proposition 43 could be used to fold a segment of reciprocal length (Figure 6): "The complements of the parallelograms, which are

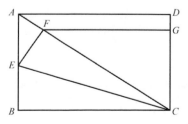

Figure 5. A solution using similar triangles.

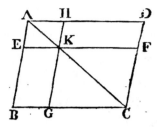

Figure 6. Euclid's *Elements*, Book I, Proposition 43.

about the diameter of any parallelogram, are equal to one another" [Playfair 45, p. 42; Kuzmarsky 10]. In other words, because $\triangle ABC$ is congruent to $\triangle ADC$, $\triangle AEK$ is congruent to $\triangle AHK$, and $\triangle KGC$ is congruent to $\triangle KFC$, then area $EKGB$ (i.e., areas $\triangle ABC - \triangle AEK - \triangle KGC$) is equal to area $HDFK$ (i.e., areas $\triangle ADC - \triangle AHK - \triangle KFC$). If we specialize to $EK = BE = 1$ and $HK = 1/\sqrt{n}$, then $KF = \sqrt{n}$. To fold this, place the given unit square $LMKE$ on an arbitrarily large piece of paper (Figure 7).

There is a straightforward iterative Euclidean construction for the square roots of the integers (Figure 8), which could be readily folded, given an arbitrarily large sheet of paper and a segment of unit length [Stein 10]. Draw a line segment AB and assign it unit length. At point B construct a perpendicular line to that segment. Draw a circle of radius AB and center B. Call the intersection of the perpendicular line and the circle C. Draw a line segment connecting points A and C. This segment has length $\sqrt{2}$. Repeat with segment AC (drawing circle of radius BC and center C) to obtain a segment of length $\sqrt{3}$ (AD), and so on.

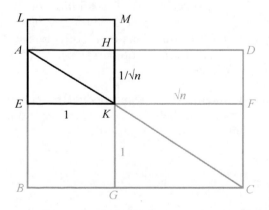

Figure 7. Folding $\sqrt{n}$ from $1/\sqrt{n}$.

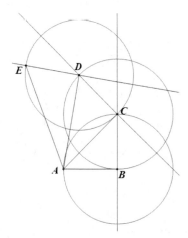

Figure 8. Three iterations of a similar Euclidean construction.

4   Conclusion

A folding method is described based on Descartes' construction of a seg-ment of length $\sqrt{n}$. It is noted that Descartes' method is a special case of Euclid's construction of the geometric mean. A method is described for folding segments the lengths of which are the square roots of the reciprocals of the integers. Finally, it is noted that one can readily (a) fold a segment of reciprocal length, following Euclid; or (b) iteratively fold segments whose lengths are the square roots of the integers, following a straightforward Euclidean construction.

It is worthwhile to ponder origami constructions in light of the rich his-tory of geometric construction. The historical examples described here are readily replicated with paper folding. The folding construction described here for the square roots of the reciprocals of the integers is simple, concise, self-contained, and iterative. I have made use of it in the design of several origami applications (manuscript in preparation). Such constructions are also pleasing in and of themselves.

Acknowledgments. I would like to acknowledge the contributions of Bob Stein, Fred Kuczmarski, and the anonymous reviewers. Without their help, this paper would have been deficient indeed.

Bibliography

[Brunton 73] James Brunton. "Mathematical Exercises in Paper Folding: I." *Mathematics in School* 2:4 (July 1973), 25–26.

[Descartes 83] René Descartes. *Geometria, à Renato des Cartes*. Amsterdam: 1683.

[Fujimoto and Nishiwaki 82] Shuzo Fujimoto and Masami Nishiwaki. *Sōzu suru origami asobi he no shōtai* (in Japanese). Osaka: Asahi Cultural Center, 1982.

[Haga 08] Kazuo Haga. *Origamics*. Singapore: World Scientific, 2008.

[Hilton and Pedersen 83] Peter Hilton and Jean Pedersen. "Approximating Any Regular Polygon by Folding Paper." *Mathematics Magazine* 56.3 (May 1983), 141–155.

[Hilton et al. 97] Peter Hilton, Derek Holton, and Jean Pedersen. *Mathematical Reflections*. New York: Springer, 1997.

[Hull 06] Thomas Hull. *Project Origami: Activities for Exploring Mathematics*. Wellesley, MA: A K Peters, 2006.

[Kuczmarski 10] Fred Kuczmarski. Private communication, November 27, 2010.

[Lang 03] Robert J. Lang. "Origami and Geometric Constructions." Available at http://www.langorigami.com/science/hha/origami_constructions.pdf, 2003.

[Playfair 45] John Playfair. *Elements of Geometry: Containing The First Six Books of Euclid*. New York: W. E. Dean, 1845.

[Row 01] T. Sundara Row. *Geometric Exercises in Paper Folding*. Chicago: The Open Court Publishing Company, 1901.

[Stein 10] Bob Stein. Private communication, November 9, 2010.

[Veenstra 09] Tamara B. Veenstra. "Fujimoto, Number Theory and A New Folding Technique?" In *Origami⁴: Fourth International Meeting of Origami Science, Mathematics, and Education*, edited by Robert J. Lang, pp. 405–416. Wellesley, MA: A K Peters, 2009.

Variations on a Theorem of Haga

Emma Frigerio

1 Introduction

Many authors have studied the problem of dividing the side of a square or rectangular piece of paper into equal parts via origami methods. Some methods are specific for one ratio only, while others, such as the Fujimoto approximation, give more general answers (see [Kasahara 88, Fujimoto and Nishiwaki 82] and, for a survey of different subdivision methods, [Lang 10]).

In its original form, Haga's theorem provides a simple and elegant way for exact division of the side of a square into thirds; its proof relies upon the Pythagorean Theorem, triangle similarity, and first-degree equations, which makes it accessible to middle school students (for an English version, see [Haga 02, Haga et al. 08, Hull 06]). The obvious generalization of Haga's theorem to other exact divisions needs only an additional (yet easy) algebraic manipulation.

I present here two algorithms for dividing the side of a square into N equal parts, both based on Haga's theorem and strongly connected with the base-2 representation of N, then compare them, discussing their pros and cons. The second method suggests a third way for equal subdivision, which works for rectangles as well. Finally, for the reader's convenience, I provide Abe's construction of a square whose area is $(1/N)$th that of a given square [Abe 03].

I have developed most of this material independently over a ten-year span, although by now parts of it have been published by other researchers.

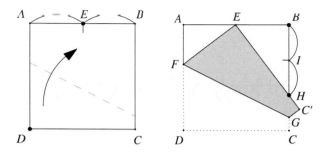

Figure 1. Going from $1/2$ to $1/3$.

2 Haga's Theorem

Assume the side of the square has length 1. As illustrated in Figure 1 (left), we first pinch the midpoint E of top edge AB, then fold D onto E. If we now pinch the midpoint I of BH (Figure 1, right), we get $BI = 1/3$. In other words, with one fold and one pinch, we pass from $1/2$ on the top side to $1/3$ on the right side (next side, for short). In his work, Haga goes on to observe that, if we start with $AE = 1/4$ and perform the same construction, then we get $BI = 1/5$, whereas if $AE = 3/4$, then $BI = 3/7$.

For a geometric proof of these results, set

$$AE = k, \quad AF = x, \quad \text{and} \quad BH = y,$$

so that

$$EB = 1 - k, \quad FE = FD = 1 - x, \quad \text{and} \quad BI = \frac{y}{2}.$$

The Pythagorean theorem for triangle AEF yields

$$x = \frac{1 - k^2}{2}.$$

From similarity between triangles AEF and BHE we get

$$x \div k = (1 - k) \div y,$$

so we obtain the main equation,

$$\frac{y}{2} = \frac{k(1 - k)}{2 \cdot x} = \frac{k}{k + 1}. \tag{$*$}$$

In particular, if we set $k = 1/2$, $1/4$, and $3/4$, then we find that $k/(k+1) = 1/3$, $1/5$, and $3/7$, respectively.

Students more used to analytic geometry than to Euclidean geometry might prefer to work out an analytic proof. This task, however, may be challenging for most of them.

3 Variation 1

In the same vein, we find that

$$k = \frac{1}{N} \Rightarrow \frac{k}{k+1} = \frac{1}{N+1};$$

hence, from the first N-subdivision point on one side, Haga's theorem (HT) constructs the first $(N+1)$-subdivision point on the next side.

This is already an algorithm for finding arbitrary divisions, albeit a "stupid" one: it would not make any sense to use it several times in order to locate the first eight-division point! So, how could we use this procedure efficiently for any natural number N?

In order to understand what is happening, let us look first at an example, say $N = 13$. Working backwards, we could obtain $1/13$ from $1/12$ by HT, $1/12$ from $1/6$ by bisection, $1/6$ from $1/3$ by bisection, $1/3$ from $1/2$ by HT, $1/2$ from 1 by bisection. Hence, working forwards, we construct

(a) $1/2$ on side AB by bisection,

(b) $1/3$ on the next side BC by HT,

(c) $1/6$ and $1/12$ on side BC by repeated bisection,

(d) $1/13$ on the next side CD by HT.

This procedure is illustrated in Figure 2.

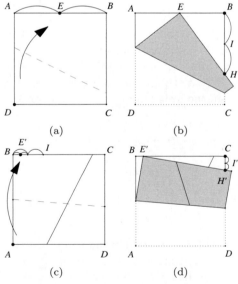

Figure 2. Construction of $1/13$.

Let b denote the operation of bisecting the left part of the top side of the square, and let h denote the application of HT as in Figure 1. Then, for $N = 13$, the above procedure is encoded by the word

$$w_{13} = b\, h\, b\, b\, h.$$

We do have to keep in mind that, in performing Haga's construction, first we fold the bottom-left vertex to the current point; second we bisect the top part of the next side; finally we open up the paper and rotate the square counterclockwise by $90°$, so that the next side becomes the new top side.

In increasing order, the sequence $s_{13} = \{a_i\}$ of denominators in the fractions above, is

$$s_{13} = \{1, 2, 3, 6, 12, 13\}.$$

Observe that we pass from a_i to a_{i+1} either by multiplying by 2 or by adding 1; hence,

$$13 = (1 \cdot 2 + 1) \cdot 2^2 + 1.$$

But this is precisely the evaluation for $x = 2$ of the polynomial

$$p_{13}(x) = x^3 + x^2 + 1,$$

whose coefficients are the digits in the base-2 representation of 13, via the Horner algorithm [Knuth 97]. In other words, as we might expect, the word w_{13} encoding the instructions for folding the fraction $1/13$ is related to $(1101)_2$, the base-2 representation of 13; moreover, the efficiency of the Horner algorithm in evaluating $p_{13}(x)$ translates into the efficiency of the procedure for folding $1/13$.

Students should be encouraged to experiment with different values of N to discover how to determine w_N just from the base-2 representation $(d_k d_{k-1} \cdots d_1 d_0)_2$ of N, without going through the sequence s_N. Possibly after some hints from the instructor, they should come up with the correct statement:

Theorem 1. *For any natural number $N \geq 2$, write N in base 2 as*

$$(d_k d_{k-1} \cdots d_1 d_0)_2.$$

Then the word w_N is obtained from $d_{k-1} \cdots d_1 d_0$ by replacing d_i with b if $d_i = 0$, and with bh if $d_i = 1$.

Proof: Here is an easy proof by induction on N. The base-2 representation of 2 is $(10)_2$ and $w_2 = b$; thus, the claim is true for $N = 2$. Now suppose the claim is true for all natural numbers less than N. If N is even, then $N = 2 \cdot M$ and $w_N = w_M b$. If $(d_{k-1} \cdots d_1 d_0)_2$ is the base-2 representation of M, then the base-2 representation of N is $(d_{k-1} \cdots d_1 d_0 0)_2$, so the claim is true also for N. If N is odd, then $N - 1 = 2 \cdot M$ is even and $w_N = w_{N-1} h = w_M bh$. If $(d_{k-1} \cdots d_1 d_0)_2$ is the base-2 representation of M, then the base-2 representations of $N - 1$ and of N are $(d_{k-1} \cdots d_1 d_0 0)_2$ and $(d_{k-1} \cdots d_1 d_0 1)_2$, respectively, so the claim is true in this case as well. $\square$

Hull hints at the idea behind the above algorithm [Hull 06, Activity 3]; likewise, it is quite feasible that others have thought of it and/or used Haga's theorem in its implementation. What is new, I believe, is its explicit connection with the base-2 representation of N expressed by Theorem 1.

4 Variation 2

Pinching midpoints and folding perpendicular bisectors are very easy origami operations; however, while pinching leaves the paper almost unmarked, perpendicular bisectors do leave creases on the paper. For instance, if $N = 15$, we would end up with three creases on the paper (see Figure 3).

Can we use HT in a different way, so as to construct just one perpendicular bisector? Going back to Equation ($*$), we can easily check that

$$k = \frac{m}{n} \Rightarrow \frac{k}{k+1} = \frac{m}{n+m}.$$

The easiest fractions to fold are the so-called *binary fractions*, fractions whose denominator n is a power of 2; this observation leads to the following construction, which, as far as I know, is new:

- Write N as $N = 2^k + m$, $0 \le m < 2^k$.

- Let E be the mth 2^k-subdivision point on the top side AB of the square (i.e., the point such that $AE = m/2^k$).

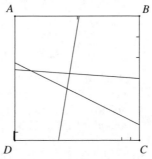

Figure 3. Construction lines for $N = 15$.

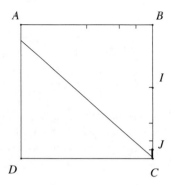

Figure 4. Using HT only once to construct $1/15$.

- Use HT and get point I on the next side BC. Thus I is the mth N-subdivision point on the "new" top side BC, which implies that

$$IC = \frac{2^k}{N}.$$

- With k repeated bisections of the right part of side BC, obtain point J, which is the last N-subdivision point on it.

Figure 4 illustrates the above construction for $N = 15$; in this case $15 = 2^3 + 7$, and hence,

$$AE = \frac{7}{8}, \quad IC = \frac{8}{15}, \quad JC = \frac{1}{15}.$$

At this point, students should be challenged to locate point E on line AB with as few pinches as possible, each of them bisecting either the left or the right part of the side. It turns out that k pinches suffice; more precisely, we need exactly k pinches if and only if m is odd (if m is even, so is N; thus m/N is not a reduced fraction). Again, this is related to the base-2 representation of N, as the following theorem shows.

Theorem 2. *For any odd number $N \geq 3$, write N in base 2 as*

$$(d_k d_{k-1} \cdots d_1 d_0)_2.$$

Let b and b' denote bisection of the left and the right parts, respectively, of a segment AB. For $i = 0, 1, \ldots, k - 1$, write b if $d_i = 0$, b' if $d_i = 1$, so as to obtain a k-letter word v_N. Then v_N encodes the instructions for locating the mth 2^k-subdivision point E on AB.

The proof is left to the reader: it can be done either by induction, as in Theorem 1, or following the method used by Veenstra [Veenstra 09,

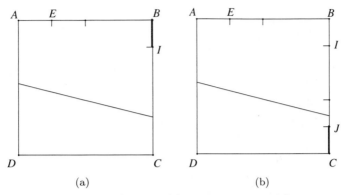

Figure 5. Constructing $1/5$ using (a) Variation 1 and (b) Variation 2.

Theorem 2]. Recall that $(0.d_{k-1} \ldots d_1 d_0)_2$ is the base-2 representation of $m/2^k$, since $(d_{k-1} \cdots d_1 d_0)_2$ is the base-2 representation of m as a k-digit number; see also [Lang 10] about binary fractions.

Finally, observe that, if $N = 2^k + 1$, the two constructions are essentially the same, as Figure 5 shows for $N = 5$, the only difference being that, with the former we get the first N-subdivision point I, whereas with the latter we obtain the last one J.

5   A Comparison

Although Variation 2 has been introduced for origami reasons, it does have two drawbacks, both of folding nature, which, unfortunately, occur together whenever m is "close enough" to 2^k:

- The two edges whose intersection is point H may form a very small angle, which makes it difficult to locate it with enough accuracy.

- Even assuming H has been accurately determined, pinching point I is awkward whenever $BH > 2 \cdot AF$.

Thus, in the end, the paper will have only one unwanted crease across, but the output may be much less accurate than when using Variation 1, where the edges determining H always form a big angle and pinching I is always easy.

This fact may lead students to an interesting discussion about the differences between theory and practice: both methods are theoretically correct, but, in the real world, results may be quite different.

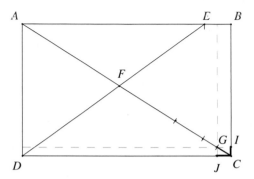

Figure 6. Constructing $1/15$ on the sides of any rectangle.

6   Variations on the Variations

This section contains two detours from Haga's theorem; in a way, we lose the main theme, but they are of some interest in spite of not being entirely new.

6.1   Rectangles

Haga presents some generalizations of HT to the silver rectangle [Haga 02]. For a more general approach, we can use the algebra in Variation 2 in a completely different subdivision method, which works for any rectangle and for any N. As before, write N as $N = 2^k + m$ and let E be the mth 2^k-subdivision point on side AB of rectangle $ABCD$. Now fold AC and DE (see Figure 6 for the case $N = 15$).

The intersection point F of these two lines is such that $AF/FC = m/2^k$, so that F is the mth N-subdivision point on AC. With a purely geometric approach, this statement can be proved by marking all mth 2^k-subdivision points on sides AB and CD; then the lines through them parallel to DE divide AC into N equal parts (otherwise, it can be obtained via the crossing-diagonals technique [Lang 10]). Next, by k repeated bisections of FC, we obtain point G, which is the last N-subdivision point on diagonal AC. Finally, from this point, we easily get the last N-subdivision points on sides BC and DC by orthogonal projection.

If applied to a square, in practice this easy geometric construction may give more precise results than Variation 2, particularly if one does not mind folding segments through two given points.

6.2   From the Strip to the Square

Let us now go back to the square and to middle school geometry. Once we have determined, by any method, $(1/N)$th of its side, we can immediately

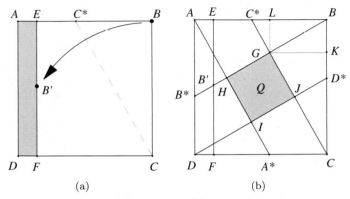

Figure 7. The shaded (a) strip and (b) square have the same area.

get a rectangular strip whose area is $(1/N)$th that of the square. But how can we construct a square with the same area? Abe presents this beautiful construction [Abe 03], which, at least in the Western world, has gone almost unnoticed, so I think it worthwhile to report it here.

As before, assume that $AB = 1$ and $AE = 1/N$. With reference to Figure 7(a), fold EF perpendicular to AB and CC^* so that B goes onto EF, reopening after each crease. Then fold BB^* perpendicular to CC^*, AA^* perpendicular to BB^* and DD^* perpendicular to AA^*, reopening after each crease (see Figure 7(b)). These lines determine square $GHIJ$, whose area is $1/N$.

In fact, since G is the midpoint of BB', L is the midpoint of EB; hence,

$$GK = LB = \frac{N-1}{2 \cdot N}.$$

Thus, the area of triangle BCG is

$$\frac{BC \cdot GK}{2} = \frac{N-1}{4 \cdot N}.$$

Square $GHIJ$ is obtained from square $ABCD$ by taking away four triangles congruent to BCG, hence its area is

$$1^2 - 4 \cdot \frac{N-1}{4 \cdot N} = 1 - \frac{N-1}{N} = \frac{1}{N}.$$

7 Conclusion

As a student, I always enjoyed subjects combining different techniques, ideas, and mathematical areas; as a teacher, I appreciate them even more:

students can use what suits them best and recognize the unity of mathematics.

I believe that Haga's theorem and its variations constitute a valuable topic for a hands-on mathematical activity, at different levels, from middle school to college, and in various courses: geometry, precalculus, abstract algebra, discrete mathematics, number theory, and introduction to proof, to name some of them.

Teachers can use different parts of this paper's content to design appropriate activities for their students. In the Italian school system, for instance, Haga's theorem (say, for $k = 1/2$, $1/4$, $1/6$) and Abe's method could be done toward the end of middle school. In high school, students could also work out the general case and discover the algorithm in Variation 1, possibly without a complete formalization of it. At a higher level, its connection with the base-2 representation of N should be emphasized in an introductory course in programming, so as to make the Horner algorithm more concrete. An abstract algebra course would be a good place for the most theoretical parts of this paper; on the contrary, the comparisons among different methods, which exemplify how data can be well- or ill-conditioned, depending on the algorithm used in solving a problem, would be appropriate in a course with more emphasis on applied mathematics. Finally, a discussion on the whole content would give prospective high school mathematics teachers a wealth of ideas about designing discovery-based activities.

Bibliography

[Abe 03] H. Abe. *Sugoizo Origami (Cool Origami)*. Tokyo: Nihon Hyoron-sha, 2003.

[Fujimoto and Nishiwaki 82] S. Fujimoto and M. Nishiwaki. *Seizo suru origami asobi no shotai (Creative Invitation to Playing with Origami)*. Tokyo: Asahi Culture Center, 1982.

[Haga 02] K. Haga. "Fold Paper and Enjoy Math: Origamics." In *Origami³: Proceedings of the Third International Meeting of Origami Science, Mathematics, and Education*, edited by T. Hull, pp. 307–328. Natick, MA: A K Peters, 2002.

[Haga et al. 08] K. Haga, J. C. Fonacier, and M. Isoda. *Origamics: Mathematical Explorations through Paper Folding*. Singapore: World Scientific Publishing Co., 2008.

[Hull 06] T. Hull. *Project Origami: Activities for Exploring Mathematics*. Natick, MA: A K Peters, 2006.

[Kasahara 88] K. Kasahara. *Origami Omnibus*. Tokyo: Japan Publications, 1988.

[Knuth 97] D. Knuth. *The Art of Computer Programming, Volume 2: Seminumerical Algorithms*, Third Edition. Reading, MA: Addison-Wesley, 1997.

[Lang 10] R. J. Lang. *Origami and Geometric Constructions*. Available at www.langorigami.com/science/hha/origami_constructions.pdf, 2010.

[Veenstra 09] T. B. Veenstra. "Fujimoto, Number Theory, and a New Folding Technique." In *Origami⁴: Fourth International Meeting of Origami Science, Mathematics, and Education*, edited by Robert J. Lang, pp. 405–416. Wellesley, MA: A K Peters, 2009.

Precise Division of Rectangular Paper into an Odd Number of Equal Parts without Tools: An Origamics Exercise

Kazuo Haga

1 Introduction

Since my first finding on scientific origami known as Haga's theorem was published by Koji Fushimi in 1979 [Fushimi 79], one of the study topics in Origamics has been how to divide a sheet into three or more odd-numbered equal divisions by bare-hand folding (i.e., with no measuring device). I already found methods of tri-, five- and seven-sections as results of the expansion of the theorem, but those divisions were incidental and fragmentary and could not be generalized [Haga 99, Haga 02, Haga 05].

In the last seven years, I have found a method to fold lines on any rectangular sheet of paper, dividing it into an odd number of equal parts, both horizontally and vertically, without any tools [Haga 04, Haga 05, Haga 08]. The dividing method starts by making a node, which is defined by the intersection of two straight lines. One of them is a constant line from the midpoint of the left side to the right upper corner of the sheet. This line is common and suitable for 3-, 5-, 7-, 9-, 11-, 13-, 15-, and 17-section methods; so, going forward, it will be called the *common line* in this paper. The second line starts from the midpoint of the upper side, which is suitable for every odd-numbered division, and ends at a designated point on the bottom side according to the particular odd number. It is called the *individual line* in this paper. Thus, our investigation of such nodes corresponding to each odd-numbered division on the common line will focus on determining the terminal point of the individual line on the bottom of the rectangular paper.

If a coordinate system is set up on the rectangular sheet with the bottom left corner as the origin, these two straight lines can be written as two linear equations, and the solution of these simultaneous equations clearly and precisely defines the desired node.

2   Preparation

In this study I often use industrial standard rectangular sheets of paper for which the ratio of length to width is $\sqrt{2}$: 1; this is the side ratio of the popular A4 size. But we can use rectangular paper of any ratio, which I now express to generalize the ratio of the adjacent sides as 1 : L (with $L \geq 1$).

Before dividing, we must make two midpoint marks as tiny folds along each of the upper and left sides of the rectangle, as shown in Figure 1. To work through all of the examples in this work, it is recommended that a number of sheets of rectangular paper of the same size be obtained and that those marks be made on every sheet. The common line (ED) will be necessary in all cases. We can make the line as a full crease from E to D by folding the paper as in Figure 1, but for practical purposes, it is better that the crease be made only in the vicinity of the desired intersection to avoid obstacles when we fold further dividing lines later.

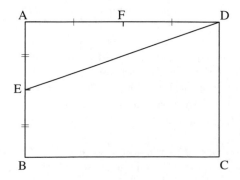

Figure 1. Mark the midpoint as a tiny crease both on the upper side and the left side, and create a line ED (the common line) as a crease.

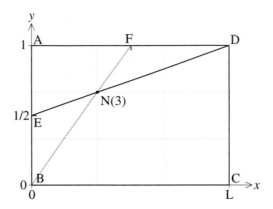

Figure 2. Draw a line FB as the individual line to obtain a node for trisection N(3). From this, the rectangle may be divided into three equal parts horizontally and vertically.

3 Trisection

The node for trisection is obtained as the intersection point on the rectangle between (a part of) the common line and (a part of) an individual line, which joins the upper side midpoint (F) to the left corner of the bottom side (B), as shown in Figure 2.

Assuming the rectangle width (AB) is 1 and length (AD = BC) is L, the coordinates of the points are as shown in Figure 2, and the equations of the lines are the following:

$$\text{Line ED (common line):} \quad y = (x + L)/(2L), \qquad (1)$$
$$\text{Line FB (individual line):} \quad y = 2x/L. \qquad (2)$$

Setting Equations (1) and (2) to be equal, we obtain the coordinates of the lines' intersection, which is the node N(3),

$$x = L/3; \quad \text{therefore,} \quad y = 2/3.$$

These coordinates show that x is $1/3$ of length L and y is $2/3$ of length 1, and the ratio of the parts into which the paper is divided is $1 : 2$ horizontally from left to right and $2 : 1$ vertically from bottom to top.

For the sake of brevity we shall call the intersection point a *#3 node (N)* for dividing the rectangle into three equal parts both horizontally and vertically as in Figure 2.

We can make the trisection folds with the following steps:

1. Align CD onto #3 node (N) and make a fold parallel to AB. Points C and D should be aligned to the sides of the rectangle.

2. Align AB onto the newly made crease from step 1 and make a fold. Now the rectangle is divided into three equal parts horizontally.

3. Align BC onto #3 node (N) and make a fold parallel to AD. Points B and C should be aligned to the left and right sides, respectively.

4. Align AD on the crease made by step 3 and make a fold.

Steps 3 and 4 divide the rectangle vertically into three equal parts, and as a result, the rectangle is divided into three equal parts horizontally and vertically.

4   Five-Section

The node for five-section (which we call #5 node, or N(5)) is obtained as the intersection point on the base rectangle with another line joining the upper side midpoint (F) with the lower right corner (C) on the opposite side, as illustrated in Figure 3.

On the 1 : L rectangle, the coordinates of the #5 node are as shown in Figure 3, and the equations of the two lines are as follows:

$$\text{Line ED (common line):} \quad y = (x + L)/(2L), \tag{3}$$
$$\text{Line FC (individual line):} \quad y = -2x/L + 2. \tag{4}$$

Again, setting the two equations equal, we obtain the coordinates of the

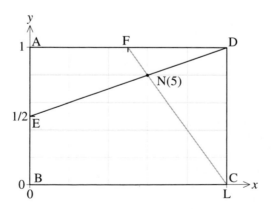

Figure 3. Draw a line FC on the rectangle of Figure 1 to obtain a node for a five-section. From this point, the rectangle may be divided into five equal parts horizontally and vertically.

node (N),

$$x = 3L/5; \quad \text{therefore,} \quad y - 4/5.$$

That is, x is 3/5 of length L, and y is 4/5 of length 1. By using the #5 node, the paper can be divided in the ratio 3:2 in the horizontal direction (left to right) and 4 : 1 in the vertical direction (bottom to top), as in Figure 3.

The procedure for a full five-section using the #5 node is the following:

1. Align CD on the #5 node (N) and make a fold parallel to AB. Points C and D should be on the sides of the rectangle.

2. Align AB on the newly made crease from step 1 and make a fold.

3. Align AB on the crease made by step 2 and make a fold, then unfold.

4. Align CD on the newly made crease from step 4 and make a fold. The rectangle is now divided into five equal parts horizontally.

5. Align BC on the #5 node (N) and make a fold parallel to AD. Points B and C should be on the left and right sides, respectively.

6. Align the bottom line of the folded part made by step 5 on the #5 node once more, and unfold.

7. Align AD on the nearest crease made by the previous step.

As a result of these steps, the rectangle is divided into five equal parts horizontally and vertically.

5 Seven-Section

The node for seven-section (#7 node) may be obtained as the intersection point on the base rectangle using an individual line that joins the upper side midpoint (F) with a point (G) situated on the bottom side, one-fourth of the way along the bottom from the left vertex, as shown in Figure 4.

On the 1 : L rectangle, the coordinates of the #7 node are as shown in Figure 4, and the equations of the two lines are as follows:

$$\text{Line ED (common line):} \quad y = (x + L)/(2L), \quad (5)$$
$$\text{Line FG (individual line):} \quad y = 4x/L - 1. \quad (6)$$

Setting the two equations equal, we obtain the coordinates of the node (N),

$$x = 3L/7; \quad \text{therefore,} \quad y = 5/7.$$

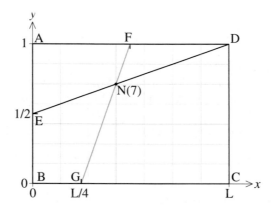

Figure 4. Draw a line from point F on the rectangle to a point on the bottom side one-fourth from corner B to get a node for seven-section.

That is, x is 3/7 of length BC, and y is 5/7 of length AB. This means that by using the #7 node, the paper can be divided in the ratio 3 : 4 in the horizontal direction (left to right) and 5 : 2 in the vertical direction (bottom to top), as in Figure 4.

The seven-section procedure using the #7 node is the following:

1. Align CD on #7 node (N) and make a fold parallel to AB. Points C and D should be on the sides of the rectangle.

2. Align AB on the newly made crease from step 1 and make a fold.

3. Align AB on the crease made by step 2 and make a fold, then unfold.

4. Align CD on the newly made crease from step 4 and make a fold. The rectangle is now divided into seven equal parts horizontally.

5. Align BC on #7 node (N) and make a fold parallel to AD. Points B and C should be on the left and right sides, respectively.

6. Align the bottom line of the folded part made in step 5 on the #7 node once more, and unfold.

7. Align AD on the nearest crease made by the previous step.

As a result of these steps, the rectangle is divided into seven equal parts horizontally and vertically.

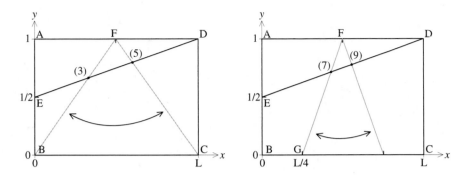

Figure 5. The individual line for trisection is swung like a pendulum to give the individual line for five-section (left). The individual line for seven-section is swung like a pendulum to give the line for nine-section (right).

6   Pendulum Symmetry

It is interesting to note that the five-section point, #5 node, may be obtained by a slight change in the procedure from that of the trisection. Imagine swinging the individual line FB to the right for the same distance from the perpendicular, as if it were a pendulum suspended from point F (Figure 5). The terminal point B of the line will fall on point C of the rectangle. The line FB makes the #3 node as an intersection with the common line ED, and the line FC makes the #5 node as an intersection with the same common line (Figure 5, top). These two individual lines (FB and FC) are similar. Their equations have the same magnitude slope, but their signs are opposite.

I have found that a similar relationship exists not only between the #3 and #5 individual lines (Figure 5, left) but also between those of #7 and #9 (Figure 5, right), between those of #11 and #13, between those of #15 and #17, and so on for all higher adjacent odd-number pairs. Therefore, when we determine a new terminal point of any individual line on the bottom side of the rectangle paper, we can easily get another point for the next odd-numbered pair.

7   Nine-Section

Let us try out this pendulum symmetry to find the #9 node. The #7 node is obtained as the intersection point between the common line and the individual line FG, which starts at the midpoint of the upper side and drops to a point one-fourth the length of the rectangle from its left corner along the bottom edge. If we swing the line FG to the right the same

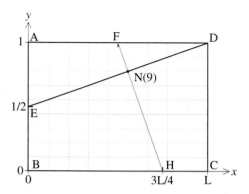

Figure 6. Draw a line from point F on the rectangle to a point on the bottom side one-fourth of the way from the right corner C to get a node for nine-section. From this node, the rectangle may be divided into nine equal parts horizontally and vertically.

distance, like a pendulum, the line drops at the same distance point from the right corner, as in Figure 6. This will be the individual line for nine-section, and so we can locate the #9 node on the common line. This node divides the length of the rectangle in proportions 5 : 4 in the horizontal direction, and divides the height in proportion 7 : 2 in the vertical direction.

On the 1 : L rectangle, the coordinates of the #9 node are as shown in Figure 6, and the equations of the lines are as follows:

$$\text{Line ED (common line):} \quad y = (x + L)/(2L), \qquad (7)$$
$$\text{Line FH (individual line):} \quad y = -4x/L + 3. \qquad (8)$$

Setting the two equations equal, we obtain the coordinates of the node (N),

$$x = 5L/9; \quad \text{therefore,} \quad y = 5/9.$$

Procedures to fully divide the paper in nine-section (and higher odd-number sections) are complicated and puzzle-like, and thus are difficult to describe briefly.

8   Individual Line for Higher-Number Sections

After various trials, I chose the starting point of the individual line at the midpoint of the rectangle's upper edge, which makes for easy calculations. The terminal (drop) point of the individual line will be somewhere on the bottom side. As previously shown, in the case of trisection, its terminal point lands at the lower left corner, and for five-section, the lower right

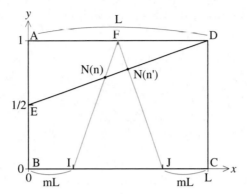

Figure 7. In general, draw an individual line FI with the lower point I placed at an arbitrary position along the bottom line (mL), and swing it like a pendulum to line FJ a distance mL from the right.

corner. For the seven-section, its terminal point lands at a point one-fourth of the distance from the left corner along the bottom side (L); and for the nine-section, the point lies at the same distance from the lower right corner.

I have determined that the corresponding distances for the 11-section and 13-section pair is 1/3 of L, and the distance for the 15-section and 17-section pair is 3/8 of L.

To generalize, for a rectangle whose side:length ratio is 1:L, we will denote the required distance for the terminal point by mL, as shown in Figure 7.

As before, the equations of two lines are:

$$\text{Common line:} \quad y = (x + L)/(2L), \tag{9}$$
$$\text{Individual line:} \quad y = 2x/(L(1 - 2m)) - 2m/(1 - 2m). \tag{10}$$

Setting the two equations equal, we obtain the coordinates of the intersection (x, y),

$$x = \frac{L(1 + 2m)}{3 + 2m}, \tag{11}$$
$$y = \frac{2 + 2m}{3 + 2m}. \tag{12}$$

Because x and y have the same denominator, the intersection divides the rectangle both horizontally and vertically.

In the case of the pendulum symmetry, the next individual line has the equation:

$$y = \frac{-2x}{L(1 - 2m)} + \frac{2 - 2m}{1 - 2m}. \tag{13}$$

Setting Equations (9) and (13) equal, we obtain the coordinates of the intersection (x, y),

$$x = \frac{L(3 - 2m)}{5 - 2m},\tag{14}$$

$$y = \frac{4 - 2m}{5 - 2m}.\tag{15}$$

These results indicate that for a pair of adjacent odd numbers, #3 and #5, #7 and #9, #11 and #13, #15 and #17, etc., the individual lines have the same slope in their equation, but the former number has a positive sign and the latter number a negative sign. Using Equations (11) and (12) and/or (14) and (15), as appropriate, we can search for suitable distances and signs of the individual line for any odd-numbered divisions.

9 Conclusions

Before my findings of the present system, Robert J. Lang reviewed several methods of proportional folding, including his own binary notation method [Lang 99]. The present system can be regarded as a variation of the crossing-diagonals technique [Lang 99]; however, I think that the system is unique in two ways. First, it has a definite line between the midpoint of the left (right) side and the upper corner of the opposite side of a rectangular paper instead of its diagonal line. Second, the system makes it simple to obtain the dividing node, which is determined as an intersection point of two lines: one is this first line, and the other line extends from the midpoint of the upper side of the paper to a designated position on the bottom side of the rectangle for each odd-numbered division. I can generally determine the position of the terminal point as a length from one or the other corner of the bottom side as a fraction for which both denominator and numerator are integers. To date, I have succeeded in making odd-number division folding from 3 to 31. Even though the technique may be used for more than 31 divisions of the paper, in practice it is difficult to fold narrowly spaced parallel lines.

In addition to the relative simplicity of construction, the previously mentioned pendulum symmetry is, I think, mathematically beautiful, and it may assist to find the desired terminal position for the second line. This system is also applicable to all right-angled quadrilateral paper, i.e., all rectangles, including squares.

Bibliography

[Fushimi 79] Koji Fushimi. "Origami Geometry, Haga's Theorem" (in Japanese). *Sugaku Seminar* 18:1 (1979), 40–41.

[Haga 99] Kazuo Haga. "Origamics Part 1: Fold a Square Piece of Paper and Make Geometrical Figures" (in Japanese). *Nihon-hyoronsha* (1999), 1–153.

[Haga 02] Kazuo Haga. "Fold Paper and Enjoy Math: Origamics." In *Origami³: Proceedings of the Third International Meeting of Origami Science, Mathematics, and Education*, edited by Thomas Hull, pp. 307–328. Natick, MA: A K Peters, 2002.

[Haga 04] Kazuo Haga. "I Can Divide a Given Rectangular Sheet of Paper into Equal Length Parts of an Optional Prime Number Vertically and Horizontally with My Empty Hands" (in Japanese). *Origami Tanteidan Magazine* 83 (2004), 14–15; 84 (2004), 14–15.

[Haga 05] Kazuo Haga. "Origamics Part 2: Fold Paper and Do Math" (in Japanese). *Nihon-hyoronsha* (2005), 1–153.

[Haga 08] Kazuo Haga. *Origamics, Mathematical Explorations through Paper Folding.* Singapore: World Scientific, 2008.

[Lang 99] Robert J. Lang. "Folding Proportions" (in Japanese). *Origami Tanteidan Magazine* 55 (1999), 14–15; 56 (1999), 14–15; 57 (1999), 14–15.

The Speed of Origami Constructions Versus Other Construction Tools

Eulàlia Tramuns

1   Introduction

Up to now, papers dealing with geometric constructions have approached the topic of constructions from a boolean point of view: a construction can or cannot be done using a tool. This focus omits important differences among constructions, such as the number of points and curves that take part in the construction, the number of steps needed to carry out the construction, and the number fields associated with the construction.

E. Lemoine, a French mathematician, defined *geometrography* as a measure of the simplicity of constructions using ruler and compass (see, e.g., [Mackay 93]). His measure allows one to compare different constructions done with ruler and compass, and such an approach could certainly be generalized to other tools. However, I feel that proceeding in this way, one mixes information. From the point of view that it is clearer to separate different aspects of a construction, I have defined new measures to characterize constructions. However, the definition of these measures needs a formalization of the mathematical concepts of *tool* and *construction*.

For this purpose, Section 2 introduces a tool. Section 3 gives a formal definition of a construction and defines measures that characterize geometric and algebraic aspects of a construction. The geometric measures are the *order* and the *level* of a construction. The algebraic measure is the *degree* of a construction. After these definitions, I present comparison criteria for constructions.

Finally, Section 4 shows an example of two optimal constructions, one done with origami and the other with both ruler and compass or conics.

2   Geometric Tools

Geretschläger [Geretschläger 95] characterized ruler-and-compass constructions by the use of *procedures*, two of which allow the construction of lines and circles, denoted by ($E1$, $E2$), and three others that allow the construction of a point as the intersection of lines and circles, denoted by ($E3$, $E4$, $E5$). Origami constructions are also characterized by procedures, one that allows the construction of the point of intersection of two lines and seven that allow the construction of lines (folds). These seven procedures correspond with six of the seven Huzita-Justin axioms [Lang 11a].

Based on this approach, I give a definition of a tool as a mathematical object.

As usual, let us identify a complex number $z = x + iy$ with the point $P = (x, y)$ in the Cartesian plane.

Let $\mathcal{P}(\mathbb{C})$ be the set of finite subsets of $\mathbb{C}$, and $\mathcal{C}(\mathbb{R})$ be the set of finite subsets of affine plane algebraic curves with real coefficients.

Definition 1 (Construction Axiom). A *construction axiom* is a map $A : \mathcal{P}(\mathbb{C}) \times \mathcal{C}(\mathbb{R}) \to \mathcal{C}(\mathbb{R})$.

Definition 2 (Intersection Axiom). An *intersection axiom* is a map $B : \mathcal{P}(\mathbb{C}) \times \mathcal{C}(\mathbb{R}) \to \mathcal{P}(\mathbb{C})$.

Definition 3 (Toolmap). Given a finite set of construction axioms $\mathcal{A}$ and a finite set of intersection axioms $\mathcal{B}$, a *toolmap* E_T is a map

$$E_T : \mathcal{P}(\mathbb{C}) \times \mathcal{C}(\mathbb{R}) \longrightarrow \mathcal{P}(\mathbb{C}) \times \mathcal{C}(\mathbb{R})$$

$$(S, C) \longmapsto \left(\bigcup_{B \in \mathcal{B}} B \left(S, \bigcup_{A \in \mathcal{A}} A(S, C) \right), \bigcup_{A \in \mathcal{A}} A(S, C) \right).$$

Definition 4 (Tool). A *tool* T is defined by the 4-tuple $(\mathcal{A}, \mathcal{B}, E_T, I_T)$, where $\mathcal{A}$ is a finite set of construction axioms, $\mathcal{B}$ is a finite set of intersection axioms, E_T is the toolmap associated with $\mathcal{A}$ and $\mathcal{B}$, and $I_T \in \mathcal{P}(\mathbb{C}) \times \mathcal{C}(\mathbb{R})$. I_T is the *initial set* of the tool.

Remark 1. Although intuitively one might think an intersection axiom should be a map $B : \mathcal{C}(\mathbb{R}) \to \mathcal{P}(\mathbb{C})$, there are tools, such as a marked ruler, that can obtain points from points and curves; see [Martin 98].

Let $(S, C) \in \mathcal{P}(\mathbb{C}) \times \mathcal{C}(\mathbb{R})$ and $(S', C') \in \mathcal{P}(\mathbb{C}) \times \mathcal{C}(\mathbb{R})$. We define the union of (S, C) and (S', C') as $(S, C) \amalg (S', C') := (S \cup S', C \cup C')$.

Definition 5 (Point Curve Sequences). Let $T = (\mathcal{A}, \mathcal{B}, E_T, I_T)$ be a tool, with $I_T = (S_0^T, C_0^T)$. The *sequences of points and curves associated with T* are $\{S_n^T\}_{n \geq 0}$ and $\{C_n^T\}_{n \geq 0}$, where

$$(S_{n+1}^T, C_{n+1}^T) := (S_n^T, C_n^T) \amalg E_T(S_n^T, C_n^T), \forall n \geq 0.$$

Definition 6 (Constructible Points and Curves). Let $T = (\mathcal{A}, \mathcal{B}, E_T, I_T)$ be a tool, with $I_T = (S_0^T, C_0^T)$. The *sets of constructible points and curves with T* are S_T and C_T, where

$$(S_T, C_T) := \amalg_{n=0}^{\infty} (S_n^T, C_n^T).$$

Remark 2. Note that, in general, S_T and C_T will not be finite nor discrete.

Origami, conics, and both ruler and compass are all tools having initial set $(S_0^T, C_0^T) := (\{0, 1\}, \emptyset)$. Let us describe their characteristics.

2.1   Origami

Let us consider origami (namely, O) as the technique of folding one unbounded piece of paper, identified as the complex plane. The construction axioms $O1,..., O7$, correspond with the seven Huzita-Justin axioms described in [Lang 11a].

Let us describe briefly the construction axioms associated with O. We express these axioms as sets, where $S \in S_O$ and $C \in C_O$.

- $O1(S, C)$ is the set of lines passing through each pair of points of S.

- $O2(S, C)$ is the set of all the perpendicular bisectors of pairs of points of S.

- $O3(S, C)$ is the set of all the angle bisectors of pairs of lines of C.

- $O4(S, C)$ is the set of lines perpendicular to each line of C passing through each point of S.

- $O5(S, C)$ is the set of all the tangents to parabolas with directrix in C and focus in S passing through a point of S.

- $O6(S, C)$ is the set of all the tangents common to two parabolas having directrix in C and focus in S.

- $O7(S, C)$ is the set of all tangents to parabolas with directrix in C and focus in S, but perpendicular to a line in C.

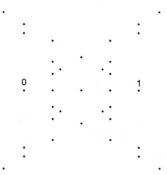

Figure 1. S_2 with origami.

There is only one intersection axiom, denoted by $O8$.

- $O8(S, C)$ is the set of all the intersection points of lines in C.

It is known that S_O is the smallest subfield of $\mathbb{C}$ closed under square roots, cube roots, and complex conjugation [Alperin 00].

Figure 1 shows the graphical representation of S_2 constructed with origami.

2.2 Ruler and Compass

A ruler and a compass (RC) consists of an infinite ruler without marks and a compass (a modern compass, with variable and unlimited opening) that allows distances to be transported. The construction and intersection axioms associated with RC correspond with the procedures described previously, $E1$, ..., $E5$. There are two construction axioms $E1$ and $E2$, and three intersection axioms $E3$, $E4$, and $E5$.

Let us now describe these axioms. Let $S \in S_{RC}$ and $C \in C_{RC}$.

- $E1(S, C)$ is the set of all lines passing through pairs of points of S.

- $E2(S, C)$ is the set of circles having center in S and radius the distance between two points of S.

- $E3(S, C)$ is the set of all intersection points of lines in C.

- $E4(S, C)$ is the set of the intersection points of each line and each circle in C.

- $E5(S, C)$ is the set of all the intersection points of circles of C.

It is known that S_{RC} is the smallest subfield of $\mathbb{C}$ closed under square roots [Cox 04].

2.3   Conics

Constructions with conics (CO) are known as *solid constructions*. A detailed description of this tool can be found in [Videla 97]. It uses three construction axioms that allow the construction of lines, circles, and conics. There is one intersection axiom that allows the construction of points obtained by the intersection of lines, circles, parabolas, ellipses, and hyperbolas. It is known that $S_{CO} = S_O$ [Alperin 00].

A comment on notation going forward: when the image of a construction axiom or an intersection axiom is a set of one element, $\{v\}$, it is written as v instead of $\{v\}$, for brevity.

3   Constructions and Measures

This section gives a formal definition of a construction and defines magnitudes that will be useful to compare constructions.

Given $A = (S, C) \in \mathcal{P}(\mathbb{C}) \times \mathcal{C}(\mathbb{R})$ and $B = (S', C') \in \mathcal{P}(\mathbb{C}) \times \mathcal{C}(\mathbb{R})$, we define $A \sqsubseteq B$ to mean $S \subseteq S'$ and $C \subseteq C'$.

Let T be a fixed tool and $I = (S, C) \sqsubseteq (S_T, C_T)$. Let v be a constructible point or a constructible curve with T such that $v \notin S$ and $v \notin C$.

Definition 7 (Construction). A *construction* $\Phi_{I,T,v}$ of v is a finite sequence of elements of $\mathcal{P}(\mathbb{C}) \times \mathcal{C}(\mathbb{R})$, with first term (S, C), such that

- if v is a point, the first element of the last term is $\{v\}$, and if v is a curve, the last term is $(\emptyset, \{v\})$;

- each term of the sequence belongs to the image of the union of the previous terms of the sequence under the toolmap E_T;

- (unicity) any other sequence beginning with (S, C), obtained by removing subsets of points or curves from the original sequence, does not satisfy one of the above conditions.

I is the *initial set* of the construction, S_F is the union of all sets of points of the construction, and C_F is the union of all sets of curves of the construction.

3.1   Measures Associated with Constructions

Let T be a tool, $I = (S, C) \sqsubseteq (S_T, C_T)$, v a constructible point or curve, and $\Phi_{I,T,v}$ a construction of v.

Definition 8 (Order of a Construction). The *order of the construction* $\Phi_{I,T,v}$ is
$$\text{ord}(\Phi_{I,T,v}) := |S_F| + |C_F|.$$

Let us denote by E_T^n the nth iterate of the function E_T.

Definition 9 (Level of a Construction). The *level of the construction* $\Phi_{I,T,v}$, denoted by $\mathrm{lv}(\Phi_{I,T,v})$, is the integer $n > 0$ such that $(S_F, C_F) \sqsubseteq E_T^n(I)$ and $(S_F, C_F) \not\sqsubseteq E_T^{n-1}(I)$. The value n is the length of the sequence $\Phi_{I,T,v}$ minus 1.

Let K be a number field, $S' \in \mathcal{P}(\mathbb{C})$ and $C' \in \mathcal{C}(\mathbb{R})$. Then $K(S')$ will be the field obtained adjoining to K each element of S' and its complex conjugate.

$K(C')$ will be the field obtained adjoining to K the coefficients of the equations of each curve in C'.

Definition 10 (Degree of a Construction). The *degree of the construction* $\Phi_{I,T,v}$ is $\mathrm{dg}(\Phi_{I,T,v}) := [\mathbb{Q}(S_F, C_F)(i) : \mathbb{Q}(S, C)(i)]$.

Definition 11 (Optimal Speed). A construction $\Phi_{I,T,v}$ has *optimal speed* if the $\mathrm{ord}(\Phi_{I,T,v})$ is minimal over the orders of the constructions of v with initial set I.

Definition 12 (Optimal Simplicity). A construction $\Phi_{I,T,v}$ has *optimal simplicity* if the $\mathrm{lv}(\Phi_{I,T,v})$ is minimal over the levels of the constructions of v with initial set I.

Definition 13 (Geometrically Optimal). A construction with optimal speed and optimal simplicity is said to be *geometrically optimal*.

Definition 14 (Algebraically Optimal). A construction $\Phi_{I,T,v}$ is *algebraically optimal* if the degree $\mathrm{dg}(\Phi_{I,T,v})$ is minimal over the degrees of the constructions of v with initial set I.

Definition 15 (Uniqueness of Construction). A construction $\Phi_{I,T,v}$ is *unique* if there is no other construction of v with initial set I, having the same order, level, degree, $|S_F|$ and $|C_F|$ as $\Phi_{I,T,v}$.

Let us illustrate with an example the measures we have defined. Consider the tool O. Let $z \in S_O$ and $I = (\{0, 1, z\}, \emptyset)$. Figure 2 shows the points and lines needed to construct $\bar{z}$ with O.

The construction is $\Phi_{I,O,\bar{z}}$. The explicit description of this sequence allows the calculation of its order, level, and degree. It is given by

$$\Phi_{I,O,\bar{z}} = \{(\{0, 1, z\}, \emptyset), (\emptyset, \{l_1\}), (\emptyset, \{l_2\}), (\emptyset, \{l_3\}), (\{P\}, \{l_4\}), (\{\bar{z}\}, \{l_5\})\}.$$

The lines and points of $\Phi_{I,O,\bar{z}}$ are $l_1 = O1(\{0, 1\}, \emptyset)$, $l_2 = O4(\{z\}, \{l_1\})$, $l_3 = O4(\{z\}, \{l_2\})$, $l_4 \in O3(\emptyset, \{l_2, l_3\})$, $P = O8(\emptyset, \{l_1, l_4\})$, $l_5 = O4(\{P\}, \{l_4\})$, and $\bar{z} = O8(\emptyset, \{l_2, l_5\})$.

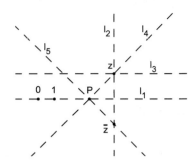

Figure 2. Complex conjugate construction with origami.

Clearly, $\mathrm{ord}(\Phi_{I,O,\bar{z}}) = 10$ and $\mathrm{lv}(\Phi_{I,O,\bar{z}}) = 5$.

Let us calculate the degree of $\Phi_{I,O,\bar{z}}$. Let $z = a + bi$. Then $a, b \in \mathbb{Q}(z, \bar{z}, i) := K$. The lines obtained from axioms $O1$ and $O4$ have their coefficients in K.

The use of the axiom $O3$ to obtain l_4 in the construction could have made the degree of $\Phi_{I,O,\bar{z}}$ greater than 1. Nevertheless, the slope of l_4 is 1, and line l_4 has its coefficients in K. The coordinates of the point P also belong to K. Thus, $\mathrm{dg}(\Phi_{I,O,\bar{z}}) = 1$.

3.2 Comparison of Constructions

Let T and T' be tools, and let v be a constructible point or curve with T and T'. Let $I \sqsubseteq (S_T, C_T)$ and $I \sqsubseteq (S_{T'}, C_{T'})$. Let $\Phi_{I,T,v}$ and $\Phi_{I,T',v}$ be two constructions of v.

Definition 16 (Faster, Simpler, Lower Degree). Construction $\Phi_{I,T,v}$ is *faster* than $\Phi_{I,T',v}$ if $\mathrm{ord}(\Phi_{I,T,v}) < \mathrm{ord}(\Phi_{I,T',v})$; $\Phi_{I,T,v}$ is *simpler* than $\Phi_{I,T',v}$, if $\mathrm{lv}(\Phi_{I,T,v}) < \mathrm{lv}(\Phi_{I,T',v})$, and $\Phi_{I,T,v}$ is *of lower degree* than $\Phi_{I,T',v}$, if $\mathrm{dg}(\Phi_{I,T,v}) < \mathrm{dg}(\Phi_{I,T',v})$.

Definition 17 (Geometric Equivalence). If $\mathrm{ord}(\Phi_{I,T,v}) = \mathrm{ord}(\Phi_{I,T',v})$ and $\mathrm{lv}(\Phi_{I,T,v}) = \mathrm{lv}(\Phi_{I,T',v})$, the constructions with T and T' are *geometrically equivalent*.

Definition 18 (Algebraic Equivalence). If $\mathrm{dg}(\Phi_{I,T,v}) = \mathrm{dg}(\Phi_{I,T',v})$, the constructions with T and T' are *algebraically equivalent*.

4 Optimal Constructions

This section proves the optimality of two constructions: the addition of two complex numbers with O and with RC. The construction with CO is the same as with RC.

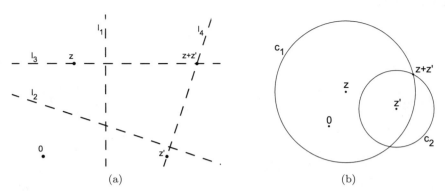

(a)   (b)

Figure 3. Addition construction: (a) origami and (b) conics and ruler and compass.

Let z and z' be two numbers constructible with O and RC, such that 0, z, and z' are not collinear and $|z| \neq |z'|$. Figure 3 shows $\Phi_{I,O,z+z'}$ and $\Phi_{I,RC,z+z'}$, where $I = (\{0, z, z'\}, \emptyset)$.

Let us describe the sequences of each construction. First, with origami (O),

$$\Phi_{I,O,z+z'} = \{(\{0, z, z'\}, \emptyset), (\emptyset, \{l_1, l_2\}), (\{z + z'\}, \{l_3, l_4\})\}.$$

At the first step, the lines are $l_1 = O2(\{0, z'\}, \emptyset)$ and $l_2 = O2(\{0, z\}, \emptyset)$. At the second step, the lines and points are $l_3 = O4(\{z\}, \{l_1\})$, $l_4 = O4(\{z'\}, \{l_2\})$, and $z + z' = O8(\emptyset, \{l_3, l_4\})$.

Now, for ruler-and-compass, let us denote by $c(A, BC)$ the circle of center A and radius given by the distance between B and C. Then,

$$\Phi_{I,RC,z+z'} = \{(\{0, z, z'\}, \emptyset), (\{z + z'\}, \{c_1, c_2\})\},$$

where $c_1 = c(z, 0z') \in E2(\{z, 0, z'\}, \emptyset)$, $c_2 = c(z', 0z) \in E2(\{z, 0, z'\}, \emptyset)$, and $z + z' \in E5(\emptyset, \{c_1, c_2\})$.

Table 1 shows the values of the measures of these constructions. The constructions with RC and CO are faster and simpler than with O and all are algebraically equivalent.

Theorem 1. *The constructions $\Phi_{I,O,z+z'}$ and $\Phi_{I,RC,z+z'}$ are geometrically and algebraically optimal and unique.*

Construction	Order	Level	Degree
$\Phi_{I,O,z+z'}$	8	2	1
$\Phi_{I,RC,z+z'}$	6	1	1

Table 1. Measures of addition constructions.

Proof: Clearly, $\Phi_{I,O,z+z'}$ and $\Phi_{I,RC,z+z'}$ are constructions. The order, level, and degree of a construction are all integers.

In general, if $\Phi_{I,T,z}$ is a construction of a point z, then $\mathrm{ord}(\Phi_{I,T,z}) \geq |S| + |C| + 1$, $\mathrm{lv}(\Phi_{I,T,z}) \geq 1$, and $\mathrm{dg}(\Phi_{I,T,z}) \geq 1$. Thus, any construction having $\mathrm{ord}(\Phi_{I,T,z}) = |S| + |C| + 1$ and $\mathrm{lv}(\Phi_{I,T,z}) = 1$ is geometrically optimal. When $\mathrm{dg}(\Phi_{I,T,z}) = [\mathbb{Q}(S,C)(i,z,\overline{z}) : \mathbb{Q}(S,C)(i)]$, the construction is algebraically optimal.

Let us consider the case of ruler and compass. Since there are no initial curves constructed, and two curves are needed to obtain a point, the lowest possible order is 6. Thus the construction is clearly geometrically and algebraically optimal.

To study the uniqueness, let us analyze all possible cases. The construction axiom $E1$ alone does not add new points and so would not allow construction of $z + z'$. With $E1$ and $E2$, we construct one line and one circle that do not intersect in $z + z'$ because the points 0, z, and z' are not collinear. Thus, the construction is unique.

If we add the conic construction axiom, we need lines constructed to be the directrix of the conic, and the level of the construction would be greater than 1. So this construction is also unique using conics.

Let us now consider the geometric optimality in the case of origami. There is no construction of z with level one. Indeed, at the first step of iteration, having three points constructed and no lines, only axioms $O1$ and $O2$ can be used. If the points are in a general position and $|z| \neq |z'|$, we cannot obtain $z+z'$ in this step, because perpendicular bisectors intersect in the circumcenter of the triangle $0zz'$, which is different from $z+z'$. Besides, perpendicular bisectors do not intersect an extended side of a triangle in the sum. So the construction has the lowest possible simplicity.

Let us now do an exhaustive search of the orders of valid constructions having level 2. Possible orders lower than 8 are 6 or 7. Thus, the pair $(|S_F|, |C_F|)$ can be $(4,2)$, $(4,3)$, $(5,1)$, $(5,2)$, and $(6,1)$. But since points are obtained as the intersection of two lines, necessarily $|C_F| \geq 2$. Moreover, with two lines, we can obtain just one intersection point, so the case $(5,2)$ does not make sense here. Thus, there are still two cases, $(4,2)$ and $(4,3)$. As we have seen before, two lines constructed in the first step of iteration do not allow the construction of $z + z'$, because $|z| \neq |z'|$. So, if the order of the construction is $(4,2)$, then one line is constructed in the first step and the other in the second step. But, since $|z| \neq |z'|$, $z + z'$ is not on any line constructible in the first step, so $z + z'$ cannot be obtained as the intersection of one line in the first step with any other line. Using the same argument, the lines of a construction satisfying $(|S_F|, |C_F|) = (4,3)$ have to be constructed, one in the first step, and the two others in later steps. Thus, after the first step, we have constructed three points and one line. $O3$ and $O7$ cannot be applied because they need two lines. $O1$ and $O2$

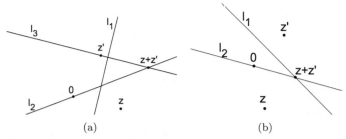

Figure 4. Addition constructions when (a) $|z| = |z'|$ and (b) $|z| = |z'|$ and $\angle z0z' = 120°$.

give the same lines that can be obtained in the first step, and we have seen that they do no intersect in the sum. Moreover, $z + z'$ is not in any line constructed by axioms $O5$ and $O6$. Thus, the only axiom that can be used is $O4$.

If we use axiom $O1$ in the first step to obtain one line, then, using $O4$ in the second step, we will construct one line, but we need two. However, by using axiom $O2$, $O4$ constructs two parallel lines that do not intersect the line constructed in the first step in the point $z + z'$ because $z + z'$ is not in any perpendicular bisector of the triangle $z0z'$.

Thus, there is no construction with lower order.

To prove uniqueness, we use the same type of arguments. Any construction with order 8 needs four points and four lines, two lines constructed in the first step of iteration and two others in the second. Other possibilities do not make sense. An exhaustive search shows that there is no other construction having these properties. Indeed, in the first step, two lines obtained by axiom $O1$ do not allow the construction of $z + z'$, so we have to use $O2$. Over the three possibilities of pairs of lines constructed with $O2$, the only one that constructs $z + z'$ is in $\Phi_{I,O,z+z'}$. At the second step, only axiom $O4$ constructs $z + z'$.   □

Remark 3. Observe that the construction of $z - z'$ is geometrically the same construction as the construction of $z + z'$. Thus, the last result also applies for the construction of $z - z'$.

Now, let z and z' be two numbers, constructible with O and RC, such that 0, z, and z' are not collinear, and $|z| = |z'|$. The construction of $z + z'$ with RC is the same as in the general case, but the origami construction can be improved.

Figure 4 shows two examples of the addition construction with O when $|z| = |z'|$.

Clearly, $z + z'$ belongs to the perpendicular bisector of zz' if and only if $|z| = |z'|$.

Suppose that $z + z'$ belongs to the perpendicular bisector of zz' but $z + z'$ is not the circumcenter of the triangle $0zz'$. Then a construction of $z + z'$ is

$$\Phi'_{I,O,z+z'} = \{(\{0, z, z'\}, \emptyset), (\emptyset, \{l_1, l_2\}), (\{z + z'\}, \{l_3\})\},$$

where $l_1 = O2(\{0, z\}, \emptyset)$, $l_2 = O2(\{z, z'\}, \emptyset)$, $l_3 = O4(\{z'\}, \{l_1\})$, and $z + z' = O8(\emptyset, \{l_2, l_3\})$; see Figure 4(a).

The order of $\Phi'_{I,O,z+z'}$ is 7, its level is 2, and its degree is 1. Following the arguments from the proof of Theorem 1, we see that the construction is geometrically and algebraically optimal. The construction is not unique because redefining l_1 to be $O2(\{0, z'\}, \emptyset)$ and l_3 to be $O4(\{z\}, \{l_1\})$, one obtains a different construction of $z + z'$.

Under the assumption that $|z| = |z'|$, $z + z'$ is the circumcenter of the triangle $0zz'$ if and only if $\angle z0z' = 120°$.

In this case, the construction is $\Phi''_{I,O,z+z'} = \{(\{0, z, z'\}, \emptyset), (\{z + z'\}, \{l_1, l_2\})\}$, where $l_1 = O2(\{0, z'\}, \emptyset)$, $l_2 = O2(\{z, z'\}, \emptyset)$, and $z + z' = O8(\emptyset, \{l_1, l_2\})$; see Figure 4 (b).

The order, level, and degree of $\Phi''_{I,O,z+z'}$ are respectively 6, 1, and 1. These are their minimal possible values. Thus, the construction is geometrically and algebraically optimal. The construction is not unique because any two of the three perpendicular bisectors intersect in the circumcenter of the triangle $0zz'$.

5   Conclusions

Clearly, the measures defined have practical applications. In fact, programs exist that simulate origami constructions as well as others that simulate ruler and compass constructions. These programs could take into account these measures to improve the construction's algorithms.

A real example of an application of a geometric measure to improve constructions is the case of Lang's ReferenceFinder program [Lang 11b]. It uses a measure similar to the order, called *rank*, which is the number of times the Huzita-Justin axioms are applied.

The order and degree of a construction are concepts that could have been defined without defining any new mathematical object. However, the formalization of the definition of a tool as a mathematical object allows us to define the level of a construction, which values the complexity of a construction. In particular, applied to origami constructions, it is a measure that takes into account the dependence of the folds that take part in a construction. A lower level means a low degree of dependence among the folds.

An algorithm to obtain constructions with the lowest possible level can be implemented. In fact, given a tool T and fixing $N \geq 1$, we can calculate $E_T^i(S_0^T, C_0^T)$, for $i = 1 \ldots N$. Extracting data from the calculations, we obtain constructions with the lowest level for the points $z \in S_i^T$ such that $z \notin S_{i-1}^T$, and for the curves $c \in C_i^T$ such that $c \notin C_{i-1}^T$, for $i = 1, \ldots, N$. All these constructions have level i. Moreover, among all the constructions with level i, one can choose the ones with lowest order. These are geometrically optimal.

Another problem related to this subject is the study of the existence of constructions being both geometrically and algebraically optimal. However, given a construction, the determination of its optimality and uniqueness in a general way remains an open problem.

Finally, it would also be interesting to establish relations satisfied by the orders, levels, and degrees of constructions.

Acknowledgements.   The author was partially supported by MTM2009-13060-C02-02 from the Spanish MEC. The author is very grateful to the reviewer and the editors for their useful comments.

Bibliography

[Alperin 00] Roger C. Alperin. "A Mathematical Theory of Origami Constructions and Numbers." *New York Journal of Mathematics* 6 (2000), 119–133.

[Cox 04] David A. Cox. *Galois Theory.* New York: Wiley, 2004.

[Geretschläger 95] Robert Geretschläger. "Euclidean Constructions and the Geometry of Origami." *Mathematics Magazine* 68:5 (1995), 357–371.

[Lang 11a] Robert J. Lang. "Huzita-Justin Axioms." Available at http://www.langorigami.com/science/hha/hha.php4, 2011.

[Lang 11b] Robert J. Lang.   "ReferenceFinder."   Available at http://www.langorigami.com/science/reffinder/reffinder.php4, 2011.

[Mackay 93] J. S. Mackay. "The Geometrography of Euclid's Problem." *Proceedings of the Edinburg Mathematical Society* 12 (1893), 2–16.

[Martin 98] George E. Martin. *Geometric Constructions.* New York: Springer-Verlag, 1998.

[Videla 97] Carlos Videla. "On Points Constructible from Conics." *The Mathematical Intelligencer* 19:2 (1997), 53–57.

A Note on Operations of Spherical Origami Construction

Toshikazu Kawasaki

1 Introduction

Geometric construction using folding methods proposed by Humiaki Huzita is based on marking figures using the tools of origami. Instead of a ruler and a compass, "creases" are used as tools to shape figures; i.e., a crease given by folding can simultaneously bring about coincidences of points and segments via reflection across the fold line. Huzita referred to these basic folding steps as "axioms" and defined seven such origami axioms [Huzita 91]. In this paper, I call them (origami) operations. There are eight operations for folding in the Euclidean plane; the eighth operation was added by Jacques Justin [Justin 91] and Koshiro Hatori. (See [Alperin and Lang 09] for details.)

In this paper, we try to create operations for spherical origami construction in order to know the essence of origami construction. We reformulate operations of planar origami construction into operations that act within spherical origami, with the condition that each point and its antipodal point on the sphere are to be set in our study. Considering the correspondence between two poles and the equator (a great circle) of the sphere, we clarify the essence of origami construction. Finally, we reduce the operations to four spherical origami operations.

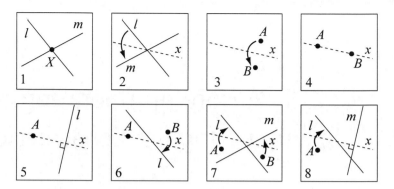

Figure 1. The eight operations of origami construction.

2   The Eight Operations of Planar Origami Construction

To begin, we review the eight operations of planar origami construction, illustrated in Figure 1.

1. Given two lines l and m, we may place a point X at that intersection.

2. Given two lines l and m, we may fold a line x to make a coincidence of l and m.

3. Given two points A and B, we may fold a line x to make a coincidence of A and B by the folding x.

4. Given two points A and B, we may fold a line x to pass through A and B simultaneously.

5. Given a point A and a line l, we may fold a line x that is simultaneously perpendicular to l and passes through A.

6. Given two points A and B and a line l, we may fold a line x that simultaneously passes through A and translates B into l.

7. Given two points A and B and two lines l and m, we may fold a line x to reflect A into l and also reflect B into m.

8. Given a point A and two lines l and m, we may fold a line x that is perpendicular to m and reflects A into l.

But these are too many operations—they must be reduced to some kind of order. When A lies on l and B lies on m in operation 7, a line x makes a coincidence of l and m. Thus, operation 2 is a special case of operation 7.

In fact, it is well known that operations 2–6 and 8 are special cases of operation 7. This means that operation 7 is a kind of "super-ruler." In order to get another reduction and to know better what constitutes origami construction, we begin to build another origami construction system, one in a spherical geometry.

3 Notation

In this section, a *line* means a straight line in a sheet of paper. If it is possible to define a line x that maps a given point A to another given point B by reflection with respect to line x, then we denote this condition by $xA = B$. Similarly, if it is possible to define a line x that maps a given line l to another given straight line m by reflection with respect to x, then we denote this condition by $xl = m$. It is evident that $xl = m$ if and only if $xl = m$ and $xA = B$ if and only if $xB = A$, as shown in Figure 2.

If a line x passes through a given point A, the element A belongs to the set x. So we can state this condition as $A \in x$, using the notation of set theory. We can also express the condition that a given line x is perpendicular to a line l as $x \perp l$. These are illustrated in Figure 3.

Using this notation, we can formulate the following eight origami operations of planar origami construction.

Proposition 1. *The origami operations of planar origami are as follows:*

PO1: $X \in l,\ X \in m$.

PO2: $xl = m$.

PO3: $xA = B$.

PO4: $A \in x,\ B \in x$.

PO5: $x \perp l,\ A \in x$.

PO6: $xB \in l,\ A \in x$.

PO7: $xA \in l,\ xB \in m$.

PO8: $x \perp l,\ xA \in m$.

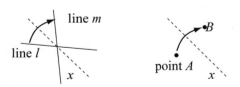

Figure 2. Conditions $xl = m$ (left) and $xA = B$ (right).

Figure 3. Conditions $A \in x$ or x passes through A (left) and $x \perp l$ or x is perpendicular to l (right).

4 Spherical Origami

We now can define a spherical form of origami [Kawasaki 90]. In this section, we consider origami construction within a spherical geometry. In this geometry, a crease is a great circle that divides the sphere into two hemispheres and reflects them onto each other, so a great circle is simply called a line.

Any point A on a sphere uniquely determines an *antipodal* point, hereafter denoted by A', as shown in Figure 4. Point A also determines a line and a disk that is perpendicular to the segment AA'. We denote them by the same symbol $a = A^*$. Conversely, a line a, or a disk a, uniquely determines two points A and A', as in Figure 5.

Proposition 2. *For any points A and B on sphere, the following conditions are equivalent, where $a = A^*$ is the disk (line) defined relative to point A, and $b = B^*$ is defined similarly for point B:*

(1) $A \in b$: A belongs to b.

(2) $A' \in b$: A' belongs to b.

(3) $a \perp b$ as disks.

(4) $AA' \perp BB'$.

(5) $B \in a$: B belongs to a.

(6) $B' \in a$: B' belongs to a.

Figure 4. Point A and its antipodal point A'.

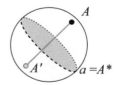

Figure 5. Point A uniquely determines the disk and the line $a = A^*$.

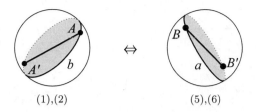

Figure 6. Condition relationships $A, A' \in b \Leftrightarrow a \ni B, B'$.

These relationships are illustrated in Figure 6.

Proof: Since segments AA' and BB' are diameters of disks a and b, respectively, (1) $\Leftrightarrow$ (2) and (5) $\Leftrightarrow$ (6) are obvious (see Figure 6).

(1) ⇒ (3): Since both A and A' belong to the same line b, segment AA' is also included in disk b. Since AA' is a normal line of disk a and any plane that includes AA' is perpendicular to a, b is perpendicular to a.

(3) ⇒ (1): We denote $a \cap b = \{N, S\}$ and put them on the north pole and the south pole. The top view (Figure 7) shows that the two disks a and b look like a cross. The normal line AA' of disk a overlaps with b. So AA' is included in the disk b. That is, $A \in b$ and $A' \in b$.

(3) ⇔ (5): This can be proved in the same way.

(3) ⇔ (4): Segments AA' and BB' are normal lines of the disks a and b, respectively. Therefore, $a \perp b \Leftrightarrow AA' \perp BB'$. □

Definition 1. Let X be a point on the sphere. For any point P on the sphere, we define an action X on point P to be the 180 degrees rotation of P about axis XX', as illustrated in Figure 8.

Proposition 3. *Let X be a point and x be the line X^*. For any points A and B and any line l, we have*

(1) $XA = xA'$ and $XA' = xA$ [Hatori 09],

(2) $Xl = xl$,

(3) $xA = B$ or $xA = B' \Leftrightarrow xa = b$, where $a = A^*$ and $b = B^*$.

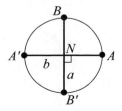

Figure 7. Top view showing $AA' \perp BB' \Leftrightarrow a \perp b$.

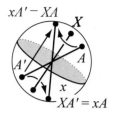

Figure 8. Illustrations of $XA = xA'$ and $XA' = xA$ (left) and $Xl = xl$ (right).

Proof: (1) See Figure 8 (keft). Line x reflects XA to A', namely $xXA = A'$. Since xx is the identity map, $XA = xxXA = x(xXA) = xA'$. In the same way, $XA' = xA$.

(2) See Figure 8 (right). For an arbitrary $P \in l$, its antipodal point P' also belongs to l. Since (1) gives $XP = xP' \in xl$, we may conclude that $Xl \subset xl$. Conversely, for an arbitrary point $Q \in xl$, its antipodal point Q' also belongs to xl because xl is a line. Condition (1) also gives $XQ = xQ' \in xxl = l$. Thus we, have $Q = XXQ \in Xl$. That is, $xl \subset Xl$.

(3) We prove $xa = b \Rightarrow xA = B$ or $xA = B'$. We can assume the lines a and b are the meridians. Figure 9 shows the equatorial plane and north pole $N \in a \cap b$. It can be considered a plane figure and is symmetrical with respect to x. Reflection through line x simultaneously reflects AA' to BB' and a to b. Thus we get $xA = B'$. Since points B and B' represent both ends of the diameter, they are interchangeable. Thus, $xA = B'$ may imply $xA = B$.

Next, we prove $xA = B$ or $xA = B' \Rightarrow xa = b$. If $xA = B'$, then $xA' = B$. So AA' and $B'B$ are symmetrical with respect to disk x. Since AA' and $B'B$ are normal lines of disks a and b, respectively, a and b are also symmetrical with respect to disk x. That is, $xa = b$. □

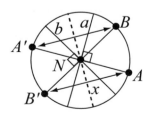

Figure 9. Illustration of $xa = b \Leftrightarrow xA = B$ or $xA = B'$.

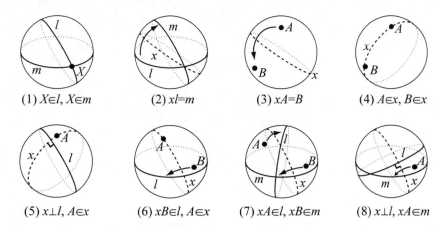

(1) $X \in l$, $X \in m$ (2) $xl=m$ (3) $xA=B$ (4) $A \in x$, $B \in x$

(5) $x \perp l$, $A \in x$ (6) $xB \in l$, $A \in x$ (7) $xA \in l$, $xB \in m$ (8) $x \perp l$, $xA \in m$

Figure 10. Spherical origami operations.

5 Operations of Spherical Origami Construction

In Section 3, the eight operations PO1–PO8 of *planar origami* construction were represented by set symbols. Therefore we can try to formally interpret them as operations of *spherical origami* construction. We can see that each planar origami operation gives an actual spherical origami operation, as shown in Figure 10.

Rewriting the letters in Figure 10, we can define corresponding spherical origami operations, as shown in Figure 11.

Here we have eight operations. But is there another operation? Koshiro Hatori has found a ninth operation S09: Given two lines a and b, we fold a

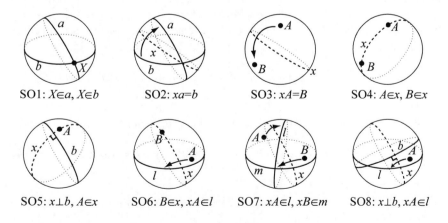

SO1: $X \in a$, $X \in b$ SO2: $xa=b$ SO3: $xA=B$ SO4: $A \in x$, $B \in x$

SO5: $x \perp b$, $A \in x$ SO6: $B \in x$, $xA \in l$ SO7: $xA \in l$, $xB \in m$ SO8: $x \perp b$, $xA \in l$

Figure 11. Spherical origami operations.

Figure 12. SO9: $x \perp a$, $x \perp b$.

line x that is perpendicular to both l and m (i.e., $x \perp a$ and $x \perp b$). This ninth operation is illustrated in Figure 12.

Proposition 4. *Operations SO1, SO4, SO5, and SO9 are the same; operations SO2 and SO3 are the same; and operations SO6 and SO8 are the same.*

Proof: Since Proposition 2 states that $X \in a \Leftrightarrow A \in x$ and $X \in b \Leftrightarrow B \in x$, SO1 is the same as SO4: $A \in x$ and $B \in x$. Since Proposition 2 also claims that $x \perp a$ and $A \in x$, $x \perp b$ and $B \in x$ are interchangeable, SO5 ($A \in x$, $x \perp b$) and Hatori's operation SO9 ($x \perp a$, $x \perp b$) are the same as SO4. That is, SO1, SO4, SO5, and SO9 are the same. In the same way, one can show that SO6 is the same as SO8.

Proposition 3(3) states that $xa = b \Leftrightarrow xA = B$ or $xA = B'$. Relations $xA = B$ and $xA = B'$ have the same structure: point meets point. Therefore, SO2 and SO3 have the same structure. □

This brings us to the main result of our paper.

Theorem 1. *Nine operations of spherical origami construction may be reduced to the following four operations (illustrated in Figure 13). For any two distinct points A and B that are not antipodal and any two distinct lines l and m,*

S1: $A \in x$, $B \in x$: we may fold a line x to simultaneously pass through A and B.

S2: $xA = B$: we may fold a line x to make a coincidence of A and B.

S3: $A \in x$, $xB \in l$: we may fold a line x that passes through A to reflect B into l.

S4: $xA \in l$, $xB \in m$: we may fold a line x to simultaneously reflect A into l and B into m.

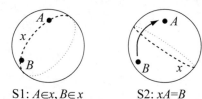 

S1: $A{\in}x, B{\in}x$ S2: $xA{=}B$ S3: $B{\in}x, xA{\in}l$ S4: $xA{\in}l, xB{\in}m$

Figure 13. Four spherical origami operations.

6 Conclusion

Eight origami operations of planar origami construction have been described using the notation of set theory. We have formally interpreted them as operations of spherical origami construction and have confirmed that they actually work on the sphere. From these, we have obtained nine origami operations, including Hatori's operation SO9. Although these operations appear different, some of them produce the same result. Thus, the nine operations may be reduced to four, namely, S1 ($A \in x$, $B \in x$), S2 ($xA = B$), S3 ($A \in x$, $xB \in l$), and S4 ($xA \in l$, $xB \in m$). We therefore see that the spherical origami operation system is simpler and more symmetrical than the planar system.

Bibliography

[Alperin and Lang 09] Roger C. Alperin and Robert J. Lang. "One-, Two-, and Multi-Fold Origami Axioms." In *Origami^4: Fourth International Meeting of Origami Science, Mathematics, and Education*, edited by Robert J. Lang, pp. 371–393. Wellesley, MA: A K Peters, 2009.

[Hatori 09] Koshiro Hatori. "Comment for Kawasaki's presentation 'Duality of Spherical Origami Construction System and Planar Origami's One.'" Presented at The Sixth Meeting on Origami Science, Mathematics, and Education, Tokyo, Japan, June 21, 2009.

[Huzita 91] Humiaki Huzita. "Axiomatic Development of Origami Geometry." In *Proceedings of the First International Meeting on Origami Science and Technology*, edited by H. Huzita, pp. 143–159. Padova, Italy: Dipartimento di Fisica dell'Università di Padova, 1991.

[Justin 91] Jacques Justin. "Resolution par le pliage de l'equation due troiseme degre et applications geometriques." In *Proceedings of the First International Meeting on Origami Science and Technology*, edited by H. Huzita, pp. 251–261. Padova, Italy: Dipartimento di Fisica dell'Università di Padova, 1991.

[Kawasaki 90] Toshikazu Kawasaki. "On Flat Origami of 2-Sphere" (in Japanese). *Report of Sasebo College of Technology* 26 (1990), 149–161.

Origami Alignments and Constructions in the Hyperbolic Plane

Roger C. Alperin

1 Introduction

Neutral geometry is the geometry made possible with the first 28 theorems of Euclid's Book 1—those results that do not rely on the parallel postulate. Hyperbolic geometry diverges from Euclidean geometry in that there are two parallels (asymptotic) to a given line through a given point and infinitely many other lines through the point that do not meet the given line (ultraparallel). In hyperbolic geometry, similar triangles are congruent; triangles have less than 180 degrees; there are no squares (regular four sided with 90 degree angles); and the area of a triangle is its defect (radian difference between π and the angle sum). As a consequence of this last remark, Bolyai (circa 1830) showed that one can square some circles (one uses a regular four sided polygon) in the hyperbolic plane using a ruler and compass [Gray 04, Curtis 90, Jagy 95]. Our aim here is to introduce origami constructions as a new method for doing geometry in the hyperbolic plane. We discuss the use of origami for making any ruler-compass construction and the possibilities for other constructions that are not ruler-compass constructions.

In the first section, we give the alignments for folding in the hyperbolic plane. These are the analogues of the classical origami axioms in the plane discussed by Huzita and Justin [Alperin and Lang 09]. There is one additional alignment that is possible. Next, we discuss the relations between the alignments in the context of neutral geometry, Euclidean geometry, and hyperbolic geometry. This leads up to the the relation of these alignments to the ruler-compass constructions in the hyperbolic plane. Basically, we show that one can do ruler-compass constructions equivalently with a subset of the folds, similar to origami in the Euclidean plane.

We use the projective model (Cayley-Klein) for the hyperbolic plane [Fishback 69]. The model is the interior of the unit disk; the bounding circle is called the *absolute*. The points and lines of this geometry are the same as points and lines of the plane that lie in the the interior of the disk. Since the model embeds in the Euclidean plane, we can make these folds using some of the Euclidean origami folds—we call this a *simulation*. We also discuss the relations of ruler-compass methods to the coordinates of the constructed points.

It is important to use the projective model. We show in Section 5 that the non-Euclidean parabola involved in axioms $\mathbb{H}_5$ and $\mathbb{H}_6$ is a conic and thus by Bezout's theorem there are at most four common tangents to a pair of these curves.

Finally, we discuss the use of the fold $\mathbb{H}_6$, which accomplishes geometrical constructions that are not ruler-compass constructions. These allow constructions that can be used to solve cubic and quartic equations. We show that the real subfield of Euclidean origami numbers [Alperin 00] can be realized with these constructions. For this, we show how to construct cube roots and trisections of angles in hyperbolic geometry.

2 Basic Alignments and Folds

2.1 Alignments

A *fold line* is the axis of a perpendicular reflection. We want to find (minimal) *alignments* of points and lines that are brought into coincidence (aligned) by one reflection. A fold line is determined by specifying its two degrees of freedom. The basic alignments and partial alignments (using only one degree of freedom) are the following:

1. *Fold two points together:* $P \leftrightarrow Q$ (uses two degrees of freedom).

   If the points are the same, $P \leftrightarrow P$ means the point must be on the fold line. There is one degree of freedom remaining.

2. *Fold two lines together:* $L \leftrightarrow M$ (uses two degrees of freedom).

   If the lines are equal, $L \leftrightarrow L$ yields a fold perpendicular to the L; this uses one degree of freedom.

3. *Fold a point and a line together:* $P \leftrightarrow L$ (uses one degree of freedom).

For the fold when P is on L, there is just one degree used; the folds produced are either perpendicular to L or fold lines pass through P.

The partial alignments are now combined with other partial alignments to give an alignment or fold line. An alignment can thus be expressed using either one or two $\leftrightarrow$ symbols.

2.2   Folds

We consider the alignments that describe a line or finite set of lines in the hyperbolic plane using the projective model of Cayley-Klein as the interior of a circle (the absolute). The lines and points are ordinary lines that meet the interior of the absolute and ordinary points interior to the absolute.

We use the notation $\mathbb{H}$ to denote the alignments in the hyperbolic plane.

2.3   Unique Alignment Folds

The following alignment rules determine at most one line.

$\mathbb{H}_0$: $L \leftrightarrow L$, $M \leftrightarrow M$. This fold is the unique perpendicular to two ultraparallel lines. (See Figure 1.)

$\mathbb{H}_1$: $P \leftrightarrow P, Q \leftrightarrow Q$. This alignment or fold is the line passing through two points. (See Figure 2.)

$\mathbb{H}_2$: $P \leftrightarrow Q$. This fold is the perpendicular bisector of PQ. (See Figure 2.)

$\mathbb{H}_3$: $P \leftrightarrow P$, $L \leftrightarrow L$. This fold is the perpendicular to L through P. (See Figure 3.)

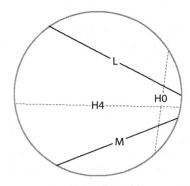

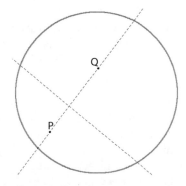

Figure 1. $\mathbb{H}_0$ is the common perpendicular, and $\mathbb{H}_4$ is the midline of L, M.

Figure 2. $\mathbb{H}_1$ is the line through P, Q, and $\mathbb{H}_2$ is the perpendicular bisector of P, Q.

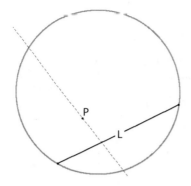 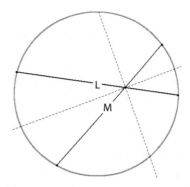

Figure 3. $\mathbb{H}_3$ is the perpendicular to L through P.

Figure 4. $\mathbb{H}_4$ gives the two perpendicular angle bisectors of L, M.

2.4 Quadratic Folds

The above alignments have two solutions in general.

$\mathbb{H}_4$: $L \leftrightarrow M$. These alignment folds are the two (perpendicular) angle bisectors. (See Figure 4.)

$\mathbb{H}_5$: $P \leftrightarrow L$, $Q \leftrightarrow Q$.

All the reflections of P in the pencil of lines at Q (using $Q \leftrightarrow Q$) form the circle centered at Q passing through P. The intersection points of this circle with L are the reflected images P_1, P_2 of P. The hyperbolic circle with center Q appears as an ellipse in the projective model, which passes through P with minor axis OQ (O is the center of the absolute). (See Figure 5.)

An alternative interpretation of these folds is that they are the tangents to a non-Euclidean parabola with focus P and directrix L, where the fold lines pass through Q. We discuss this in more detail in Section 5.

$\mathbb{H}_{5'}$: $P \leftrightarrow L$, $M \leftrightarrow M$.

The reflections of P in all the perpendiculars to M (using $M \leftrightarrow M$) gives the equidistant curve to M passing through P. The intersections of this equidistant curve with L gives at most two reflected images whose perpendicular bisectors with P are the fold lines. The line L may meet the equidistant curve in 0, 1, or 2 points on the same side of M as P. (See Figure 6.)

The equidistant curve appears in the projective model as a branch of a conic tangent to the absolute at the ends of M and passes through P.

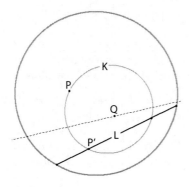

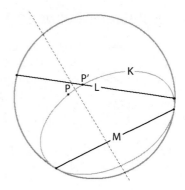

Figure 5. $\mathbb{H}_5$ with circle K having center Q and radius QP.

Figure 6. $\mathbb{H}_{5'}$ with equidistant curve K having axis M and passing through P.

2.5 Quartic Folds

These alignments determine four lines in general.

$\mathbb{H}_6$: $P \leftrightarrow L$, $Q \leftrightarrow M$.

The reflection of Q in the tangents to the non-Euclidean parabola, enveloped by the folding $P \leftrightarrow L$, give a quartic curve. It is a rational singular quartic related to the constructions of pedals of conics [Alperin 04]. One of the singularities of this quartic is at Q. Intersections of the line M with this quartic give four possible solutions. In the Euclidean case, one of the four common tangents is at infinity so there are at most three fold lines (see Figure 7). In the non-Euclidean case we may have some of these four possible fold lines outside of the absolute (see Figure 8).

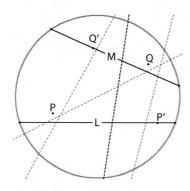

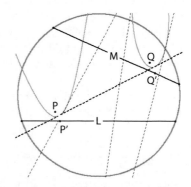

Figure 7. $\mathbb{H}_6$ with fold line P to P', Q to Q'.

Figure 8. $\mathbb{H}_6$ with non-Euclidean parabolas shown.

3 Relations between the Alignment Axioms

We first investigate relations between these alignments in the context of neutral geometry. We use the notation O_i (for neutral geometry) rather than $\mathbb{H}_i$, indicating that we are not using any properties of parallels.

Theorem 1. *In neutral geometry with O_1, $O_6 \to O_5$, $O_5 \to O_4$, $O_5 \to O_3$, $O_5 \to O_2$, $O_4 \to O_2$, and $O_6 \to O_{5'}$, $O_{5'} \to O_3$, $O_{5'} \to O_2$.*

Proof: $O_6 \to O_5$: Let $M = PQ$. Folding Q onto M means the fold passes through Q or is perpendicular to M. The folds through Q are those which satisfy O_5.

$O_5 \to O_4$: Fold P onto $L = QR$ (passing through Q) gives the angle bisectors of $\angle PQR$. If the lines don't meet, then use Bolyai's construction: place two points P, P' on L and Q, Q' on M. Construct, using O_1, the line PQ and $P'Q'$. Construct angle bisects at P, Q meeting at R and P', Q' meeting at R'. The line RR' is the desired fold line.

$O_5 \to O_3$: Choose point Q on L (and different from P); now fold, using O_5, the line that reflects Q to L and passes through P. This line must be perpendicular to L and pass through P.

$O_5 \to O_2$: Make perpendiculars at P and Q on PQ; bisect these right angles using O_4; now use O_1 to connect intersections of bisectors giving the perpendicular bisector.

$O_4 \to O_2$: Fold P onto L. The bisection of the $180°$ angle at P is the same as the perpendicular to P at L. For two points P, Q we construct the perpendiculars at P, Q to the line $L = PQ$. Now bisect these right angles, and then construct the perpendicular bisector of PQ in the standard way using a line through the intersections of the corresponding bisectors.

$O_6 \to O_{5'}$: Fold P on L and Q on M when Q is on M; this is same as folding P to L and perpendicular to M.

$O_{5'} \to O_3$: To see this, we first place a point Q on M and create $L = PQ$ using O_1. Now, by $O_{5'}$, fold P to L so that the fold line is perpendicular to M. The fold line must pass through P or be perpendicular to L, since P is incident to L. The fold line cannot be perpendicular to both L and M, since they meet. Thus, the fold line passes through P and is perpendicular to M.

$O_{5'} \to O_2$: Using O_1, make $M = PQ$. Now, using $O_{5'} \to O_3$, create the line L perpendicular to M at Q. Folding P to L perpendicular to M gives the fold line, which is the perpendicular bisector of PQ.   □

3.1 Totally Real Constructions

In the early seventeenth century, Van Schooten discussed constructions in Euclidean geometry that depend on the transfer of lengths. In Euclidean geometry, this is equivalent to the use of angle bisectors, which gives the

Pythagorean or totally real constructions of Hilbert [Hilbert 38, Alperin 00]. Here we show the constructive power of $O_1 - O_4$ in neutral geometry.

Theorem 2. *With O_1, O_3, O_4, given center P, we can construct the symmetry or $180°$ rotation about P of any point A.*

Proof: Construct $M = PA$ and L the perpendicular to M at P; create the perpendicular from A to the angle bisectors of L, M. Construct the perpendicular from A to first bisector meeting L at A_1; construct the perpendicular from A_1 to second bisector meeting PA at A'. □

Corollary 1. *With O_1, O_3, O_4, given a line L and a point P not on L, we can construct the reflection P' of P across L. Hence the reflection of a segment PQ across L can be constructed.*

Proof: Construct the perpendicular line M from P to L, meeting it at Z; now by symmetry, using Theorem 2 about Z, we can move P to P' on M, which is the same as the reflection of P across L. For the segment PQ, we use the same construction as above for each point P and Q. □

Corollary 2. *With O_1, O_3, O_4, we can move segment of length AB to any point P on the line $L = AB$.*

Proof: Construct the perpendicular bisector M of P and A. Reflect B across M by Corollary 1 to B'. Then PB' has same length as AB. □

Corollary 3. *With O_1, O_3, O_4, we can move a segment AB to begin at any given point on any given line L through P.*

Proof: Construct M, the perpendicular bisector of PA. Reflect B across M to B'. Construct angle bisectors of L and PB'. Construct a perpendicular to the angle bisector through B_1 meeting L at B'. Then the length of PB' is the same as the length of AB. □

Corollary 4. *With O_1, O_2, O_3, the following constructions are equivalent:*

(a) a given length can be marked on any constructed ray;

(b) the angle bisector axiom O_4;

Proof: (a)$\rightarrow$ (b): Without loss of generality, we can assume the angle with vertex O is less than $180°$. We mark the given length on each ray of the angle say at A and B. Construct the midpoint of AB, then OC is an angle bisector.

(b)$\rightarrow$ (a): This is Corollary 3. □

3.2   Radical Axis and Ruler-Compass Constructions

The following construction of the radical axis works in neutral geometry; it is similar to a construction in [Handest 56]. Given two circles with centers A and B, construct the line $L = AB$. The perpendiculars to L at A and B meet the respective circles at C, D. These points are constructed using Theorem 4. The perpendicular bisector of CD meets L at P. The midpoint of AB is Z. The radical axis N is perpendicular to L passing through Q, where $ZQ = ZP$. This last can be accomplished by using a reflection, as in Theorem 2. Thus, this all can be accomplished by O_1, O_5.

Theorem 3. *All ruler-compass constructions in neutral geometry can be done using O_1, O_5.*

Proof: To see that we can do all ruler-compass constructions, we need only to show that the intersection of two circles can be constructed. By the remarks above, we can construct the radical of the two circles, and by O_5 we can construct the intersection of the radical with either of the circles to get the intersections of the two circles.   □

Theorem 4. *With O_1, O_5, we may construct the intersection of a line and circle. With O_1, $O_{5'}$, we may construct the intersection of a line and an equidistant curve.*

Proof: From the previous result, we can use O_5 to create O_3. Given line L and the circle with center Q and passing through P, we use O_5 to construct the two folds reflecting P onto L with fold lines passing through Q. Now, using O_3, we drop a perpendicular from P to these fold lines which meet L at the reflected images of P.

From the previous result, we can use $O_{5'}$ to create O_3. Given line L and an equidistant curve with axis M and passing through P, we use O_5 to construct the two folds of P on L with fold lines perpendicular to M. Now, using O_3, we drop a perpendicular from P to these fold lines, which meet L at the desired points.   □

3.3   Saccheri Quadrilaterals

Now we restrict our study to hyperbolic geometry. We can prove that $\mathbb{H}_0$ follows from the axioms $\mathbb{H}_1$–$\mathbb{H}_4$ using Saccheri quadrilaterals. A quadrilateral $ABCD$ with $AC \equiv BD$ and angles at A and B that are right angles is called a Saccheri quadrilateral.

Theorem 5. *In a Saccheri quadrilateral, the angles at B and D are equal (in neutral geometry) and less than $90°$ in $\mathbb{H}^2$.*

Proof: By the side-angle-side theorem (SAS), $\triangle ABC \equiv \triangle ABD$, so the diagonals are equal, $AD \equiv BC$. Congruence of diagonals then yields $\triangle ACD \equiv \triangle BCD$ by the side-side-side theorem (SSS); hence the angles at C and D are equal. In $\mathbb{H}^2$, triangles have less than 180 degrees so the last result now follows. $\square$

Theorem 6. *In a Saccheri quadrilateral, the midpoint line EF is perpendicular to sides CD and AB.*

Proof: By construction, $\triangle AFC \equiv \triangle BDF$, so $CF \equiv DF$. Thus $\triangle CEF \equiv \triangle DEF$; hence, $\angle CEF = \angle DEF$ and are $90°$ degrees because they are on a line. $\square$

This will give a construction of the common perpendicular to two (ultra) parallel lines once we establish a Saccheri quadrilateral.

Theorem 7. $\mathbb{H}_4, \mathbb{H}_3, \mathbb{H}_1 \to \mathbb{H}_0$.

Proof: Let CD be on one line and AB be the feet of the perpendiculars on the second line. In the case $AC \equiv BD$, the midpoint line EF is the common perpendicular. Use $\mathbb{H}_2$ to construct midpoints. Otherwise, one side is longer than the other. We move one in until we get equality, using the ability to transfer distances and angles, Corollary 3; now the result follows from the first case. $\square$

3.4   Ruler-Compass Constructions

As we have shown in Section 2.4, ruler-compass constructions can be done with $\mathbb{H}_1$ and $\mathbb{H}_5$; also, it is easy to see that these can be done with ruler and compass by Theorem 3, so the constructive power of ruler-compass is equivalent to $\mathbb{H}_1, \mathbb{H}_5$. The parallel ruler can do all ruler-compass constructions by Handest's results [Handest 56], so it is equivalent to using $\mathbb{H}_5, \mathbb{H}_1$.

However, there are other circle like objects or cycles in the hyperbolic plane, the equidistant curve, and also the horocycle. The equidistant curve is related to the use of $\mathbb{H}_{5'}$; it is a cycle where the center is outside the absolute, whereas the horocycle is a cycle with its center on the absolute.

The famous results of Nesterovich [Coxeter 47] show that the extra cycles do not add new constructive power.

Theorem 8 (Nestorovich). *Usage of other compasses (horocompass and hypercompass, which have centers on and outside the absolute, respectively, and pass through an interior point) adds no new information; that is, these constructions can be done using an ordinary (hyperbolic) compass.*

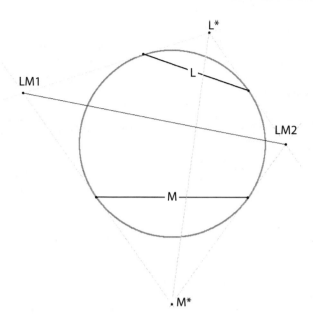

Figure 9. $\mathbb{H}_0$: common perpendiculars.

3.5   Simulation of Constructions with $\mathbb{H}_0$–$\mathbb{H}_4$

Allowing the Euclidean construction of the two tangents to the absolute at the ends of a line L, which meet at L^*, gives the facility for construction of perpendiculars to L, since any perpendicular to L passes through L^*. Note that these tangents are also perpendicular to the radial lines from the center of the absolute. We also allow the construction of the intersection of a line with the absolute.

With this extra ability, we explain how to make the non-Euclidean folds of $\mathbb{H}_0$–$\mathbb{H}_4$ on an ordinary flat piece of paper. Of course, $\mathbb{H}_1$ is the folding of a line through two given points.

$\mathbb{H}_0$: Perpendicular to two lines, L, M, we construct the line L^*M^*. The other two points of intersection of these four tangent lines are LM_1, LM_2; these can be used to give the midline of L, M, thereby solving $\mathbb{H}_4$ when the lines do not meet. (See Figure 9.)

$\mathbb{H}_2$: To make the perpendicular bisector to given points A, B, we construct AL^*, BL^*, meeting the absolute in P, Q, R, S on opposite sides of L. The lines PS, QR meet L at the midpoint M of AB, and then ML^* is the perpendicular bisector. (See Figure 10.)

$\mathbb{H}_3$: Make the perpendicular to L through P by constructing the line PL^*. (See Figure 11.)

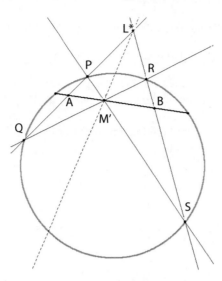

Figure 10. $\mathbb{H}_2$: perpendicular bisector.

$\mathbb{H}_4$: To make the angle bisectors of lines L, M meeting at O, first construct the ends P, Q, R, S of L, M, respectively; the lines PS, QR, PR, QS meet at O_1^*, O_2^*, respectively. The lines OO_1^*, OO_2^* are the angle bisectors. (See Figure 12.)

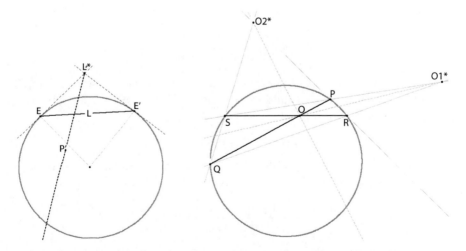

Figure 11. $\mathbb{H}_3$: perpendicular to L through P.

Figure 12. $\mathbb{H}_4$: angle bisectors.

4   Trigonometry and More Folding in $\mathbb{H}^2$

4.1   Hyperbolic Coordinates, Distances, Angles

For the projective model of hyperbolic geometry, the interior of the unit circle, we measure distance ρ from the origin and angle θ with respect to the x-axis.

These are the hyperbolic coordinates of $Q = (u, v)$. The distance from the origin satisfies

$$\cosh(\rho) = \frac{1}{\sqrt{1 - u^2 - v^2}};$$

hence,

$$\sinh(\rho) = \frac{\sqrt{u^2 + v^2}}{\sqrt{1 - u^2 - v^2}}, \quad \tanh(\rho) = \sqrt{u^2 + v^2};$$

also

$$\cos(\theta) = \frac{u}{\sqrt{u^2 + v^2}}, \quad \sin(\theta) = \frac{v}{\sqrt{u^2 + v^2}},$$

so in terms of the hyperbolic coordinates

$$Q = (u, v) = \tanh(\rho)(\cos(\theta), \sin(\theta)).$$

More generally, if $P = (r, s)$ and $Q = (u, v)$ are two points inside the circle, the hyperbolic distance $\rho(P, Q)$ between them satisfies

$$\cosh(\rho(P, Q)) = \frac{1 - ru - sv}{\sqrt{(1 - r^2 - s^2)(1 - u^2 - v^2)}}.$$

If line L has equation $ux + vy = 1$ and M has equation $rx + sy = 1$, then

$$\cos(\angle LM) = \frac{ur + sv - 1}{\sqrt{(u^2 + v^2 - 1)(r^2 + s^2 - 1)}}.$$

Equation of the circle. The equation of the circle centered at $O = (c, d)$ passing through $P = (r, s)$ and general point $X = (x, y)$ can be expressed simply by an algebraic relation using

$$\cosh(\rho(O, P))^2 = \cosh(\rho(O, X))^2.$$

Expanding this equation using the definition of the distance given above yields the equation of a (projective) conic,

$$x^2 + y^2 - 1 + \frac{(1 - r^2 - s^2)}{(1 - cr - ds)^2}(cx + dy - 1)^2 = 0.$$

Hence, the non-Euclidean circle with center at the origin $(0,0)$ appears as an ordinary Euclidean circle in the projective model (where the dual of the center of the absolute is the line at infinity).

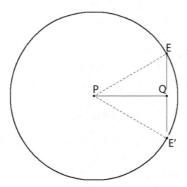

Figure 13. Asymptotic angle of $30°$.

Hyperbolic trigonometry. For a right triangle with included angle θ, opposite side length a, adjacent side b, and hypotenuse c, we have

$$\sin\theta = \frac{\sinh a}{\sinh c}, \quad \cos\theta = \frac{\tanh b}{\tanh c}.$$

More complicated formulas exist for other trigonometric functions of the angles and ratios of hyperbolic trigonometric functions of the side lengths [Carslaw 16]. The analogue of the Pythagorean theorem is that

$$\cosh c = \cosh a \cosh b.$$

Angle of parallelism. The famous formula of Bolyai,

$$\cos\Pi(x) = \tanh(x),$$

relates the perpendicular distance $x = PQ$ from P to a line L (at Q) with the angle of parallelism $\Pi(x) = \angle QPE$ between PQ and the line PE where E is an end of L.

One can regard the asymptotic lines from the origin as having an asymptotic angle of $30°$ to the perpendicular line to the x-axis at $Q(\sqrt{3}/2, 0)$ as in Figure 13. The distance ρ is about 1.317, whereas the distance to the boundary point $(1,0)$ from $P(0,0)$ is infinite.

Using a method of Bolyai, we may construct the asymptotic line through a given point.

Theorem 9 (Bolyai). *Given a point P not on a line L, we may construct, using $\mathbb{H}_5, \mathbb{H}_1$, the asymptotic lines through P to L.*

Proof: Construct Q on line L and construct perpendicular line M to L through P. Mark another point R on L. Construct L', the perpendicular line to M at P. Transfer distance QR to line L' as PR'. Construct M' perpendicular L at R.

Using Theorem 4, we may construct the points T, T' on the line M' which are also on the circle centered at P passing through R'. The lines PT, PT' are asymptotic (parallel) to L. $\qquad\qquad\qquad\qquad\square$

4.2 Parallel Ruler and Its Simulation

The parallel ruler is an instrument discussed by Handest [Handest 56]. The parallel ruler constructs a line though a point P which is also parallel to a given line L, i.e. it constructs an asymptotic line (see Figure 14). In the hyperbolic plane there are two such lines. Using the result from Theorem 9, we can construct these lines using $\mathbb{H}_5$ and $\mathbb{H}_1$. Another interesting tool for ruler compass constructions is the hyperbolic ruler [Al'Dhahir 62].

It is interesting to notice that this parallel ruler is like a fold line using a modified $\mathbb{H}_1$ where we allow one of the points to be on the absolute.

We can easily simulate the parallel ruler using Euclidean constructions by intersection of the line L with the circle absolute, and then construct the lines from P to the ends E, E' of L.

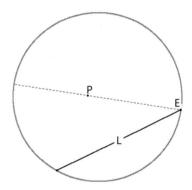

Figure 14. Parallel ruler folds a line through P parallel to L.

4.3 Theorem of Mordukhai-Boltovskoi

Theorem 10, from Mordukhai-Boltovskoi [Curtis 90], says that a segment of length x is ruler-compass constructible if adn only if (iff) $\sinh(x)$ lies in the real subfield $\sqrt{\mathbb{Q}}$, the square root closure of $\mathbb{Q}$. From this and the trigonometrical results, we see that an angle of parallelism, or any constructible angle θ, is ruler-compass constructible iff $\cos(\theta)$ is ruler-compass constructible (this is the same as the Euclidean condition).

A unit length is not constructible, since $\tanh(1)$ is transcendental. But for the points $O = (0, 0)$ and $P = (1/2, 1/2)$, we see that the distance is ruler-compass constructible since $\tanh(\rho) = 1/\sqrt{2}$ and $\cos(\theta) = 1/\sqrt{2}$. For $Q = (1/2, 0)$, since $\cosh(\rho) = 2/\sqrt{3}$, then $\tanh(\rho) = 1/2$ and $\cos(\theta) = 1$ and hence this point is also constructible.

We assume that the center O of the absolute is given.

Theorem 10. *Points in the hyperbolic plane are ruler-compass constructible iff* $\tanh(\rho), \cos(\theta) \in \sqrt{\mathbb{Q}}$, *the field of surds. Lengths are constructible iff* $\tanh(\rho)$ *is a surd. Angles are constructible iff they are Euclidean constructible; i.e.,* $\cos(\theta)$ *is obtained by using repeated square roots and field operations.*

Without using the result of Nesterovich, we can directly show how to deduce $\mathbb{H}_{5'}$ by using ruler-compass constructions (i.e., $\mathbb{H}_1, \mathbb{H}_5$).

Proposition 1. $\mathbb{H}_1, \mathbb{H}_5 \to \mathbb{H}_{5'}$.

Proof: We are given P, L, M, and we want to fold P onto L at P' with a reflection line perpendicular to M. First, construct the perpendicular from P to M meeting at Q, with PQ of length a.

Suppose that L, M meet at R, making an angle ω, and $P'R$ of length c. From Bolyai's formula for the right triangle $\triangle P'Q'R$, we have $\sin \omega = \sinh a / \sinh c$, so we can mark the distance $\sinh c$ starting from R along L to meet the equidistant curve at P'. The perpendicular bisector of PP' (or QQ') is the fold line of $\mathbb{H}_{5'}$. (Note that, to do this construction, we know that c is a constructible length since $\sinh c$ lies in the field containing $\sinh a, \sin \omega$, which is of degree a power of 2, since a and ω are constructible.)

If L, M do not meet, then first construct the intersections A and B of the common perpendicular N to lines L and M. From the unknown point P' on L construct the perpendicular meeting M at Q'. The quadrilateral $BAQ'P'$ is a Lambert quadrilateral; that is, it has three right angles. Using trigonometric formulas, we can solve for $\cosh b$, $b = BQ'$. Since we know $\sinh a$, $a = P'Q' = PQ$, and $\tanh a'$, $a' = AB$, using the 90° angle at B, we get $\sinh a / \sinh c = \tanh a' / \tanh c$ for $c = BP'$; hence, $\cosh c = \sinh a / \tanh a'$. Hence, by the Pythagorean theorem, we can determine $\cosh b$, so b is constructible. The perpendicular bisector of QQ' is thus the fold line needed for $\mathbb{H}_{5'}$ and can be constructed using the ruler-and-compass constructions (i.e., by Theorems 10 and 3 using $\mathbb{H}_1, \mathbb{H}_5$). □

4.4 Construction of Regular Tessellations

Suppose that we want to construct, by ruler and compass, the tessellation generated by the triangle group (p, q, r) inside the hyperbolic plane. We must have that the angles are constructible. Thus, the sines and cosines of the angles $A = \pi/p, B = \pi/q, C = \pi/r$ lie in a quadratic field. Thus,

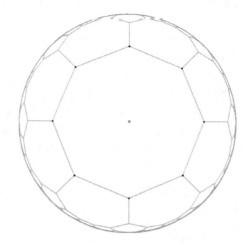

Figure 15. Tessellation of $\mathbb{H}^2$ by regular octagons.

p, q, r can be products of the Fermat primes $3 = 2 + 1, 5 = 2^2 + 1, 17 = 2^4 + 1, 257 = 2^8 + 1$, etc. and arbitrary powers of 2. These angles determine a unique triangle, so it remains only to compute the side lengths. This is easily accomplished with the hyperbolic dual triangle formula:

$$\cos C = -\cos A \cos(B) + \sin A \sin B \cosh c.$$

Hence, it follows that the cosh of the side length is constructible since it lies in the field generated from the angles. For example to construct a $(8,3,2)$ triangle and hence an octagon that tessellates the plane, we have $\cosh(c) = \cot A \cot B = (1+\sqrt{2})/\sqrt{3}$, which satisfies the polynomial $9x^4 - 18x^2 + 1 = 0$. The triangle tessellation contains a tessellation by octagons that meet three at a vertex (Figure 15), so the octagons have the interior angle of $120° = 2\pi/3$. If we use other right triangles $(p, q, 2)$, we can construct tessellations by regular polygons with p sides meeting q to a vertex.

For the tessellation by right-angled pentagons, we use the $(5, 4, 2)$ triangle group. In this case, we have $\cos(\pi/5) = \sin(\pi/2)\sin(\pi/4)\cosh(c)$ and then $\cosh(2c) = (1 + \sqrt{5})/2$ corresponds to the edge length of the right-angled pentagon.

5   The Non-Euclidean Parabola

Modifying $\mathbb{H}_5$ by folding the point F onto the line L in all possible ways, we create the tangent lines to a curve. We call this the non-Euclidean parabola with focus F and directrix L (Figure 16).

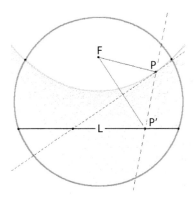

Figure 16. Non-Euclidean parabola with directrix L and focus F.

We show now that the locus $\mathcal{K}$ of points $P = (x, y)$, whose distance to the focus $F = (u, v)$ is the same as the (non-Euclidean) perpendicular distance to the directrix $L : Ax + By = 1$, is a part of a conic in the ambient plane of the absolute. We show that this curve $\mathcal{K}$ is also the envelope of perpendicular bisectors of F with points on the directrix. The tangent line to $\mathcal{K}$ at P is also the perpendicular bisector of FP', where P' is the foot of the perpendicular from P on L. The perpendicular bisector passes through P, since $PF = PP'$; any other point Q of the curve that is also on the perpendicular bisector has $QF = QQ' < QP'$, since Q' is shortest distance from Q to L. But the perpendicular bisector is the locus of X with $XF = XP'$. Therefore, there can be only one point of the perpendicular bisector on this conic; hence, it is a tangent. Thus, the locus $\mathcal{K}$ is the curve enveloped by the folds of F on L. This shows that our non-Euclidean parabola is the same as the curve discussed in [Henle 98].

Since the closest point on the directrix $Ax + By = 1$ is the intersection of the line PL^* (the perpendicular from P) and the directrix L, we use the formula for $\cosh^2$ of the distance to obtain the equation for $\mathcal{K}$:

$$
\begin{aligned}
K = {} & 1 - (u^2 + v^2)(A^2 + B^2) + 2x(A(v^2 + u^2 - 1) + u(A^2 + B^2 - 1)) \\
& + 2y(B(u^2 + v^2 - 1) + v(A^2 + B^2 - 1)) \\
& + x^2((1 - B^2)(1 - v^2) - u^2 A^2) + y^2((1 - A^2)(1 - u^2) - v^2 B^2) \\
& + 2xy(AB(1 - u^2 - v^2) + uv(1 - A^2 - B^2)).
\end{aligned}
\tag{1}
$$

Choosing special points and lines $(u, v) = (0, a)$ and $(A, B) = (0, -\frac{1}{a})$, we obtain the simplified form $4ay = (1 - a^2)x^2$. The tangents from $(b, 0)$ have a slope given by points on the dual curve, which in this case has the equation $y = (a/(a^2 - 1))x^2$.

6 $\mathbb{H}_6$

Certainly if we realize a fold line N by $\mathbb{H}_6$, $P \leftrightarrow L, Q \leftrightarrow M$, then P folds to P' on L and Q folds to Q' on M, so P', Q' are in the hyperbolic plane, since they are reflections of P, Q in N.

We know that $\mathbb{H}_6 \to \mathbb{H}_{5'}$, $\mathbb{H}_6 \to \mathbb{H}_5$ by Theorem 1 so this axiom apparently allows more than ruler-compass constructions.

Can we construct one-third of a constructible length using these origami axioms? The Euclidean origami method uses similar triangles, but they are not available in hyperbolic geometry [Ludwig 78]. The problem can be restated: suppose $\sinh(x)$ is constructible, then is $\sinh(x/3)$ also? There is a familiar cubic relationship $\sinh(3x) = 4\sinh(x)^3 + 3\sinh(x)$.

Can we trisect angles? The Euclidean trisection method of Abe uses results about alternate interior angles for parallel lines, so the argument does not work in $\mathbb{H}^2$. The trisection of an angle gives a well-known cubic relationship between $\cos(\theta)$ and $\cos(\theta/3)$, coming from the identity $\cos 3\theta = 4\cos^3 \theta - 3\cos\theta$.

Can we solve these cubic equations $4x^3 \pm 3x - a = 0$ with the aid of origami in the hyperbolic plane? This problem is discussed in the next section.

6.1 Higher Origami Constructions

Axiom $\mathbb{H}_6$ permits folds of the common tangents to two non-Euclidean parabolas. Since these curves are conics, there are four possible tangents; these can be determined as the common points to the two adjoint or dual curves.

The equation for the dual curve of K is $K^d = 0$, where

$$
\begin{aligned}
K^d =& 1 + vB + uA - 2(A + u)x - 2(B + v)y \\
& + 2(Bu + vA)yx + (1 - vB + uA)x^2 + (1 + vB - uA)y^2.
\end{aligned}
\tag{2}
$$

To locate common tangents of two parabolas, we use the common points of the two dual curves or the resultant of the two dual curves, which is a polynomial of degree 4. The solutions to this polynomial give the information to recover the equation of the tangent lines. We can take an intersection of the adjoint or dual curves and construct the tangent lines to the given conics by using inversion. The perpendicular projection $P = \pi(L) = \tanh(\rho)(\cos\theta, \sin\theta)$ of the origin on a line L is given by inversion of the dual of L.

Now, since the intersection points of conics have their coordinates lying in the field of Vietans, $\mathbb{V}$, the real subfield of origami numbers discussed in [Alperin 00], the coordinates of P also belong to $\mathbb{V}$; hence, the sum of squares of these coordinates belongs to $\mathbb{V}$, so $\tanh\rho(O, P) \in \mathbb{V}$, and then it also follows that $\cos\theta \in \mathbb{V}$.

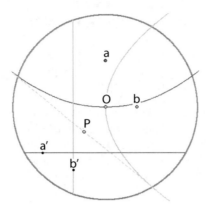

Figure 17. Using $\mathbb{H}_6$ to create $\sqrt[3]{6}$.

Theorem 11. *Any constructible length ρ or constructible angle θ using the hyperbolic origami axioms $\mathbb{H}_0 - \mathbb{H}_6$ has $\tanh \rho$ and $\cos \theta$ in the real subfield $\mathbb{V}$ of origami numbers.*

By Theorem 10, the ruler-compass constructions give the square root closure of $\mathbb{Q}$ as a subfield of the non-Euclidean origami constructible numbers. In general, the four possible tangents of two non-Euclidean parabolas have line coordinates belonging to the field $\mathbb{V}$ of Euclidean origami numbers. Thus, we can realize all Vietans if we can construct all cube roots and trisections of angles by our hyperbolic origami constructions.

We now give methods to obtain resultant equations for trisections or real cube roots.

In order to accomplish the cube roots of elements, we can use two parabolas, as in the example of Section 5. The dual equation is also the equation of a parabola. Let $(A, B) = (0, -1/a)$ and $(u, v) = (0, a)$ for the first parabola, and $(A, B) = (-1/b, 0)$ and $(u, v) = (b, 0)$ for the second parabola; then the resultant has the cubic factor $y^3 = g(a, b)$, and we can solve for values of a, b that give any positive value of $g(a, b) \in \mathbb{V}$ using $a, b \in \mathbb{V}$. For example with $a = 1/2$, $b = 1/3$, the projection of the origin on the fold line, as in Figure 17, gives a point $P = (u, v)$ with $u/(u^2 + v^2) = \sqrt[3]{6}$, where $u = \tanh \alpha$, $v = \tanh \beta$, and α, β are the lengths of the projections of P on the coordinate axes.

Because the adjoint (or dual) equations can be put in the form $X = Y^2, Y = \alpha X^2$, by suitable constructible choices of a, b, we have common solutions when $X^3 = \alpha^{-2}$; thus, we obtain the following important result.

Theorem 12. *Using the hyperbolic origami axioms, we can construct $\sqrt[3]{\tanh \rho}$ for any hyperbolic constructible length ρ.*

Using the two non-Euclidean parabolas with focus-directrix coordinates, $(u, v), (A, B)$,

$$(0, -a), \quad \left(0, \frac{1}{a}\right), \quad \left(-\frac{A}{A^2 - B^2}, \frac{B}{A^2 - B^2}\right), \quad (A, B),$$

the resultant enables us to solve cubic equations, since these dual curves by construction have an intersection at the origin.

For the cubic $4x^3 + 3x - m$ for m in the Vietens, we use $B = 2A, a = \frac{1}{A}, A = \sqrt{r}$ and first solve for an auxiliary cube root. Namely, we solve using real cube roots and real square roots for r, so that

$$m = -\frac{3\sqrt{6}}{8}(3r + 1)\sqrt{r - 1}\sqrt{(r + 1)^3}.$$

Thus, we can trisect any segment since $\sinh(3\rho) = 4\sinh(\rho)^3 + 3\sinh(\rho)$.

For the choice of parameters $B = A/2, a = 1/A, A = -\sqrt{r}$, the resultant of the dual conics simplifies to (cf. [Alperin 05]) $4x^3 - 3x - m$ for any $m \in [.64, .84]$ in the Vietens by first solving for an auxiliary cube root, to realize m. Thus, we can also trisect angles. The somewhat intricate details of this are beyond the scope of this discussion. Basically, we need to justify only that these conic duals meet outside the absolute. For this, we develop some further properties of the dual curves; for example, when the directrices meet inside the absolute, then the conics meet only once inside the absolute; hence, there is an intersection outside the absolute and hence a solution to $\mathbb{H}_6$ that is a valid hyperbolic line.

Theorem 13. *Any point in the hyperbolic plane with coordinates in $\mathbb{V}$ can be constructed using the axioms $\mathbb{H}_0 - \mathbb{H}_6$. Hence, by hyperbolic origami, any length ρ or angle θ with $\tanh(\rho) \in \mathbb{V}, \cos(\theta) \in \mathbb{V}$ can be constructed.*

Bibliography

[Al'Dhahir 62] M. W. Al'Dhahir. "An Instrument in Hyperbolic Geometry." *Proceedings of the American Mathematical Society* 13:2 (1962), 298–304.

[Alperin 00] Roger C. Alperin. "A Mathematical Theory of Origami Constructions and Numbers.", *New York J. Math* 6 (2000) 119–133. (Available at http://nyjm.albany.edu.)

[Alperin 04] Roger C. Alperin. "A Grand Tour of Pedals of Conics." *Forum Geometricorum* 4 (2004), 143–151. (Available at http://forumgeom.fau.edu.)

[Alperin 05] Roger C. Alperin. "Trisections and Totally Real Origami." *American Mathematical Monthly* 112:3 (2005), 200–211.

[Alperin and Lang 09] Roger C. Alperin and Robert J. Lang. "One-, Two-, and Multi-Fold Origami Axioms." In *Origami⁴: Fourth International Meeting of Origami Science, Mathematics, and Education*, edited by Robert J. Lang, pp. 371–393. Wellesley, MA: A K Peters, 2009.

[Carslaw 16] H. S. Carslaw. *The Elements of Non-Euclidean Plane Geometry and Trigonometry*. London: Longman, Green, and Co., 1916. (Reprint, New York: Chelsea publishing Co., 1960.)

[Coxeter 47] H. S. M. Coxeter. *Non-Euclidean Geometry*. Toronto: University of Toronto, 1947.

[Curtis 90] Robert R. Curtis. "Duplicating the Cube and Other Notes on Constructions in the Hyperbolic Plane." *Journal of Geometry* 39 (1990), 38–59.

[Fishback 69] W. T. Fishback. *Projective and Euclidean Geometry*. New York: Wiley, 1969.

[Gray 04] Jeremy Gray. *János Bolyai, Non-Euclidean Geometry and the Nature of Space*. Cambridge, MA: Burndy Library, 2004.

[Handest 56] Frans Handest. "Constructions in Hyperbolic Geometry." *Canadian Journal of Mathematics* 8 (1956), 389–394.

[Henle 98] M. Henle. "Will the Real Non-Euclidean Parabola Please Stand Up?" *Mathematics Magazine* 71:5 (1998), 369–376.

[Hilbert 38] David Hilbert. *Foundations of Geometry*. La Salle, IL: Open Court, 1938.

[Jagy 95] William C. Jagy. "Squaring Circles in the Hyperbolic Plane." *Mathematical Intelligencer* 17:2 (1995), 31–36.

[Ludwig 78] Hubert J. Ludwig. "Segment Trisection in Absolute Geometry." *Mathematics Magazine* 51:2 (1978), 124–125.

A Combinatorial Definition
of 1D Flat-Folding

Hidefumi Kawasaki

1   Introduction

An interesting theorem on flat-foldability of one-dimensional (1D) origami says, "Any flat-foldable 1D origami can be folded by local operations (crimp and end fold)" [Arkin et al. 04, Demaine and O'Rourke 07]. The aim of this paper is to present a combinatorial definition of 1D flat-folding in order to give a more rigorous proof of this theorem.

2   Flat-Foldable 1D Origami

A *1D origami* is a closed interval $[v_0, v_n]$ with nodes $v_0 < v_1 < \cdots < v_n$. Any 1D origami is assumed to be oriented from v_0 to v_n. We denote by e_i the edge (v_{i-1}, v_i). Its length is denoted by $|e_i|$. We also denote the 1D origami by $e_1 \cdots e_n$.

When an assignment of mountain = right (valley = left) is given to node v_i, the 1D origami is folded right (left) with a double right angle at node v_i. Such a 1D origami is called a *1D origami with assignments*.

A 1D origami $e_1 \cdots e_n$ with assignments is said to be *flat-foldable* if there exists a permutation $\sigma \in \mathfrak{S}_n$, which indicates the layer distribution, such that the graph constructed in the following manner is *connected* and *doesn't self-cross* (see Figure 1). This construction is as follows:

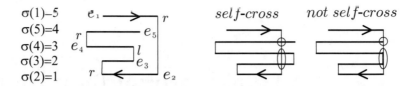

Figure 1. Flat-foldability, illustrating self-crossing.

1. Put e_i horizontally in the $\sigma(i)$th layer.

2. When i is odd and the assignment of v_i is right (left), connect right endpoints of e_i and e_{i+1} by a vertical downward (upward) line segment, if possible.

3. When i is even and the assignment of v_i is right (left), connect left endpoints of e_i and e_{i+1} by a vertical upward (downward) line segment, if possible.

Throughout this discussion, $\sigma(i)$ denotes the layer number of edge e_i. The bottom layer's number is defined to be 1, and the top layer's number is some number n. A flat-foldable 1D origami is illustrated in Figure 2; the 1D origami in Figure 3 is not flat-foldable.

To make our definition strict, we have to define *non-self-crossing* and *connectivity*. We define the *relative coordinate* of the nodes by

$$v_0 = 0, \quad v_i = v_{i-1} - (-1)^i |e_i| \quad (i = 1, \ldots, n). \tag{1}$$

For simplicity, we denote by v_i the relative coordinate of node v_i (Figure 4); there should be no confusion which is meant in a given context.

We say edge e_j is placed between e_i and e_{i+1} if either $\sigma(i) < \sigma(j) < \sigma(i+1)$ or $\sigma(i) > \sigma(j) > \sigma(i+1)$ holds, and simply denote this condition by

$$\sigma(i) \lessgtr \sigma(j) \lessgtr \sigma(i+1). \tag{2}$$

It is clear that Equation (2) is equivalent to $\{\sigma(i)-\sigma(j)\}\{\sigma(i+1)-\sigma(j)\} < 0$.

Next, we say that edge e_j doesn't cross node v_i if either $v_j, v_{j-1} \leq v_i$ or $v_j, v_{j-1} \geq v_i$ holds, and simply denote this condition by

$$v_j, v_{j-1} \lessgtr v_i. \tag{3}$$

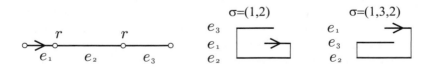

Figure 2. Both $\sigma = (1, 2)$ and $\sigma = (1, 3, 2)$ achieve flat-folding.

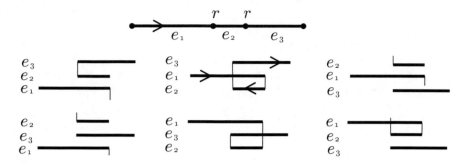

Figure 3. A 1D origami that is not flat-foldable.

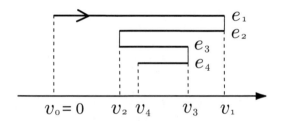

Figure 4. Relative coordinate of the nodes.

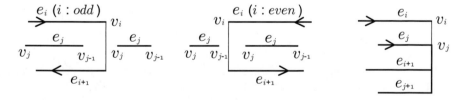

Figure 5. Non-self-crossing (left and middle) and self-crossing (right).

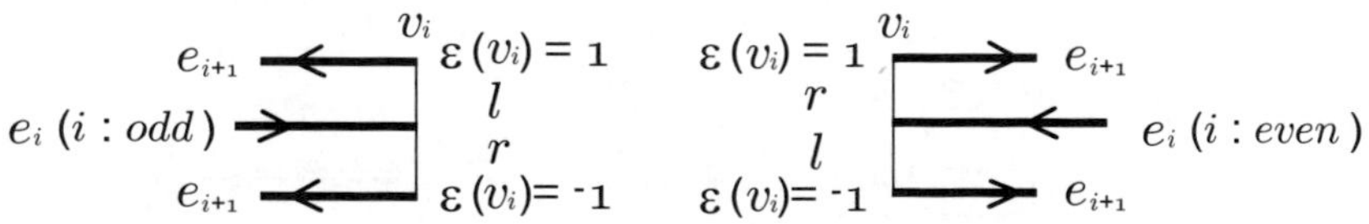

Figure 6. Edges e_i and e_{i+1} are connectable if $\{\sigma(i+1) - \sigma(i)\}\varepsilon(v_i) > 0$.

It is evident that Equation (3) is equivalent to $(v_j - v_i)(v_{j-1} - v_i) \geq 0$, and the first condition of non-self-crossing is

$$\sigma(i) \lessgtr \sigma(j) \lessgtr \sigma(i+1) \quad \Rightarrow \quad v_j, \; v_{j-1} \gtrless v_i. \tag{4}$$

However, when $v_i = v_j$, Equation (3) is trivially satisfied, so we need an additional condition to eliminate the possibility of Figure 5 (right), which is

$$\left\{ \begin{array}{l} \sigma(i) \lessgtr \sigma(j) \lessgtr \sigma(i+1) \\ v_i = v_j, \; (-1)^i = (-1)^j \end{array} \right. \quad \Rightarrow \quad \sigma(i) \lessgtr \sigma(j+1) \lessgtr \sigma(i+1). \tag{5}$$

Finally, we say that edges e_i and e_{i+1}, placed in the $\sigma(i)$th and $\sigma(i+1)$th layers, respectively, are *connectable* if

$$\{\sigma(i+1) - \sigma(i)\}\varepsilon(v_i) > 0, \tag{6}$$

where $\varepsilon(v_i)$ is a sign indicating consistency of the assignment at v_i and connectivity of e_i and e_{i+1}:

$$\varepsilon(v_i) := \left\{ \begin{array}{rl} 1 & (i \text{ is odd, } l) \text{ or } (i \text{ is even, } r), \\ -1 & (i \text{ is odd, } r) \text{ or } (i \text{ is even, } l). \end{array} \right. \tag{7}$$

Summarizing the two non-self-crossing conditions and the one connectivity condition (Figure 6), we obtain our definition of a flat-foldable 1D origami:

Definition 1. A 1D origami is said to be *flat-foldable* if there exists a permutation $\sigma \in \mathfrak{S}_n$ satisfying Equations (4)–(6).

3   Mingling

This section is devoted to defining important concepts: *crimp*, *end-fold*, and *mingling* (see [Arkin et al. 04, Demaine and O'Rourke 07]).

A pair of consecutive nodes $\{v_i, v_{i+1}\}$ is called a *crimpable pair* if they have opposite assignments and $|e_i| \geq |e_{i+1}| \leq |e_{i+2}|$. We call the operation to make a new edge from a crimpable pair a *crimp*. We also call the zigzag in Figure 7 (left) a *crimp*.

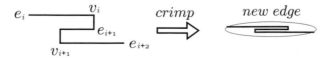

Figure 7. Crimp.

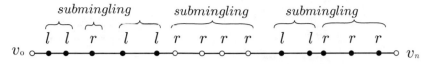

Figure 8. Mingling.

When $|e_1| \leq |e_2|$ (or $|e_{n-1}| \geq |e_n|$), we can flat-fold e_1 and e_2 at v_1 (or e_{n-1} and e_n at v_{n-1}) and create a new edge. We call such an end a *foldable end*, and call this operation an *end-fold*. Crimps and end-folds are called *local operations*.

A maximal sequence of consecutive nodes v_i, v_{i+1}, $\ldots$, v_j with a same assignment is called a *submingling* if either $|e_{i-1}| \leq |e_i|$ or $|e_j| \geq |e_{j+1}|$ holds. When any maximal sequence of consecutive nodes with a same assignment is a submingling, the 1D origami is called a *mingling*. See Figure 8.

Lemma 1. [Arkin et al. 04, Demaine and O'Rourke 07] *Let v_i, v_{i+1}, $\ldots$, v_j be a maximal sequence of consecutive nodes with a same assignment. If 1D origami $[v_{i-1}, v_{j+1}]$ is flat-foldable, then one of the following conditions holds:*

$$|e_i| \leq |e_{i+1}| \leq \cdots \leq |e_{j+1}|,$$
$$|e_i| \geq |e_{i+1}| \geq \cdots \geq |e_{j+1}|,$$
$$|e_i| \leq \cdots \leq |e_{l-1}| \leq |e_l| \geq |e_{l+1}| \geq \cdots \geq |e_{j+1}| \text{ for some } i < l \leq j.$$

Lemma 2. [Arkin et al. 04, Demaine and O'Rourke 07] *Any flat-foldable 1D origami is a mingling. Any mingling has either a crimpable pair or a foldable end.*

4   Proof of the Flat-Foldability Theorem

Theorem 1. [Arkin et al. 04, Demaine and O'Rourke 07] *Any local operation preserves flat-foldability.*

Proof: Assume that 1D origami $e_1 e_2 \cdots e_n$ is flat-foldable. Then there exists $\sigma \in \mathfrak{S}_n$ satisfying Equations (4)–(6).

End-fold: The result of end-fold at v_{n-1} is nothing but the 1D origami $e_1 e_2 \cdots e_{n-1}$. Define $\rho \in \mathfrak{S}_{n-1}$ by

$$\rho(i) = \begin{cases} \sigma(i) & \sigma(i) < \sigma(n), \\ \sigma(i) - 1 & \sigma(i) > \sigma(n). \end{cases}$$

Since $\rho(i) < \rho(j)$ if and only if $\sigma(i) < \sigma(j)$, ρ satisfies Equations (4)–(6), so 1D origami $e_1 e_2 \cdots e_{n-1}$ is flat-foldable.

Figure 9. Moving the edges out of the crimp.

Figure 10. Four kinds of crimps.

Crimp: (This part of the proof is the main concern of this paper.) The point is to prove that we can move the edges out of the crimp, as in Figure 9. We construct a permutation τ of the layer distribution of Figure 9 (right) from the permutation σ of the layer distribution of Figure 9 (left).

There are four kinds of crimps, as shown in Figure 10, and it's enough to prove this for the leftmost crimp. Now we divide the set of edges into four groups (Figure 11):

$E_- := \{e_j \text{ is on the left-hand side of the crimp} \mid \sigma(i+2) < \sigma(j) < \sigma(i)\}$,

$E_0^1 := \{e_j \mid \sigma(j) < \sigma(i+2)\}$,

$E_0^2 := \{e_j \mid \sigma(i) \leq \sigma(j)\}$,

$E_+ := \{e_j \text{ is on the right-hand side of the crimp} \mid \sigma(i+2) \leq \sigma(j) < \sigma(i)\}$.

The edges in E_- are shifted downward right under e_{i+2}, keeping their hierarchical relation so that their layer numbers decrease (Figure 12, left). Any edge in $E_0^1 \cup E_0^-$ doesn't change its layer.

Although any edge in E_+ doesn't move, its layer number increases with the shift of E_- (Figure 12, right). At this point, we don't care about non-

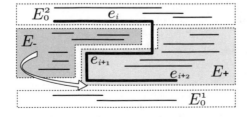

Figure 11. Four groups of edges.

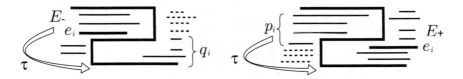

Figure 12. The edges in E_- are shifted downward (left), so the edges in E_+ are relatively shifted upward (right).

self-crossing and connectivity, so we can shift the edges. Indeed, this shift is achieved by the following permutation τ:

$$\tau(j) := \begin{cases} \sigma(j) + p_j, & e_j \in E_+, \\ \sigma(j) - q_j, & e_j \in E_-, \\ \sigma(j), & e_j \in E_0^1 \cup E_0^2, \end{cases} \tag{8}$$

where $p_j := |\{e_k \in E_- \mid \sigma(k) > \sigma(j)\}|$ and $q_j := |\{e_k \in E_+ \mid \sigma(k) < \sigma(j)\}|$.

Hence, our proof is completed if we show that τ satisfies Equations (4)–(6). We address each in turn:

Equation (4): Suppose that there exist l, m such that $\tau(l) < \tau(m) < \tau(l+1)$ and $v_m < v_l < v_{m-1}$. Then, since σ satisfies Equation (4), $\sigma(l) \lessgtr \sigma(m) \lessgtr \sigma(l+1)$ doesn't hold. Hence, the hierarchical relation of $\{e_l, e_m, e_{l+1}\}$ changes. By the definition of τ, such a change is possible only when one of these edges belongs to E_- and another one belongs to E_+. Since E_- and E_+ are separated by the crimp, there are four cases. By considering each of the four cases, we get a contradiction in each case. For example, in Figure 13, which shows the case where $e_l \in E_-$, $e_m \in E_+$, and $\sigma(l) > \sigma(m)$, since $\tau(m) < \tau(l+1)$, we get $e_{l+1} \in E_0^2$. Since $\sigma(l) < \sigma(i) < \sigma(l+1)$, we have $v_l \leq v_{i-1}, v_i$. However, since $e_m \in E_+$, we get $v_{i-1} \leq v_{m-1}, v_m$. So $v_l \leq v_{m-1}, v_m$, which contradicts $v_m < v_l < v_{m-1}$.

Similarly, we can prove that there exist no l, m such that $\tau(l) > \tau(m) > \tau(l+1)$ and $v_m < v_l < v_{m-1}$. Therefore, τ satisfies Equation (4).

Equation (5): Suppose that there exist l, m such that $v_l = v_m$, $(-1)^l = (-1)^m$, $\tau(l) < \tau(m) < \tau(l+1)$, and not $\{\tau(l) \lessgtr \tau(m+1) \lessgtr \tau(l+1)\}$. If $\sigma(l) < \sigma(m) < \sigma(l+1)$, then by Equation (5) $\sigma(l) < \sigma(m+1) <$

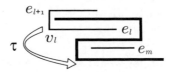

Figure 13. The case where $e_l \in E_-$, $e_m \in E_+$, and $\sigma(l) > \sigma(m)$.

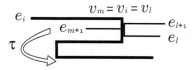

Figure 14. The case where $e_{m+1} \in E_-$, $e_l \in E_+$, and $\sigma(m+1) > \sigma(l)$.

$\sigma(l + 1)$. Hence, the hierarchical relation of $\{e_l, e_{m+1}, e_{l+1}\}$ changes. By the definition of τ, such a change is possible only when one of the edges belongs to E_- and another one belongs to E_+. Since E_- and E_+ are separated by the crimp, there are four cases. By case analysis, we get a contradiction in each of these four cases as well.

For example, in Figure 14, in the case where $e_{m+1} \in E_-$, $e_l \in E_+$, and $\sigma(m + 1) > \sigma(l)$, since $\tau(m + 1) < \tau(l) < \tau(l + 1)$, we get $e_{l+1} \in E_+ \cup E_0^2$. Since e_{m+1} and e_l are separated by the crimp, it holds that $v_m \leq v_i \leq v_l$. By assumption, we have $v_m = v_i = v_l$. Hence, l is even, so m is also even. Thus, edges e_m and e_{m+1} are separated by the crimp, which contradicts connectivity. Therefore, τ satisfies Equation (5).

Equation (6): Assume that there exists m such that $\{\tau(m + 1) - \tau(m)\}\varepsilon(v_m) < 0$. Since σ satisfies Equation (6), $\{\sigma(m+1)-\sigma(m)\}\varepsilon(v_m) > 0$, so

$$\{\sigma(m + 1) - \sigma(m)\}\{\tau(m + 1) - \tau(m)\} < 0.$$

Hence, one of $\{e_m, e_{m+1}\}$ belongs to E_-, and the other belongs to E_+, which contradicts that e_m and e_{m+1} are connected.

In addition to the edges in E_-, we can shift the edges in E_+, except e_{i+1} and e_{i+2}, upward right above e_i. Then we can apply the crimp operation to crimp $e_i e_{i+1} e_{i+2}$, and the result is flat-foldable.   □

We now come to the result of our paper; it follows from Lemma 2 and Theorem 1.

Theorem 2. [Arkin et al. 04, Demaine and O'Rourke 07] *Any flat-foldable 1D origami can be folded by a sequence of local operations.*

Our definition of 1D flat-foldability is combinatorial, rather than continuous; we note that Jacques Justin [Justin 97] gave a combinatorial definition of flat-folding of 2D origami.

Acknowledgments. The author would like to thank Prof. R. Uehara for his valuable comment on the definition of flat-foldability. The author would like to thank the referees for their important comments.

Bibliography

[Arkin et al. 04] E. M. Arkin, M. A. Bender, E. Demaine, M. Demaine, J. S. B. Mitchell, S. Sethia, and S. S. Skiena. "When Can You Fold a Map?" *Computational Geometry: Theory and Applications* 29 (2004), 23–46.

[Demaine and O'Rourke 07] E. Demaine and J. O'Rourke. *Geometric Folding Algorithms.* Cambridge, UK: Cambridge University Press, 2007.

[Justin 97] J. Justin. "Towards a Mathematical Theory of Origami." In *Origami Science and Art: Proceedings of the Second International Meeting of Origami Science and Scientific Origami*, edited by K. Miura, pp. 15–29. Shiga, Japan: Seian University of Art and Design, 1997.

Stamp Foldings with a Given Mountain-Valley Assignment

Ryuhei Uehara

1   Introduction

What is the best way to fold an origami model? Origamists around the world struggle with this problem daily, searching for more clever, more accurate, or faster folding sequences and techniques. Even if you have a good origami model with its crease pattern, this is not the end of the challenge.

We focus here on a simple kind of one-dimensional creasing, where the piece of paper is a long rectangular strip, which can be abstracted into a line segment, and the creases uniformly subdivide the strip. A mountain-valley pattern can then be represented simply as a binary string over the alphabet $\{M, V\}$ (M for mountain, V for valley), which we call a *mountain-valley string*.

Of particular interest in origami is the *pleat*, whose binary string alternates $MVMVMV\cdots$; see Figure 1. As will be shown in Section 2, the pleat folding is unique in the sense that its folded state is unique; that is, there is only one folded state consistent with the string, and only pleat folding has this property.

In general, this uniqueness is not the case. For example, surprisingly, 100 distinct folded states are consistent with the string

$$MMVMMVMVVV,$$

which raises the question: among them, what is the *best* folded state?

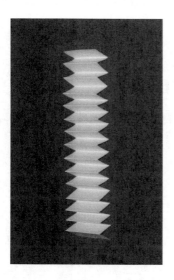

Figure 1. Simple pleats (left) and an origami angel (right) with many pleats folded by Takashi Hojyo (reproduced with his kind permission).

From practical considerations, we would like to minimize the number of paper layers between a pair of paper segments hinged at a crease. If we have many layers between the hinged papers, it becomes difficult to fold with accuracy, and if we have too many layers, we cannot fold any more. This is a typical problem encountered when folding recent complex origami models.

For a folded state, we define the *crease width* at a crease by the number of paper layers between the papers hinged at the crease. Then, we can consider two kinds of optimization problems:

> Given a paper of length $n + 1$ with a mountain-valley string s in $\{M, V\}^n$, among the folded states consistent with s, find a folded state that

1. minimizes the maximum crease width at a crease in the folded state, or

2. minimizes the total crease width for all creases in the folded state.

We note that it is possible to consider the minimization problem for the average crease width, but this is equivalent to the second optimization problem (by dividing by n).

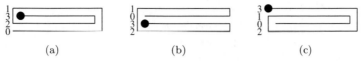

Figure 2. Three foldings for the mountain-valley string VVV.

For example, suppose that we are given a strip of length 4 with a mountain-valley string $s = VVV$. Then we have three different folding choices (Figure 2). For the folded state in Figure 2(a), the maximum crease width is 2 (the rightmost crease in the figure has two layers at the crease), and the total crease width is $2 + 1 + 0 = 3$. All of the folded states in Figure 2 have a minimum maximum crease width 2, but the minimum total crease width 2 is achieved by only the folded state in Figure 2(b).

Interestingly, in general, the two minimization problems have different solutions for a given binary string. For example, among the 100 folded states for the string $MMVMMVMVVVV$, the minimum maximum crease width is 3, which is achieved by the folded state $[4|3|2|5|6|0|1|7|9|11|10|8]$ (the meaning of this notation will be described in Section 2), and the minimum total crease width is 11, achieved by the state $[4|3|2|0|1|5|6|7|9|11|10|8]$. Moreover, these solutions are unique for this string. (The solutions and their uniqueness were checked by an exhaustive search program that I developed.)

Here we state an open problem:

> Determine the computational complexity of the minimization problems of the crease width of the folded state for a given string s in $\{M, V\}^n$.

We first show that the problem is well defined, even in a simple folding model. The simple folding model is one of the basic origami models, which was introduced by Arkin et al. [Arkin et al. 04]. We show that, even in the simple folding model, every folded state consistent with any given mountain-valley string can be folded. This result is related to the locked chain problem, which has a long and rich history [Demaine and O'Rourke 07, Chapter 6], and one-dimensional flat foldings [Demaine and O'Rourke 07, Section 12.1].

The open problem seems to be hard in general. We next prove this intuition from a mathematical viewpoint. A straightforward algorithm for solving the problem is the so-called exhaustive search; for any given mountain-valley string, generate all possible folded states and find the minimum one. If this algorithm runs efficiently, we have hope that the open problem can be solved positively. However, this is not the case.

In this paper, we state the following negative results for the exhaustive search approach.

Theorem 1. *Let s be a mountain-valley string of length n taken uniformly at random, and let $f(n)$ be the expected number of folded states consistent with s. Then $f(n) = \Omega(1.53^n)$, and $f(n) = O(2^n)$.*

That is, we give the lower bound and the upper bound of the number of folded states for a random string. The results guarantee that $f(n)$ is an exponential function, and hence the exhaustive search approach has no hope, in general. For example, suppose that your computer has a CPU running with a 5-GHz clock, and your nice exhaustive search program checks each folding way in a single clock cycle. Then, if you have a random string of length 100, your program would run at least $1.53^{100}/(5 \times 10^9) = 5.89 \times 10^8$ seconds on average, which is equal to 1120 years!

Theorem 1 comes from the more general counting problem:

Theorem 2. *Let $F(n)$ be the number of folded states of a paper of length $n + 1$. Then $F(n) = \Omega(3.06^n)$, and $F(n) = O(4^n)$.*

From Theorem 1, the open problem seems to be difficult to solve positively in general. Unfortunately, we have no idea how to show its hardness (e.g., NP-hardness), up to now.

2   Preliminaries

A *paper strip* is a one-dimensional line with creases at every integer position. The paper is rigid except for the creases on the integer positions; that is, we are allowed to fold only at these integer positions, and the direction of a crease (in $\{M, V\}$) at the end of the foldings follows the letter. To simplify, the ith letter of the mountain-valley string indicates the final folded state of the crease at integer point i in $[1 \ldots n]$. That is, a paper of length $n + 1$ with a string of length n is placed at the interval $[0 \ldots n + 1]$ at first. (We refer to this state as an *initial state*.) We call each paper segment between i and $i + 1$ the ith *segment*. Each (final) folded state of unit length can be represented by the ordering of the segments; for example, a pleat folding $MVMV$ is described by $[0|1|2|3|4]$ or $[4|3|2|1|0]$, and a crease string VVV can produce $[1|3|2|0]$, $[1|0|3|2]$, $[3|1|0|2]$, or their reverses (Figure 2). We distinguish between the left and right ends of the strip, but we sometimes ignore the reverse of one folded state because they are essentially the same. In fact, the side of a folded state is sometimes changed when we fold all paper layers at a crease from right to left or from left to right.

We employ the simple fold model of Arkin et al. [Arkin et al. 04] (see also [Demaine and O'Rourke 07, Section 14.1] and [Cardinal et al. 09]). Put precisely, each simple folding is the folding from a flat-folded state to another flat folded state by the following operations:

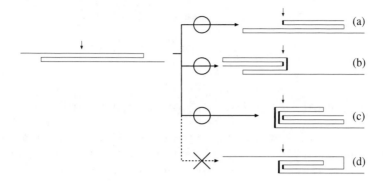

Figure 3. Simple foldings (a)–(c); (d) is not allowed.

1. Position the flat(-folded) paper (on the reverse side, if necessary).

2. Choose an integer point to fold.

3. Valley-fold consecutive inner layer paper segments at the crease.

We note that a simple folding of a set of layers has the restriction that all segments between two folded layers have to be folded together. That is, in Figure 3, (a), (b), and (c) are simple foldings, but (d) is not allowed. When we unfold the paper, we can use a rewind move of a simple folding; that is, we can unfold a folded state a to a folded state b if and only if a can be obtained from b by a simple folding.

For a mountain-valley string s, we call a folded state *legal for s* if it follows the string. A mountain-valley string $MVMVMV\cdots$ is called *pleats*. The pleat folding is a special folding in the following sense:

Lemma 1. *For a mountain-valley string s in $\{M,V\}^n$, the legal folded state is unique (up to reversal of the paper) if and only if s is pleats.*

Proof: It is easy to see that a pleat folding has no other folded state. On the other hand, we suppose that a mountain-valley string has a unique folded state. Without loss of generality, we can assume that the first crease is mountain. If the second crease is also mountain, we have two distinct folded states of the segments 0, 1, and 2; it can be $[0|2|1]$ and $[2|1|0]$. Hence the second crease should be valley if the strip has a unique folded state. We can repeat this argument for each crease, and by forcing uniqueness at each fold, obtain the pleats. □

Figure 4. Two legal folded states for a string that cannot be exchanged by local simple foldings.

3   Universality of the Simple Folding Model

In this section, we show that the simple folding model is strong enough to address the paper strip with equidistant creases. More precisely, we show that every legal folded state for any string can be made by a sequence of simple foldings.

We first observe that any string has some legal folded state:

Proposition 1. *For any given mountain-valley string s in $\{M, V\}^n$, there exists a legal folded state.*

Proof: We can fold a paper for any given string by using the idea of an "end fold" [Demaine and O'Rourke 07, p. 192]. More precisely, we repeatedly fold n times at the leftmost crease according to s. After the ith folding, the leftmost parts of length i are all folded according to s, and they are stacked on the leftmost unit length area. Thus, the ith folding has no influence on the previously folded area. Therefore, after n foldings, we fold the strip into a unit length, and obtain a legal folded state for s.   □

Next, we show that any legal folded state can be folded from the initial state, which is the flat unfolded paper, by a sequence of simple foldings.

Theorem 3. *Let P be any legal folded state for a mountain-valley string s in $\{M, V\}^n$. Then P can be folded from the initial state by a sequence of simple foldings.*

Before proving Theorem 3, we comment on the claim of the theorem. One may think that Theorem 3 is "trivial." But it is not so trivial. A typical counterexample for this intuition is shown in Figure 4; these two folded states are legal for the same mountain-valley string, but they cannot be exchanged by just local simple foldings. (In fact, folding the left folded state is not so trivial to make by a sequence of simple foldings.) This fact implies that folding of these states from the initial state requires some global strategy.

By the definition of unfolding, a folded state P can be folded from the initial state by simple foldings if and only if P can be unfolded to the initial state. Hence, we prove Theorem 3 by showing how to unfold any folded state P to the initial state. This is strongly related to two well-investigated problems in origami science.

First, this is a kind of "(un)locked chain problem in 2D," which has a long and rich history [Demaine and O'Rourke 07, Chapter 6]. It is known that there is no locked chain in 2D [Demaine and O'Rourke 07, Section 6.6]. However, this fact does not imply Theorem 3, since the operations are restricted to simple unfoldings in our theorem, whereas the general unlocked chain may require multiple simultaneous motions.

Second, our problem is also related to "one-dimensional flat foldings" [Demaine and O'Rourke 07, Section 12.1]. In the problem, we aim to determine whether there *exists* a flat folded state, and the known result says that we can find *some* flat folded state by repeating basic folding operations if it exists. (In fact, Proposition 1 is a special case of this problem.) Thus, the known algorithm cannot construct a given *specified* folded state from the initial state. (In contrast with Theorem 3, this is not always possible for non-unit case; see Figure 7 in Section 5.)

Thus, in a sense, our problem is more difficult than these two problems; the folded state is specified, and we can use only simple foldings to make it. But, all links in our "linkage" have unit length. We strongly rely on this property.

In the following proof, we do not use the fact that P is folded into unit length. (We use only that we can fold at every integer point.) Hence, we prove the following stronger claim than Theorem 3:

Theorem 4. *Let P be any folded state of a paper of length $n + 1$ such that every folded point is placed at an integer point in $[1 \ldots n]$ in the initial state. Then P can be folded from the initial state by a sequence of simple foldings that are made at each integer point. Moreover, the total number of simple foldings (or unfoldings) is bounded above by $2n$.*

Proof: As mentioned previously, we prove the theorem by unfolding any folded state P to the initial state. In general, two endpoints of P may not be seen from the outside. Thus, we first peel off the papers covering the last segment and make it appear. After that, we arrange the last consecutive segments to form a straight line. To describe in detail, we let p be the last endpoint of the paper, which will be placed at integer point $n + 1$ in the initial state. We use the symbol P to denote the current (flat-)folded state. We here define *visibility* of a point on P: a point is *visible* if and only if it appears on an exterior surface of P. All visible points are drawn in thick lines in Figure 5. We note that a crease can be visible even if it is between two invisible segments (e.g., the crease point q in Figure 5(a) is a visible point of length 0). According to the visibility of the last end p, we have two cases. (In the context of the algorithm, we have two "phases.")

Case 1: The point p is not visible in the folded state P. (In Figure 5, p is not visible in cases (a), (b), and (c).) Let q be the closest visible crease to p. That is, all points $r > q$ (including p) are invisible. We note that q

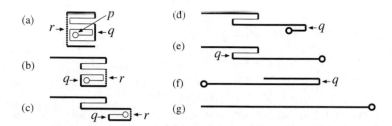

Figure 5. Simple (un)foldings. Thick lines describe the points appearing on the surface. (a)–(c) The first phase (when the endpoint p is not visible from outside). (d)–(g) The second phase (when p is visible).

can be flat. Let q' be the closest folded crease to q between p and q with $q' \neq q$. If there is no folded crease between p and q, set $q' = p$.

We first suppose that the crease q is flat. Then, by the visibility of q, the papers on the visible side of q can be flipped by a simple folding (or a simple unfolding) at the crease point q'. Then the closest visible crease to p is updated from q to q', and q' is properly closer to p than q.

Next, we suppose that q is a folded crease. Without loss of generality, the crease $q+1$ may be placed left of q, as in Figure 5(a). Then, the papers on the opposite side of $q-1$ with respect to the segment $q\ (= [q, q+1])$ cover the point $q+1$ but do not cover $q-1$ since q is visible. This fact implies that these papers can be flipped by a simple folding (or a simple unfolding) at the crease point q' that is closer to p than q. (In Figure 5(a)–(c), the bottom paper is flipped by (un)folding the crease r. The point r covers the crease point q', where q' is closer to p than q, q' is a folded crease, and q' is not visible until (un)folding at r.)

In any case, the closest visible crease is updated from q to q' by one (un)folding. We repeat this process until point p becomes visible. The number of repeatings is at most n, and hence the total number of (un)foldings in case 1 is at most n.

Case 2: Point p is visible in the folded state P. (In Figure 5, p is visible in cases (d), (e), and (f).) Let q be the closest folded crease to p. If q is not visible, since there is no folded crease between p and q, and p is visible, we can make q visible by just one simple (un)folding at the point q by using the same technique as in case 1. Now we can assume that all points in $[q, p]$ are visible. Then, these points can be seen from one side. (For example, suppose that segments $[q, r]$ are visible from the top and $[r, p]$ are visible from the bottom. In this case, since p is the endpoint of the paper, the paper becomes disconnected.) Hence we can unfold at the folded crease q and make it flat. This does not change the visibility of p. Thus, we can repeat this process until the entire set of creases becomes flat. We can observe that these two (un)foldings (to make q visible if necessary, and to make q flat) can be done at once. Hence, the total number of (un)foldings in this case is at most n.

From these two arguments, we have Theorem 4 and hence Theorem 3. $\Box$

By Proposition 1, the optimization problems are well defined for any mountain-valley string. Moreover, by Theorem 3, the problem is worth considering on the simple folding model. Furthermore, if we have an optimal solution, it can be folded in linear time by Theorem 4.

4 The Number of Folded States

In this section, we prove Theorems 1 and 2. Using Theorem 2, Theorem 1 follows easily. Hence, we first focus on Theorem 2. Before giving theoretical bounds, it is worth observing numerical experimental results.

Recall that $F(n)$ is the number of folded states of a paper of length $n + 1$. The author developed a program that produces all possible folded states for any given mountain-valley string, but we can find the correct values up to large n in *The On-Line Encyclopedia of Integer Sequences* (OEIS).[1] (According to the site, this sequence is "the number of ways of folding a strip of n labeled stamps," which describes our problem.) Plotting the sequence by $+$ and the line 3.3^x in Figure 6, we can observe that $F(n) = \Theta(3.3^n)$, and hence $f(n) = \Theta(1.65^n)$. Now we turn to theoretical upper and lower bounds.

Lemma 2. $F(n) = O(4^n)$.

Proof: We first assume that n is even, say $n = 2k$, and each folded state of unit length is placed on the interval $[0 \ldots 1]$. We see the relationship among the papers at point 0. The papers should not be penetrated through each other. That is, at point 0, k creases with one end (of the left end of segment 0) make a nested structure. The number of nested structure with k pairs is given by the Catalan number $C_k = (1/(k+1))\binom{2k}{k} = (2k)!/((k+1)!k!)$ (see, e.g., [Stanley 97]). Once the left end is connected to the right nested structure at point 1, the paper order is automatically determined. The number of possible connections of the left end to the right nested structure is k. Hence, the number of folding ways can be bounded above by kC_kC_k.

Next, let us assume that n is odd, say $n = 2k + 1$. Then, using the same argument, we have an upper bound of $(k + 1)C_{k+1}C_k$. Since $C_k \sim 4^k/k^{3/2}\pi = O(4^k)$, the lemma follows. $\Box$

---

[1]With id:A000136, http://www.research.att.com/~njas/sequences/A000136.

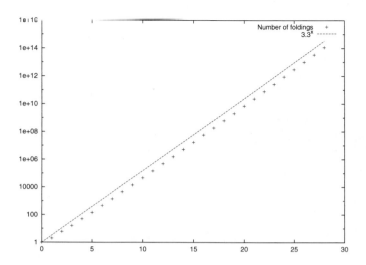

Figure 6. Experimental results of the number of folded states: each $+$ describes the number of ways of folding a strip of n labeled stamps, and the dotted line describes the function 3.3^x.

We note that we do not include the connectivity of the paper segments in the proof. We would have to consider this property to improve the upper bound.

Lemma 3. $F(n) = \Omega(3.065^n)$.

Proof: We imagine folding the last k creases for some $k \ll n$. After folding the last k creases into unit length, we glue it, and obtain a paper of length $n - k + 1$ with $n - k$ creases. Let $G(k)$ be the number of the folding ways of this last k creases under the constraint that the $(n - k)$th crease is not *covered*, which means the segments $(n-k-1)$ and $(n-k)$ are not separated by the other papers in $[n - k \ldots n + 1]$. Repeating this process, we have a lower bound: $F(n) > (G(k))^{\frac{n}{k}} = (G(k)^{\frac{1}{k}})^n$. This function $G(k)$ is also listed in *The On-Line Encyclopedia of Integer Sequences.*[2] (According to the OEIS, this sequence is "the number of ways a semi-infinite directed curve can cross a straight line n times." This may not seem to fit our problem, but the semi-infinite directed curve corresponds to the paper strip itself, and the straight line corresponds to the point $(n - k + \frac{1}{2})$.) Since the function $G(k)$ is a monotone increasing function for k, we use the largest value $G(43) = 830776205506531894760$ on the list, and obtain the lower bound $F(n) > (830776205506531894760^{\frac{1}{43}})^n = 3.06549^n$ for sufficiently large n. $\qquad\square$

---

[2]With id:A000682, http://www.research.att.com/~njas/sequences/A000682.

By Lemmas 2 and 3, we have Theorem 2 immediately.

Next we turn to Theorem 1. The number of mountain-valley strings of length n is 2^n. Hence, dividing the values in Theorem 2 by 2^n, we have Theorem 1.

Corollary 1. *We suppose a mountain-valley string of length n is given uniformly at random. Then, the exhaustive search algorithm for the minimization problems for the crease width of a paper strip runs in an exponential time on average.*

5   Concluding Remarks

In this paper, we show that the number of folding ways for a given mountain-valley string is extremely large, in general. This fact does not directly imply that finding an optimal folding way is impossible. However, it seems likely to be computationally complex. I conjecture that finding an optimal folding is NP-hard, which will be the topic of future work.[3]

Only the structure known as pleats has a unique (and optimal, of course) folded state. Hence, if a string is "close" to pleats, the number of folding ways may be small. However, the size of this "closeness" is not necessarily clear. The string[4] $(MV)^{i-1}MM(VM)^{i-1}$ seems to be close to pleats except for the center MM, but the left pleats and right pleats may be combined in any order so as to fold into unit length. As a result, this string has $\binom{2i}{i} \sim 2^i$ distinct legal folded states, which are also exponentially large. (This string is called the "shuffle pattern" and plays an important role to prove NP-completeness of a strongly restricted version of a puzzle "Kaboozle: The Labyrinth Puzzle" [Asano et al. 10].) Hence, the characterization of strings that are "close" to pleats represents other further work.

From the algorithmic viewpoint, it is also interesting to develop a polynomial time algorithm that solves the crease width problem for small crease width. For example, is there a polynomial time algorithm that determines if a given mountain-valley string has a folded state of crease width at most k? From Figure 2, it is easy to see that the mountain-valley string V^i has only i foldings. Then what happens if you add k mountains into this string for small k?

In Section 3, we mentioned that Theorem 3 is strongly related to the (un)locked chain problem in 2D [Demaine and O'Rourke 07, Chapter 6], and the one-dimensional flat foldings [Demaine and O'Rourke 07, Section 12.1]. Extending the proof of Theorem 3, for the one-dimensional

---

[3]In 2011, NP-completeness of the minimization problem of the maximum crease width was shown by Takuya Umesato, Toshiki Saitoh, Ryuhei Uehara, and Hiro Ito.

[4]Here we use the standard notation of string repetition; e.g., $(MV)^3MM(VM)^3 = MVMVMVMMVMVMVM$.

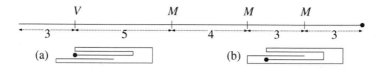

Figure 7. Two legal folded states: (a) foldable by simple foldings, and (b) not foldable by simple foldings.

flat foldings, one might wonder if any specified legal folded state can be folded from the initial state even if we allow nonuniform intervals. It turns out that this is not the case. In Figure 7, both (a) and (b) are legal folded states for the mountain-valley string $VMMM$. It is not difficult to see that Figure 7(a) is foldable by a sequence of simple foldings but Figure 7(b) is not. In fact, Figure 7(b) cannot be unfolded at all from this position by a simple unfolding. For consideration of industrial applications, the characterization of folded states that can be folded by a sequence of simple foldings seems to be a nice topic for future work.

Finally, we compare our stamp folding with *meanders*, which correspond to the systems formed by the intersections of two curves in the plane, with equivalence up to homeomorphism within the plane. There are multiple variants of meanders, classified according to their endpoints. An *ordinary meander* is a closed curve, but an *open meander* of order n is defined by a non-self-crossing oriented (open) curve in R^2, which transversally intersects another line at n points for some positive integer n. Two meanders are said to be equivalent if they are homeomorphic in the plane. Meanders are a widely investigated notion, and the interested reader can find a good survey online [Croix 03]. It seems that a folded state of a paper strip folded to unit length corresponds to an open meander of order n, as seen in the proof of Lemma 3. But the open meander number of n gives us only a lower bound of $F(n)$. The key difference is that two open meanders are equivalent if they are homeomorphic. But our paper strip has a direction and two kinds of distinguished labels M/V. For example, VVV and MMM are equivalent as meanders, and there is no other equivalent string. But VMM has three other equivalent strings VVM, MVV, and VMM.

Acknowledgement. The author is grateful to the reviewer of this article for his/her fruitful comments.

Bibliography

[Arkin et al. 04] Esther M. Arkin, Michael A. Bender, Erik D. Demaine, Martin L. Demaine, Joseph S. B. Mitchell, Saurabh Sethia, and Steven S. Skiena.

"When Can You Fold a Map?" *Computational Geometry: Theory and Applications* 29:1 (2004), 23–46.

[Asano et al. 10] T. Asano, E. Demaine, M. Demaine, and R. Uehara. "Kaboozle Is NP-complete, Even in a Strip Form." In *Fun with Algorithms: 5th International Conference, FUN 2010, Ischia, Italy, June 2–4, 2010, Proceedings*, Lecture Notes in Computer Science 6099, pp. 28–36. Berlin: Springer-Verlag, 2010.

[Cardinal et al. 09] J. Cardinal, E. D. Demaine, M. L. Demaine, S. Imahori, S. Langerman, and R. Uehara. "Algorithmic Folding Complexity." In *Algorithms and Computation: 20th International Symposium, ISAAC 2009, Honolulu, Hawaii, USA, December 16–18, 2009, Proceedings*, Lecture Notes in Computer Science 5856, pp. 452–461. Berlin: Springer-Verlag, 2009.

[Croix 03] M. La Croix. "Approaches to the Enumerative Theory of Meanders." Manuscript, 2003. (Available at http://www.math.uwaterloo.ca/~malacroi/Latex/Meanders.pdf.)

[Demaine and O'Rourke 07] E. D. Demaine and J. O'Rourke. *Geometric Folding Algorithms: Linkages, Origami, Polyhedra.* Cambridge, UK: Cambridge University Press, 2007.

[Stanley 97] R. P. Stanley. *Enumerative Combinatorics*, I. Cambridge, UK: Cambridge University Press, 1997.

Flat Vertex Fold Sequences

Thomas C. Hull and Eric Chang

1   Introduction

A *flat vertex fold* is a single vertex from a flat origami crease pattern. Almost everything is known about flat vertex folds, but in this paper we investigate an aspect that is not known. In particular, a flat vertex fold is completely determined by the sequence of consecutive angles

$$\vec{v} = (\alpha_1, \alpha_2, ..., \alpha_{2n})$$

about the vertex, which we may think of as a vector. In order to fold flat, however, mountain and valley directions need to be assigned to the creases between these angles, and there are various restrictions on such mountain-valley (MV) assignments. We can then compute the number of *valid* MV assignments for a given vertex $\vec{v}$ and call this number $C(\vec{v})$.

We then ask, "What are the possible values for $C(\vec{v})$ over all flat vertex folds $\vec{v}$ of degree $2n$?" Answering this question leads to some interesting sequences of numbers that shed light into the possibilities of folding a single vertex flat.

In Section 2 we describe, briefly, the basics of flat vertex folds. In Section 3 we outline how to calculate the possible values of $C(\vec{v})$ and begin an initial analysis of the sequences they generate. We conclude the paper with some open questions for further work.

2   The Basics of Flat Vertex Folds

The two most fundamental results concerning flat vertex folds are known as Maekawa's theorem and Kawasaki's theorem.

Theorem 1 (Maekawa). *Let M and V denote the number of mountain and valley creases, respectively, in a flat vertex fold. Then $M - V = \pm 2$.*

Maekawa's theorem was independently discovered by Jun Maekawa and Jacques Justin in the 1980s [Kasahara and Takahama 85, Justin 86]. Note that it can be used to prove that every flat vertex fold must have an even number of creases.

Theorem 2 (Kawasaki). *Let $\vec{v} = (\alpha_1, ..., \alpha_{2n})$ be a sequence of consecutive angles between creases meeting at a vertex. Then $\vec{v}$ is a flat vertex fold if and only if*

$$\alpha_1 - \alpha_2 + \alpha_3 - \cdots - \alpha_{2n} = 0.$$

Kawasaki's theorem was independently discovered in the 1980s by Toshikazu Kawasaki and Jacques Justin [Kasahara and Takahama 85, Justin 86]. However, the first reference (and proof) for this result was given by S. A. Robertson [Robertson 78] in 1977, but only for the necessary direction of the theorem.

It is also interesting to note that these two theorems are actually special cases of a more general result of Justin's, although it requires us to expand our view to flat-foldable crease patterns with more than one vertex.

Theorem 3 (Justin). *Given a flat origami crease pattern on a simply connected piece of paper, let γ be a vertex-avoiding, simple closed curve on the paper that crosses crease lines $l_1, ..., l_{2n}$, in order. Let α_i be the (oriented) angle between crease lines l_i and l_{i+1} (where we take $l_{2n+1} = l_1$), and let M and V denote the number of mountain and valley creases, respectively, among the creases l_i. Then we have*

$$\alpha_1 - \alpha_2 + \alpha_3 - \cdots - \alpha_{2n} = (M - V)\pi \quad \bmod 2\pi.$$

The proof of this is just an easy application of winding numbers (or, if we take some liberties, just the Gauss-Bonnet theorem). Note, though, that sufficiency is lost; determining whether a crease pattern will fold flat globally is notoriously difficult. Also, Justin proved a more general version of this theorem [Justin 97], showing that it holds for paper with holes as well, although this requires a more complicated proof. We won't be using Justin's theorem specifically in this paper, but it will be relevant for one of the open problems at the end.

Using Kawasaki's and Maekawa's theorems as main tools, we can begin an attack on the problem of counting valid MV assignments for flat vertex

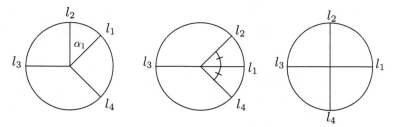

Figure 1. The cases for a degree-4 flat vertex fold.

folds of a specified even degree. (See [Hull 03] for more details.) To offer a motivating example, consider the flat vertex folds in Figure 1. They make up the three canonical cases for a degree-4 flat vertex fold.

In the first one (Figure 1, left), creases l_1 and l_2 cannot have the same MV parity, for if they did, the two angles surrounding α_1 would both cover it on the same side of the paper, forcing the paper to self-intersect (or forcing another crease to be made). Maekawa's theorem then tells us that l_3 and l_4 have to have the same MV parity, and thus we have four possible MV assignments for the creases $l_1, ..., l_4$ ($C(\vec{v}) = 4$ for this vertex).

For the second flat vertex fold in Figure 1 (middle), crease lines l_1, l_2, and l_4 cannot all be M or all be V. (If they were all the same, then Maekawa's Theorem would imply that l_3 is different from them, and the two small, equal angles would have to contain the two big angles, which is impossible.) This gives us $C(\vec{v}) = 6$.

In the third example (Figure 1, right), all the angles are equal. Here there are no restrictions on where the Ms and Vs can be—we only have to obey Maekawa's theorem. This gives us $C(\vec{v}) = 8$.

Notice what is happening in these examples. Whenever we have a sequence of small, equal angles in a row surrounded by bigger angles, the creases among those small angles have restrictions placed on them. What is not immediately apparent from the examples in Figure 1 is that we can think of this process as being recursive. If we have a smallest angle (or sequence of consecutive smallest angles), then we can assign MVs to them and fold those creases only, turning our paper into a cone. The creases remaining on this cone can then be looked at by themselves as a separate flat vertex fold (they still must follow Maekawa's and Kawasaki's theorems, though).

This insight provides the following bounds:

Theorem 4. *Let $\vec{v} = (\alpha_1, ..., \alpha_{2n})$ be a flat vertex fold, on either a flat piece of paper or a cone. Then*

$$2^n \leq C(\vec{v}) \leq 2\binom{2n}{n-1}$$

are sharp bounds.

The upper bound is achieved by the all equal angles case with Maekawa's theorem applied to it. The lower bound is generated by recursion in the case where there is always one smallest angle to find and assign either MV or VM to its creases, giving us 2^n for a degree $2n$ flat vertex fold.

The question then becomes: what values can $C(\vec{v})$ attain between these bounds? Our recursive process can, surprisingly enough, generate recursive formulas.

Theorem 5. [Hull 03] *Let* $\vec{v} = (\alpha_1, ..., \alpha_{2n})$ *be a flat vertex fold in either a piece of paper or a cone, and suppose we have* $\alpha_i = \alpha_{i+1} = \alpha_{i+2} = \cdots = \alpha_{i+k}$ *and* $\alpha_{i-1} > \alpha_i$ *and* $\alpha_{i+k+1} > \alpha_{i+k}$ *for some* i *and* k. *Then*

$$C(\alpha_1, ..., \alpha_{2n}) = \binom{k+2}{\frac{k+2}{2}} C(\alpha_1, ..., \alpha_{i-2}, \alpha_{i-1}-\alpha_i+\alpha_{i+k+1}, \alpha_{i+k+2}, ..., \alpha_{2n})$$

if k *is even, and*

$$C(\alpha_1, ..., \alpha_{2n}) = \binom{k+2}{\frac{k+1}{2}} C(\alpha_1, ..., \alpha_{i-1}, \alpha_{i+k+1}, ..., \alpha_{2n})$$

if k *is odd.*

These recursions allow us to compute the various values of $C(\vec{v})$, as will be detailed in the next section.

3   Flat Vertex Fold Sequences

For a given vertex degree $2n$, the set of possible vectors $\vec{v}$ of dimension $2n$ that could represent a flat vertex fold are restricted by Kawasaki's theorem. Hull previously described the configuration space P_{2n} of all flat vertex folds of degree $2n$ [Hull 09].

We define

$$SC(n) = \{C(\vec{v}) \mid \vec{v} \in P_{2n}\}.$$

That is, $SC(n)$ is the set of all possible values that $C(\vec{v})$ can attain for flat vertex folds of degree $2n$. We can translate the recursion in Theorem 5 into the sets $SC(n)$, and to do this, we employ the following notation: If x is an integer, then let us denote by $xSC(k)$ the set of all elements of $SC(k)$ multiplied by x. (That is, $xSC(k) = \{xa \mid a \in SC(k)\}$.) Then Theorem 5 may be translated as follows:

Theorem 6. $SC(1) = \{2\}$, *and for* $n \geq 2$ *we have*

$$SC(n) = \left(\bigcup_{k=1}^{n-1} \left(\binom{2n-2k}{n-k} SC(k) \cup \binom{2n-2k+1}{n-k} SC(k)\right)\right) \cup \left\{2\binom{2n}{n-1}\right\}.$$

This allows us to easily create the sets $SC(n)$. Below are the first five such sets.

$$SC(1) = \{2\}$$
$$SC(2) = \{4, 6, 8\}$$
$$SC(3) = \{8, 12, 16, 18, 20, 24, 30\}$$
$$SC(4) = \{16, 24, 32, 36, 40, 48, 54, 60, 70, 72, 80, 90, 112\}$$
$$SC(5) = \{32, 48, 64, 72, 80, 96, 108, 120, 140, 144, 160, 162, 180, 200, 210,$$
$$216, 224, 240, 252, 270, 280, 300, 336, 420\}$$

We define the sequence $N_n = |SC(n)|$, the sequence of sizes of the sets $SC(n)$. Using Mathematica we computed

$$(N_n) = (1, 3, 7, 13, 24, 39, 62, 97, 147, 215, 312, 440, 617, 851, 1161, \ldots).$$

This gives us two patterns to investigate: the sets $SC(n)$ and the sizes N_n of the sets. First, we discuss some bounds on N_n.

Theorem 6 states that to create the set $SC(n)$, we take all previous sets $SC(k)$ for $k < n$, multiply each of them by a binomial coefficient, union them together, and union one more number. If the recursion resulted in all distinct values for $SC(n)$, then the size of the union of all the sets would be equal to the sum of the sizes of the sets being unioned. It turns out, however, that the binomial coefficients and the elements in all previous sets $SC(k)$ for $k < n$ multiply to many repeated elements.

For example, in $SC(3)$, the recursion gives us

$$SC(3) = \binom{2}{1}SC(2) \cup \binom{3}{1}SC(2) \cup \binom{4}{2}SC(1) \cup \binom{5}{2}SC(1) \cup \left\{2\binom{6}{2}\right\}$$
$$= \{8, 12, 16\} \cup \{12, 18, 24\} \cup \{12\} \cup \{20\} \cup \{30\}$$
$$= \{8, 12, 16, 18, 20, 24, 30\}.$$

We define $SC(k)_i$ to be the ith element in $SC(k)$ ordered least to greatest. In this case, note that $\binom{2}{1}SC(2)_2 = \binom{3}{1}SC(2)_1 = \binom{4}{2}SC(1)_1$; the element 12 has been triple counted. Still, if we disregard the repeats for now, we can use Theorem 6 to easily find an upper bound for N_n. Let a_n be a sequence defined by the recursion

$$a_n = 1 + \sum_{k=1}^{n-1} 2a_k, \qquad \text{where } a_1 = 1. \tag{1}$$

Then a_n is counting exactly what $|SC(n)|$ would be if there were no duplicate numbers ever appearing in the Theorem 6 recursion. Thus a_n is over-counting and we have $N_n \le a_n$ for all n. After expanding the recursion in Equation (1), we have

$$a_n = 2a_{n-1} + a_{n-1} = 3a_{n-1},$$

and therefore $a_n = 3^{n-1}$. Thus we have that $N_n \le 3^{n-1}$ for all n.

Note that this bound loses accuracy very quickly. Let's compare N_n and a_n for $n = 1, 2, ..., 10$:

$$N_n \ : 1, 3, 7, 13, 24, 39, 62, 97, 147, 215, ...$$
$$a_n \ : 1, 3, 9, 27, 81, 243, 729, 2187, 6561, 19683, ...$$

By $n = 10$, $a_n - N_n = 19468$. Fortunately, we can do better. Let us look for some patterns in the generation of the sets $SC(n)$, using the recursion in Theorem 6. Look again at the recursion for computing $SC(3)$ that we detailed above, and compare it to the recursion for $SC(4)$:

$$SC(4) = \binom{2}{1}SC(3) \cup \binom{3}{1}SC(3) \cup$$
$$\binom{4}{2}SC(2) \cup \binom{5}{2}SC(2) \cup \binom{6}{3}SC(1) \cup \binom{7}{3}SC(1) \cup \left\{2\binom{8}{3}\right\}$$
$$= \{16, 24, 32, 36, 40, 48, 60\} \cup \{24, 36, 48, 54, 60, 72, 90\} \cup$$
$$\{24, 36, 48\} \cup \{40, 60, 80\} \cup \{40\} \cup \{70\} \cup \{112\}$$
$$= \{16, 24, 32, 36, 40, 48, 54, 60, 70, 72, 80, 90, 112\}.$$

You can see a pattern in these examples. In each level of the recursion, the first set is a subset of the second set on the previous line. For example, $\binom{6}{3}SC(1) \subset \binom{5}{2}SC(2)$ and $\binom{4}{2}SC(2) \subset \binom{3}{1}SC(3)$ in the computation of $SC(4)$. If we look at the way the recursion generates these binomial coefficients, we can see that some repeats will always happen.

Looking at the generation of $SC(4)$: The set $\{24, 36, 48\}$ comes from multiplying the elements of $SC(2)$ by $\binom{4}{2} = 6$. Note, though, that $\{24, 36, 48\} \subset \{24, 36, 48, 54, 60, 72, 90\}$. That is because the set $\{24, 36, 48, 54, 60, 72, 90\}$ comes from multiplying the elements of $SC(3)$ by $\binom{3}{1} = 3$. If you consider the fact that some of the elements in $SC(3)$ come from $2 * SC(2)$, then what you have are $3 * (2 * SC(2))$, meaning the elements of $6 * SC(3)$ are purely repeats, mainly because $\binom{4}{2} = 6 = 2\binom{3}{1}$. Similarly, the fact that $\binom{5}{2} = 10$, and $\binom{6}{3} = 20 = 2\binom{5}{2}$ explains why $\binom{6}{3}SC(1) \subset \binom{5}{2}SC(2)$.

This persists throughout the recursion beyond this one example, and we can prove it!

Lemma 1. *Using the notation of Theorem 6, where $1 \le k \le n - 1$, we have*

$$\binom{2n - 2k + 2}{n - k + 1}SC(k - 1) \subset \binom{2n - 2k + 1}{n - k}SC(k).$$

Proof: Note that

$$\binom{2n-2k+2}{n-k+1} = \frac{(2n-2k+2)(2n-2k+1)!}{(n-k+1)(n-k)!(n-k+1)!}$$

$$= 2\frac{(2n-2k+1)!}{(n-k)!(n-k+1)!}$$

$$= 2\binom{2n-2k+1}{n-k}.$$

Also, by the recursion in Theorem 6, we have that $SC(k) = 2SC(k-1) \cup$ (other sets), so $2SC(k-1) \subset SC(k)$ is always true. Therefore,

$$\binom{2n-2k+2}{n-k+1}SC(k-1) = 2\binom{2n-2k+1}{n-k}SC(k-1)$$

$$\subset \binom{2n-2k+1}{n-k}SC(k). \qquad \square$$

This means that the recurrence in Theorem 6 is partly redundant. We can express this as the following:

Theorem 7. *If $SC(n)$ is defined as above, then an equivalent recursion to generate the sets $SC(n)$ is*

$$SC(n) = 2SC(n-1)\cup 3SC(n-1)\cup \left(\bigcup_{k=1}^{n-2}\binom{2n-2k+1}{n-k}SC(k)\right)\cup\left\{2\binom{2n}{n-1}\right\}$$

for $n \geq 2$ and $SC(1) = \{2\}$.

We can use this to obtain a better bound on $N_n = |SC(n)|$. Recall that the sequence given by $a_n = 3a_{n-1}$ gave us a very simple bound on N_n by supposing that all the binomial coefficient terms in Theorem 6 were distinct. Theorem 7 tells us that after the first pair, half of all the other terms will be redundant. So a better bounding sequence $a_n \geq N_n$ would be given by

$$a_n = 3a_{n-1} - a_{n-2} - a_{n-3} - \cdots - a_1.$$

This may be simplified to

$$a_n = 3a_{n-1} - 4a_{n-2} + 3a_{n-2} - a_{n-3} - \cdots a_1$$

$$= 3a_{n-1} - 4a_{n-2} + a_{n-1}$$

$$= 4a_{n-1} - 4a_{n-2},$$

where $a_1 = 1$ and $a_2 = 3$. Solving this homogeneous recurrence gives us $a_n = (n+1)2^{n-2}$. Thus we have proven the following:

Theorem 8. $N_n \leq (n+1)2^{n-2}$ for all $n \geq 1$.

This bound can certainly be improved, as there are plenty of other repeated numbers in the $SC(n)$ recursion. However, the exponential upper bound makes one suspect that the sequence N_n might be exponential as well. An exponential lower bound would be needed to more accurately characterize the nature of N_n, but finding a decent lower bound appears to be more difficult. The authors have made only preliminary progress on this and hope to develop it in a future report.

4 Conclusion

Readers may wonder whether an explicit formula for N_n can be found, but this seems highly unlikely. Doing this from the recursions in Theorems 5 and 6 would require knowing the prime factorizations of $\binom{2n}{n}$ and $\binom{2n}{n-1}$. Such prime factorizations are unknown, although they have been studied extensively (see, for example, [Erdös et al. 75]). However, it is entirely possible that what knowledge does exist about the prime factorization of $\binom{2n}{n}$ could lead to a more comprehensive strategy for avoiding repeated numbers in our recursions, and thus result in better bounds.

Another avenue for future work might be in trying to generalize from the single vertex case toward cases for which Justin's theorem would apply. That is, Justin defines a *crown of faces* C_γ to be all the faces in a flat origami crease pattern that are crossed by a simple, closed, vertex-avoiding curve γ [Justin 97]. Such a crown of faces would form a ring and thus can be viewed as similar to a flat vertex fold, but two problems arise: (1) the creases no longer meet at a point, so the results of "equal angles in a row" in the single vertex case that allowed us to count valid MV assignments will no longer apply; and (2) the distances between consecutive creases in our crown of faces will further restrict MV assignments. No results on counting valid MV assignments for crowns of faces are known to us, so this could be an interesting and fruitful open area to pursue.

Bibliography

[Erdös et al. 75] P. Erdös, R. L. Graham, I. Z. Ruzsa, and E. G. Straus. "On the Prime Factors of $\binom{2n}{n}$." *Mathematics of Computation* 29:129 (1975), 83–92.

[Hull 03] Thomas C. Hull. "Counting Mountain-Valley Assignments for Flat Folds." *Ars Combinatoria* 67 (2003), 175–188.

[Hull 09] Thomas C. Hull. "Configuration Spaces for Flat Vertex Folds." In *Origami⁴: Fourth International Meeting of Origami Science, Mathematics, and Education*, edited by Robert J. Lang, pp. 361–370. Wellesley: A K Peters, 2009.

[Justin 86] Jacques Justin. "Mathematics of Origami, Part 9." *British Origami* 118 (1986), 28–30.

[Justin 97] Jacques Justin. "Toward a Mathematical Theory of Origami." In *Origami Science and Art: Proceedings of the Second International Meeting of Origami Science and Scientific Origami*, edited by K. Miura, pp. 15–29. Shiga, Japan: Seian University of Art and Design, 1997.

[Kasahara and Takahama 85] Kunihiko Kasahara and Toshi Takahama. *Top Origami*. Tokyo: Japan Publications, 1985.

[Robertson 78] S. A. Robertson. "Isometric Folding of Riemannian Manifolds." *Proceedings of the Royal Society of Edinburgh* 79:3–4 (1977/78), 275–284.

Circle Packing for Origami Design Is Hard

Erik D. Demaine, Sándor P. Fekete, and Robert J. Lang

1   Introduction

Over the past 20 years, the world of origami has been changed by the introduction of design algorithms that bear a close relationship to, if not outright ancestry from, computational geometry. One of the first robust algorithms for origami design was the circle/river method (also called the tree method) developed independently by Lang [Lang 94, Lang 97, Lang 96] and Meguro [Meguro 92, Meguro 94]. This algorithm and its variants provide a systematic method for folding any structure that topologically resembles a graph theoretic weighted tree. Other algorithms followed, notably one by Tachi [Tachi 09] that gives the crease pattern to fold an arbitrary 3D surface.

Hopes of a general approach for efficiently solving all origami design problems were dashed early on, however, when Bern and Hayes showed in 1996 that the general problem of crease assignment—given an arbitrary crease pattern, determine whether each fold is mountain or valley—was NP-complete [Bern and Hayes 96]. In fact, they showed more: given a complete crease assignment, simply determining the stacking order of the layers of paper was also NP-complete. Fortunately, even though crease assignment in the general case is hard, the crease patterns generated by the various design algorithms carry with them significant extra information associated with each crease—enough extra information that the problem of crease assignment is typically only polynomial in difficulty. This is certainly the case for the tree method of design [Demaine and Lang 09].

Designing a model using the tree method (or one of its variants) is a two-step process: the first step involves solving an optimization problem in which one solves for certain key vertices of the crease pattern, and the second step constructs creases following a geometric prescription and assigns their status as mountain, valley, or unfolded. The process of constructing the creases and assigning them is definitely polynomial in complexity; but, up to now, the computational complexity of the optimization has not been established.

There were reasons for believing that the optimization was, in principle, computationally intractable. The conditions on the vertex coordinates in the optimization can be expressed as a packing problem, in which the packing objects are circles and "rivers," (which are curves of constant width) of varying size. It is known that many packing problems are, in fact, NP-hard, and our intuition suggested that this might be the case for the tree method optimization problem.

In this paper, we show that this is, in fact, the case. The general tree method optimization problem is NP-hard. In the usual way with such problems, we show that any example of a 3-Partition can be expressed as a tree method problem. At the same time, we show that deciding whether a given set of circles can be packed into a rectangle, an equilateral triangle, or a unit square are NP-hard problems, settling the complexity of these natural packing problems. On the positive side, we show that any set of circles of total area 1 can be packed into a square of edge length $4/\sqrt{\pi} = 2.2567\ldots$.

2   Circle-River Design

The basic circle-river method of origami has been described elsewhere [Lang 96, Demaine and Lang 09]; we briefly recapitulate it here. As shown in Figure 1, we are presented with a polygon P', which represents the paper to be folded, and an edge-weighted tree, T, which describes the topology of the desired folded shape. The design problem is to find the crease pattern that folds P' (or some convex subset) into an origami figure whose perpendicular projection has the topology of the desired tree T and whose edge lengths are proportional to the edge weights of T. The coefficient of proportionality m between the dimensions of the resulting folded form and the specified edge weights is called the *scale* of the crease pattern. The optimization form of the problem is to find the crease pattern that has the desired topology and that maximizes the scale m.

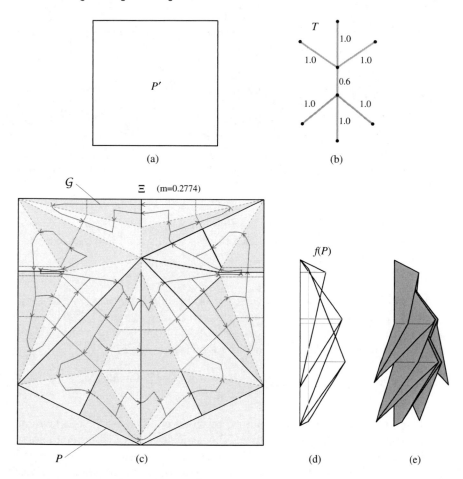

Figure 1. Schematic of the problem. (a) P' is the paper to be folded. (b) T is an edge-weighted tree that describes the desired shape. (c) A solution to the optimization problem, showing creases and the ordering graph on the facets. (d) An x-ray view of the folded form. (e) A visual representation of the folded form.

Formally, the problem can be expressed as follows. There is a one-to-one correspondence between leaf nodes $\{n_i\}$ of the tree T and *leaf vertices* $\{v_i\}$ of the crease pattern whose projections map to the leaf nodes. We denote the edges of T by $\{e_j\}$ with edge weights $w(e_j)$. For any two leaf nodes $n_i, n_j \in T$, there is a unique path $p_{i,j}$ between them; this allows us to define the *path length* $l_{i,j}$ between them as

$$l_{i,j} \equiv \sum_{e_k \in p_{i,j}} w(e_k). \tag{1}$$

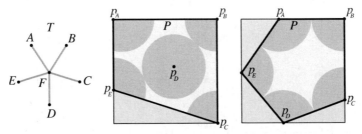

Figure 2. A star tree (left) and two possible solutions for the leaf vertices (middle and right). Each solution corresponds to a packing of the circles centered on the leaf vertices.

We showed previously [Lang 96] that a necessary condition for the existence of a crease pattern with scale m was that, for all leaf vertices,

$$|v_i - v_j| \geq ml_{i,j}, \tag{2}$$

and subsequently, that with a few extra conditions, Equation (2) was *sufficient* for the existence of a full crease pattern (and we gave an algorithm for its construction). The largest possible crease pattern for a given polygon P', then, can be found by solving the following problem:

$$\text{optimize } m \text{ subject to } \left\{ \begin{array}{l} |v_i - v_j| \geq ml_{i,j} \text{ for all } i, j, \\ v_i \in P' \text{ for all } i. \end{array} \right. \tag{3}$$

There is a simple physical picture of these conditions: if we surround each vertex by a circle whose radius is the scaled length of the edge incident to its corresponding leaf node and, for each branch edge of the tree, we insert into the crease pattern a curve of constant width (called a *river*) whose width is the scale length of the corresponding edge, then Equation (3) corresponds exactly to the problem of packing the circles and rivers in a non-overlapping way so that the centers of the circles are confined to the polygon P' and the incidences between touching circles and rivers match the incidences of their corresponding edges in the tree T.

A special case arises when there are no rivers, such as in the case of a star tree with only a single branch node, as illustrated in Figure 2. In this case there are no rivers, and the optimization problem reduces to a single packing of circles, one for each leaf node, whose radius is given by the length of the edge incident to the corresponding node.

Thus, several problems in origami design can be reduced to finding an optimum packing of some number of circles of specified radii within a square (or other convex polygon). Several examples of such problems (and their solutions) are described by Lang [Lang 03].

We now show that this circle-packing problem is NP-complete.

3 Packing and Complexity

Problems of packing a given set of objects into a specific container appear in a large variety of applied and theoretical contexts. Many one-dimensional variants are known to be NP-complete (e.g., *Bin Packing*, where the objective is to pack a set of intervals of given lengths into as few unit-sized containers as possible). A special case of Bin Packing that is still NP-hard is *3-Partition*, for which an instance is given by $3n$ numbers x_i with $1/4 < x_i < 1/2$, and $\sum_{i=1}^{3n} x_i = n$. Clearly, n unit-sized containers suffice for packing the object, iff there is a partition of the x_i into n triples that each have combined weight 1; hence the name *3-Partition*. An important property of the problem is that it is *strongly* NP-complete: it remains hard even if there is only a constant number of different values x_i [Garey and Johnson 79].

Like their one-dimensional counterparts, higher-dimensional packing problems tend to be hard. Typically, the difficulty arises from complicated container shapes (e.g., a nonsimple polygon to be filled with a large number of unit squares) or complicated objects (e.g., rectangles of many different sizes to be filled into a square, which is a generalization of Bin Packing.) This does not mean that packing simple objects into simple containers is necessarily easy: for some such problems, it is not even known whether they belong to the class NP. One example is the problem of *Pallet Loading*, deciding whether n rectangles of dimensions $a \times b$ can be packed into a larger rectangle of dimensions $A \times B$, for positive integers n, a, b, A, B; it is open whether the existence of any feasible solution implies the existence of a packing that can be described in space polynomial in the input size $\log n + \log a + \log b + \log A + \log B$, as the two different orientations of the small rectangles may give rise to complicated patterns. (See Problem #55 in The Open Problems Project, [Demaine et al. 04].)

None of these difficulties arises when a limited number of simple shapes without rotation, in particular, different squares or circles, are to be packed into a unit square. Leung et al. [Leung et al. 90] managed to prove that the problem of *Square Packing*, deciding whether a given set of squares can be packed into a unit square, is an NP-complete problem. Their proof is based on a reduction of the 3-Partition problem mentioned above: any 3-Partition instance Π_{3p} can be encoded as an instance Π_{sp} of Square Packing, such that Π_{sp} is solvable iff Π_{3p} is, and the encoding size of Π_{sp} is polynomial in the encoding size of Π_{3p}. Membership in NP is not an issue, as coordinates of a feasible packing are integers of a description size polynomial in the encoding size of Π_{sp}.

In the context of circle/river origami design, we are particularly interested in the problem of *Circle Packing*: given a set of n circles of a limited number of different sizes, decide whether they can be packed into a unit square. More precisely, we are interested in *Circle Placement*: given a set of n circles, place the circle centers on the paper, such that the overall circle layout is nonoverlapping. Clearly, this feels closely related to Square Packing, so it is natural to suspect NP-completeness. However, when packing circles, another issue arises: tight packings may give rise to complicated coordinates. In fact, the minimum size C_n of a $C_n \times C_n$ square that can accommodate n unit circles is known only for relatively moderate values of n; consequently, the membership of Circle Packing in NP is wide open. (At this point, $n = 36$ is the largest n for which the exact value of C_n is known; see [Specht 10] for the current status of upper and lower bounds for $n \leq 10,000$.)

Paradoxically, this additional difficulty has also constituted a major roadblock for establishing the NP-hardness of Circle Packing, which requires encoding desired combinatorial structures as appropriate packings: this is hard to do when little is known about the structure of optimal packings.

The main result of this paper is to describe an NP-hardness proof of Circle Placement, based on a reduction of 3-Partition; it is straightforward to see that this also implies NP-hardness of Circle Packing. In Section 4, we describe the key idea of using *symmetric 3-pockets* for this reduction: a triple of small "shim" circles C_{i_1}, C_{i_2}, C_{i_3} and a medium-sized "plug" circle can be packed into such a pocket iff the corresponding triple of numbers x_{i_1}, x_{i_2}, x_{i_3} add up to at most 1. In the following sections, we show how symmetric 3-pockets can be forced for triangular paper (Section 5), for rectangular paper (Section 6), and for square paper (Section 7). The technical details for the proof of NP-hardness are wrapped up in Sections 8 and 9, in which we sketch additional aspects of filling undesired holes in the resulting packings, approximating the involved irrational coordinates, and the polynomial size of the overall construction. On the positive side, we show in Section 10 that circle packing becomes a lot easier if one is willing to compromise on the size of the piece of paper; we prove that any given set of circles of total area at most 1 can easily and recursively be packed into a square of edge length $4/\sqrt{\pi} = 2.2567\ldots$.

4   Symmetric 3-Pockets

Our reduction is based on the simple construction shown in Figure 3. It consists of a *symmetric 3-pocket* as the container, which is the area bounded by three congruent touching circles. Into each pocket, we pack a medium-sized circle (*plug*) that fits into the center, and three small identical circles (*shims*) that fit into the three corners left by the plug. Straightforward trigonometry (or the use of Proposition 2 in Section 8) shows that for a

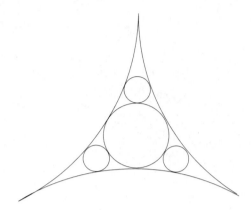

Figure 3. A symmetric 3-pocket with plug and shims.

pocket formed by three unit circles, the corresponding size is $2/\sqrt{3} - 1 = 0.1547...$ for the plug; the value for the shims works out to $1/(5 + \sqrt{3} + 2\sqrt{7 + 4\sqrt{3}}) = 0.07044....$

Clearly, this packing is unique, and the basic layout of the solution does not change when the plug is reduced in size by a tiny amount, say, $\varepsilon = 1/N$ for a suitably big N, while each shim is increased by a corresponding amount that keeps the overall packing tight. This results in a radius of r_p for each plug, and a radius of r_s for each shim.

Now consider the numbers x_i for $i = 1,\ldots,3n$, constituting an instance of 3-Partition. We get a feasible partition iff all triples (i_1, i_2, i_3) are feasible, i.e., $\sum_{j=1}^{3} x_{i_j} = 1$. By introducing $x_i' = 1/3 - x_i$ and using $\sum_{i=1}^{3n} x_i = n$, it is easy to see that a partition is feasible iff $\sum_{j=1}^{3} x_{i_j}' \leq 0$ for all triples (i_1, i_2, i_3). Note that a 3-Partition instance involves only a constant number of different sizes, so there is some $\delta > 0$ such that any infeasible triple (i_1, i_2, i_3) incurs $\sum_{j=1}^{3} x_{i_j}' \geq \delta$. By picking N large enough, we may assume $\delta > \varepsilon$.

As a next step, map each x_i to a slightly modified shim S_i by picking the shim radius to be $r_i = r_s - x_i'/N^2$. We make use of the following elementary lemma; see Figure 4.

Lemma 1. *Refer to Figure 4. Consider an equilateral triangle $\Delta = (v_1, v_2, v_3)$ bounded by the lines ℓ_1, ℓ_2, ℓ_3 through the triangle edges e_1, e_2, e_3. For an arbitrary point p, let d_j be the distance of p from ℓ_j. Define $y_j = d_j$ if p is on the same side of ℓ_j as Δ, and $y_j = -d_j$ if p is separated from Δ by ℓ_j. Then $\sum_{j=1}^{3} y_j$ is independent of the position of p.*

Proof: Without loss of generality, let $e_i = 1$. Consider the three triangles (v_1, v_2, p), (v_2, v_3, p), (v_3, v_1, p). Their areas are $d_3/2$, $d_1/2$, $d_2/2$, hence $d_1/2 + d_2/2 + d_3/2$ is always equal to the area of Δ, i.e., a constant.   □

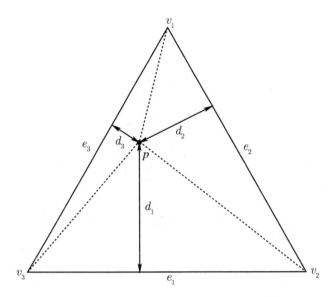

Figure 4. For an equilateral triangle, the sum of distances from the three sides is a constant.

The crucial argument for our reduction is the following:

Lemma 2. *A set of three shims S_{i_1}, S_{i_2}, S_{i_3} and a plug P of radius r_p can be packed into a 3-pocket, iff $\sum_{j=1}^{3} x'_{i_j} \leq 0$, i.e., if (i_1, i_2, i_3) is feasible.*

Proof: Refer to Figure 5. Let c be the center point of the pocket. For each of the three corners of the pocket, consider the two tangents $T_{i_j}^1$ and $T_{i_j}^2$ between an unmodified shim of radius r_s and the touching pocket boundary; let $2\phi \in]0, \pi[$ be the angle enclosed by those two tangents. (If the pocket were an equilateral triangle, we would get $\phi = \pi/6$; the exact value for pockets with circular boundaries can be computed, but the exact value does not matter.)

Now consider the shim motion arising by modifying r_s by x'_{i_j}/N^2, while keeping the shim tightly wedged into the corner. This moves its center point p_{i_j} along the bisector b between $T_{i_j}^1$ and $T_{i_j}^2$. Let $c = 1/\sin\phi$. Considering the first-order expansion of the shim motion, we conclude that p_{i_j} moves by $c \times x'_{i_j}/N^2 + \Theta(1/N^4)$ along b, to a position q_j.

Finally, refer to Figure 6 and consider the possible placement of a plug after placing the modified shims S_{i_1}, S_{i_2}, S_{i_3} into the the three corners; this requires finding a point within the pocket that has distance at least $r_p + r_s - x'_{i_j}/N^2$ from each p_{i_j}. For this purpose, consider the circle C_{i_j} of radius $r_p + r_s - x'_{i_j}/N^2$ around each p_{i_j}. As shown in Figure 7, let t_{i_j}

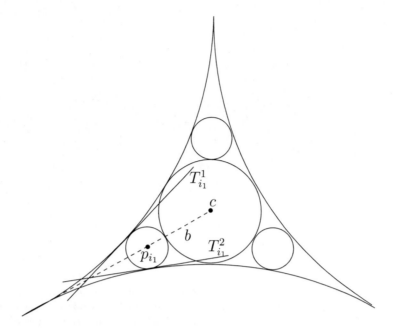

Figure 5. Changing the size of a shim.

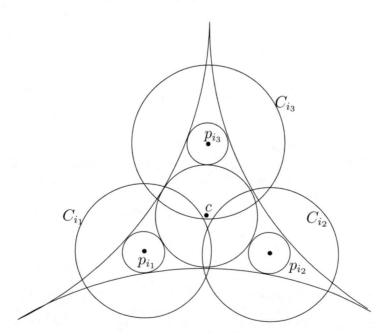

Figure 6. Finding a feasible placement for the plug.

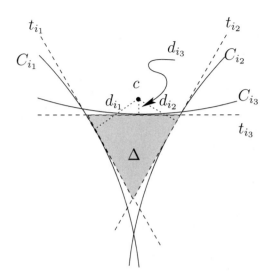

Figure 7. The existence of a feasible placement for the plug depends on the sum of distances of c from the sides of Δ. (Distances are not drawn to scale so that circles and tangents can be distinguished; in reality, they are much closer.)

be the tangent to C_{i_j} at the point closest to c; let d_{i_j} be the distance of c to t_{i_j}. If we define $y_{i_j} = d_{i_j}$ for c is outside of C_{i_j}, and $y_{i_j} = -d_{i_j}$ for c is inside of C_{i_j}, then $y_{i_j} = -((c+1) \times x'_{i_j}/N^2 + \Theta(1/N^4))$. Now consider the set Δ of points separated by t_{i_1} from p_{i_1}, t_{i_2} from p_{i_2}, t_{i_3} from p_{i_3}. Making use of Lemma 1, we conclude that Δ is a nonempty isosceles triangle, iff $\sum y_{i_j} \geq 0$, i.e., iff $\sum x'_{i_j}/N^2 - \Theta(1/N^4) \leq 0$. Given that $\sum x'_{i_j} > 0$ implies $\sum x'_{i_j} \geq \delta > 1/N$, we conclude that $\sum x'_{i_j} \leq 0$ implies the existence of a feasible packing.

Conversely, consider $\sum x'_{i_j} > 0$. Given that each t_{i_1} has a distance $\Theta(1/N^2)$ from c, we observe that the corners of the triangle formed by t_{i_1}, t_{i_2}, t_{i_3} are within $\Theta(1/N^4)$ from C_{i_1}, C_{i_2}, C_{i_3}. However, because $\sum x'_{i_j} \geq \delta > 1/N$, we conclude that any point of Δ is at least $\Theta(1/N^3)$ from being feasible. This implies that there is no feasible placement for the plug, concluding the proof. $\square$

5   Triangular Paper

For making use of Lemma 2 and completing the reduction, we need to define a set of circles (called *rocks*) that can only be packed in a way that results in a suitable number of 3-pockets. In the case of triangular paper,

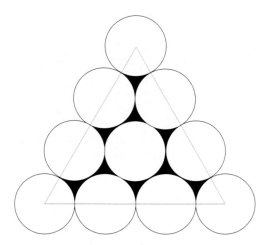

Figure 8. The unique packing of $(k+2)(k+1)/2$ unit circles into an equilateral triangle of edge length $2k$ leaves k^2 identical symmetric 3-pockets.

this is relatively easy by making use of a result by Graham [Folkman and Graham 69]. (See Figure 8.)

Proposition 1. *An equilateral triangle of edge length $2k$ has a unique packing of $(k+2)(k+1)/2$ unit circles; this uses a hexagonal grid pattern, placing circles on the corners of the triangle.*

This creates $\sum_{i=1}^{k}(2i-1) = k^2$ symmetric 3-pockets. After handling some issues of accuracy and approximation (which are discussed in Section 9), we get the desired result.

Theorem 1. *Circle/river origami design for triangular paper is NP-hard.*

As a corollary, we get the following:

Corollary 1. *It is NP-hard to decide whether a given set of circles can be packed into an equilateral triangle.*

6   Rectangular Paper

Similar to triangular paper, it is easy to force a suitable number of symmetric 3-pockets for the case of rectangular paper; see Figure 9. Disregarding symmetries, $2k$ unit circles can be packed into an $2k-1$ by $\sqrt{3}$ rectangle only in the manner shown. This creates $2k-2$ symmetric 3-pockets, which can be used for the hardness proof.

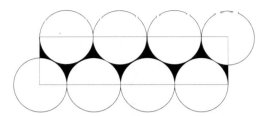

Figure 9. Packing $2k$ unit circles into a rectangle of dimensions $2k - 1$ and $\sqrt{3}$ leaves $2k - 2$ identical symmetric 3-pockets.

Because the input created for encoding an instance Π_{3p} of 3-Partition needs to be a set of rationals whose size is bounded by a polynomial in the encoding size of Π_{3p}, the irrational numbers need to be suitably approximated without compromising the overall structure. This is discussed in Section 9. As a consequence, we get the following:

Theorem 2. *Circle/river origami design for rectangular paper is NP-hard.*

This yields the following easy corollary:

Corollary 2. *It is NP-hard to decide whether a given set of circles can be packed into a given rectangle.*

7   Square Paper

Setting up a sufficient number of symmetric 3-pockets for square paper is slightly trickier: there is no infinite family of positive integers n for which the optimal patterns of packing n unit circles into a minimum-size square are known. As a consequence, we make use of a different construction; without loss of generality, our piece of paper is a unit square.

As a first step, we use four large circles of radius $1/2$, creating a symmetric *4-pocket*, as shown in Figure 10. A circle of radius $(\sqrt{2} - 1)/2$ has a unique feasible placement in the center of the pocket, leaving four smaller auxiliary pockets, as shown. Now we use 12 identical "plug" circles and four slightly smaller "fixation" circles, such that three plugs and one shim have a tight packing, as shown in the figure. For these, it is not hard to argue that not more than three plugs fit into an auxiliary pocket, ensuring that precisely three must be placed into each pocket. Moreover, it can be shown that at most one additional shim can be packed along with the three plugs; this admits precisely the packing shown in Figure 10, creating a symmetric 3-pocket in each auxiliary pocket. In addition, we get a number of undesired asymmetric pockets, which must be used for accommodating

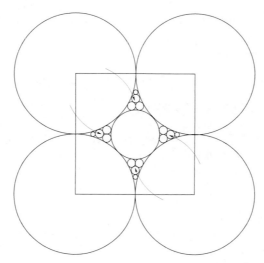

Figure 10. A gadget for creating identical triangular pockets. This set of 13 circles has a unique packing into a symmetric 4-pocket, creating four smaller symmetric 3-pockets, indicated by arrows.

appropriate sets of "filling" circles, leaving only small gaps that cannot be used for packing the circles that are relevant for the reduction.

As shown in Figure 11, we can use a similar auxiliary construction (consisting of 13 circles) for the 3-pockets in a recursive manner in order to replace each symmetric 3-pocket by three smaller symmetric 3-pockets.

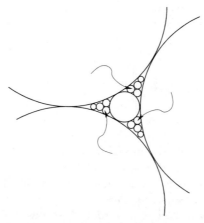

Figure 11. A gadget for creating multiple identical triangular pockets. This set of 13 circles has a unique packing into a symmetric 3-pocket, creating three smaller symmetric 3-pockets, indicated by arrows.

The argument is analogous to the one for 4-pockets. Again, additional filling circles are used; these do not compromise the overall structure of the packing, as the overall argument holds.

Theorem 3. *Circle/river origami design for square paper is NP-hard.*

This yields the following easy corollary.

Corollary 3. *It is NP-hard to decide whether a given set of circles can be packed into a given square.*

8 Filling Gaps

The constructions shown create a number of additional gaps in the form of asymmetric 3-pockets. Each is bounded by three touching circles, say, of radius r_1, r_2, r_3. By adding appropriate sets of "filler" circles that precisely fit into these pockets, we can ensure that they cannot be exploited for sidestepping the desired packing structure of the reduction. Computing the necessary radii can simply be done by using the following formula.

Proposition 2. *The radius r of a largest circle inscribed into a pocket formed by three mutually touching circles with radii r_1, r_2, r_3 satisfies*

$$1/r = 1/r_1 + 1/r_2 + 1/r_3 + 2\sqrt{1/r_1 r_2 + 1/r_1 r_3 + 1/r_2 r_3}.$$

Note that the resulting r is smaller than the smallest r_i, and at least a factor of 3 smaller than the largest of the circles. Therefore, computing the filler circles by decreasing magnitude guarantees that all gaps are filled in the desired fashion, and that only a polynomial number of such circles is needed.

9 Encoding the Input

To complete our NP-hardness proof for Circle Placement, we still need to ensure that the description size of the resulting Circle Placement is polynomial in the size of the input for the original 3-Partition. It is easy to see from the previous discussions that the total number of circles remains polynomial. This leaves the issue of encoding the radii themselves: if we insist on tightness of all packings, we get irrational numbers that can be described as nested square roots. As described in Section 4, the key mechanisms of our construction still work if we use a sufficiently close approximation. This allows us to use sufficiently tight approximations of the involved square roots in other parts of the construction, provided the involved computations are fast and easy to carry out. For our purposes, even Heron's quadratically converging method (which doubles the number of correct digits in each simple iteration step) suffices.

10   A Positive Result

Our NP-hardness results imply that there is little hope for a polynomial-time algorithm that computes the smallest possible triangle, rectangle or square for placing or packing a given set of circles. However, it is possible to guarantee the existence of a feasible solution, if one is willing to use larger paper. In fact, we show that a square of edge length $4/\sqrt{\pi} = 2.2567\ldots$ suffices for packing any set of circles that have total area 1.

Theorem 4. *Consider a set $\mathcal{S}_C$ of circles of total area 1, and a square S of edge length $4/\sqrt{\pi}$. Then $\mathcal{S}_C$ can be packed into S.*

Proof: Refer to Figure 12. For each circle C_i of radius r_i, let n_i be chosen such that $\gamma/2^{n_i+1} < r_i \leq \gamma/2^{n_i}$. Hence, replacing each C_i by a square S_i of size $\gamma/2^{n_i}$ increases the edge length by a factor of at most $\gamma = 4/\sqrt{\pi}$. Now a recursive subdivision of S into sub-squares of progressively smaller size can be used to pack all squares S_i, showing that all circles C_i can be packed. □

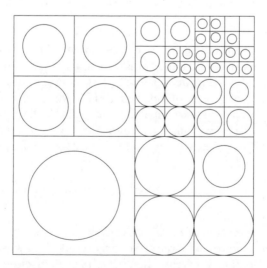

Figure 12. A quad-tree packing guarantees that any set of circles of total area at most 1 can be packed into a square of edge length $\gamma = 4/\sqrt{\pi} = 2.2567\ldots$.

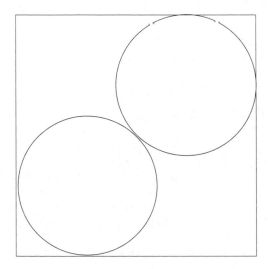

Figure 13. A lower-bound example for packing circles: two circles of area $1/2$ require a square of edge length at least $1.362\ldots$.

11   Conclusions

In this paper, we have proven that even the aspect of circle packing in circle/river origami design is NP-hard. On the positive side, we showed that the size of a smallest sufficient square for accommodating a given set of circles can easily be approximated within a factor $2.2567\ldots$. A number of interesting open questions remain:

- Our 2.2567-approximation is quite simple. The performance guarantee is based on a simple area argument. This gives rise to the following question: what is the smallest square that suffices for packing any set of circles of total area 1? We believe the worst case may very well be shown in Figure 13, which yields a lower bound of $(1 + \sqrt{2})/\sqrt{\pi} = 1.362\ldots$. There are ways to improve the upper bound; at this point, we can establish $2\sqrt{2}/\sqrt{\pi} = 1.5957\ldots$ [Demaine and Fekete 11].

- The same question can be posed for *placing* circles instead of packing them.

- The approximation of circle packing does not produce a "clustered" layout, as required by circle/river origami design, where objects that are close in the hierarchy should be placed in close proximity. In the absence of rivers, we can reproduce the quad-tree packing in this context by making use of a space-filling curve.

- We do not know yet, however, how to approximate the necessary paper size in the presence of rivers of positive width.

Acknowledgments. We thank Ron Graham for several helpful hints concerning the state of the art on packing circles. We also thank Vinayak Pathak for pointing out a numerical typo related to Figure 13.

Bibliography

[Bern and Hayes 96] Marshall Bern and Barry Hayes. "On the Complexity of Flat Origami." In *Proceedings of the 7th ACM-SIAM Symposium on Discrete Algorithms*, pp. 175–183. New York: ACM Press, 1996.

[Demaine and Fekete 11] Erik Demaine and Sándor P. Fekete. "Worst-Case Bounds for Dense Packings." Manuscript, 2011.

[Demaine and Lang 09] Erik D. Demaine and Robert J. Lang. "Facet Ordering and Crease Assignment in Uniaxial Bases." In *Origami⁴: Fourth International Meeting of Origami Science, Mathematics, and Education*, edited by Robert J. Lang, pp. 259 – 272. Wellesley, MA: A K Peters, 2009.

[Demaine et al. 04] Erik D. Demaine, Joseph S.B. Mitchell, and Joseph O'Rourke. "The Open Problems Project." Availalbe at http://cs.smith. edu/~orourke/TOPP/Welcome.html, 2004.

[Folkman and Graham 69] Jon H. Folkman and Ronald L. Graham. "A Packing Inequality for Compact Convex Subsets of the Plane." *Canadian Mathematics Bulletin* 12 (1969), 745–752.

[Garey and Johnson 79] Michael R. Garey and David S. Johnson. *Computers and Intractability: A Guide to the Theory of NP-Completeness.* New York: W. H. Freeman, 1979.

[Lang 94] Robert J. Lang. "Mathematical Algorithms for Origami Design." *Symmetry: Culture and Science* 5:2 (1994), 115–152.

[Lang 96] Robert J. Lang. "A Computational Algorithm for Origami Design." In *Proceedings of the Twelfth Annual Symposium on Computational Geometry*, pp. 98–105. New York: ACM Press, 1996.

[Lang 97] Robert J. Lang. "The Tree Method of Origami Design." In *Origami Science and Art: Proceedings of the Second International Meeting of Origami Science and Scientific Origami*, edited by Koryo Miura, pp. 73–82. Shiga, Japan: Seian University of Art and Design, 1997.

[Lang 03] Robert J. Lang. *Origami Design Secrets: Mathematical Methods for an Ancient Art.* Natick, MA: A K Peters, 2003.

[Leung et al. 90] Joseph Y.-T. Leung, Tommy W. Tam, C. S. Wing, Gilbert H. Young, and Francis Y.L. Chin. "Packing Squares into a Square." *J. Parallel Distrib. Comput.* 10:3 (1990), 271–275.

[Meguro 92] Toshiyuki Meguro. "Jitsuyou Origami Sekkeihou (Practical Methods of Origami Designs)." *Origami Tanteidan Shinbun* 2 (1991–1992), 7–14.

[Meguro 94] Toshiyuki Meguro. "Tobu Kuwagatamushi to Ryoikienbunshiho (Flying Stag Beetle and the Circular Area Molecule Method)." In *Oru*, pp. 92–95. Tokyo: Sojusha, 1994.

[Specht 10] Eckard Specht. "The Best Known Packings of Equal Circles in a Square." Available at http://hydra.nat.uni-magdeburg.de/packing/csq/csq.html, 2010.

[Tachi 09] Tomohiro Tachi. "3D Origami Design Based on Tucking Molecules." In *Origami⁴: Fourth International Meeting of Origami Science, Mathematics, and Education*, edited by Robert J. Lang, pp. 259–272. Wellesley, MA: A K Peters, 2009.

Contributors

- **Hugo Alves Akitaya**, Computer Science Department, Univesity of Brasília, Campus Darcy Ribeiro, ICC Norte Subsolo Módulo 18, 70510-500, Brasilia, Brazil,
  hugoakitaya@gmail.com

- **Roger C. Alperin**, Department of Mathematics, San Jose State University, San Jose, CA 95192, USA,
  alperin@math.sjsu.edu

- **Nana Y. Atuobi**, 3421 University Place, Baltimore, MD 21218, USA,
  natuobi1@jhu.edu

- **Noy Bassik**, Medical Scientist Training Program, Johns Hopkins University, 409 Range Road, Towson, MD 21204, USA,
  noy@jhu.edu

- **Alex Bateman**, Wellcome Trust Sanger Institute, Valley Farm Cottage London Road, Balsham, Cambridgeshire CB21 4, UK,
  agb@sanger.ac.uk

- **Nadia M. Benbernou**, Massachusetts Institute of Technology, 1301 Delaware Avenue, Wilmington, DE 19806, USA,
  nbenbern@mit.edu

- **Norma J. Boakes**, Richard Stockton College of New Jersey, PO Box 195, Pomona, NJ 08240, USA,
  Norma.Boakes@stockton.edu

- Hans Georg Bock, Interdisciplinary Center for Scientific Computing (IWR), University of Heidelberg, Im Neuenheimer Feld 368, 69120 Heidelberg, Germany, scicom@iwr.uni-heidelberg.de

- Alla Brafman, 34-15 Morlot Avenue, Fair Lawn, NJ 07410, USA, abrafma1@jhu.edu

- Krystyna Burczyk, ul. Konwaliowa 22A, 32-080 Zabierzow, Poland, burczyk@mail.zetosa.com.pl

- Wojciech Burczyk, ul. Konwaliowa 22A 32-080 Zabierzow Poland, burczyk@mail.zetosa.com.pl

- Eric Chang, 75 Brookwood Drive, Longmeadow, MA 01106, USA, changeric.teach@gmail.com

- Yan Chen, School of Mechanical and Aeorspace Engineering, Nanyang Technological University, Singapore 639798, Singapore, chenyan@ntu.edu.sg

- Herng Yi Cheng, NUS High School of Mathematics and Science, 20 Clementi Avenue 1, Singapore 129957, Singapore, herngyi@gmail.com

- Neus Dasquens, Escola Font de l'Alba, Carrer Consell de Cent, 146, 08226 Terrassa, Barcelona, Spain, neusdasq@gmail.com

- Erik D. Demaine, Massachusetts Institute of Technology, 32 Vassas Street, Cambridge, MA 02139, USA, edemaine@mit.edu

- Martin L. Demaine, Massachusetts Institute of Technology, 32 Vassas Street, Cambridge, MA 02139, USA, mdemaine@mit.edu

- Klaus Drechsler, Institute of Aircraft Design, Institut fuer Flugzeugbau (IFB), University of Stuttgart, Pfaffenwaldring 31, 70569 Stuttgart, Germany, drechsler@ifb.uni-stuttgart.de

- Christine E. Edison, 13295 Clinton Avenue, Berwyn, IL 60402, USA, cedison@prodigy.net

- Sándor Fekete, Technische Universität Braunschweig, Department of Computer Science, Mühlenpfordtstr. 23, D-38106 Braunschweig, Germany, s.fekete@tu-bs.de

- Maria Lluïsa Fiol, Facultat de Ciències de l'Educació, Universitat Autònoma de Barcelona (UAB), 08193, Cerdanyola del Vallès, Spain, marialluisa.fiol@uab.es

- Emma Frigerio, Università di Milano, Dipartimento di Matematica, Via Saldini, 50, 20133 Milano, Italy, emma.frigerio@unimi.it

- Yukio Fukui, University of Tsukuba, Department of Computer Science, Tennodai 1-1-1, Tsukuba, Ibaraki, 305-8573, Japan,
  fukui@cs.tsukuba.ac.jp

- Matthew Gardiner, Ars Electronica Futurelab, Ars Electronica Strasse 1, Linz A-4040, Austria,
  mg@matthewgardiner.net

- Miri Golan, Israeli Origami Center, Aba Hillel 146, Ramat Gan 52572, Israel,
  origami@netvision.net.il

- Faye Goldman, 616 Valley View Road, Ardmore, PA 19003, USA,
  fayeG@ix.netcom.com

- David H. Gracias, Department of Chemical and Biomolecular Engineering, Johns Hopkins University, 3400 N Charles Street, 221 Maryland Hall, Baltimore, MD 21218, USA,
  dgracias@jhu.edu

- Steven Gray, University of Pennsylvania, Department of Mechanical Engineering and Applied Mechanics, School of Engineering and Applied Science, University of Pennsylvania, Towne 229, 220 S. 33rd Street, Philadelphia, PA 19104-6315, USA,
  stgray@seas.upenn.edu

- Simon D. Guest, Department of Engineering, University of Cambridge, Trumpington Street, Cambridge CB2 1PZ, UK,
  sdg13@cam.ac.uk

- Kazuo Haga, Haga's Laboratory for Science Education, 16-2-514, Higashiarai, Tsukuba, Ibaraki, 305-0033, Japan,
  hagak@hi-ho.ne.jp

- Koshiro Hatori, 3-14-9 Futaba #301, Shinagawa-ku, Tokyo 142-0043, Japan,
  origami@ousaan.com

- Andrew Hudson, 2803 Grinnel Dr. Davis, CA 95618, USA,
  andrew.hudson13@gmail.com

- Thomas C. Hull, Western New England College, Mailbox H-5174, 1215 Wilbraham Road, Springfield, MA 01119, USA,
  thull@wne.edu

- Yoshihiro Kanamori, University of Tsukuba, Department of Computer Science, Tennodai 1-1-1, Tsukuba, Ibaraki, 305-8573, Japan,
  kanamori@cs.tsukuba.ac.jp

- Miyuki Kawamura, 3-10-521, Yaemizo Saga, Saga 849-0935, Japan,
  myu3@beige.plala.or.jp

- Hidefumi Kawasaki, Faculty of Mathematics, Kyushu University, Motooka 744, Nishi-ku, Fukuoka 819-0395, Japan,
  kawasaki@math.kyushu-u.ac.jp

- Toshikazu Kawasaki, Anan National College of Technology, Aoki, Minobayashi, Anan City, Tokushima Pref. 774-0017, Japan,
  kawasaki@anan-nct.ac.jp

- Yves Klett, Institute of Aircraft Design, Institut fuer Flugzeugbau (IFB), University of Stuttgart, Pfaffenwaldring 31, 70569 Stuttgart, Germany, klett@ifb.uni-stuttgart.de

- Carla Koike, Computer Science Department, Univesity of Brasília, Campus Darcy Ribeiro, ICC Norte Subsolo Módulo 18, 70510-500, Brasilia, Brazil, ckoike@cic.unb.br

- Duks Koschitz, Design and Computation Group, Massachusetts Institute of Technology, 168 Webster Avenue, Cambridge, MA 02141, USA, duks@mit.edu

- Jason S. Ku, 3D Optical Systems Group, Massachusetts Institute of Technology, KBL #312, 129 Franklin Street, Cambridge, MA 02139, USA, jasonku@MIT.EDU

- Vijay Kumar, University of Pennsylvania, Department of Mechanical Engineering and Applied Mechanics, School of Engineering and Applied Science, University of Pennsylvania, Towne 229, 220 S. 33rd Street, Philadelphia, PA 19104-6315, USA, kumar@grasp.upenn.edu

- Kaori Kuribayashi-Shigetomi, Institute of Industrial Science, The University of Tokyo; Takeuchi Biohybrid Innovation Project, Exploratory Research for Advanced Technology (ERATO), Japan Science and Technology Agency (JST), 4-6-1 Komaba, Meguro-ku, Tokyo 153-8505, Japan, kaorik@iis.u-tokyo.ac.jp

- Shi-Pui Kwan, The Hong Kong Institute of Education, Department of Mathematics and Information Technology, D3-2/F-12, 10 Lo Ping Road, Tai Po, New Territories, Hong Kong, spkwan@ied.edu.hk

- Tung Ken Lam, University of Cumbria, Bowerham Road, Lancaster, LA1 3JD, UK, tklorigami@yahoo.co.uk

- Robert J. Lang, 899 Forest Lane, Alamo, CA 94507, USA, robert@langorigami.com

- Cheng Chit Leong, 480 Segar Road #05-378, Segar Gardens, Singapore 670480, Singapore, leongccr@singnet.com.sg

- Jiayao Ma, University of Oxford, Department of Engineering Science, Parks Road, Oxford, OX1 3PJ, UK, jiayao.ma@balliol.ox.ac.uk

- Jun Maekawa, National Astronomical Observatory, Japan 1-1-21-805 Tobitakyu, Chofu, Tokyo, 182-036, Japan, maekawa@nro.nao.ac.jp

- Crystal Elaine Mills, 8687 Great Horned Owl Lane, Blaine, WA 98230, USA crysmills@mac.com

- Jun Mitani, University of Tsukuba, Department of Computer Science, Tennodai 1-1-1, Tsukuba, Ibaraki, 305-8573, Japan,
  mitani@cs.tsukuba.ac.jp

- Hiroyuki Moriwaki, Department of Information Design, Tama Art University, 2-1723 Yarimizu, Hachiojo, Tokyo, 192-0394, Japan,
  moriwaki@tamabi.ac.jp

- Charlene Morrow, SummerMath, Mount Holyoke College, 50 College Street, South Hadley, MA 01075-1441, USA,
  cmorrow@mtholyoke.edu

- James Morrow, SummerMath, Mount Holyoke College, 50 College Street, South Hadley, MA 01075-1441, USA,
  jmorrow@mtholyoke.edu

- Robert Orndorff, P.O. Box 15266, Seattle, WA 98115, USA,
  orndorff@alum.mit.edu

- Aviv Ovadya, Massachusetts Institute of Technology, 2785 West 5th Street, Apt 23C, Brooklyn, NY 11224, USA,
  avivo@mit.edu

- Sue Pope, Liverpool Hope University, Hope Park, Liverpool L16 9JD, UK,
  popes@hope.ac.uk

- Montserrat Prat, Facultat de Ciències de l'Educació, Universitat Autònoma de Barcelona (UAB), 08193, Cerdanyola del Vallès, Spain,
  montserrat.prat@uab.es

- Jose Ralha, University of Brasilia, Campus Universitrio Darcy Ribeiro, ASA Norte, 70510-500 Brasilia, Brasil,
  ralha@cic.unb.br

- Matheus Ribeiro, Computer Science Department, Univesity of Brasília, Campus Darcy Ribeiro, ICC Norte Subsolo Módulo 18, 70510-500, Brasilia, Brazil,
  matruskan@gmail.com

- Mark Schenk, University of Cambridge, Clare College, Cambridge, CB2 1TL, UK,
  me@markschenk.com

- George M. Stern, Master's of Science in Engineering, John Hopkins University, 514 Ridge Road, Wilmette, IL 60091, USA,
  gstern4@gmail.com

- Tomohiro Tachi, University of Tokyo, 3-8-1 Komaba, Meguro-ku, Tokyo 153-8902, Japan,
  tachi@idea.c.u-tokyo.ac.jp

- Shoji Takeuchi, Institute of Industrial Science, The University of Tokyo; Takeuchi Biohybrid Innovation Project, Exploratory Research for Advanced Technology (ERATO), Japan Science and Technology Agency (JST), 4-6-1 Komaba, Meguro-ku, Tokyo 153-8505, Japan,
  takeuchi@iis.u-tokyo.ac.jp

- Koichi Tateishi, Department of English, Kobe College, 4-1 Okadayama, Nishi-
  nomiya, Hyogo 662-8505, Japan,
  tateishi@mail.kobe-c.ac.jp

- Eulàlia Tramuns, Universitat Politècnia de Catalunya EETAC, Edifici C3,
  Despatx 013, Esteve Terradas, 7 - 08860 Castelldefels, Spain,
  etramuns@ma4.upc.edu

- Naoya Tsuruta, University of Tsukuba, Department of Computer Science,
  SB0925, Laboratory of Advanced Research B, Tennodai 1-1-1, Tsukuba,
  Ibaraki, 305-8573, Japan,
  tsuruta@npal.cs.tsukuba.ac.jp

- Arnold Tubis, 8099 Paseo Arrayan, Carlsbad, CA 92009-6963, USA,
  tubisa@aol.com

- Ryuhei Uehara, School of Information Science, Japan Advanced Institute of
  Science and Technology (JAIST), Asahidai 1-1, Nomi, Ishikawa 923-1292,
  Japan,
  uehara@jaist.ac.jp

- Kunfeng Wang, Block 657B, Jurong West Street 65 #07-652, Singapore
  642657, Singapore,
  wangkunfeng@gmail.com

- Michael J. Winckler, Interdisciplinary Center for Scientific Computing, Uni-
  versity of Heidelberg, HGS MathComp, Im Neuenheimer Feld 368, 69120
  Heidelberg, Germany,
  Michael.Winckler@iwr.uni-heidelberg.de

- Kathrin D. Wolf, Zum Wasserschloss 9, 74821 Mosbach-Lohrbach, Germany,
  kathrin.d.wolf@web.de

- Mark Yim, University of Pennsylvania, Department of Mechanical Engineering and Applied Mechanics, School of Engineering and Applied Science,
  Towne 229, 220 S. 33rd Street, Philadelphia, PA 19104-6315, USA,
  yim@grasp.upenn.edu

- Zhong You, University of Oxford, Department of Engineering Science, Parks
  Road, Oxford, OX1 3PJ, UK,
  zhong.you@eng.ox.ac.uk

- Nathan J. Zeichner, University of Pennsylvania, 418 West 51st Street, New
  York, NY 10019, USA,
  nathanz@seas.upenn.edu

Index